Radioanalytical Chemistry

Bernd Kahn, Editor

Radioanalytical Chemistry

 Springer

Library of Congress Control Number: 2006925171

ISBN 10: 0-387-34122-6   Printed on acid-free paper.
ISBN 13: 978-0387-34122-4

9 8 7 6 5 4 3 2 1

springer.com

Contents

Acknowledgments

The following members of the Editorial Advisory Board (EAB) contributed enormously to this textbook by suggesting format, reviewing successive drafts, and some even writing chapters (as indicated in chapter headings). It was a great pleasure to work with all contributors, and they have my heartfelt thanks for their participation.

- Moses Attrep, Jr., LANL-retired
- Darleane Hoffman, LBL
- Kenneth Inn, NIST
- John Keller, ORNL
- Harry Miley, PNNL
- Stan Morton, RESL-IDO-retired
- Glenn Murphy, UGA
- Dick Perkins, PNNL
- Charles Porter, EPA-retired
- Jake Sedlet, ANL
- John Wacker, PNNL

Isabel Fisenne and A. L. Boni were members of the EAB for several years and then resigned.

My deepest sympathy goes to the family of Jake Sedlet, who died while this work was in progress.

I appreciate the efforts of others who contributed by writing a chapter:

- Craig Aalseth, PNNL
- Gregory Eiden, PNNL
- Pam Greenlaw, EML
- Jeffrey Lahr, GT
- Scott Lehn, PNNL
- Keith McCroan, EPA
- Robert Rosson, GT
- Paul Schlumper, GT
- Linda Selvig, Boise ID

- Liz Thompson, GT
- Arthur Wickman, GT

Each chapter is credited to the main author or authors, but many portions of their original manuscripts were moved to other chapters for better text integration.

This effort was supported by NNSA–DOE under grant DE-FG07-01ID14224. Dan Griggs and Stephen Chase were successive project officers. I am grateful to them for their active involvement in the preparation of this text, and to Dale Perry and Terry Creque of NNSA for participating in the EAB meetings.

I thank the Georgia Institute of Technology, my professional home when the book was written, and my coworkers there. Liz Thompson was the editorial writer who untiringly worked on every aspect of the text to prepare it for publication with great insight and competence. Robert Rosson and Jeff Lahr are radiochemists who were most helpful in working with me and writing specific chapters. Jean Gunter was the skilled coordinator of all EAB meetings, who made attendance a pleasure.

I wish to express my deepest respect and thanks to the two radiochemists who taught me, William S. Lyon at ORNL and Charles D. Coryell at MIT.

This book is dedicated with love to my wife, Gail, and my parents, Alice and Eric Kahn.

1
Introduction

BERND KAHN

1.1. Background

Radioanalytical chemistry is devoted to analyzing samples for their radionuclide content. For this purpose, the strategies of identifying and purifying the radioelements of interest by chemical methods, and of identifying and measuring the disintegration rate ("activity") of radionuclides by nuclear methods, are combined. Radioanalytical chemistry can be considered to be a specialty in the subdiscipline of nuclear and radiochemistry.

This textbook was written to teach radioanalytical chemistry in the classroom and support its application in the laboratory. Its emphasis is on the practical aspects of the specialty, notably setting up the laboratory, training its staff, and operating it reliably. The information presented herein, outlined in Section 1.4, is the accumulated product of a century of nuclear chemistry and radiochemistry practice.

Radioanalytical chemistry was first developed by Mme. M. Curie, with contributions by many other distinguished researchers, notably E. Rutherford and F. Soddy. These pioneers performed chemical separations and radiation measurements on terrestrial radioactive substances during the 20 years following 1897 and in the process created the very concept of radionuclides. Their investigations defined the three major radiation types, confirmed the emission of these radiations by the nucleus and the associated atomic transformations, established the periodic table between bismuth and uranium, and demonstrated the distinction between stable and radioactive isotopes.

Thereafter, cosmic rays were observed and explained, and many cosmic-ray-produced radionuclides were identified. The number of known and characterized radionuclides increased dramatically with the development and application of nuclear-particle accelerators in the 1930s and nuclear-fission reactors in the 1940s.

Since Mme. Curie's time, applications of radioanalytical chemistry have proliferated. Modern practitioners of nuclear and radiochemistry have applied chemical and nuclear procedures to elucidate nuclear properties and reactions, used

Environmental Radiation Branch, Georgia Tech Research Institute, Georgia Institute of Technology, Atlanta, GA 30332

1

radioactive substances as tracers for solving problems, and measured radionuclides in many types of samples. The work plays an integral part in research related to chemistry, physics, medicine, pharmacology, biology, ecology, hydrology, geology, forensics, atmospheric sciences, health protection, archeology, and engineering. Applications that generate great interest include forming and characterizing new elements, determining the age of materials, and creating radioactive reagents for specific tracer use, notably targeting selected tissues and organs with radionuclides for diagnosis and treatment. New and clever analytical methods are developed to monitor ever-more radionuclides at ever-lower concentrations in persons and in the environment.

The increasingly common use of radioactive materials has engendered an entire industry dedicated to monitoring the locations and amounts of those radionuclides. Radionuclides are monitored in the environment, in effluent and process streams, and in workers by nuclear research laboratories, nuclear fuel cycle facilities, radiological measurement contractors, and government agencies. Ongoing monitoring efforts examine the products of routine operation, research projects, waste processing, and storage; additional monitoring may be ordered in response to contamination incidents, nuclear site closures, or terrorist actions. Throughout the world, radioanalytical chemists collaborate in investigations of contamination from nuclear weapon tests and nuclear accidents.

Because many of the required skills are the same, a competent nuclear chemist or radiochemist can move from one of the above-mentioned fields to another. Radioanalytical chemists combine separation techniques of classical chemical qualitative and quantitative analysis with the radiation detection techniques of nuclear physics. Practitioners polish techniques for processing distinctive sample matrices, separating various sets of radionuclides, and improving detection sensitivity. In some instances, informed use of radiation detection techniques eliminates any need for chemical separations, or mass spectrometry replaces radiation detection.

Despite the continuing interest and expanding applications, the number of practitioners in nuclear chemistry and radiochemistry and its radioanalytical chemistry specialty has decreased to an alarmingly low level from its peak in the 1950s and 1960s. Two causes have combined to produce the situation. For a variety of reasons, many academic institutions no longer teach the discipline, and so the number of pertinent degree programs and the students who graduate from them have declined markedly. At the same time, the senior members of the radiochemistry community are leaving the profession by retirement, and so many of the practitioners who contributed to the flourishing of radioanalytical chemistry no longer are active.

International and U.S. agencies recognize the seriousness of this problem and have taken initial steps to reverse the trend. This textbook was written to contribute to these efforts. First, it provides a newly prepared text for expanded programs in teaching students the fundamentals of radioanalytical chemistry. Secondly, it attempts to preserve and transmit the practical aspects of this specialty for training professionals that are newly assigned to radioanalytical chemistry tasks.

1.2. Information Sources for Radioanalytical Chemistry

Nuclear chemistry and radiochemistry are described in many excellent texts. Two of the more widely used ones are by Friedlander *et al.* (1981) and Choppin *et al.* (1995). A five-volume *Handbook of Nuclear Chemistry*, edited by Vertes *et al.* (2003), has been published recently. It includes pertinent applications such as activation analysis and tracer use, and an excellent brief history by Friedlander and Herrman (2003). Radioanalytical chemists can obtain information from these texts and others on such vital aspects of the work as the sources of radionuclides, radiation detection, radiation interactions, and applications to varied fields. Other useful books on these topics were written within the past 15 years by Ehmann and Vance (1991), Navratil *et al.* (1992), and Adloff and Guillaumont (1993).

Several books that address certain aspects of radioanalytical chemistry are valuable sources of information for the radiochemist. The *CRC Handbook of Radioanalytical Chemistry* (Tolgyessy and Bujdoso 1991) is a two-volume set that devotes the majority of its more than 1700 pages to physical constants, mathematical tables, radionuclide data, and activation data. It includes charts and tables with relevant data for solvent extraction, ion exchange, masking reactions, isotope dilution analysis, radiometric titrations, and organic and inorganic radioreagent methods. A table of radiochemical analysis methods is reproduced from NCRP (1976b). This handbook is an updated and condensed version of *Nuclear Analytical Chemistry*, Tolgyessy's five-volume work of 20 years earlier (Tolgyessy 1971).

A text with a scope similar to this book is *Radiochemical Methods in Analysis* (Coomber 1975). The text contains such relevant chapters as "Separation methods for inorganic species" and "The use of tracers in inorganic analysis." A chapter titled "Determination of radioactivity present in the environment" contains information geared toward sample collection.

Elementary Practical Radiochemistry (Ladd and Lee 1964) contains 20 brief experiments that illustrate detection techniques such as measurement of ingrowth and decay, as well as ion exchange, extraction, and coprecipitation. The text *Radioisotope Laboratory Techniques* (Faires and Boswell 1981) primarily addresses nuclear physics, radionuclide production, and counting techniques. It briefly mentions laboratory apparatus but omits discussion of separation techniques.

The section "Radioactive Methods" in volume 9 of the *Treatise on Analytical Chemistry* (Kolthoff and Elving 1971) discusses radioactive decay, radiation detection, tracer techniques, and activation analysis. It has a brief but informative chapter on radiochemical separations. A more recent text, *Nuclear and Radiochemistry: Fundamentals and Applications* (Lieser 2001), discusses radioelements, decay, counting instruments, nuclear reactions, radioisotope production, and activation analysis in detail. It includes a brief chapter on the chemistry of radionuclides and a few pages on the properties of the actinides and transactinides.

Radioanalytical chemistry methods have been published for a wide variety of samples. Compilations of methods developed in the United States during World War II were published in the *National Nuclear Energy Series* (Coryell and

Sugarman 1951). A set of monographs published by the National Research Council over a period of several years, entitled *The Radiochemistry of [Element]* (NAS-NRC 1960a), traverses the entire periodic table. Another set (NAS-NRC 1960b) of monographs is on radiochemical techniques. Laue and Nash (2003) edited symposium presentations of recently developed chemical and radiation detection methods, together with overviews of historical developments and current needs.

New methods were published in the United States in the journal *Analytical Chemistry* and the *Journal of the American Chemical Society*, and corresponding publications in other countries. From 1949 to 1986, *Analytical Chemistry* issued biennial reviews of nucleonics articles; a similar set of reviews in the field of water chemistry also held articles of interest to the radioanalytical chemist.

Journals such as *Radiochemistry, Radioactivity and Radiochemistry*, and the *Journal of Radioanalytical and Nuclear Chemistry* currently are vehicles for publishing radioanalytical chemistry methods. Groups such as the American Society for Testing and Materials (ASTM) and the American Public Health Association (APHA) publish standard methods for radiochemistry (see Section 6.5), among other topics.

A manual devoted to radioanalytical chemistry methods was published by the U.S. Environmental Protection Agency (EPA 1984) and another one was updated periodically by the Environmental Measurements Laboratory of the U.S. Department of Energy (Chieco 1997). An intergovernmental task group recently prepared the Multi-Agency Radiological Laboratory Analytical Protocols (MARLAP) manual (EPA 2004) to guide radiation survey and site investigations, this detailed work addresses many aspects of radioanalytical chemistry and is on the Internet at www.eml.doe.gov/marlap. Additional sources of methods are the manuals published at major DOE facilities—Oak Ridge National Laboratory (ORNL), Savannah River Site (SRS), Los Alamos National Laboratory (LANL), and Idaho National Engineering Laboratory (INEL), among others. Each has a set of manuals to describe every method employed at the site, including radioanalytical chemistry methods.

No modern textbook is complete without the use of electronic information sources to augment hard-copy sources. Online resources have several advantages: they are always accessible (never closed and rarely restricted), unfettered by the bounds of printing (and so usually fairly extensive in content and illustration), and generally up to date (if maintained by professionals). This format, however, is impermanent and subject to change with time. Online references are current as of the viewing date noted with each Web address cited, but the Web site content may change, the address may change, or the site may not be maintained.

1.3. Radioanalytical Chemistry Program Elements

This textbook was written to build the knowledge base that a competent radioanalytical chemist should possess. From these chapters, the student should be able to attain the knowledge necessary to perform skillfully the following duties:

- Operate the radioanalytical chemistry laboratory safely and in accord with regulations, with assured quality and cost-effectiveness;
- Plan and perform identification and measurement of radionuclides through a combination of chemical separation and radiation detection methods;
- Select sample size, sample processing, radiation detection instruments with peripherals, and measurement period to match detection sensitivity specifications;
- Identify radionuclides by their chemical behavior and radiation, and resolve ambiguities by expanding or revising separations and measurements;
- Calculate radionuclide concentration values and their uncertainty with consideration of the radionuclide decay scheme, sample processing losses, radiation detection efficiency, and interfering radiation.

These integrated activities require both professional knowledge and hands-on experience. Hence, this textbook is designed for a radioanalytical chemistry lecture course and an associated laboratory course. A laboratory manual and an instructor's guide are in preparation to support the text. The prerequisite study program should include (1) analytical chemistry lectures and associated wet-chemistry laboratory and (2) nuclear physics lectures and associated radiation detection laboratory.

1.4. Use of This Text

The book is intended as a guide to the functioning radioanalytical chemistry laboratory. It presents a one-semester course at the senior or graduate level to train students to work in, manage, or interact with a laboratory devoted to radionuclide measurement. Such a laboratory has two components: the wet laboratory to prepare samples for measuring radionuclides, and the counting room to perform the measurements. The samples being considered here are primarily for environmental monitoring in media such as air, water, vegetation, foodstuff, wildlife, soil, and sediment. Samples may also be for bioassay, process control, and various research and application uses.

Nuclear physics as applied to radionuclide identification and measurement is introduced in Chapter 2 to refresh the reader's memory. Detailed information pertinent to the main radiation detectors routinely used in the radioanalytical chemistry counting room is given in Chapter 8. Chapter 9 discusses radionuclide identification by decay scheme and gives examples.

Chapter 3 presents general information on the purification processes in the wet laboratory that underlie radioanalytical chemistry, as well as the background information in analytical chemistry necessary to apply those processes. Information specifically associated with the behavior of radionuclides in aspects such as their low concentration and the effect of radiation emission is given in Chapter 4. Chapter 5 describes the form of the usual samples submitted to the radioanalytical chemistry laboratory and the treatment of various sample matrices to indicate the context in which analytical results are considered.

The chemistry work in the radioanalytical laboratory is described in Chapters 6 and 7. Chapter 6 discusses the radioanalytical chemistry process from initial sample preparation to complete purification and gives examples of separation methods for the commonly encountered radionuclides. Chapter 7 describes preparation of the counting source that is submitted for the measurements by radiation detection instruments presented in Chapter 8.

The next four chapters are devoted to various aspects of data interpretation, data presentation, and quality assurance. Chapter 9 considers interpretation of data for radionuclide identification by decay scheme. Chapter 10 reviews the important topics of data calculation, measurement uncertainty, data evaluation, and reporting the results. Chapter 11 describes the quality assurance plan that must govern all laboratory operations. Chapter 12 discusses methods diagnostics to correct the analytical and measurement problems that can be expected to plague every laboratory.

The following two chapters address laboratory operation. Chapter 13 describes laboratory practice, design, and management in terms of a model radioanalytical chemistry laboratory and its staffing, maintenance requirements, and costs of operation. Chapter 14 discusses the practice of laboratory safety and the management elements that reinforce the safety culture. Although the placement of these chapters in the text is intended to maintain learning continuity, their contents should be considered at the very beginning of laboratory practice.

The last three chapters consider special aspects of radioanalytical chemistry that have become increasingly important and visible. Chapter 15 describes the automated systems that are used to measure radionuclides in the counting room and in the environment. Chapter 16 is devoted to identification and measurement of the radionuclides beyond the actinides. These are research projects at the cutting edge of radiochemistry that apply novel rapid separations in order to measure a few radioactive atoms before they decay. Much must be inferred from limited observations. In Chapter 17, several versions of mass spectrometers combined with sample preparation devices are described. The mass spectrometer, applied in the past as a research tool to detect a small number of radioactive atoms per sample, is now so improved that it serves as a reliable alternative to radiation detection for radionuclides with half-lives as short as a few thousand years.

In total, it is anticipated that these chapters will allow the student to begin to understand the practice of radioanalytical chemistry. The authors wish you a good journey and look forward to your success.

2
Radiation Detection Principles

Moses Attrep, Jr.[1,2]

2.1. Introduction

As noted in Chapter 1, a background in *analytical* chemistry is essential to the practice of *radioanalytical* chemistry. The parallels between the two fields will become evident as analytical and radioanalytical chemistry principles are covered in Chapters 3 and 4, and culminate in the discussion of applied radioanalytical chemistry in Chapter 6. However, the theoretical underpinnings of the two disciplines are markedly different.

The primary distinction between analytical chemistry and radioanalytical chemistry is the nature of the transformations being examined. The analytical chemist is concerned with *chemical* transformations, brought on by the interaction of an atom's valence electrons with its physical environment. The radioanalytical chemist, on the other hand, is primarily interested in the *nuclear* transformation of a given atom. For practical purposes, the physical environment of the atom has no effect on the nuclear event. Consequently, many of the instrumental methods of detection most widely utilized in the normal course of analytical characterization have little use in the radioanalytical laboratory.

Instead, the radioanalytical chemist focuses on the detection of radiation, the by-product of a nuclear transformation. The analyst must understand the types of radiation that may be encountered and the way that each interacts with matter. With this knowledge, the analyst can adapt the method of detection to the particular radionuclide of interest. The goal of this chapter is to provide a brief review of nuclear chemistry as it relates to the principles of radiation detection. Next, an overview of the operating principles of commonly used detectors is provided as a basis for understanding the material presented in Chapter 8.

2.2. Radioactive Decay and Types of Radiation

When the unstable nucleus of a radioactive isotope undergoes a nuclear transformation, i.e., decays, a new atomic species is formed with the concomitant emission

[1]344 Kimberly Lane, Los Alamos NM 87544.
[2]Los Alamos National Laboratory (retired), Los Alamos, NM 87545.

TABLE 2.1. The three common types of radiation emitted from unstable nuclei

Types of radiation	Symbol	Description
Alpha	α	A helium-4 nucleus composed of two protons and two neutrons; mass is approximately 4 Da; charge +2; no spin
Beta	β	An electron; mass $\sim 1/1822$ Da; charge -1 or $+1$; spin 1/2
Gamma	γ	Electromagnetic radiation; no mass; no charge

of one or more forms of radiation. The three basic types of radiation are alpha (α), beta (β) and gamma (γ) radiation (see Table 2.1). Three primary modes of nuclear decay are alpha-particle emission, beta-particle emission, and gamma-ray emission. Above atomic number 90, spontaneous fission (SF), a natural decay mode in which the nucleus spontaneously divides into two large fragments, becomes an increasingly important mode of decay. During all of these processes, mass, charge, and energy must be conserved. Thus, the sum of the mass numbers A and the atomic numbers Z of the products (or daughters) must equal those of the initial radionuclide (the parent). Additionally, the mass of the parent must equal the masses of the daughter and emitted particles plus mass equivalents of the kinetic energy of the products. The known radioactive nuclides and their radiation are given in tables and charts of nuclides together with the stable isotopes (Firestone 1996).

The specific mode of decay listed in Table 2.1 depends upon the nature of the parent's instability relative to the lower energy states to which it may decay, and functions to transform the radioactive nucleus into a more stable nucleus. Alpha emission (α) occurs in the neutron-deficient lanthanides and is the major mode of decay in many heavy radionuclides of the actinides and transactinides. Beta radiation, which can take the form of either negatron (β^-) or positron (β^+) emission, results from an imbalance in the neutron-to-proton (N/Z) ratio of the parent and occurs in the isotopes of elements throughout the periodic table. Gamma radiation (γ) is the result of an excited nucleus de-exciting to lower energy levels. A fourth mode of decay is SF, where the parent nucleus fissions into two large, neutron-rich fission fragments (FF). The FF may be stable or may de-excite by neutron emission to lower mass nuclei and by β^- emission to higher atomic number nuclides until further decay is no longer energetically possible.

Each of these decay process is explained in more detail in the following sections. All processes described below are exoergic; i.e., they give off energy.

2.2.1. Alpha-Particle Decay

Alpha decay is characterized by the emission of an alpha garticle, which is equivalent to a ^{4}He nucleus. This mode of decay decreases the original nucleus by two protons and two neutrons for a total mass loss of about 4 Da (or atomic mass units, amu), and a reduction in nuclear charge by 2. That mass and charge is carried away by the alpha particle, as shown in Eq. (2.1):

$$^A_Z X \rightarrow {}^{A-4}_{Z-2} Y + {}^4_2 \alpha \qquad (2.1)$$

Because the momentum and energy evolved in the decay must be conserved, they are distributed between the product nucleus (sometimes called the recoil nucleus) and the emittedg alpha particle. The unit of energy commonly used in describing nuclear decay and radiation is the electron volt and its multiples. One electronvolt equals 1.6×10^{-19} J, which is numerically equal to the electron charge e in coulombs.

Given the energy difference Q between isotopes X and Y, the kinetic energy of the alpha particle is $QY/(Y + \alpha)$ (where Y and α are masses) and the kinetic energy of the recoiling isotope Y is $Q\alpha/(Y + \alpha)$. The lighter alpha particle is quite energetic between 2.5 and 9 MeV, although more commonly between 4 and 6 MeV. In contrast, the recoil energy of the heavier product nucleus typically is about 0.1 MeV.

A radionuclide may emit several alpha-particle groups of different energies and intensities, but every alpha particle in a given group is of the same energy. As shown in the example of ^{241}Am in Section 9.3.7, the most energetic alpha-particle group decays to the ground state. Less energetic groups decay to states slightly more energetic than the ground state and then decay promptly by conversion electron (CE) and gamma-ray emission (Section 2.2.3) to the ground state.

2.2.2. Beta-Particle and Electron-Capture Decay

Negatron (β^-) emission occurs when a nucleus has an excess of neutrons with respect to protons, as compared to the stable nucleus isobar (same A value, where $A = N + Z$). This type of transition in effect converts a neutron to a proton. Conversely, positron emission (β^+) in effect converts a proton to a neutron to attain greater stability. It occurs when the radionuclide neutron-to-proton ratio is deficient as compared to the stable nucleus of the same A value. These processes are represented in Eqs. (2.2) and (2.3):

$$\begin{aligned} {}_Z^A X &\to {}_{Z+1}^A Y + \bar{\nu} + \beta^- & (2.2) \\ {}_Z^A X &\to {}_{Z-1}^A Y + \nu + \beta^+ & (2.3) \end{aligned}$$

The $\bar{\nu}$ represents the antineutrino; ν is the neutrino. Neutrino and antineutrino emissions serve to balance the energy and rotation before and after decay. Neutrinos have no charge and little mass; as a result, they interact to a vanishingly small degree with matter and are difficult to detect without elaborate apparatus. The neutrino (or antineutrino) must be included in the decay equation to conserve energy, angular momentum, and spin. The neutron, proton, beta particle, and neutrino all have a nuclear spin of $1/2$. A fuller discussion of this topic is in nuclear chemistry texts such as Choppin et al. (1995).

The neutrino and antineutrino groups carry somewhat more than half of the decay energy, while the beta-particle group carries somewhat less than half. The kinetic energy of the product nuclide is very small because of the several-thousand-fold smaller beta-particle mass. This recoil energy, in the range of a few electron volts, nevertheless may cause chemical change such as displacement of an atom from a crystal lattice.

A radionuclide can decay by emitting one or several energy groups of beta particles. The beta-particle energy in a given group ranges from nearly zero to the maximum energy E_{max}. The value of E_{max} is characteristic of each radionuclide and is generally between a few kiloelectron volts and 4 MeV.

Positron emission occurs only when the energy difference between the parent radionuclide and the products exceeds 1.02 MeV (the energy equivalent of the sum of the masses of an electron and a positron). The atom's recoil, as for beta-particle emission, is a few electron volts. At lesser energy differences, a proton in the nucleus can be converted to a neutron by electron capture, i.e., the capture by the nucleus of an atomic electron from, most probably, an inner electron shell (see discussion below of CEs). The process of electron capture parallels positron emission and may occur in the same isotope. It is accompanied by emission of a neutrino and characteristic X rays due to the rearrangement of atomic electrons. Electron capture may also be signaled by the subsequent emission of gamma rays. Examples of these decays are given in Sections 9.3.4 and 9.3.6.

2.2.3. Gamma-Ray and Conversion Electron Emission

Gamma-ray emission follows the above-mentioned modes of decay when the decay leaves the product nucleus in an excited state. The nucleus is capable of further de-excitation to a lower energy state by the release of electromagnetic energy. No change in N or Z accompanies this release, only a change in the energy and spin of the nuclide. The excited nucleus is said to have a significant lifetime when it remains excited for longer than a nanosecond (10^{-9} s) without releasing its energy. In this case, it is called a metastable nucleus, and marked with an "m" as in ^{238m}U.

The state-to-state transition energy for gamma-ray emission ranges from a few kiloelectron volts to generally less than 3 MeV. The energy of the gamma ray is dictated by the nuclear structure of the emitting radionuclide, and thus is characteristic of that radionuclide.

An excited nucleus may also de-excite by emitting a conversion electron (CE). This is *not* a two-step mechanism, where a gamma-ray leaves the nucleus and then transfers energy by "kicking" the electron out of an atomic orbital. Instead, electron location probability calculations indicate that orbital electrons are on occasion located *inside* the nucleus. When an orbital electron encroaches on the nuclear space, the excitation energy of the nucleus may be transferred not to a gamma ray but to that electron, in a process called internal conversion. The resultant CE is ejected with a kinetic energy equal to the excitation energy of the nucleus minus the binding energy of the electron.

Conversion electron fractions ε_K, ε_L, etc., are tabulated as the fractions from the respective K, L, etc., atomic shells relative to the number of gamma rays of the pertinent energy (Firestone 1996). The CE and gamma-ray emissions are often observed together, especially in the decay of high-Z radionuclides and those that emit weaker gamma rays. Little CE emission occurs in low-Z radionuclides or with energetic gamma rays because the CE/gamma-ray fraction is proportional to Z^3 and inversely proportional to Q, the energy of the transition process.

Ejection of a CE (as well as the process of electron capture) leaves behind an inner shell electron vacancy. To fill this vacancy, an electron from an outer shell may fill the inner shell (moving from, say, an L shell to the K shell). X-ray emission commonly accompanies this rearrangement. The X-ray energy reflects the difference in binding energies of the two shells involved in the transition. Several transitions are usually possible (L3 to K, L2 to K, etc.), so that X rays of several different energies appear, with designations such as $K_{\alpha 1}$, $K_{\alpha 2}$, and K_β. Essentially instantaneously, the electron vacancy is filled by a cascade accompanied by X rays with ever-lower energies. The energy depends on the electron structure of the individual element, so that each X-ray energy is characteristic of the element from which it is emitted. The X-ray energies observed for each element are tabulated (Firestone 1996) together with their fractional emissions.

Auger electron emission is an alternative to X-ray emission. The process consists of one electron from the outer shell filling the inner shell, but instead of X-ray emission, a second electron, the Auger electron, is emitted from the outer shell. It has a kinetic energy that is equal to the difference in binding energies for the two shells minus the additional binding energy of the Auger electron. This leaves behind two electron vacancies in the higher energy shell that are, in turn, filled instantaneously.

X-ray emission and Auger electron ejection are competitive processes, especially in low-Z elements. As the Z value of the material becomes higher, X-ray emission becomes the more likely process. This is reflected in tabulations of fluorescence yield, which is the ratio of X-ray transition to Auger electron plus X-ray transitions; this value increases with Z from 0.01 at fluorine to 0.97 at and above polonium (Firestone 1996).

2.2.4. Spontaneous Fission Decay

Decay by SF has been observed in the elements heavier than thorium ($Z = 90$). At californium ($Z = 98$), SF begins to compete favorably with alpha-particle emission as a mode of decay, and becomes the primary decay mode for many of the higher atomic number actinides and the transactinides. As with all fission, SF releases neutrons. Unlike the induced fission process described in Section 2.3.2, SF takes place without addition of energy. Radioisotopes for which SF is an important decay mode may be used as neutron sources.

The SF process that results in two nearly equal mass fragments (a process called "symmetric fission") has been observed in ^{259}Fm (1.5 s). More commonly, SF occurs as "asymmetric fission," a split of the parent radionuclide into two unequal large FF. As in neutron-induced fission, many different asymmetric mass (and charge) divisions with varying yields can result, with mass numbers from about 70 to 170, each with many isotopes. Hundreds of different nuclides can be produced. Figure 2.1 displays the predominantly asymmetric mass yields as a function of mass number (dubbed "mass–yield curves") that have been measured for several SF and neutron-induced fission nuclides.

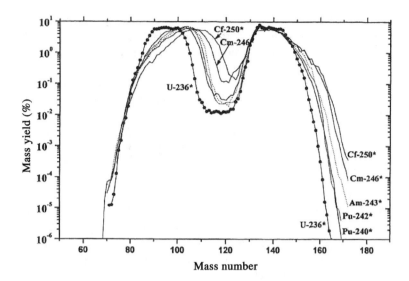

FIGURE 2.1. Mass–yield curves for some fission products, including ^{235}U and ^{239}Pu. The curve labels refer to the compound nuclei, as in ^{235}U $+ {}^1$n $= {}^{236}$U. (From Vertes *et al.* 2003, p. 222.)

2.2.5. Radioactive Decay

Radioactive decay is a random first-order process described by

$$\frac{dN}{dt} = -N\lambda \tag{2.4}$$

where dN/dt is the disintegration rate of a given radionuclide in which dN is the change in the number of its atoms (N) that undergo nuclear transformation in the time (t) interval, dt. The proportionality constant λ is the decay constant; it is in units of inverse time, such as s^{-1}. The solution to this differential equation is

$$N_f = N_0 \, e^{-\lambda t} \tag{2.5}$$

The subscripts f and 0 respectively represent the final and initial number of atoms. As shown in Eq. (2.4), the disintegration rate is proportional to the number of atoms; hence the equation can be written as

$$\left(\frac{dN}{dt}\right)_f = \left(\frac{dN}{dt}\right)_0 e^{-\lambda t} \tag{2.6}$$

If the rate is expressed as activity A in units of net disintegrations per unit time, then successive measurements performed in the same detector without any change in detection characteristics or sample location can be related by

$$A_f = A_0 \, e^{-\lambda t} \tag{2.6a}$$

The equation defines a straight line in semilogarithmic coordinates with slope $-\lambda$,

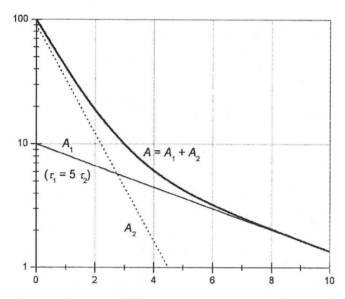

FIGURE 2.2. Radioactive decay. The ordinate is in count rate and the abscissa is the time in hours. (From Vertes *et al.* 2003, p. 263.)

as displayed graphically in Fig. 2.2. Note that the two straight lines pertain to two radionuclides that are measured in the same sample. The curves may be plotted in terms of the net count rate, $R_G - R_B$, shown in Eq. (8.1) as the measured count rate from which the background count rate has been subtracted, if the counting efficiency—the number of observed counts per disintegration—remains the same for all measurements. The curved line in Fig. 2.2 is the total net count rate, the sum of the two individual net count rates.

Measurements of the slope of the line in Fig. 2.2 or the disintegration rate at two separate times can be used to calculate the half-life $t_{1/2}$, i.e., the time period during which the radionuclide decays to one-half of a previous value. The half-life is $\ln 2/\lambda$, i.e., $t_{1/2} = 0.693/\lambda$.

The disintegration rate can be determined from the measured values of radionuclide mass m in grams and the half-life $t_{1/2}$ in seconds. Equation (2.7) relates the decay rate to the mass in terms of Avogadro's number A_v of 6.02×10^{23} atoms per mol and the isotope mass number A in g/mol:

$$\frac{dN}{dt} = \frac{0.693 m A_v}{A t_{1/2}} \tag{2.7}$$

The mass at radionuclide disintegration rates commonly encountered in the laboratory is extremely small: at a typical disintegration rate of 1 disintegration s^{-1} or Bq, the mass may be about 1×10^{-15} g. Weighable amounts of radionuclides occur for radionuclides that have half-lives of billions of years. For such long-lived radionuclides, it has been commonplace to calculate the disintegration rate from the known half-life and measured mass, or the half-life from the measured mass and disintegration rate. Recently, masses have been measured by mass spectrometer for radionuclides with half-lives of as short as few thousand years (see Chapter 17).

In some instances, a radionuclide decays not to a stable nuclide but to a second radionuclide (or even an entire chain of radionuclides) before a stable nuclide is finally reached. In that case, when the second radionuclide is separated from the first, it immediately begins to grow into the first radionuclide while decaying in the separated portion. For two successive radionuclides, the disintegration rate of the daughter (subscript 2) is related to the disintegration rate at the time of separation, $t = 0$, of the parent (subscript 1) by

$$\left(\frac{dN}{dt}\right)_2 = \left(\frac{dN}{dt}\right)_{1(t=0)} \left[\frac{\lambda_2}{(\lambda_2 - \lambda_1)}\right] (e^{-\lambda_1 t} - e^{-\lambda_2 t}) + \left(\frac{dN}{dt}\right)_{2(t=0)} (e^{-\lambda_2 t})$$

(2.8)

The equation indicates that for a daughter that has a much larger value of λ than its parent, i.e., a short-lived daughter of a long-lived parent, the second part of the first term on the right approaches 1 and the third part approaches $(1 - e^{-\lambda_2 t})$. This is called secular equilibrium; i.e., the daughter reaches the same disintegration rate as the parent (see Fig. 9.10). The second term on the right describes the decay of the separated daughter. Two other distinctive cases can occur (see Section 9.3.1). Transient equilibrium refers to the value of λ_2 that is only somewhat larger than that of λ_1, so that the daughter disintegration rate after some time exceeds at an almost constant ratio the disintegration rate of the parent (see Fig. 9.11). In the case of no equilibrium, λ_2 is less than λ_1; in this scenario, the daughter decays at a slower rate than the parent after initial ingrowth, as depicted in Fig. 9.12.

Calculations for longer decay chains under some conditions can be simplified by assuming that short-lived daughters and long-lived parents have the same disintegration rate. In some complex chains, the Bateman equation (in the same form as Eq. (2.8), but with terms added to describe further decays) can be used to determine the ingrowth and decay pattern of three or four successive radionuclides (Evans 1955).

2.3. Formation of Radionuclides by Nuclear Reactions

Some primordial radionuclides are still with us because their half-lives are near or exceed 1 billion years or because they are shorter-lived progeny of these long-lived radionuclides. All other radionuclides are formed by reaction of a nucleus with atomic or subatomic particles or radiation. Reactions with high-energy cosmic rays form some radionuclides in nature. Others are man-made, mostly in accelerators, nuclear reactors, and nuclear explosions. The hundreds of such radionuclides are shown in the Chart of Nuclides (Parrington et al. 1996) on both sides of the stable elements and at masses heavier than the stable elements.

2.3.1. Energy Requirements for Nuclear Reactions

Common nuclear reactions are viewed as an interaction of a subatomic particle such as a neutron or proton with the nucleus of an atom to form a compound nucleus in an excited, or higher energy, state. One or more reactions then occur that result in a different nucleus and emission of a nuclear particle or a gamma ray, or even nuclear spallation to form two smaller atoms. The path of the reaction depends on the energy available during the reaction as well as the energy states in the nucleus. The reactions must conserve mass, energy, momentum, spin, nucleon number, and charge. An example is the following reaction:

$$^{27}\text{Al} + {}^{1}\text{H} \rightarrow {}^{27}\text{Si} + {}^{1}n + Q \qquad (2.9)$$

Aluminum-27 reacts with a proton (^{1}H or p) to form silicon-27 and a neutron. The symbol Q represents the energy of the reaction, in million electron volts. The reaction may be written in a shorthand format as ^{27}Al(p,n)^{27}Si. The 13 protons of aluminum plus 1 proton for ^{1}H yield a total of 14 protons on the left side of the equation, and silicon has 14 protons on the right side of the equation. A total of 28 protons plus neutrons are on both the left and right sides of the equation, thus balancing the reaction.

The energy Q related to the nuclear reaction is determined from the differences in the masses M of the reactants and the products converted to million electron volts so that, for the example reaction, $Q = [M_{^{27}\text{Al}} + M_{^{1}\text{H}} - (M_{^{27}\text{Si}} + M_{^{1}n})] \times 931.5$. The masses are expressed in atomic mass units as neutral atoms and the conversion factor is 931.5, in units of million electron volts per atomic mass unit. A more convenient calculation is to use, instead of M, the commonly tabulated mass excess or defect Δ. The quantity Δ is the atomic mass minus the mass number (A) for the nuclide, expressed in million electron volts. These quantities for the individual reactants and products can be substituted in the calculation of Q. For this example, $Q = \Delta(^{27}\text{Al}) + \Delta(^{1}\text{H}) - \Delta(^{27}\text{Si}) - \Delta(^{1}n) = -17.194 + 7.289 + 12.385 - 8.071$ MeV $= -5.591$ MeV. The negative value of Q shows that the kinetic energy of the proton is required for the reaction.

Before the proton energy can be specified, the energy required to overcome the coulombic barrier at the nucleus of the ^{27}Al target must be estimated. The maximum repulsion that the proton will experience is when the distance between the center of the proton and the center of the ^{27}Al is at the minimum defined by the radii. The coulombic barrier is estimated to be about 3.1 MeV. Hence, the energy of 5.591 MeV calculated above is sufficient to overcome the coulombic barrier. A momentum correction is needed to adjust for dissipation of the projectile kinetic energy by both products. The required energy of the proton must be at least 5.591 $(28/27) = 5.8$ MeV for the reaction to occur with reasonable yield.

2.3.2. Production of Radionuclides

The rate of production of atoms by induced reactions is proportional to the number of target atoms, the rate of incoming projectiles, and the probability that the reaction will occur. For a thin target irradiated by a well-defined particle beam, the number of target atoms is expressed by the term nx, where n is the number of atoms of target nuclei per cubic centimeters and x is the thickness of the target material in centimeters. The beam intensity is I, the number of incoming particles per second. The probability of the reaction is given by the cross section with symbol σ and units of centimeter square per atom. The cross section is tabulated in units of barns (b), where 1 b is 10^{-24} cm^2.

The rate of production, P, in s^{-1} of atoms by reactions such as that shown in Eq. (2.9) is

$$P = nxI\sigma \qquad (2.10)$$

In the case of neutron activation in a nuclear reactor, the rate of formation of atoms is given by

$$P = N\Phi\sigma \qquad (2.10a)$$

where Φ is the neutron flux expressed in neutrons s^{-1} cm^{-2} and N is the number of atoms in the sample, i.e., the product mA_v/A in Eq. (2.7).

At constant irradiation intensity, the accumulated number of product atoms is the integral of P over the time of production. In a simple case, the result is Pt. It is less than this value when the projectile flux applicable to this reaction is depleted significantly by parallel reactions; the number of target atoms decreases significantly as a result of the reactions; and the number of product atoms decreases significantly because of radioactive decay or further nuclear interactions.

If the produced atom is radioactive, the rate of radionuclide production in terms of the disintegration rate [shown in Eq. (2.4)] is $R\lambda$. The disintegration rate of the accumulated atoms, balancing the production and decay rates, is then

$$\frac{dN}{dt} = N\Phi\sigma(1 - e^{-\lambda t}) \qquad (2.11)$$

in units of disintegrations per second.

2.3.3. Sources of Neutrons

The most common example of neutron activation is the capture of a neutron by a nucleus, followed by the emission of a gamma ray, designated as (n,γ). This reaction produces an isotope of the target nuclei that has a mass number increased by 1. Competing reactions can occur, such as (n,2n), (n,p), and (n,α), that depend on the energy of the neutron and the relative stability of the products. Neutrons exist in variety of energy ranges, from slow thermal neutrons (0.025 eV) to the high-energy neutrons produced in cosmic showers and accelerators (about 14 MeV). Listed below are some typical neutron sources and their associated energy range.

- A low-Z element (e.g., Be) mixed with a high-specific activity alpha-particle emitter (such as ^{226}Ra or ^{239}Pu) produces neutrons by ^{9}Be(α,n)^{12}C with an energy range of 3–5 MeV. The neutron generation rate depends on the amounts of low-Z element and alpha-particle emitter and is relatively low.

- Some spontaneously fissioning radionuclides, such as ^{252}Cf, produce 3–4 neutrons per fission with an energy spectrum of average energy of about 2–3 MeV. The neutron generation rate depends on the amount of ^{252}Cf, which emits $\sim$(2–3) $\times$ 10^{12} neutrons s^{-1} g^{-1}.
- Accelerators provide a variety of nuclear reactions for production of neutrons. Cockcroft–Walton accelerators can generate 14.8 MeV neutrons by accelerating deuterons (^{2}H) onto a tritium target to produce 10^8–10^{11} neutrons s^{-1}. Cyclotrons and linear accelerators can produce high-energy neutrons with a broad spectrum of energies in spallation reactions that result from the bombardment of heavy elements by charged particles.
- Nuclear reactors are the most common source of neutrons. Inside the reactor, a sustained nuclear reaction of fissile material produces fast neutrons. When ^{235}U is used as the reactor fuel, 2–3 MeV neutrons are produced, along with other neutrons at other energy ranges. A neutron moderator slows the fast neutrons to reduce their energies to the thermal level. This is required to continue the chain reaction by further absorption of neutrons by surrounding atoms of ^{235}U. Other neutron energy ranges are the epithermal between 0.1 and 1 eV and resonances between 1 eV and 1 keV.

2.3.4. Nuclear Fission

As discussed in Section 2.2, SF is a "natural" nuclear decay mode. Nuclear fission can also be induced in numerous nuclides by irradiation with projectiles such as neutrons, protons, and deuterons.

The fission process can be described in terms of the liquid drop model. The drop is capable of undergoing various deformations considered as vibrational modes. As the drop becomes more distorted, it eventually breaks into two primary fragments. The two fragments are usually unequal in mass, with a mass distribution that depends on the manner in which the drop splits. In nuclear fission, absorption of a neutron induces similar oscillations in the target radionuclide that distort its shape until it splits into the two primary FF. A wide array of possible mass combinations exists for the FF. Shown below are three such possibilities for ^{235}U.

$$\begin{aligned}
{}^{1}_{0}\text{n} + {}^{235}_{92}\text{U} &\rightarrow {}^{142}_{54}\text{Xe} + {}^{90}_{38}\text{Sr} + 4\,{}^{1}_{0}\text{n} \\
{}^{1}_{0}\text{n} + {}^{235}_{92}\text{U} &\rightarrow {}^{139}_{56}\text{Ba} + {}^{94}_{36}\text{Kr} + 3\,{}^{1}_{0}\text{n} \\
{}^{1}_{0}\text{n} + {}^{235}_{92}\text{U} &\rightarrow {}^{144}_{55}\text{Cs} + {}^{90}_{37}\text{Rb} + 2\,{}^{1}_{0}\text{n}
\end{aligned} \tag{2.12}$$

The sum of the atomic masses of the heavy fragment, the light fragment, some additional free neutrons, and the released energy equals the mass of the fissioning atom. The Q value for such neutron-induced fission reactions is in excess of 200 MeV per fission.

For a given radionuclide, some fission-fragment pairs are more common than others. This is easily seen in the mass–yield curves of Fig. 2.1, which show the spectrum of atomic masses given off by the fission of several radionuclides. Atom fission yields range up to 6–7% for an individual fission product at the heavy and light mass peaks. These peaks represent the mass fragments that are most common when the parent nucleus splits. The particular shape of the spectrum depends on both the identity of the fissioning nucleus and the energy of the projectile. Fission product yields, in the form of tables that give the yield of each fragment nuclide at each mass number (from approximately 66 to 171), have been compiled over the course of several decades, and are available online at http://ie.lbl.gov/fission/endf349.pdf (January 2006); these are also published in a series of reports by England and Rider (1994).

Fission products are neutron-rich and thus are negatron emitters. At each mass, a decay series may consist of as many as five radionuclides that decay, one into the other, until the chain stops at a stable element. In a decay series, the fission yield may begin at a smaller value for the initial short-lived radionuclides and increase for subsequent radionuclides to the maximum shown in the figure. Among these fission products are ones with half-lives from days to years that are readily measured for monitoring nuclear reactors and nuclear weapon tests.

The fission process releases the large amounts of energy used in nuclear reactors and weapons and also several neutrons. The number of these neutrons depends on both the process and the fissioning atom, as indicated in the ^{235}U decay equations above. When the number of neutrons exceeds one per fission and the energy of the neutrons is suitably moderated to induce further fission, a chain reaction is induced that can be used for energy production.

2.4. The Basis for Detecting Radiation

Detection of radionuclides based on their radiations—alpha, beta, or gamma—commonly occurs by one of two processes. Electrons bound to the atoms or molecules of the material through which charged particles such as alpha and beta particles pass are released in a process called ionization. The separated electrons and ions can be collected at two electrodes by imposing a potential difference across the detector space; their presence is then measured as pulses or a current.

Alternatively, the transfer of the radionuclides' energy to these bound electrons raises them to an excited state in the atom or molecule. When the excited species returns to its ground state energy level, the excited atom or molecule may emit electromagnetic energy in the ultraviolet to visible (UV/Vis) region. This light can be detected by a photomultiplier tube (PMT) in a scintillation counter system.

Less common detection of electrons is by photons emitted during their deceleration in a medium. The two mechanisms are creation of bremsstrahlung (continuous X-ray spectrum) and Cherenkov radiation (visible light). Cherenkov radiation is detected with a PMT; bremsstrahlung X rays are detected as discussed below for gamma rays.

Detection of gamma or X rays by ionization or excitation is a secondary process where the first step is the formation of a free electron by interaction with matter.

A less common process is the creation of a positron–electron pair, as discussed below.

2.4.1. Alpha Particles and Ion Formation

The energy per ion formation, w, is a measured value that has been tabulated for many gases (ICRU 1979). Average values are 35 eV in air and 26 eV in argon; w is in the same energy range for many other gases. Hence, a 5-MeV alpha particle theoretically has enough energy to create 192,000 ion pairs in argon. That the average pair formation energy greatly exceeds the first ionization potential for the gas—e.g., 15.7 eV is needed to remove the first electron from argon—suggests that some of the required pair formation energy is devoted to effects other than ioniziation, such as the excitation discussed above.

Ionization per unit path length is not constant when a charged particle loses energy. As the alpha-particle energy loses energy along its path, the stopping power of the medium, defined as the average energy loss per path length, gradually increases to a maximum as the probability of interaction of the alpha particle with the atoms or molecules along its path increases. The peak value at about 0.7 MeV in air is $2.0 \, \text{MeV} \, \text{mg}^{-1} \, \text{cm}^{-2}$, compared to $0.8 \, \text{MeV} \, \text{mg}^{-1} \, \text{cm}^{-2}$ at 5.0 MeV. Below 0.7 MeV, the stopping power decreases to 0 MeV at zero energy (ICRU 1993).

The alpha particle loses most energy in interactions with electrons. Below 0.1 MeV, collision with the nucleus of the material along its track contributes significantly to energy loss. The unit of mass per area (or distance times density) is used instead of distance for stopping power to obtain similar values for the entire range of elements. In fact, stopping power values increase by about an order of magnitude as a function of the atomic mass of the element with which the alpha particle interacts.

Because alpha particles are monoenergetic, all in a given energy group can be expected to stop at the same distance. A small deviation occurs because the energy of the alpha particles after multiple interactions has a normal distribution.

The range ($\Re$) of an alpha particle does not vary linearly with energy because of the above-cited pattern of variation in stopping power. An empirical relation of range to its initial energy E_α is given by Eq. (2.13) (Friedlander et al. 1981):

$$\Re = k \, (E_\alpha \,)^a \tag{2.13}$$

The constants k and a are determined experimentally. For example, the range of an alpha particle in air at STP (standard temperature and pressure conditions) is given by Eq. (2.13a) (Eichholz and Poston 1979):

$$\Re = 0.31 \, (E_\alpha \,)^{3/2} \tag{2.14}$$

where the alpha-particle energy is in million electron volts and the range is in centimeters.

Figure 2.3 is a graph of the range calculated by the continuous-slowing-down approximation (CSDA) of an alpha particle, in both air and aluminum. The data that generated the graph (ICRU 1993) indicate that a 5-MeV alpha particle has a

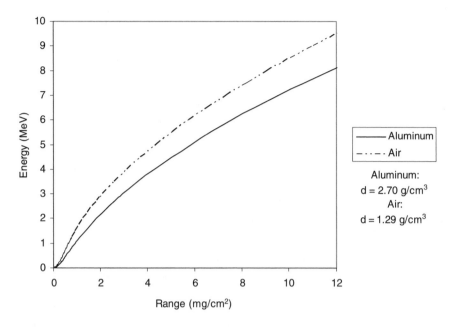

FIGURE 2.3. Alpha-particle range in air and aluminum. (Data from ICRU 1993, pp. 192 and 213.)

range of 3.4 cm (CSDA range 4.37 mg cm^{-2} divided by the density of air, 1.29 mg cm^{-3}) in air. Equation (2.13) gives a value of 3.47. The Bethe equation for the energy loss of charged particles is complex but indicates that the loss is related to the electron density, i.e., the atom density times the atomic number Z. The empirically derived Bragg–Kleeman rule suggests that the range in one material relative to another is proportional to $\rho A^{0.5}$ (Evans 1955), where ρ is the density of the attenuating material.

2.4.2. Beta Particles and Ion Formation

At relatively low energies, the ion formation process by beta particles and CE is similar to that of alpha particles but their range is much greater because of the 7000-fold smaller mass and 2-fold smaller charge. For example, the stopping power for electrons in aluminum is 1.5 keV mg^{-1} cm^{-2} at 1000 keV, reaches a maximum of 300 keV mg^{-1} cm^{-2} at 0.1 keV, and then decrease to 0 at 0 keV. Energy transfer is basically by electron–electron interaction, but bremsstrahlung contributes significantly at beta-particle energies above about 500 keV. The relativistic velocity of higher energy electrons results in little change in the stopping power between about 700 and 4000 keV, and a gradual increase in stopping power above that energy.

As for alpha particles, the relation of range to energy of electrons (or beta particles) is not linear. All monoenergetic electrons in a group have the same

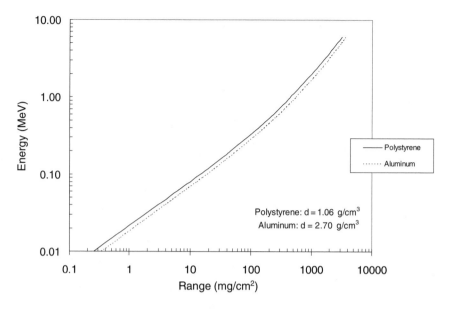

FIGURE 2.4. Beta-particle range–energy curve (log/log) in aluminum and polystyrene. (Data from ICRU 1984.)

range. A set of range–energy curves is shown in Fig. 2.4, with range in units of mg cm^{-2} in aluminum and polystyrene. Range values in terms of this unit can be applied to various substances within a restricted range of Z. An equation for beta-particle range similar to the equation for alpha particles above is

$$\Re = 412 \left(E_\beta \right)^n \tag{2.14}$$

where $n = 1.265 - 0.0954 \ln E$, and E refers to the E_{max}, maximum energy of the beta-particle group. Simpler equations have been proposed for portions of the range–energy curve for which linear approximations are possible (Evans 1955).

The energy spectrum of a beta-particle group takes the form of a continuum that has different shapes for different radionuclides; three are shown in Fig. 2.5. Because of the energy distribution of beta particles, the relationships observed for the interaction of beta particles with matter are not as simple as those of alpha particles.

The typical curve of beta-particle attenuation in aluminum absorbers, shown in Fig. 2.6, at lower energies resembles the exponential attenuation observed for gamma rays (see Section 2.4.4). The final part of the line curves downward to reach the distinct range associated with E_{max}. Attenuation curves for the various beta-decay radionuclides differ because of the different beta-particle energy spectra, but both the characteristic range and the approximately exponential attenuation have been used to estimate maximum beta-particle energies (Evans 1955).

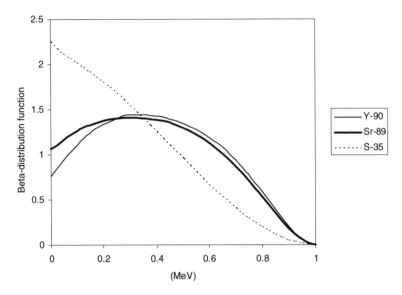

FIGURE 2.5. Three typical beta-particle spectra. (Data from ICRU 1997, pp. 107–108.)

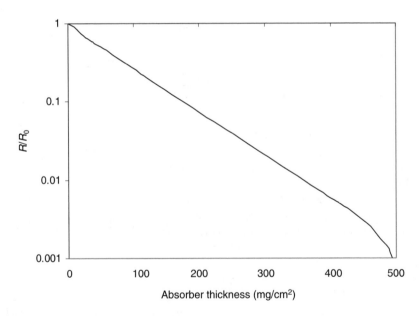

FIGURE 2.6. Beta-particle attenuation curve: ^{210}Bi in aluminum absorbers. (Based on Zumwalt 1950, p. 40.)

2.4.3. Bremsstrahlung and Cherenkov Radiation

These two forms of electromagnetic radiation are produced by charged particles—electrons in the present context—that slow down in matter ("bremsstrahlung" in German means "braking radiation"). An electron moving with decreasing velocity in the electric field of atoms emits part of the energy associated with this decrease in velocity as continuous X rays. The maximum-energy X ray equals the energy of the electron (or beta particle or positron). The X-ray intensity is very small at this maximum energy but increases exponentially with decreasing energy. The fraction of beta-particle energy converted to bremsstrahlung is approximately (3.5×10^{-4}) ZE_{max}. Thus, in the passage of a 1-MeV beta-particle group through lead ($Z = 82$), approximately 3% of the energy is converted to bremsstrahlung, while through beryllium ($Z = 4$) it is 0.14%. Most of the X rays are at energies below 0.1 MeV.

Cherenkov radiation (named after its discoverer, P.A. Cherenkov) is bluish light emitted forward when a very energetic charged particle, traveling in a transparent medium at a velocity faster than light could in the same medium, slows to the velocity of light in that medium. This situation occurs when the refractive index of the medium is well above 1, such as 1.33 in water. A low-energy cut-off for Cherenkov radiation decreases with the refractive index. In water, beta particles below an energy of 0.265 MeV do not stimulate Cherenkov radiation. The number of photons per electron is about 200 at 1 MeV and about 600 at 2 MeV (Knoll 1989). The Cherenkov light at an energy of about 2.5 eV is detected by a liquid scintillation counter. Less than 0.1% of the electron kinetic energy is transferred to Cherenkov radiation.

Bremsstrahlung is of interest in radioanalytical chemistry because some of the energy of electrons stopped in detector-shielding material is converted to X rays that can penetrate the shield. Cherenkov radiation permits scintillation counting of radionuclides in plain water samples if the electron energy is sufficiently high and the detection system is sufficiently sensitive.

2.4.4. Gamma Radiation Interactions

Gamma rays are electromagnetic radiation without mass and charge. They are far more penetrating than electrons of the same energy, and *much more* penetrating than alpha particles. Three major gamma-ray interactions produce free electrons that can be detected: Compton scattering, photoelectric effect, and pair formation. The interactions have been given respective interaction coefficients σ, τ, and κ that define the exponential attenuation of gamma rays in matter and are used to calculate the extent of shielding (i.e., gamma ray removal from a beam), interaction (i.e., energy deposition), and radiation detector efficiency. The total attenuation coefficient μ for gamma rays is the sum of the individual coefficients. The ratio of the gamma-ray flux after attenuation, I_f, to that before attenuation, I_0, is a function of the distance traveled x:

$$\frac{I_f}{I_0} = e^{-\mu x} \tag{2.15}$$

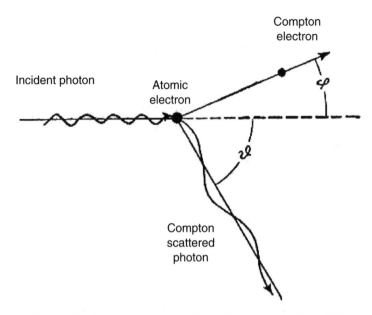

FIGURE 2.7. Compton scattering. (Figure from Evans 1955, p. 675.)

The units of μ can be cm^2 g^{-1}, cm^2 atom^{-1}, or cm^{-1}; respective units of x are g cm^{-2}, atom cm^{-2}, or cm. The interaction fraction with the medium is $I_0 - I_f$,

$$(I_0 - I_f)/I_0 = 1 - e^{-\mu x} \tag{2.15a}$$

Compton scattering occurs when a gamma ray interacts with an electron bound weakly (i.e., an outer orbital electron) to an atom or molecule and transfers a fraction of its energy to the electron. The resulting weaker gamma ray departs at an angle, as illustrated in Fig. 2.7. The energy distribution between the electron and the produced gamma ray depends on the angles at which the two move relative to the initial gamma ray. Much or all of the energy transferred from the initial gamma ray to the electron is deposited in the absorbing material in which the electron slows down.

The electron energy is distributed between zero and a maximum energy that is somewhat less than the energy of the incoming gamma ray, E_γ. The maximum electron energy is

$$E_{max} = E_\gamma \left[\frac{2E_\gamma/0.51}{1 + (2E_\gamma/0.51)} \right] \tag{2.16}$$

For example, if the incoming gamma-ray energy is 1.0 MeV, the maximum electron energy is 0.8 MeV. The energy of the scattered gamma ray is the energy of the initial gamma ray minus that of the ejected electron. That gamma ray

TABLE 2.2. Photon interaction and energy absorption coefficients for common materials[a]

Energy (keV)	Mass attenuation (cm^2g^{-1})			Mass energy absorption (cm^2g^{-1})			
	Water	SiO$_2$	CaCO$_3$	Water	Air	Muscle	Bone
50	0.208	0.282	0.485	0.0394	0.0384	0.0409	0.158
100	0.165	0.158	0.181	0.0252	0.0231	0.0252	0.0386
200	0.136	0.135	0.125	0.0300	0.0268	0.0297	0.0302
300	0.118	0.106	0.107	0.0320	0.0288	0.0317	0.0311
500	0.096	0.087	0.087	0.0330	0.0297	0.0327	0.0316
1000	0.0707	0.0636	0.0636	0.0311	0.0280	0.0308	0.0297
2000	0.0494	0.0447	0.0448	0.0260	0.0234	0.0257	0.0248
3000	0.0397	0.0363	0.0366	0.0227	0.0205	0.0225	0.0219

[a] Data from Shleien (1992).

usually has sufficient energy to interact further, by more Compton scattering or otherwise.

The two solid curves in Figs. 2.8 and 2.9 show the distinction between gamma-ray attenuation, represented by μ_0, due in part to Compton attenuation σ_0, and gamma-ray energy absorption, represented by μ_a, due in part to Compton energy absorption σ_a. Fractional attenuation refers to the number of gamma rays that interact in the medium relative to the number of gamma rays that enter the medium. Fractional energy absorption refers to the average amount of energy absorbed per interaction relative to the energy involved in each interaction. Because some gamma rays are scattered and carry some energy away from the point of interaction, the energy absorption fraction is always less than the attenuation fraction, as shown in Table 2.2.

Values from the upper solid curves (labeled "Total attenuation") in Figs. 2.8 and 2.9 are applied with Eq. (2.15) to calculate the fractional transmission at a thin absorber of gamma rays in a collimated beam. Equation (2.15a) is applied to values from the lower solid curves (marked "Total absorption") to calculate the fractional deposition of energy from gamma rays in the absorbing material; more generally, these calculations give the radiation dose, in energy per unit mass, for radiation protection and material response purposes. Table 2.2 gives the values of these coefficients for some typical materials of interest; many others are also available.

In practical applications, when the beam is broad and the attenuating material is thick, fractional transmission of gamma rays is greater than the values calculated from the upper curve in Figs. 2.8 and 2.9. While the gamma rays scattered in a very thin attenuator are removed from the beam and do not reach a point in the beam behind the scattering material, in a thick attenuator, some of the scattered gamma rays again are scattered and will reach the point in the beam behind the scattering material. The fraction of increased radiation to which this point is exposed had earlier been determined empirically for various materials shapes, and thicknesses,

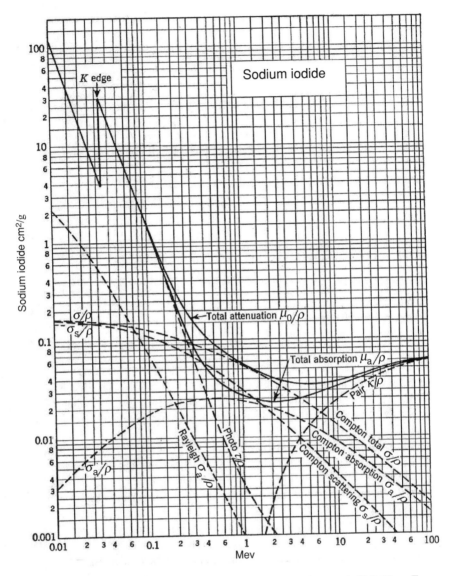

FIGURE 2.8. Gamma-ray interaction coefficients as function of energy in NaI. (From Evans 1955, p. 716.)

and is now calculated by Monte Carlo simulation. This "buildup factor" B (Shleien 1992) exceeds 1.0 and is used to multiply I_f in Eq. (2.15). The buildup factor depends on gamma-ray energy, source-shield configuration, and shield thickness. It usually is reported in terms of the dimensionless relaxation length $\mu_0 x$.

The second mechanism by which gamma radiation loses energy is the photo-electric effect. In this interaction, all of the energy is transferred from the gamma

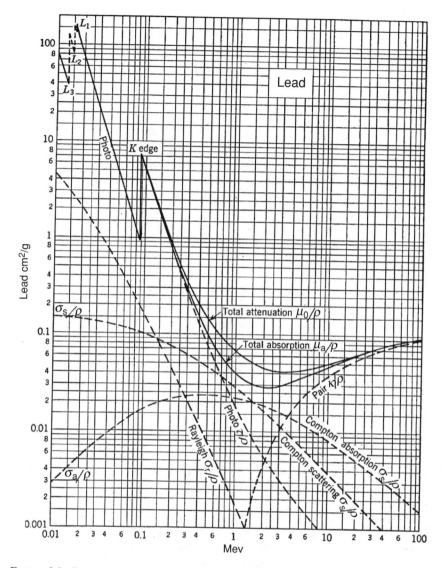

FIGURE 2.9. Gamma-ray interaction coefficients as function of energy in Pb. (From Evans 1955, p. 717.)

ray to a strongly bound interior orbital electron. When the energy transmitted by the gamma ray exceeds the electron binding energy E_{BE}, the electron is ejected from the atom or molecule with energy E_e according to Eq. (2.17):

$$E_e = E_\gamma - E_{BE} \qquad (2.17)$$

The photoelectric effect becomes more likely with lower E_γ and as electrons are more tightly bound in their shells (i.e., K > L > M). The missing electron is replaced instantaneously by another electron with the emission of an X ray or Auger electron as described in Section 2.2.3. The electrons stopped in the absorbent material plus the stopped associated X rays and Auger electrons deposit energy E_γ. In relatively small absorbers, when the X ray escapes without being stopped, the energy deposited is E_γ minus the energy for the escaped X ray.

The third mechanism of gamma-ray interaction is pair formation. The gamma ray transforms itself near a nucleus into a positron and an electron. For pair formation to occur, the energy of the gamma ray must be at least twice that of the electron rest energy, mc^2, of 0.511 MeV, so that the energy must be equal to or greater than 1.022 MeV.

After pair formation, the electron and positron move in opposite directions so as to conserve energy and momentum. Each has a kinetic energy of $0.5(E_\gamma - 1.022)$ MeV. The moving electron causes ionization and excitation. The positron behaves similarly until it slows down near an electron and both are annihilated, creating two gamma rays that are directed away from each other, each with an energy of 0.511 MeV. These two 0.511-MeV gamma rays then proceed to interact by Compton scattering or the photoelectric effect. This annihilation process occurs within about 0.1 cm in solid materials, but can be at a greater distance in air. Because of the potential for interaction at a distance from the source and for Compton scattering, the fraction of energy deposition is less than that for interaction, as discussed for Compton scattering.

The interaction coefficients for the three modes of gamma interaction with sodium iodide and lead are shown in Figs. 2.8 and 2.9. The photoelectric effect declines as the energy increases, as does Compton scattering. The jagged edge in the photoelectric effect curve occurs at the binding energy of K orbital electrons. A fine structure that is not shown at this edge arises from the various X-ray energies discussed in Section 2.2.3. Pair formation starts at 1.022 MeV and increases as the energy increases. The total interaction at any given energy is the sum of the three mechanisms. The electrons generated by these mechanisms will act as described above for beta particles and CEs to ionize or excite molecules in systems used for radiation detection.

Gamma rays also interact with matter by other reactions, two of which occasionally may be considered. One interaction is coherent (Rayleigh) scattering at low energies. The gamma ray is absorbed by an atom and reemitted with unchanged energy. The direction of the gamma ray is changed due to scattering from bound electrons. The scattered gamma ray commonly proceeds to interact by the photoelectric effect. Because coherent scattering produces interference patterns, the process is used in crystallographic analysis to make structural determinations. Its attenuation coefficient decreases with energy and accounts for only a few percent of attenuation below 0.2 MeV in lower Z elements, though it is still significant in lead to 1 MeV.

Another interaction is photodisintegration, generally a (γ, n) reaction. This reaction can occur if photons have energies of 8 MeV or higher. Exceptions occur at energy thresholds of 1.67 and 2.23 MeV for ^{9}Be and ^{2}H, respectively, to produce ^{8}Be and ^{1}H plus neutrons.

2.5. Detectors

2.5.1. Gas-Filled Detectors

Gas-filled detector systems collect and record the electrons freed from gaseous atoms and molecules by the interaction of radiation with these atoms and molecules. The systems have been classified into the primary categories of ionization chambers, proportional counters, and Geiger–Mueller (G-M) counters. The detectors have a variety of designs, but essentially consist of a chamber, which serves as the cathode, and a center wire, which serves as the anode. The electrical field is characterized by chamber shape and radius, the wire radius, and the applied voltage (Knoll 1989). The chamber is filled with a gas and a potential is applied between the two electrodes. The configuration of the system is shown in Fig. 2.10.

Ionized electrons in the gas are collected onto the anode. Figure 2.11 shows the response of the system in terms of the number of electrons or ions produced in the gas when the applied voltage is increased. The curves show six regions, each of which has different properties. The higher and lower curves pertain to particles that deposit more and less energy, respectively, in the detector.

In Region 1, many electrons and ions produced in the gas recombine because the voltage applied between cathode and anode is not large enough to collect all electrons. This region is not useful for counting radiation.

Region 2 is the saturation region. The potential difference is sufficient to collect all freed electrons. The resulting very weak current or pulse over brief periods is measured with extremely sensitive electrometer devices. A detector working in the saturation region is called an ionization chamber. Its output is proportional to the deposited radiation energy. Internal or thin-window ionization chambers are used as alpha-particle and fission-fragment detectors. External samples are measured for beta particles and gamma rays with ionization detectors; the latter are larger and contain counting gas at elevated pressure.

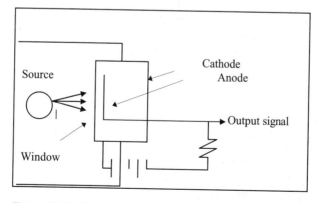

FIGURE 2.10. Configuration of a gas-flow proportional counter.

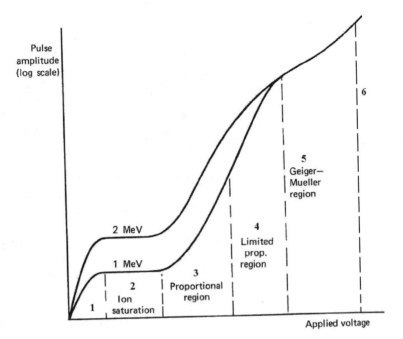

FIGURE 2.11. Variation of pulse height with applied voltage in a counter. (From Knoll 1989, p. 162.)

In Region 3, the proportional region, the applied voltage is strong enough that the electrons freed by the initial radiation are accelerated, so that they, in turn, ionize additional atoms or molecules (secondary ionization) to free more electrons. This electron multiplication generates an avalanche toward the anode for each primary electron that was freed. The applied voltage domain is called the proportional region because each avalanche is characterized by the same electron multiplication at a given applied voltage. The output signal is directly proportional to the deposited energy, although each pulse is many times larger than in the ionization region. As the applied voltage is increased, the amplification increases uniformly for the entire range of deposited energy.

A problem in proportional counters operating at these higher voltages is production of a secondary avalanche due to molecular excitation, which interferes with the detection of subsequent pulses. To prevent excitation that causes this secondary avalanche, a quenching agent is added to the fill gas (see Section 8.5.1). The multiplication factor in this region is about 10^4, which requires the measurement device to be far less sensitive than that for the ionization chambers. The proportional counter system has a preamplifier and a linear amplifier.

In Region 4, the proportionality of the output signal to the deposited energy at a given applied voltage no longer applies. Amplification of the greater deposited energy reaches its limit while that for the lesser deposited energy continues to increase. This region is called the region of limited proportionality and is usually avoided as a detection region.

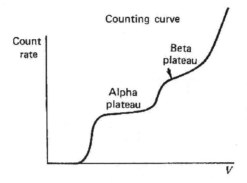

FIGURE 2.12. Plateau regions for both alpha and beta particles in a proportional counter.

The third useful region for detection of radiation is Region 5, the G-M region. This applied voltage is high enough that any deposited energy produces sufficient secondary electrons to discharge the entire counting gas. The linear amplifier is no longer needed. One no longer can distinguish between a small and a large deposition of energy. This discharge must be quenched so that the next pulse can be detected. Either the applied voltage must be removed briefly or quenching gases (see Section 8.5.1) must be added.

In Region 6, an electrical discharge occurs between the electrodes. This voltage region has been used for some purposes but generally is avoided because the discharge can disable the detector.

Although gas-filled detectors are operated in the second, third, and fifth voltage regions, it would be a mistake to assume that a particular detector can be used in all of them simply by changing the applied voltage. Detector components are designed to be used in a single voltage domain. The operating voltage then is selected at a plateau region on the basis of the curve of count rate vs. applied voltage, shown in Fig. 2.12.

In proportional and G-M detectors, this curve begins with zero count rate at a relatively low voltage, reaches a plateau at intermediate voltage, and then increases when a continuous electrical discharge occurs. The initial increasing count rate represents the increasing number of pulses that are sufficiently amplified to pass the lower discriminator. At the plateau, all pulses are detected. In a proportional counter, a first plateau is reached for the much larger alpha-particle pulses, and a second plateau at higher applied voltages is then reached for beta particles as well as alpha particles. Detectors are operated near the middle of the plateau to avoid erroneous data due to small fluctuations in applied voltage. In proportional counters, the applied voltage is reduced to measure alpha particles but not beta particles.

2.5.2. Solid-State Detectors

The operating principle for solid-state detectors is analogous to that of the gas-filled detector except that collection at the electrodes of electrons and electron holes

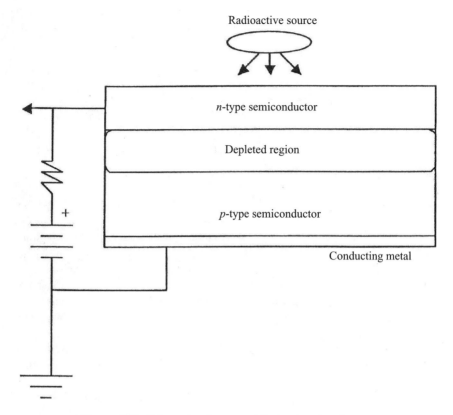

FIGURE 2.13. Schematic of a reverse bias p-n junction detector.

replaces electrons and positive ions. One advantage is the thousandfold greater electron density in solids for the much greater stopping power that is desirable for detecting gamma radiation. Another advantage is the 10-fold smaller value of w, the energy absorption per ionization, which yields better energy resolution. The solid medium currently is hyperpure germanium or silicon that has been prepared to function as semiconductor material. A schematic of a reverse bias p-n junction detector is shown in Fig. 2.13.

A p-type material is one with positive holes and an n-type material is one with electrons in excess. To create these types of matrices, small amounts of impurities are incorporated into some host materials. For example, if the host material is silicon, which has four valence electrons, it may be "doped" with boron, which has three valence electrons. This creates regions in the matrix where there are "holes" for electrons to occupy; it becomes a p-type material. The energy for creating an electron–hole pair is 3.7 eV in silicon and 3.0 eV in germanium (Knoll 1989). Thin wafers of silicon diode surface barrier detectors that have a very thin layer of gold on the surface are used for alpha-particle spectroscopy. Hyperpure germanium detectors, typically a closed-end coaxial, 6 cm in diameter and 6 cm or more in length, are currently used for gamma-ray spectroscopy.

Conversely, if phosphorus or arsenic, each with five outer shell electrons, is inserted into the silicon crystalline material, then the resulting material will have an "excess" of electrons in the lattice, producing the n-type material. The placement together of an n-type and p-type creates a p-n junction.

When a positive charge is applied to the n-type semiconductor and a negative charge is applied to the p-type material, the positive holes are attracted to the negative electrode and the electrons are attracted toward the p-n junction. This creates the depletion layer. Radiation enters the n-type side where the depleted region serves as the radiation-sensitive volume. There, electron–hole pairs are created that will be rapidly collected to create the pulse for amplification. Silicon semiconductors of this type are generally used for beta-particle and CE spectroscopy.

Surface-barrier detectors for measuring alpha particles are formed from n-type silicon with an oxidized p-type surface. A very thin gold layer is evaporated onto this surface to function as one of the electrodes. The sample that emits alpha particles faces this side.

2.5.3. Scintillation Detection Systems

Certain materials that become excited by the absorption of radiation are able to de-excite through emission of light. These scintillators—or phosphors—enable radiation detection through the observation of flashes of luminescence. The amount of light observed is proportional to the amount of radiation energy absorbed by the medium. A scintillation detector utilizes a phosphor (to absorb radiation and produce light), optically coupled to a photomultiplier tube (PMT), which detects the light and converts it into an electrical signal. A computer then collects the signal produced by the PMT.

The phosphor can be an inorganic crystal or an organic solid, liquid, or gas. Materials function well as a scintillator for radiation if they efficiently absorb and convert radiated energy to light, exhibit good transparency for the light to escape the material, and emit a frequency of light that matches the photon-detection system. The photons emitted by the phosphors and detected by PMTs typically are in the 2.5–3.6-eV UV/V is energy range. Some common scintillation detectors are given in Table 2.3.

One effective medium in common use for gamma-ray measurements is thallium-activated sodium iodide [NaI(Tl)]. The relatively high-Z iodine atom provides a high attenuation coefficient for interacting with energetic gamma radiation. It

TABLE 2.3. Some common scintillators

Scintillation material	Physical state	Radiation primarily detected	Conversion efficiency (%)
Anthracene	Solid	Beta	4.5
NaI(Tl)	Solid	Gamma	10
Liquid scintillators	Liquid	Alpha, beta	2–4
Plastic scintillators	Solid	Alpha, beta, gamma	3.0

results in an appreciable fraction of counts in the full-energy peak, which is useful for spectral analysis (see following section). The crystal is sealed in an aluminum can because it is hygroscopic, with a light pipe at one end for coupling to the PMT.

The inclusion of thallium in the NaI(Tl) detector provides energy levels that are somewhat lower than the conduction band. According to the band theory of crystalline solids, the valence electrons commonly occupy the highest filled atomic electron band. Above this energy is a band gap, and higher still is the conduction band. Incoming radiation (if sufficiently energetic to surmount the band gap) can excite the valence electrons to the conduction band; from here the electrons drop back, de-exciting by the emission of light. The thallium impurity facilitates the de-excitation of electrons to the valence band. The benefit of a smaller energy gap is that the excited electron will shed its excess energy more quickly for more efficient detector response. Second, the energy emitted will not be the same as that which caused the initial excitation; consequently, photon emission will not cause re-excitation (i.e., the detector is transparent to its radiation).

Some other scintillation materials, such as cesium iodide and bismuth germanate, have characteristics that are less favorable than NaI(Tl) for general use, but recommend them for some special measurements. For example, CsI and NaI(Tl) can be combined for coincidence or anticoincidence counting by distinguishing between output from the two detectors by their pulse shapes.

Silver-activated zinc sulfide [ZnS(Ag)] has been used since radioactivity was first measured to detect alpha particles. It is relatively insensitive to electrons and gamma rays because it is not transparent to its own radiation. Radiation interactions within the detector are not recorded; only its surface, where alpha particles interact, emits scintillations. The ZnS is doped with silver to shift its scintillations to a longer wavelength for better PMT response.

Organic scintillators function similarly to inorganic crystals. One major difference is that impurities are not necessary to produce energy levels for de-excitation because the selected organic substances quickly shed energy by passing through multiple vibrational states. Organic scintillators that have resonance structures, such as anthracene, are excellent scintillators. The effect is similar to doping; the de-excitation photon is rarely of the same energy as the absorbed radiation, so that the organic scintillator is transparent to its own emissions.

Organic scintillators have relatively low counting efficiency for gamma rays because of their low-Z carbon, oxygen, and hydrogen atoms. Any spectral analysis has to be based on Compton-edge energy measurements because of the low full-energy peak intensity. Solid organic scintillators can be useful for beta-particle detection because they can be exposed without a container. They can also be machined to large volume in many shapes.

In a liquid scintillation (LS) system, the sample is mixed with a cocktail that consists of an organic scintillator dissolved in an organic solvent. The cocktail and the usual aqueous sample form an emulsion. The radiation emitted by the intimately mixed radionuclide deposits its energy in the solvent, which transfers it to the scintillator. The scintillations are then detected by the PMT. The LS counter is useful for detecting alpha particles and low-energy beta particles from samples that are dispersed throughout the LS detector, whereas in other detectors the radiation would be absorbed in the sample.

2.5.4. Silicon and Germanium Spectrometers

Solid-state detectors—notably silicon for alpha particles and germanium for gamma rays—are particularly effective for spectral analysis. Spectral analysis is used both to identify the radiation by its characteristic energy peak and to determine its activity based on the count rate within this peak. Further advantages of restricting analysis to the narrow band of the characteristic energy peak are reducing interference from other radionuclides in the sample and from background radiation. Other materials, such as cadmium telluride and cadmium zinc telluride, have been developed for specific advantages such as inclusion of a high-Z material, but are not in routine use.

The detection systems for alpha-particle and gamma-ray spectrometers are distinctly different. For alpha particles, a thin silicon detector and a thin sample are placed in a small vacuum chamber to eliminate energy loss in air. No further shielding is needed because the chamber is a shield for ambient background alpha particles while other ambient radiation is not detected in the thin solid to any significant extent. For gamma rays, a large germanium detector and samples as large as necessary are placed in a radiation shield of lead or steel.

The spectrometer systems require highly stable power supplies and amplifiers to support analysis at high-energy resolution. The germanium detector has a cryostat for cooling during application. A Dewar flask that contains liquid nitrogen coolant is commonly used, but electromechanical cooling can be substituted. Various geometric arrangements enable the germanium detector to view the sample vertically or horizontally.

The detected radiation is displayed to show near-Gaussian peaks (see Fig. 2.14) in a 1024-, 2048-, or 4096-channel spectrometer. The spectrometer is calibrated for energy by matching the channel at the midpoint of the peak to the known energy of gamma radiation emitted by a set of radioactivity sources. Typically, the channel width is set to approximately one-fourth of the detector peak resolution, but some compromise is needed because the resolution changes with energy.

The energy resolution of a recorded set of many identical pulses is the extent to which individual values deviate from the expected single value. If a normal, Gaussian energy distribution can describe the deviation from the expectation value that results in the observed peaks, a simple view (see Section 10.3.4) suggests that the standard deviation σ for this peak equals the square root of the number of events for each peak. The number of events is E/w, where E is the deposited energy and w is the energy required to form an electron–hole pair. The standard deviation in units of eV is σ_w; relative to the energy it is σ_w/E; hence

$$\frac{\sigma_w}{E} = \frac{w}{E}\sqrt{\frac{E}{w}} = \sqrt{\frac{w}{E}} \tag{2.18}$$

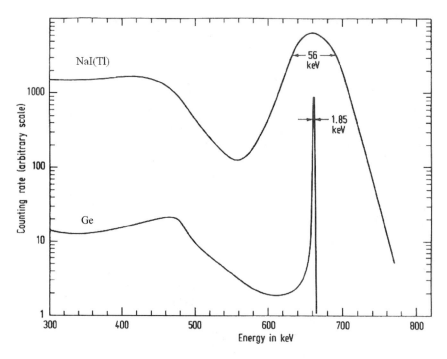

FIGURE 2.14. Peak resolution in gamma-ray spectra from Ge and NaI(Tl) detectors. (From Friedlander *et al.* 1981, p. 259.)

The resolution of a peak is commonly described by the full width at half maximum, FWHM. Since FWHM equals 2.36σ, to describe the resolution in these terms,

$$\text{FWHM} = 2.36E\sqrt{\frac{w}{E}} = 2.36\sqrt{wE} \qquad (2.19)$$

At the w value of 3.0 eV for germanium and 3.7 eV for silicon, the equation yields the FWHM of 4.7 keV for germanium at 1.33 MeV and of 9 keV for silicon at 4 MeV. The resolution measured at these energies is about 3-fold better in germanium and 1.5-fold worse in silicon.

Two sets of factors that work in opposite directions are not considered in Eq. (2.19). The equation overestimates the FWHM by $\sqrt{F}$, where F is the empirically observed Fano factor. The existence of this factor is attributed to the circumstance that the generated electrons do not necessarily act independently in producing the ionization pulse, and so the peak is narrower than when attributed to random events. On the other hand, detector drift, noise, and incomplete carrier collection each contributes to widening the FWHM (Knoll 1989).

The counting efficiency of these detectors is calibrated with radionuclide standards or Monte Carlo simulation (Briesmeister 1990). Typically, the alpha-particle detector has the same efficiency for thin samples at all commonly encountered energies, while the gamma-ray detector efficiency has a maximum near 100 keV for samples of various dimensions.

Commercially available spectrometer systems include a computer for operating the system, storing data, analyzing data, and providing information output. These functions are used to set counting times and periods, identify peaks by channel and energy, store and subtract background, calculate gross and net counts in multiple peak regions, calculate count rates, and convert count rates to disintegration rates on the basis of stored decay scheme data. Statistics software programs are available to estimate output data uncertainty.

2.5.5. Other Spectrometers

Alpha-particle spectral analysis before development of solid-state detectors was accomplished with the Frisch grid chamber. In this large ionization detector, the effect on pulse height due to sample location and ion drift is minimized. A collimator restricts incoming alpha particles to a limited volume, and a grid that is as transparent to electrons as possible shields the sensitive area from the positive ions that are generated. Although the silicon diode is far more convenient because of its stability, large-area Frisch grid chambers continue to be used for alpha-particle spectral analysis of surfaces larger than about 20 cm^2.

When designed for optimum detection efficiency, gas-filled proportional counters are effective as X-ray spectrometers for photons of energy below about 80 keV. The important criterion is good counting efficiency, achieved by absorbing a large fraction of the X rays entering the detector. These detectors are larger than the detectors used for alpha- and beta-particle counting, filled with a gas that has higher electron density such as xenon, and at a pressure of several atmospheres. The window is a beryllium disk to maintain the pressure while minimizing attenuation.

Gamma-ray spectral analysis before development of solid-state detectors was by the NaI(Tl) detector. Some of these detectors are much more efficient than germanium detectors, both because iodine has a higher Z value than germanium and also because larger sodium iodide detectors can be formed. No cooling is necessary, although fairly uniform temperature is needed to eliminate calibration drift. Because the resolution is controlled by the counting statistics of energy conversion at the PMT, the resolution is much worse for NaI(Tl) detectors than for germanium detectors, typically by a factor of 30 near 0.66 MeV, as shown in Fig. 2.14. This spectrometer system is still used in selected situations where good resolution is not crucial, i.e., if only a few radionuclides are to be measured at one time, and if ease of application and portability are important factors.

Older LS counter systems (see Section 8.3.2) have three channels for distinguishing energies; new ones have full energy spectrometers. These are directly applicable for distinguishing alpha particles by energy. Because beta particles are emitted as a spectrum from zero to maximum energy, and the spectra have various shapes, identification of beta-particle emitters by energy is less feasible.

The energy spectrum presented by the LS system does not exactly reflect the beta-particle spectrum because some energy is not deposited or is lost in transmission. When multiple beta-particle energy groups are detected, only the more energetic beta particles in the group with the highest maximum energy can be seen by themselves because the other groups overlap.

The advantage of an LS counter as alpha-particle spectrometer is the minimal energy loss because the sample is integral to the detector medium. The disadvantages are the relatively poor resolution associated with the PMT, some edge effects where the alpha-particle emitter is near the vial wall, and a higher radiation background count rate. Typical FWHM is around 500 keV at 5 MeV. About 2-fold better resolution is achieved by special pulse timing discrimination (McDowell 1992).

3
Analytical Chemistry Principles

JEFFREY LAHR and BERND KAHN

3.1. Introduction

To separate and purify the radionuclide of interest in the sample, the analyst can depend on the similar behavior of the stable element and its radioisotopes. Chemical reactions involving the radionuclide will proceed with essentially the equilibrium and rate constants known for the stable element in the same chemical form. Slight differences result from small differences between the isotopic mass of the radionuclide and the atomic mass (i.e., the weighted average of the stable isotopic masses) of the stable element. Because of this similarity in chemical behavior, many radioanalytical chemistry procedures were adapted from classical quantitative and qualitative analysis. For the same reason, new methods published for separating chemical substances by processes such as precipitation, ion-exchange, solvent extraction, or distillation are adapted for and applied to radionuclides. One exception occurs when the radionuclides to be separated are two or more isotopes of the same element. Here, effective separation can be accomplished by mass spectrometer (see Chapter 17).

This chapter reviews classical separation techniques and their roles in the radioanalytical chemistry laboratory. Practical application to specific radionuclides and sample types is discussed in Chapter 6.

Chemical separation of a radionuclide from other radionuclides is intended to recover most of the radionuclide while removing most of the accompanying contaminants. Purification specifications usually can be relaxed by selecting a radiation detector that measures the radionuclide of interest without detecting some of the contaminant radionuclides, either by discrimination against certain types of radiations or by spectral energy analysis.

The measure of purification is the decontamination factor, DF, defined as the concentration of the interfering radionuclide before separation divided by its concentration after separation. The required DF depends on the initial concentrations of the interfering radionuclide and the radionuclide of interest, and the extent of acceptable contamination when counting the emitted radiation. The DF must be large if the radionuclide of interest is a small fraction of the total initial radionuclide content. The interfering radionuclide that remains in the source that is counted

Environmental Radiation Branch, Georgia Tech Research Institute, Georgia Institute of Technology, Atlanta, GA 30332

should contribute not more than a few percent to the count rate of the radionuclide of interest. Evaluation of the fractional contribution of the radiation from interfering radionuclides must consider the radioactive decay of the radionuclide of interest and the interfering radionuclides.

The separation method generally is selected from available methods that are described as purifying the radionuclide of interest from the identified contaminant radionuclides. Scavenging steps can be inserted that separate the major contaminant radionuclides from the radionuclide of interest. The selected method must be tested to determine all DF values for the contaminant radionuclides or at least to demonstrate for the sample under consideration that none interfere with the radiation measurement. A separation step will have to be repeated or additional separation steps will have to be added when a single step does not achieve the required DF.

Radionuclide analysis methods are published in analytical chemistry and radiochemistry journals, and in methods manuals issued by nuclear facilities such as government laboratories. For example, the *Environmental Measurements Laboratory Procedures* manual, HASL-300 (Chieco 1997), is an excellent source. Standard methods for radionuclide analysis (see Section 6.7) are available, and should be used whenever appropriate. If conditions differ from those to which published methods have been applied, radionuclide recovery and decontamination must be tested and additional process steps may have to be inserted.

If no applicable analytical method is found, then individual separation steps have to be selected and combined sequentially, each step to remove one or more contaminants until all are removed to the extent necessary. The first step must match the sample form, each subsequent purification step must match the preceding step, and the final step must produce the counting source in its specified form. Each step must give high recovery of the radionuclide of interest and the required removal of interfering radionuclides. Suitably designed tracer tests can provide otherwise unavailable information.

Each separation method also must be evaluated for suitability with the stable (nonradioactive) dissolved substances in the sample. These substances may interfere in separations by competing or blocking reactions. Separation or scavenging steps must be introduced to remove these interfering stable substances or reduce them to acceptable amounts.

The techniques for separating and purifying radionuclides as part of the radioanalytical chemistry process are discussed in the following sections. Although separate sections present the different techniques, the analyst is expected to combine separation techniques that produce optimum analytical efficacy.

3.2. Precipitation Separation

3.2.1. Principle

Whether and to what extent a salt precipitates is characterized by the solubility product K_{sp}. The solubility product should be recalled from general chemistry texts as the equilibrium constant describing the formation of a slightly soluble (or nearly insoluble) ionic compound from its component ions in solution. Consider the chemical reaction

$$A_a B_b \text{ (s)} = a A^{b+}(\text{aq}) + b B^{a-}(\text{aq}) \tag{3.1}$$

In Eq. (3.1), A^{b+} is the cation with charge $b+$ and B^{a-} is the anion with charge $a-$. The solubility product is the product of the thermodynamic activities of the component ions in solution at equilibrium:

$$K_{sp} = [A^{b+}]^a [B^{a-}]^b \tag{3.2}$$

Here, the term "activity" is used differently than in other chapters of this textbook; it stands for the thermodynamic value that replaces the molar concentration of a substance. The units of K_{sp} are mol/l to the power $(a + b)$. Values are listed in handbooks, such as the *CRC* (Weast 1985) and *Lange's Handbook of Chemistry* (Dean 1999), as well as in general chemistry textbooks and online.

The solubility product may be calculated from the free energy of formation G of the solids and aqueous forms (see Weast 1985) in Eq. (3.3), with the absolute temperature T and the gas constant R in consistent units:

$$\ln K_{sp} = \frac{-\Delta G}{RT} \tag{3.3}$$

If the product of the activities of the two components initially in solution is equal to or less than the K_{sp} of the product compound, then no precipitate is formed. When the product of the known initial activities of the two components exceeds the K_{sp} value, a precipitate is formed. One can then estimate the activity of the residual ion of interest (say B^{a-}) left in solution from the known value of K_{sp}. Consider for simplicity that concentrations are so low that they can be substituted for activity, and designate the initial concentrations of the two components by a zero subscript,

$$(B^{a-})^b = \frac{K_{sp}}{(A^{b+})^a} = \frac{K_{sp}}{[(A^{b+})_0 - [(a/b)(B^{a-})_0] + [(a/b)(B^{a-})]]^a} \tag{3.4}$$

The fraction of B^{a-} remaining in solution after precipitation, $(B^{a-})/(B^{a-})_0$, can be estimated by dividing both sides of Eq. (3.4) by $(B^{a-})_0$, so that

$$\left[\frac{(B^{a-})^b}{(B^{a-})_0^b} \right] = \frac{K_{sp}}{(B^{a-})_0^b [(A^{b+})_0 - [(a/b)(B^{a-})_0^b] + [(a/b)(B^{a-})^b]]^a} \tag{3.5}$$

For example, if the iodide reagent added as carrier for radioactive iodine is precipitated as AgI by adding silver nitrate to solution, where $a = b = 1$, the fraction of iodine remaining in solution after the precipitation, $(I^{1-})/(I^{1-})_0$, is

$$\left[\frac{(I^{1-})}{(I^{1-})_0} \right] = \frac{K_{sp}}{(I^{1-})_0 [(Ag^{1+})_0 - (I^{1-})_0 + (I^{1-})]} \tag{3.6}$$

The last term in the denominator of Eq. (3.6) can be ignored when it is very small compared to the difference between the other two terms within the brackets. The

K_{sp} of silver iodide is 8.49×10^{-17} $(mol/l)^2$ (Weast 1985). Typically in analyzing radioiodine, the amount of iodide carrier added to 0.035-L solution is 10 mg or 0.078 mmol, and the amount of silver added is 1 ml of 1 M-solution, i.e., 1 mmol, hence the initial concentrations are 0.0021 M iodine and 0.027 M silver. As a result, the fraction of iodide that remains in solution after precipitation is 1.6×10^{-12}.

The result may be modified by the influence of relatively high concentrations of silver nitrate and possibly other substances in solution on the activity coefficients for silver and iodide, but precipitation of iodide essentially is complete. Simple calculations of solubility from K_{sp} values are only approximations of solubility behavior due to factors such as incomplete dissociation, complex ion formation, and ion pair formation.

3.2.2. Practice

A first separation step for a radionuclide may be coprecipitation or scavenging removal with reagents suggested by group separations applied in the past for qualitative and quantitative analysis. One incentive is to scavenge, as soon as possible, any relatively intense contaminant radionuclides to retain a sample that can be handled with less radiation exposure and contamination potential. Group precipitation can reduce the sample volume by carrying radionuclides from a large initial water sample on a solid that is then dissolved in a relatively small volume. This strategy can also serve to place the radionuclides into a solution that no longer contains interfering substances and is more amenable to subsequent chemical processing. For example, individual rare earth elements then can be separated from each other on an ion-exchange column with a complexing agent under closely controlled conditions such as precise pH values.

Carriers that have proved effective for coprecipitation are the hydrous oxides of the metals, particularly of iron, other transition metals, and aluminum. Their efficacy is due to their large surface area, gelatinous character, and ability to coagulate. Some distinction among carried ions can be achieved because, as indicated by Table 3.1, ions are precipitated as hydroxides at various pH thresholds.

TABLE 3.1. Precipitation of ions as hydrous oxides at various pH values

pH threshold	Species precipitated
pH 3	$Sn^{+2}, Zr^{+4}, Fe^{+3}$
pH 4	U^{+6}, Th^{+4}
pH 5	Al^{+3}
pH 6	$Zn^{+2}, Be^{+2}, Cu^{+2}, Cr^{+3}$
pH 7	$Sm^{+3}, Fe^{+2}, Pb^{+2}$
pH 8	$Ce^{+3}, Co^{+2}, Ni^{+2}, Cd^{+2}, Pr^{+3}, Nd^{+3}, Y^{+3}$
pH 9	$Ag^{+1}, Mn^{+2}, La^{+3}, Hg^{+2}$
pH 11	Mg^{+2}

TABLE 3.2. Group precipitation

Type of precipitate	Carrier example	Representative elements precipitated
Hydroxide	Ferric hydroxide—$Fe(OH)_3$ and other hydrated oxides like $Mn(OH)_2$ and $Al(OH)_3$	Coprecipitates rare earth and actinide elements without difficulty. Used in many procedures as first step to concentrate from water, urine, and dissolved solid samples.
Phosphate	Alkaline earths, especially calcium phosphate—$Ca_3(PO_4)_2$	Ra, U, Th, Ac, Pd, Ba, Al, Bi, In, Zr, plus rare earth elements
Oxalate	Calcium oxalate—CaC_2O_4	Precipitates actinides and rare earth elements. Can separate Ra from Pb, Bi, Po, and Ca at pH 2. Th at pH 3.5. Sr, Ba, and Y carriers precipitate as oxalates. Precipitates actinides from urine and leaves behind organics
Sulfate	Barium sulfate—$BaSO_4$ Lead sulfate—$PbSO_4$	Coprecipitate actinides for counting. Carries radium for counting. Precipitates essentially all ter- and quadrivalent cations; i.e., all elements from Pb to Cf in addition to Ba, La, and the light lanthanides (Sill and Williams 1969).
Fluoride	Neodymium fluoride—NdF_3	Precipitates actinides for counting by alpha spectrometry.
Sulfide	Lead sulfide—PbS	Pb sulfide at pH 3.5 to 4 carries Po; can separate Pb, Bi, Po from Ca.
Carbonate	Strontium carbonate—$SrCO_3$	Sr, Ba carriers
Nitrate	Barium nitrate—$Ba(NO_3)_2$ Strontium nitrate—$Sr(NO_3)_2$	Fuming HNO_3 at 60% conc. will precipitate Sr and Ba but not most Ca.
Chromate	Barium chromate—$BaCrO_4$	Carries Ra, Pb, and Ba—separates from Sr; Final precipitate for Pb.
Iodate	Cerium iodate—$Ce(IO_3)_4$	Carries Th, Po, Pt, Pa, and Ce—not Am^{+6}, U^{+6}, and Pu^{+6} (Hindman 1986)

Under favorable circumstances, an initial precipitation reaction can be selected that will separate the radionuclide of interest from most contaminants. An example from the group separations in Table 3.2 is the precipitation of strontium nitrate in concentrated nitric acid. The element of interest may be accompanied by several other elements, usually in the same periodic group, as indicated in the table. Barium and radium nitrate, for example, are also insoluble. If such specific precipitation reactions can be applied directly, the only additional processes will be separation from these similar elements and preparation of the counting source.

Otherwise, procedures can be selected to separate specific radionuclides in a series of steps performed in different chemical environments to free the radionuclide of interest from all accompanying radionuclides and nonradioactive substances that would interfere with the radionuclide measurement. Each separation is effective if it recovers most of the radionuclide of interest while carrying little of the contaminants. Special techniques, such as homogeneous precipitation, have been suggested to enhance separation (Gordon *et al.* 1959).

3.3. Ion-Exchange Separation

3.3.1. Principle

Radionuclides in an aqueous solution can be retained on a solid ion-exchange medium by stirring the solid in the solution or passing the solution through a column filled with the solid. In the ideal case, the radionuclide of interest is collected almost entirely on the ion-exchange medium after sufficient time to reach equilibrium while the contaminants remain almost entirely in solution, or *vice versa*. If the purified radionuclide retained on the ion-exchange medium emits gamma rays, it is then counted on the ion-exchange medium. If not, or if the yield (see Section 4.7) must be determined, the radionuclide is eluted from the ion-exchange medium for further processing of the elutriant solution to prepare a counting source. Column techniques can achieve more complex results, such as collecting several radionuclides on the ion-exchange medium and then selectively eluting each radionuclide.

The mass-action concept of ion-exchange between an ion A^{b+} initially in solution and an ion B^{a+} on the ion-exchange (ix) medium, at equilibrium, is represented by Eq. (3.7):

$$a A^{b+} \text{ (aq) } + b B^{a+} \text{ (ix) } = a A^{b+} \text{(ix)} + b B^{a+} \text{(aq)} \tag{3.7}$$

The equilibrium or selectivity constant $E_{A/B}$ for aqueous ions A^{b+} that replace some ions fixed on the ion-exchange medium, B^{a+} is

$$E_{A/B} = \frac{[B^{a+}]^b [A^{b+}]^a_{ix}}{[A^{b+}]^a [B^{a+}]^b_{ix}} \tag{3.8}$$

The brackets represent the thermodynamic activity. At low concentrations in water, the concentration value can be used for activity; at higher concentrations or with other ions in solution, the concentration value must be multiplied by the activity coefficient. The activity of the ions on the ion-exchange medium is not readily available; the mole fraction, defined as the moles of an individual component divided by the total moles of all components in the phase, has been used in its place (Rieman and Walton 1970).

One way of bypassing calculation of $E_{A/B}$ to estimate the selectivity of a specific ion-exchange resin for various ions is to measure the distribution coefficient D_V for individual ions, including radionuclides. (The mass distribution coefficient D_M is also used for this purpose.) The volumetric distribution coefficient is the ratio of the ion concentration on the ion-exchange medium relative to the ion concentration in solution at equilibrium (both values in units of mol/l or Bq/L:

$$D_V = \frac{[A^{b+}]_{ix}}{[A^{b+}]_{aq}} \tag{3.9}$$

The larger the value of D_V, the greater is the selectivity of the resin for that ion. On ion-exchange resins (see below), trivalent ions are bound more strongly than divalent ions, which are in turn bound more strongly than monovalent ions. Among monovalent ions, the order of selectivity generally is Cs > Rb > K > Na > Li (Walton and Rocklin 1990), which is in the order of the ionic radius. The value of D_V can vary with the saturation of the ion-exchange resin by the backing ion; if the concentration of A^{b+} is low, the concentration of B^{a+} will be relatively high in both the solution and the resin phase.

The value of D_V is determined by shaking or stirring a specified amount of solution to which the ion of interest, such as the radionuclide, had been added with a specified amount of ion-exchange medium. When the radionuclide distribution between solid and liquid phase reaches equilibrium, the concentration of the radionuclide in each phase is measured. The batch of ion-exchange medium must be saturated initially with the specified backing ion and the initial solution must contain the same ion at a specified concentration. For example, if the radionuclide is a cationic radionuclide such as $^{42}K^+$ and the system for comparison is a sodium salt, then the ion-exchange medium must be in the sodium form and the solution that contains $^{42}K^+$ must be at a specified sodium backing-ion concentration. The concentration of nonradioactive potassium ion must also be specified. Any other radionuclides from which the radionuclide of interest is to be separated must be equilibrated under identical conditions of known volume ratios, ionic concentration, temperature, and equilibration period.

A column experiment can also yield the value of D_V. The radionuclide-bearing solution is added at one end of an ion-exchange medium column and is then eluted and collected in incremental volumes that are measured for radionuclide content. Ideally, the concentration of the eluted radionuclide is distributed in the incremental volumes as a Gaussian curve. For a distance of movement along the column (in this case, the length of the resin column), l, cross-sectional area a, interstitial column volume occupied by solution, i, and elutriant volume V_e measured to the peak of the radionuclide concentration curve, Eq. (3.10) applies

$$D_V = \frac{V_e}{l\,a - i} \tag{3.10}$$

Because D_V is dimensionless, the numerator and denominator on the right-hand side of Eq. (3.10) must be in the same units of volume, say cm^3. If the radionuclide can be measured while it is moving in the column (for example, with a gamma-ray detector behind a shield that has a slit so that only a narrow width of column is observed), then Eq. (3.10) can be applied for movement through the column without monitoring the effluent. As indicated in the preceding paragraph, the same backing ion and identical conditions must be used in all comparisons.

Column separations are more efficient than batch separations because a column can be viewed as a series of sequential batch separations. For column separation of two radionuclides, the difference between corresponding values of D_V represents the number of column volumes separating the two elution peaks; if the Gaussian curves are sufficiently narrow, little contamination may occur. For batch separations, in contrast, the value of D_V indicates that some contaminant always remains, and sequential batch separations may be needed to reduce such contamination to an acceptable amount. Batch separations are useful if the value of D_V differs by, say, 2 orders of magnitude and the solid:liquid volume ratio is selected for acceptable discrimination. The easiest separation is a cation from anions or *vice versa*, so that one form can be essentially completely retained while the other form is only in interstitial water.

3.3.2. Practice

The process of ion-exchange was first discovered and studied in natural inorganic compounds, of which the most abundant are the clay minerals, especially zeolites. The latter are microporous, crystalline aluminosilicate minerals. Numerous naturally occurring and synthetic zeolites exist, each with a unique three-dimensional structure that can host cations, water, or other molecules in its void space (cavities or channels).

Several inorganic compounds were found to be useful for specific separations involved in radioanalytical work (Amphlett 1964). These include hydrous oxides of chromium(III), zirconium(IV), tin(IV), and thorium(IV); aluminum molybdophosphate $[(NH_4)_3PO_4 \cdot 12MoO_3 \cdot 3H_2O]$; zirconium phosphate $[Zr(HPO_4)_2H_2O$ or $ZrO_2P_2O_5]$; molecular sieves (activated synthetic crystalline zeolites); and silica gel. These inorganic substances are amorphous and eventually degrade to fine powders. They are useful for batch operations but not for high performance chromatography.

Paper can also function as an ion-exchange medium. It has very low capacity but is suitable for separating radionuclides at their low concentrations. Typically, paper chromatography is performed on strips through which a selected solvent flows, and distinguishes radionuclides by their path lengths along the strip.

Once ion-exchange resins were synthesized in 1935, these organic exchangers replaced zeolites in applications (industrial and analytical) and in scientific investigation. The advent of nuclear power in mid-twentieth century stimulated renewed interest in inorganic exchangers that were stable at high temperatures and could withstand the effects of high doses of radiation.

The ion-exchange resins commonly used for separations in a radioanalytical chemistry laboratory are solid organic structures with ion-exchange sites, or with attached substances that function as exchange sites. Some common organic ion-exchange media are listed in Table 3.3. The nature of the functional group defines the properties of the resin. Resins with fixed positive charge groups can exchange

TABLE 3.3. Types of ion-exchange resins

Cation-exchange resins	Functional groups	Chemical formula
Strongly acidic	Sulfonic acid groups	$RSO_3^- H^+$
Moderately strong acidic	Substituted phosphoric acids	$PO(OH)_2$
Weakly acidic	Carboxylic acid group	$RCOOH$

Anion-exchange resins	Functional Groups	Chemical Formula
Strongly basic	Tetraalkylammonium groups	$[RN(CH_3)_3]^+ Cl^-$
Weakly basic	Tertiary amine groups	$[RN(CH_3)_3]^+ Cl^-$
	Secondary amine groups	$[RNH(CH_3)_2]^+ Cl^-$

anions and are thus called anion-exchange resins; resins with fixed negatively charged groups exchange cations and are called cation-exchange resins.

A few commercially available ion-exchange resins are used for most published separations. They are defined by functional group, ion-exchange capacity (in meq g^{-1}), mesh size, cross-linkage, density, and ionic form. Each descriptor affects the value of D_V in qualitatively understood ways (Korkisch 1989), so that the analyst may select a resin with optimum characteristics for the intended separation. The ionic form of the purchased resin can be replaced by washing the resin with a solution that contains the new ion until complete conversion is demonstrated by tests of the effluent solution.

Selection of ion-exchange column dimensions is based on the amount of the resin needed to hold an amount of ion related to the sample (its capacity) and to achieve separation of the radionuclide. A longer and narrower column permits better separation, but the cross-sectional area must be sufficient to minimize wall effects that lead to the liquid flowing along the wall in preference to passing through the resin. Although the analyst may desire the most rapid flow permitted by column flow resistance, the flow must be sufficiently slow to permit local equilibrium and to avoid disrupting column structure by channeling or introducing air bubbles. Flow may be either upward or downward; the former tends to decrease disruption of column structure by air bubbles.

Columns must be filled carefully for uniformity of the resin bed without voids, and maintained to prevent drying and the resulting separation within the column. In many applications, columns can be reused for a limited number of times after being washed with water and then regenerated with the ionic solution that returns them to the original form. Resin reuse is not possible for applications that partially destroy resin, clog the column with solids, coat the resin with emulsions, or load the ion-exchange sites with various ill-defined ions. Commercial ion-exchange resins usually are relatively stable but decompose slowly with time, or more rapidly when attacked by strong oxidizing agents (Rieman and Walton 1970).

The separation typically is performed by adding from a few milliliters to a few liters of the radionuclide-bearing sample to a resin column of about 5–50 ml, and then washing the radionuclides sorbed on the column with water or a dilute reagent. In the sorption phase, radionuclides of interest either are retained at the inflow end of the resin or move through the column but do not break through. The radionuclide is then eluted with a selected reagent of experimentally determined number of column volumes. Several radionuclides may be eluted in succession with different types or strengths of elutriant. Used in the scavenging mode, the radionuclide of interest passes through the column while contaminants remain on the resin.

Researchers systematically examined values of the distribution coefficient for various resins and solutions across the entire periodic table. These distribution coefficients are available in tables or graphs of the distribution coefficient *vs.* acid concentration. One well-known example, in Fig. 3.1, shows ln D_V *vs.* concentration of HCl (Kraus and Nelson 1956). Similar figures are available for HBr, HNO_3, $HClO_4$, and H_2SO_4, as well as selected acid/alcohol combinations (Korkisch 1989).

The distinction among oxidation states in Fig. 3.1 suggests oxidation or reduction as a convenient technique for eluting a radionuclide that is strongly retained on the ion-exchange resin at its original oxidation state. Taken to its extreme, a cation held on the ion-exchange resin is removed completely by being converted to an anion, and *vice versa,* or to a nonionic state. The same principle applies when a metal ion, by addition of a complex-forming agent, is retained on or removed from the ion-exchange resin. Such complexes may be uncharged, of the same charge as the original ion, or of the opposite charge. The curves in Fig. 3.1 show the effect of complex formation on retention by ion-exchange in the metal–chloride system. Such complexes carry a negative charge when fully coordinated, and are adsorbed by the anion exchanger. The analyte may be eluted by changing the concentration to cause dissociation of the anionic complexes, or by changing the oxidation state of the metal ion.

Use of complex-forming agents in solution can increase the selectivity of an ion-exchange procedure. An example is given in Fig. 3.2 for separating rare earth ions complexed with ethylene dinitrilotetraacetic acid (EDTA). The rare earth ions are held on the cation-exchange resin when not complexed and are released by changing the pH value so that they become complexed. The curves in Fig. 3.2 indicate that the radionuclide represented by the curve on the extreme right (La^{3+}) can be completely retained on the column under the study conditions at pH from 0.4 to 2.8, and eluted completely above pH 3.9. The La^{+3} is separated completely at pH 2.8 from the radionuclide represented by the fifth curve (Zn^{+2}) from the right, and from all ions to the left of the Zn^{+2} curve. It is partially separated from Sm^{+3} and UO_2^{+2} ions represented by the other two curves on the right. By operating at pH 3.4, approximately one-half of La^{+3} is retained, and is separated from all radionuclides to the left except UO_2^{+2}.

Ion-exchange resins have also been loaded with counter ions for *in situ* precipitation; for example, an anion-exchange resin in the sulfate form can collect ^{226}Ra and ^{90}Sr, or one in the tetraphenyl borate form can collect ^{137}Cs (Cesarano *et al.* 1965). The radionuclide of interest is then eluted with a solution in which it dissolves.

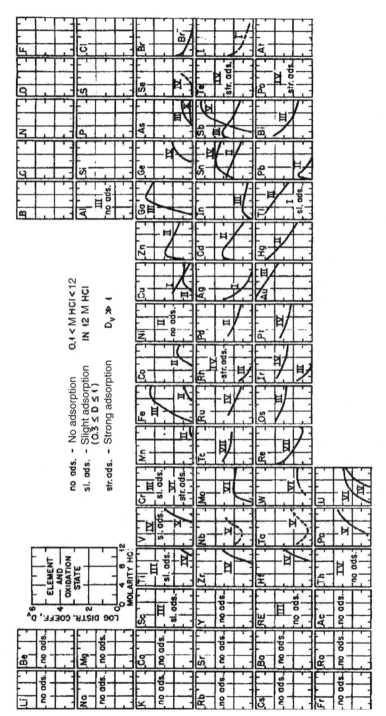

FIGURE 3.1. D_V on anion-exchange resin as a function of HCl concentration. (Figure from Kraus and Nelson 1956.)

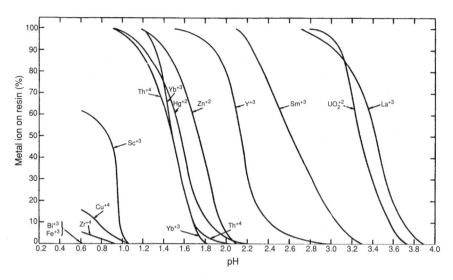

FIGURE 3.2. Elution of rare earths from a Dowex-50 cation-exchange resin, as a function of EDTA solution pH. (Figure from Fritz and Umbreit 1958, p. 513.)

3.4. Liquid–Liquid Extraction

3.4.1. Principle

In liquid–liquid extraction, an immiscible liquid—usually an organic solution—is combined with the sample in aqueous solution in an extraction flask and shaken to achieve good contact between the liquids. A reagent that functions as an extractant may have been added to one phase or the other. Information on the distribution ratio D and the extraction yield E that indicate the extent of purification from specified contaminants is available from many studies (Sekine and Hasegaw 1977). The information should describe the extractant, the organic solvent and the conditions of purity, reagent concentrations, volumes, required time, and temperature. The value of D reflects the ratio of the radioelement solubility in the organic phase to that in the aqueous phase, hence the type of solvent and the chemical form of the radionuclide to be extracted may be inferred from radioelement solubility data. If the initial conditions of the extraction procedure are not identical to those for reported extractions, the extent of extraction must be tested.

The two liquids must be as mutually immiscible and nonreactive as possible. The interface between the two liquids should be reasonably clean to avoid carryover of one phase in the other as part of an emulsion. Some time may be required for phase separation after mixing.

The distribution ratio is defined in analogy to Eq. (3.9) as

$$D = \frac{C_{\text{org}}}{C_{\text{aq}}} \tag{3.11}$$

The subscripts org and aq refer to the organic and aqueous phase, respectively, and C is the concentration of the radioelement at equilibrium.

The extraction yield E is defined as the ratio of the amount of solute extracted to its initial total amount. Relative to the volume of solution V,

$$E = \frac{D}{\left(D + \dfrac{V_{aq}}{V_{org}}\right)}$$ (3.12)

For extractions repeated n times with fresh portions of the organic phase at constant V_{aq} and V_{org}, the total recovered fraction E_r is defined as

$$E_r = 1 - \left[1 + D\left(\frac{V_{org}}{V_{aq}}\right)\right]^{-n}$$ (3.13)

From Eq. (3.13), the fraction of solute remaining in the aqueous phase is $[1 + D(V_{org}/V_{aq})]^{-n}$.

3.4.2. Practice

Extractant reagents and solvents are selected for optimum separation of the radionuclide of interest from contaminants, and also with regard to chemical stability, purity, and minimal hazard potential. The reagents and solvents may have to be stored under conditions that avoid degradation, and purified to remove minor contaminants that can interfere with separations. Use of chemicals listed as hazardous material may be feasible with appropriate care, but later disposal may be difficult.

Equilibration usually is reached in a few minutes, but should be checked for each method. Typically, the radionuclide in the aqueous phase is extracted into the organic phase under one set of conditions, and is then back-extracted under a second set of conditions. Washing steps commonly are inserted after extraction to improve the specificity of the radionuclide transfer. A single extraction and back-extraction cycle may suffice for purification or several cycles may be necessary. In analogy to ion-exchange systems, column separations have been developed with countercurrent flow of the two liquids.

Solvent extraction systems may be characterized by type of reagent used as extractant or by chemical species extracted. Data for the extraction of elements is in tabular form listed by element, plots of D $vs.$ pH curves in periodic table format, and distribution coefficient values for multiple elements plotted at specific phase compositions for a given extraction system. The many solvent extraction systems that have been reported provide many options to select a suitable system to solve a given problem. A brief description of extractant categories is given here with examples, noting that many reagents will fit more than one category according to the environment in which they are used (Marcus and Kertes 1969, Sekine and Hasegaw 1977).

Nonsolvating solvents may be used for nonelectrolyte molecular species such as rare gases, halogens, interhalogen compounds, and some metal halide complexes.

Examples include benzene, hexane, chloroform, carbon tetrachloride, toluene, and carbon disulfide. A well-known early example is the extraction of halogen molecules such as I_2 and Br_2, which are only sparingly soluble in water, into carbon tetrachloride or chloroform. The reduced form (the halides, I^- and Br^-) is then back extracted into water.

Ion-pair-forming extractants may be used for anions, anionic metal complexes, and weak acids and bases. Examples include quaternary ammonium salts such as tricaprylmonomethyl ammonium chloride (Aliquat 336); polyphenyloniums such as tetraphenylarsonium chloride; cationic dyes such as rhodamine B and alizarin blue; large anionic extractants such as tetraphenylborate (TPB); and heterocyclic polyamines such as 2,2'-dipyridyl (DIP).

Chelating extractants may be used for ionic salts. Examples include beta diketones such as thenoyltrifluoroacetone (TTA), acetylacetone, benzoylacetone, and dibenzoylmethane; 8-hydroxyquinoline (oxine); oximes such as dimethylglyoxime; nitrosophenyl compounds and nitrosohydroxylamines such as *N*-nitrosophenyl hydroxylamine (cupferron); and diphenylthiocarbazone (dithizone).

Alkylphosphoric acids are acidic extractants sometimes called liquid cation exchangers due to exchange of H^+ ions for the extracted cations. They have been used for many metal ions. Examples include mono-(2-ethylhexyl)phosphoric acid (MEHP), dibutylphosphoric acid (DBP), and di-(2-ethylhexyl)phosphoric acid (DEHP).

High molecular weight amines are basic extractants sometimes called liquid anion exchangers. They have been used for anionic metal complexes. Examples include tetraphenylborate (TPB), trioctylamine (TOA), triisooctylamine (TIOA), and trilaurylamine (TLA).

The high-molecular-weight amines R_3N can be considered liquid ion-exchange media because of the reaction

$$R_3NH^+ A^-(org) + B^-(aq) = R_3NH^+ B^-(org) + A^-(aq) \qquad (3.14)$$

This equation is analogous to Eq. (3.7). The amine and its inert solvent—for example, benzene, toluene, xylene, or kerosene—strongly extract the mineral acids *HA* and then extract anions B^- that are retained by ion-exchange.

Figure 3.3 shows ln D with triisooctyl amine of many polyvalent metal ions that form anionic chloro complexes in hydrochloric acid. Additional distribution data are available for metal nitrates, sulfates, and oxy anions that can be extracted. These ions can then be back-extracted into less acid systems in which they are no longer anionic complexes.

Neutral extractants may be used for uncharged metal complexes, ionic salts, and strong acids. Examples include ethers such as diethylether and diisopropylether; ketones such as methylisobutyl ketone (MIBK or hexone); and neutral organophosphorous compounds such as tri-*n*-butylphosphate (TBP), tri-*n*-octylphosphine oxide (TOPO), triphenylphosphate (TPP), and triphenyphosphine oxide (TPPO). Such neutral extractants have long been used for extracting actinides and lanthanides from nitric acid solutions.

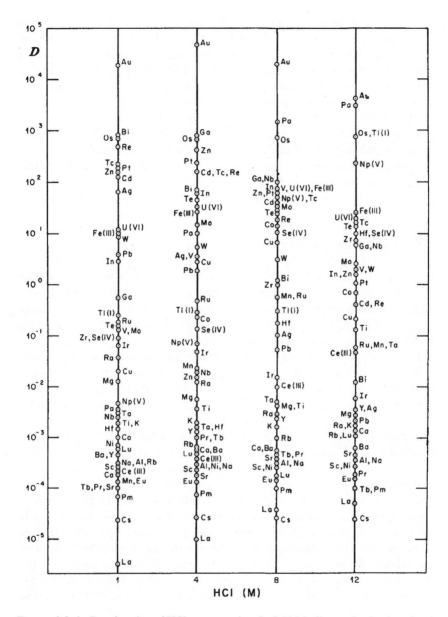

FIGURE 3.3. ln D as function of HCl concentration (in 0.11 M triisooctylamine in xylene) for extraction by amines. [Figure from Marcus and Kertes (1969), p. 960.]

One example is diethyl ether as an extractant for iron as the chloro complex $HFeCl_4$ or the cyano complex $NH_4Fe(SCN)_4$. Many of the radionuclides that form these molecules can be separated from each other under different conditions related to redox potential or pH.

Another example is the use of TOPO in cyclohexane solution. It combines with various metals in the form $M^{a+}Cl_a(TOPO)_2$ (De *et al.* 1970). Plots of D for several elements are shown in Figs. 3.4 and 3.5 for extraction from aqueous solutions as a function of HNO_3 and HCl strength. Trends can be predicted but the value of D is difficult to predict because of the many factors that influence it. Beyond the obvious ones of oxidation state and relative liquid volumes are the type and concentration of organic solvent, concentration of TOPO, temperature, and concentrations of aqueous phase contents. Acids are also extracted and compete with radionuclide extraction, although one can achieve equilibrium by first washing the organic phase with successive fractions of the aqueous phase without the radionuclide until the organic phase is saturated with the acid.

3.5. Solid Phase Extraction

A separation technique that shares some benefits of ion-exchange and solvent extraction processes is solid phase extraction (SPE), also called reversed phase partition chromatography or extraction chromatography. This technique immobilizes extractants on an inert polymeric support to retain specific radionuclides from a contacting solution (Cerrai and Ghersini 1970). Experimental applications of SPE began 50 years ago, and development for sample purification began in the 1970s. As in the case of ion-exchange separation, a liquid sample passes through a column, a cartridge, a tube, or a disc that contains a sorbent to retain the analyte. After the sample has passed through the sorbent, the retained analyte is washed and then eluted.

Extractants from liquid–liquid systems such as HDEHP [di(2-ethylhexyl) orthophosphoric acid] and CMPO/TBP (carbamoylmethylphosphine oxide derivative and tri-*n*-butyl phosphate) are supported on the solid material, as are newer ion-selective crown ethers (such as 4,4'(5')-di-*t*-butylcyclohexano 18-crown-6 for Sr). Various SPE columns available commercially from Eichrom Industries have proven useful to separate radionuclides such as Sr, Tc, Ra, Ni, Pb, Am, Pu, Th, U, Np, Cm, and lanthanides. These columns usually are small (approximately 2 ml resin bed). Their effectiveness depends on their specificity for the ion that includes the radionuclide of interest, but the small volume limits the amount (i.e., less than 10 mg) of carrier that can be retained. The specificity of each product shows promise for development of procedures for sequential radionuclide analyses from a single sample aliquot. (Burnett *et al.* 1997, Horowitz *et al.* 1998)

Analogous systems are commercially available as 3 M Empore™ Rad Disks. These come in the form of impregnated PTFE (polytetrafluoroethylene) membranes, which are used as filters for aqueous samples (Schmitt *et al.* 1990). Filter dimensions constrain this type of system to carrier-free separations. These filters with the retained radionuclide may then be washed, dried, and counted directly or the radionuclide may be eluted for further processing. Current products include filters for Ra, Sr, Tc, and Cs.

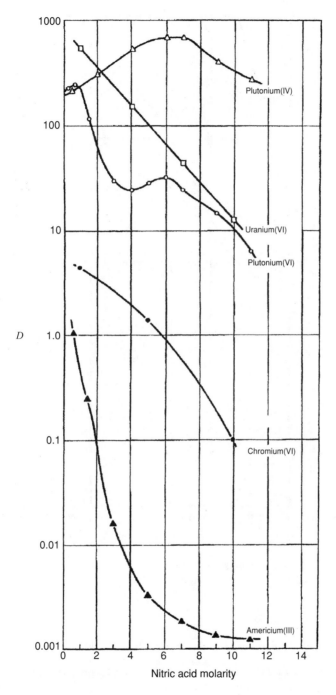

FIGURE 3.4. D as a function of nitric acid pH for extraction of metal ion into 0.1 M TOPO in cyclohexane. (Figure from Martin et al. 1961, p. 99.)

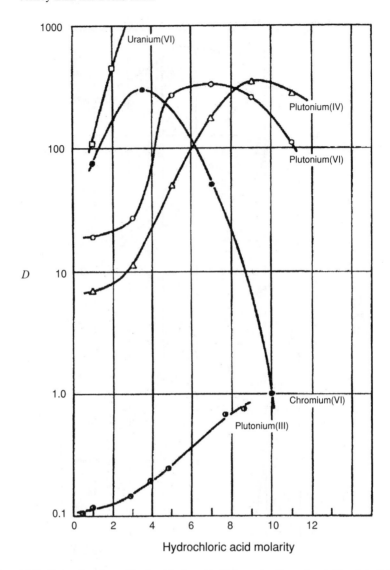

FIGURE 3.5. *D* as a function of hydrochloric acid pH for extraction of metal ion into 0.1 M TOPO in cyclohexane. (Figure from Martin et al. 1961, p. 102.)

3.6. Gas Separations

3.6.1. Distillation and Vaporization

Purification by distillation is one of the oldest separation techniques. Raoult explained the partition of components between liquid and vapor in terms of their vapor pressure. The total pressure of a vapor is equal to the sum of the partial pressures exerted by each component. Each component partial pressure is proportional to the fractional concentration of the substance in the liquid:

$$P_1 = X_1 P_1^0 \tag{3.15}$$

where P_1 is the vapor pressure of substance 1 over the solution, X_1 is the mole fraction of substance 1 in solution and P_1^0 is the vapor pressure of the pure substance at the same temperature. This is a colligative property related to the number of atoms and not to any characteristics of the substance.

A component at high concentration, such as the solvent, ideally behaves as postulated by Raoult's law. Gaseous and liquid solutes at relatively low concentrations often do not. This deviation from ideal behavior is attributed to the circumstance that the nearest neighbors of atoms at low concentration usually are not identical atoms, but rather solvent atoms.

For such solutes, Henry's law is applicable:

$$C_a = k P_a \tag{3.16}$$

where P_a is the partial pressure of substance a, C_a is the concentration of the substance in solution and k is a constant characteristic of the particular gas–liquid system. The gaseous atom concentration is proportional to the solute atom concentration, but the constant is not necessarily associated with the mole fraction. Each component of the system is independent of all others. As temperature rises, so do the partial pressures in the system (Wilson et al. 1968).

Many elements have chemical forms with sufficiently high vapor pressures at low temperatures—i.e., low boiling points—to permit separation as a gas from contaminants with higher boiling points. Figure 3.6 summarizes the elements that can be separated by various aspects of this technique.

The noble gases dissolved in water, as well as the halogens and hydrogen, carbon, nitrogen, and oxygen (as elements or in a number of compounds) can be flushed from the solution with an inert gas for collection. Radon is carried from previously sealed radium-containing solutions to measure the activity of its ^{226}Ra parent (Curtiss and Davis 1943), as discussed in Section 6.4.1. Fission-produced xenon and krypton radioisotopes and neutron-activated ^{41}Ar are flushed from samples of reactor coolant water, collected as the gas, and counted. Similarly, ^{14}C in the form of dissolved CO_2 or CO is flushed with an inert gas through the distillation system and collected in NaOH or Na_2CO_3 solution. Mercury salts can be reduced to the metal, processed by steam distillation, and collected as liquid mercury on a cool surface.

Volatile oxidation states of ionic radionuclides in solution that have been separated by distillation include group IVA, VA, and VIA halides ($GeCl_4$, $AsCl_3$, $SeBr_4$, $SnCl_4$, $SbCl_3$), and group VIII oxides (RuO_4, OsO_4, Re_2O_7, Tc_2O_7). Other volatile solutes include SiF_4, SO_2, $HMnO_4$, CrO_2Cl_2, VCl_4, and $TiCl_4$ (DeVoe 1962), and hydrides of some of the cited elements (Bachmann 1982).

A simple radioanalytical chemistry distillation apparatus is shown in Fig. 3.7. The boiling flask contains the sample solution with any needed reagents to maintain the radionuclide of interest in its volatile form and any contaminants in nonvolatile forms. An inflow tube with mouth submerged beneath the solution is available for

1	2	3	4	5	6	7	8	9	10	11	12	13	14	15	16	17	18
H abcd																	He a
Li a	Be											B bdf	C bcd	N abcd	O abcd	F abcd	Ne a
Na a	Mg											Al d	Si bd	P abcd	S abcd	Cl abcd	Ar a
K a	Ca	Sc	Ti d	V d	Cr g	Mn e	Fe d	Co	Ni	Cu	Zn	Ga bd	Ge bd	As abcd	Se bcd	Br abd	Kr ad
Rb a	Sr d	Y	Zr d	Nb d	Mo d	Tc cd	Ru cd	Rh a	Pd	Ag a	Cd a	In a	Sn bd	Sb bd	Te bcd	I abd	Xe ad
Cs a	Ba a	La	Hf d	Ta d	W d	Re cd	Os cd	Ir d	Pt	Au a	Hg ad	Tl a	Pb	Bi ab	Po ad	At ab	Rn ad
Fr a	Ra	Ac															

Ce	Pr	Nd	Pm	Sm	Eu	Gd	Tb	Dy	Ho	Er	Tm	Yb	Lu
Th	Pa d	U d	Np d	Pu d	Am	Cm	Bk	Cf	Es	Fm	Md	No	Lr

Form in which elements are volatile
(a) Element
(b) Hydride
(c) Oxide
(d) Halides
(e) As permanganic acid
(f) As boric acid
(g) As chromyl chloride

FIGURE 3.6. Distillation separations by species. (Figure from Coomber 1975, p. 307.)

carrier gas flow or reagent addition. The condenser tube is cooled with water and leads into the collector, which may either be dry or contain a solution to collect the condensate. For optimum decontamination, the distillation process should minimize liquid-droplet carry-over.

Tritium as tritiated water undoubtedly is the most common radionuclide purified by distillation. For separating tritiated water from samples such as biological material or vegetation, azeotropic distillation can be more effective (Moghissi *et al.* 1973). In this method, an organic solvent not miscible with water is mixed intimately with the tritiated aqueous sample. An excellent website on azeotropic distillation can be found at http://www.chemstations.net/documents/DISTILLATION. PDF. Cyclohexane, which forms a constant-boiling mixture at 94°C is used, as are other organic liquids, such as *para*-xylene and toluene. The distillation apparatus is assembled with the azeotropic distillation receiver shown in Fig. 3.8, rather than the usual collection flask. During distillation, the azeotrope separates into an aqueous and an organic phase in the receiver. The aqueous phase is used for analysis while the organic phase returns to the distilling flask. Distillation can continue until essentially all of the water has been collected, achieving complete distillation. This is particularly important in tritium purification, where incomplete distillation will enrich the residue in ^{3}H (see Section 4.8) and yield results that are underestimates by about 10%.

Many of the elements footnoted in Fig. 3.6 by the letters c or d have been separated from compounds that have higher boiling points by heating the salts in a volatilization apparatus at a relatively elevated temperature. The vapor is

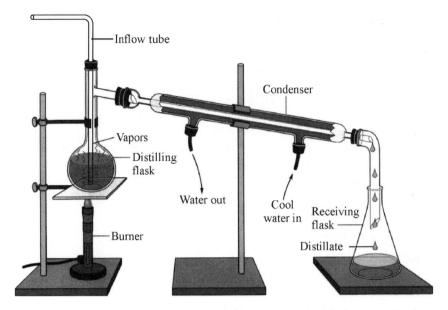

FIGURE 3.7. Simple distillation apparatus. (Figure modified from Chemweb 2005.)

collected on a cold plate suitable for counting or subsequent separations. Several radionuclides can be separated from the same source by successively increasing the temperature at the sample to each boiling point and replacing the cooled target for

FIGURE 3.8. Azeotropic distillation receiver. (Figure modified from Chemweb 2005.)

each radionuclide. The technique is occasionally used for rapidly separating a more volatile radionuclide from an irradiated target, but is limited in the radioanalytical chemistry laboratory to preparations of purified compounds in cases where the apparatus and skill are available. An example of this separation is the volatilization of plutonium chloride from soil in a gas chromatograph (Dienstbach and Bachmann 1980). Another example is the collection of ^{90}Sr as the chloride salt on a cooled target when the source is heated between 900 and 1400°C; when heated above 1400°C, its daughter ^{90}Y is volatilized (Sherwin 1951).

Radionuclides that are volatilized during ashing or freeze-drying, can be collected in a commercially available apparatus. For freeze-drying, the sample is placed into a vacuum distillation apparatus and the vaporized radionuclide is collected at a cooled surface or on a sorbent material.

3.6.2. Separation of Gases

Noble gases in air can be collected and concentrated by sorption at very low temperature on material such as charcoal, silica gel, or molecular sieve. Sorbed radioactive gases can be counted directly for gamma rays or flushed from the collectors for separation and measurement by application at elevated temperature of a vacuum or a stream of inert gas such as helium. The conditions for separation by sorption typically are selected from reported applications and adjusted empirically for the samples at hand. The components of the sample matrix can have major influence on the degree of uptake and the saturation capacity of the sorption medium available for the gas of interest.

Noble gases collected from air or water (see Section 3.6.1) may be separated from each other in the carrier gas by selective condensation on sorption media, followed by selective volatilization at temperatures near their boiling points of the separated gases (see Section 15.4.2). Radon can be sorbed from air or water and then measured directly by counting gamma rays emitted by ^{222}Rn or ^{220}Rn progeny.

Water vapor in air is collected on these sorption media at measured flow rates and then removed from the sorbent by distillation. The tritium activity is reported relative to the amount of water. Collection of various gaseous forms of tritium, ^{14}C, and radioiodine is discussed in Section 5.3.2.

Binding of molecules in nonionic states to these materials is attributed to van der Waal's forces at specific sorption sites. Calculations have been performed to match theoretical predictions to measured values with respect to rate of sorption (Blue and Jarzemba 1992), approach to equilibrium, and relative binding for different molecules (Underhill 1996). Rates of sorption appear to depend on molecular diffusion in at least three regions: the air or water medium, the immediate environment at the sorbent surface, and within the sorbent material (Kovach 1968). Commercially available sorption media may be rated by capacity for a specified substance.

In molecular sieves, the dimensions of molecules considered for sorption can be matched to the sorbent network, rated by active site diameter (Karge and Weitkamp 1998). Most molecular sieves are stable to 500°C. Molecular sieves can be selected to separate molecules based on size, shape, polarity, and degree of unsaturation. The range of pore sizes commercially available is approximately 0.3–0.8 nm. A mixture of gases or vapors of different molecular sizes can be separated by passage through a series of molecular sieves arranged so that only the smallest molecules are sorbed by the first zeolite and only the smallest of the remaining compounds by a second zeolite.

3.7. Electrodeposition Separation

In electrodeposition, an ion is transferred from solution to the surface of an electrode and retained as a solid. Typically, a metallic ion is reduced and deposited as the metal at the cathode. Deposition of oxy-ions at the cathode and oxidation at the anode also are employed. The rate of deposition depends on the applied voltage and factors such as the rate of stirring, volume and nature of the solution, and the material and area of the electrode. Little deposition occurs until a critical voltage is reached, after which the rate of deposition increases rapidly with increasing voltage. Consider the half reaction in which n is the number of transferred electrons per atom of A:

$$A^{n+}_{(aq)} + ne^- \Leftrightarrow A_{(s)} \tag{3.17}$$

The energy of this reaction E, in volts, is defined as

$$E = E^\circ + \frac{RT}{nF} \ln \frac{1}{[A^{n+}_{(aq)}]} \tag{3.18}$$

E° is the voltage, relative to a standard hydrogen electrode (SHE), that is required for a reduction half reaction at extreme dilution; this value is tabulated for redox couples in handbooks (Weast 1985). The term $[A^{n+}]$ is formally defined as the thermodynamic activity of species A at 25°C (298 K) and 1 atmosphere, but often is applied as concentration when radionuclides are at very low concentrations. The term R is the gas constant, 8.314 volt-coulomb/Kelvin-mol. The Faraday constant F is 96,485 coulomb/mol, and T is the temperature in degrees Kelvin. To write the equation with logarithms to the base 10, these terms are multiplied by 2.303; for $n = 1$ and $T = 298$ K, the coefficient is 0.05916 V.

As an example, for applying Eq. (3.17) to the reduction of silver ion to metallic silver,

$$Ag^+_{(aq)} + e^- \Leftrightarrow Ag_{(s)} \tag{3.19}$$

The voltage required for electrodeposition is

$$E = E^\circ + \frac{0.05916}{1} \log \frac{[Ag_{(s)}]}{\left[Ag^+_{(aq)}\right]} - E_{over}$$

$$= E^\circ + 0.05916 \log \frac{1}{\left[Ag^+_{(aq)}\right]} - E_{over} \tag{3.20}$$

H																	He
Li	Be											B	C	N	O	F	Ne
Na	Mg											Al	Si	P	S	Cl	Ar
K	Ca	Sc	Ti	V	Cr X	Mn	Fe X	Co X	Ni X	Cu X	Zn X	Ga	Ge	As	Se	Br	Kr
Rb	Sr	Y	Zr	Nb	Mo	Tc	Ru X	Rh X	Pd	Ag X	Cd X	In X	Sn X	Sb X	Te X	I	Xe
Cs	Ba	La	Hf	Ta	W	Re X	Os	Ir X	Pt	Au X	Hg X	Tl X	Pb X	Bi X	Po X	At X	Rn
Fr	Ra X	Ac X															

Ce	Pr	Nd	Pm	Sm	Eu	Gd	Tb	Dy	Ho	Er	Tm	Yb	Lu
Th X	Pa X	U X	Np X	Pu X	Am X	Cm X	Bk	Cf	Es	Fm	Md	No	Lr

FIGURE 3.9. Electrodeposition separations (shown with X). (Data from Blanchard et al. 1960.)

The overvoltage or overpotential E_{over} is inserted in Eq. (3.20) to adjust for other processes that compete in the system and make electrodeposition less than ideally efficient. These processes are irreversible and include the effects of the decomposition of water, other solutes, and imperfections in the electrode surface. Because of these processes, a greater potential difference than calculated from the reference potential and the ionic concentration must be applied in order to achieve deposition. For the same reason, spontaneous deposition, inferred from a positive value of $E°$, may not occur if the overvoltage exceeds it. Overvoltage effects occur at both the cathode and the anode.

The positive value of $E°$ for many metals (for Ag^+, $E° = 0.799$ V) and the additional effect of low concentration from the second term on the right in Eq. (3.20) suggest that these metals will deposit readily at the cathode to which a low negative voltage, i.e., a few tenths of a volt, is applied, or even spontaneously. The equilibrium voltage $E°$ applies to an electrode—in this case, the cathode–uniformly constructed of the same metal as the ion; the initial voltage required for deposition is different for a cathode of a different material. Only when a coating of the same metal has been deposited on the cathode surface, the value of $E°$ will apply.

The current, which is proportional to the amount deposited, can be calculated to be small at low concentrations. Practical requirements for applied voltage and

TABLE 3.4. Electrodeposition or reduction to metal: grouping in classes (Lindsey 1964)

Group	Name	Range
A	Precious metal group	Gold to mercury
B	Copper group	Copper, bismuth, antimony, arsenic
C	Lead group	Lead to tin
D	Zinc group	Nickel to zinc
E	Mercury cathode group	Alkali metals, lanthanons, etc.
F	Anodic group	Metals anodically deposited as oxides

current will differ because of the above-cited competing reactions by the water and other substances in solution.

Most metals except the alkalis and alkaline earths can be electroplated at the cathode with suitable applied voltage from acid solutions. Relatively early experience with electrodeposition of various metals is summarized in Fig. 3.9. The process typically is applied for carrier-free or low-concentration samples to prepare sources for alpha-particle spectral analysis. It is also useful for depositing thin sources for counting radionuclides that emit beta-particles with low maximum energy.

Because electrodeposition usually is the last procedural step, the composition of the solution can be controlled for ease of electrodeposition. The reagents that are present may consist of an acid to avoid hydrolysis and a redox agent to retain the radionuclide in a state suitable for final reduction. As shown in Table 3.4, metals can be grouped for electrodeposition analysis based on their standard electrode potentials.

Some elements are not suitable for electrodeposition from aqueous solution as the metal. Among these are the radionuclides plutonium, uranium, and thorium, which are prepared for alpha-particle spectral analysis by deposition of oxides. Other metals, such as lead, can also be deposited as oxides under empirically derived conditions (Laitinen and Watkins 1975).

4
Radioanalytical Chemistry Principles and Practices

JEFFREY LAHR,[1] BERND KAHN,[1] and STAN MORTON[2]

4.1. Introduction

The preceding chapter describes methods that can be applied to the analysis of both radionuclides and their stable element counterparts. One important difference to consider is the extremely low amounts of radionuclides generally analyzed in the radioanalytical chemistry laboratory. As noted in Section 13.1, the environmental samples in a radioanalytical chemistry laboratory typically are in the picocurie to nanocurie (0.037 to 37 Bq) range; this corresponds to a radionuclide sample mass of around 10^{-15} g, depending on half-life and molecular weight of the radionuclide. Precise measurement of such a low amount can be achieved because radionuclides emit energetic radiation.

The low concentration of the typical radionuclide sample affects some aspects of its behavior as a solute. One cause is believed to be interactions with surfaces and other solutes that are not taken into consideration for species at the more usual concentrations. The chemical status of a radionuclide sample may also be affected by the energy liberated when a radionuclide is created and the energy deposited when the emitted radiation passes through the solvent medium (Vertes *et al.* 2003).

The peculiarities of radionuclides at extremely low concentration were of great interest to radiochemists during the first half of the twentieth century and are described in excellent texts such as Wahl and Bonner (1951) and Haissinsky (1964). Some of these effects still are not clearly understood. Considered in this chapter are issues pertinent to radioanalytical chemistry, including:

- Production of radionuclides
- Conventional equilibria values extended to very low concentrations
- Low-concentration "radiocolloidal" behavior
- Sample preservation

[1]Environmental Radiation Branch, Georgia Tech Research Institute, Georgia Institute of Technology, Atlanta, GA 30332
[2]Radiobioassay Programs, General Engineering Laboratories, 1111 North Mission Park Blvd. Suite #1070, Chandler, AZ 85224

- Sample dissolution
- Carrier or tracer addition
- Separation for purification
- Source preparation for radiation measurement

The term "radiocolloidal" was applied in the early days of radiochemistry to describe certain behavior patterns of radionuclides in aqueous solution that did not conform to expectations based on the normally higher concentrations of the chemical forms assigned to them. Because many of these deviations were related to failure to pass through filters or to remain uniformly distributed in solution, colloidal behavior was inferred. This behavior was observed to affect in different ways many of the steps in the radioanalytical chemistry process itemized above, with the potential of invalidating analytical results because the radioisotope did not consistently follow the known path of its stable isotopes.

On the other hand, the low concentration of a radionuclide provides opportunities for use as tracer in chemical and physical studies. In radioanalytical chemistry, one benefit is that the addition of a stable-element "carrier" permits analysis without the requirement of quantitative analyte recovery. Another benefit is the opportunity to deposit very thin sources that minimize self-absorption in a source of alpha- and beta-particle radiation.

Radionuclides with extremely long half-lives or in very large amounts will not be at such extremely low concentrations, as indicated by Eq. (2.7). Radionuclides with half-lives in excess of 10^9 years, such as ^{238}U and ^{232}Th, can be measured by some of the conventional techniques of analytical chemistry as alternatives to radiation measurement. Radionuclides at concentrations much above 10^4 Bq/l usually are not submitted to a low-level radioanalytical chemistry laboratory because of the threat of contaminating other samples.

Another distinction pertains to radionuclides that have no stable isotopes. Chemical analysis for these radionuclides has no basis in conventional analytical chemistry except as studies performed with the usual small amounts, based on similarities in chemical behavior to homologous stable elements according to their location in the periodic table. When sufficiently large amounts of these radionuclides are produced and purified to permit observation by microchemical manipulations, any conclusions must consider the impact of the intense radiation on the observed chemical reactions.

The atomic mass difference between the radionuclide and the mix of its stable isotopes in nature, although minor in terms of its effect on chemical equilibrium and reaction rates, provides opportunities for separation, identification, and quantification at low concentration by mass spectrometer, as discussed in Chapter 17. The mass difference ratio is at its extreme for tritium (T or ^{3}H) relative to the stable isotopes ^{1}H and ^{2}H. This distinction causes minor separation between ordinary water with molecular mass 18 and tritiated water (HTO) with molecular mass 20 during distillation, and can be applied to enriching tritiated water in the laboratory by electrolysis.

4.2. Effects in Producing Radionuclides

When a radionuclide is produced by fission or activation processes or in radioactive decay of a radionuclide to its progeny, the product has a different form than its antecedent and moves from its original site. That is, the nuclear reaction may rupture the chemical bond between the radioactive atom and the molecule of which the atom was a part, and the newly created radioactive atom may have several new electron configurations. This result is described as the Szilard–Chalmers effect.

The Szilard–Chalmers effect permits separation of radionuclides at high specific activity and purity from the matrix in which they are produced. Preparation of the pure product is an empirical process because of the complex interaction of the three sequential steps: producing the free radioactive ion or atom, maintaining the product in its new form in the sample matrix, and separating the product from the matrix. Success of the process is evaluated empirically in terms of the specific activity of the product relative to the matrix or, alternately, the fractional yield of the product.

As discussed in Sections 2.2.1, the recoil energy of a product nuclide following alpha-particle emission is about 0.1 MeV; in Section 2.2.2, it was noted that the recoil energy observed following beta-particle emission is of the order of a few kiloelectron volts. Recoil is also observed during gamma-ray emission and electron capture and when an atom is produced in reactions such as (n,γ) or (n,α) (Vertes *et al.* 2003). The recoil energy of the produced radioactive atom generally is sufficient to perturb its original position in a molecule or crystal.

The recoil energy is not the only factor that controls the degree to which the new radionuclide separates itself from the original matrix, and some of the recoil energy it is dissipated by other mechanisms. Moreover, the recoiling atom may return to its original molecule or crystal site, especially when the product remains the same element as its antecedent, as in an (n,γ) reaction or after an internal transition. The product nuclide and its antecedent nuclides may rapidly interchange to reduce the specific activity of the product.

The last step in the collection process is to separate the product. Coprecipitation and solvent extraction are common techniques for producing pure, high specific activity, radionuclides such as chlorine or bromine. To keep the produced radionuclide separated at high specific activity, a chemical form is selected for the product that has a slow rate of interchange with the product, such as certain oxyions or organic compounds. Separation techniques that have been found especially convenient are sweeping a gaseous radionuclide from the medium and collecting a recoiling radionuclide on a surface that faces a deposited solid.

4.3. Anticipated Low-Concentration Behavior

Analytical processes at the conventional concentrations discussed in Chapter 3 involve such concepts as the mass action law and the solubility product; these tools can be applied to the analysis of radionuclides at extremely low, "trace-level," concentrations *with some precautions*. For example, competing reactions with a second substance, at too low a concentration to be identified but at a much higher concentration than the radionuclide, can totally overshadow the expected reaction of the radionuclide.

Some of the early studies of radionuclide behavior in solution and as solids resulted in findings that could be explained by extrapolating the behavior of the ions or molecules from the conventional to "trace-level" amounts or concentrations. Other findings could be attributed to the unusual environment faced by the relatively few radioactive atoms that were detected only because they emitted radiation.

Various anomalies in low-concentration behavior have been reported. In precipitation, a trace-level radionuclide that does not exceed its solubility product may coprecipitate with a second substance. In electrodeposition, despite adjustment for concentration (or more accurately, thermodynamic activity) by the Nernst equation, the predicted voltage may not be suitable because of the factors discussed in Section 3.7 pertaining to overvoltage and an electrode surface not fully covered by the element of the radioisotope (Haissinsky 1964). In volatilization or sublimation of a radionuclide, distinctive effects have been attributed to the situation that atoms at trace-level are not bonded to each other but to the other atoms in the medium or on a surface. A radioactive cation sorbed on a negatively charged solid particle that it does not fully neutralize will behave like a negative particle.

A radionuclide at trace level can be incorporated in a precipitate, either by intention or inadvertently. In the absence of a visible precipitate, the radionuclide may appear to act as a colloid (i.e., act insoluble and stratify or settle), as suggested by the term "radiocolloid."

Such precipitation or colloidal behavior occurs under conditions that would be expected if the radionuclide hydrolyzed to form a hydrous oxide at higher concentration than inferred from radiation measurements. At a very small solubility product K_{sp} (see Section 3.1), any unknown small amount of stable ion in solution may be sufficient to cause such an effect. Coprecipitation of trace-level radionuclides with another insoluble ion, such as Ra^{+2} with $BaSO_4$ and Pu^{+3} with LaF_3, incorporates the radioactive atoms within the crystal structure in various ways or sorbs it on particle surfaces (Kolthoff 1932). A precipitate such as $Fe(OH)_3$ in neutral or slightly basic solution can scavenge from solution many tracer-level radionuclides that hydrolyze under the conditions of the procedure (see Table 3.1 for the effect of pH).

In practice, tests are needed to determine the fraction of a radionuclide that is carried under the conditions of the process. For example, radioactive strontium tracer is carried by $BaCO_3$ in slightly basic solution and by $Y(OH)_3$ in more strongly basic solution. Such carrying generally is intended for the former but not for the latter. Carrying by $Y(OH)_3$ can be prevented by adding even a small amount, such as 1 mg, of stable strontium ion. When no stable element is available, as for plutonium, behavior due to hydrolysis must be avoided by strongly acidifying the sample, possibly to as low pH as 2.

Much trace-level behavior has been understood only after detailed studies. In the absence of visible evidence, behavior must be inferred from radiation measurements during the reaction. In some cases, an observed effect can be attributed to various mechanisms, as in the above-cited processes of coprecipitation as mixed crystals or by surface sorption, colloid formation, or sorption on suspended material, and retention on filters by conventional filtration or ion exchange on cellulose. In each instance, the second effect has been termed "radiocolloidal" at one time or another.

4.4. Sample Loss by "Radiocolloidal" Behavior

Samples submitted for analysis that contain radionuclides at extremely low concentration present a challenge to the analyst because these radioactive atoms unexpectedly can deposit on surfaces. Such behavior has been widely observed and was termed "radiocolloidal" (Hahn 1936). Radionuclides at trace levels are lost from solution by sorption on container walls (Korenman 1968), suspended solids (Wahl and Bonner 1951), immersed glass slides (Eichholz *et al*. 1965), metal foils (Belloni *et al*. 1959), and filters (Granstrom and Kahn 1955). Filters of cellulose, cotton wool, and glass wool retain various radionuclides at trace levels.

This loss is of greatest concern at the very beginning of the sample path, as when a water sample is poured into a sampling bottle. Observations of losses stimulated studies of radionuclide sorption on container walls of different types of glass and plastics, pretreated or coated with reagents, and from aqueous solutions either unmodified or with added carriers, acids, or complexing agents (Kepak 1971). Washing containers with various solutions was tested for recovering some of the sorbed radionuclides (Hensley *et al*. 1949).

A simple explanation for this sorption, although qualitative and incomplete, is that a significant fraction of the radionuclide in solution is retained at a few container and filter surface sites by ionic or electrostatic sorption (Kepak 1971). Fractional retention varies among radionuclides, presumably reflecting their affinities for the sites. Even a few available sites can be detected by radiation measurement if they are of the same magnitude as the number of radioactive atoms in solution.

According to this explanation, the fractional sorption at these few sites becomes immeasurably small when carrier is added to the sample. Various reagents can also compete for the sites. Conversely, the fewer the competing ions, the greater the fractional sorption (Thiers 1957); this is the case with deionized water. Pretreatment of glass or plastic container surfaces can reduce radionuclide sorption by saturating the sites; treatment by heating or washing glass with a basic solution, on the other hand, can increase sorption.

Tests have shown that some glasses and plastics are better than others in minimizing surface sorption losses for specific radionuclides (Eichholz *et al*. 1965), but the results are so variable that actual samples should be tested to assure that loss is minimal for the samples and containers under consideration. Unfortunately, test results for actual samples can be ambiguous at the low count rates of typical samples, while tests with radioactive tracers at higher concentrations may not apply unless the tracers are in the same chemical form as the radionuclides in the sample. Alternate methods that counteract the effect of sorptive behavior are listed in Section 12.3.2.

4.5. Sample Preservation

A liquid sample must be preserved so that the radionuclide in solution remains at the same concentration during transportation and storage by preventing loss or fractionation of the radionuclide. The only acceptable loss is by radioactive decay for which Eq. (2.6) can compensate. Problematic sources of radionuclide loss include volatile forms, the above-cited deposition on container walls, and incorporation in suspended solids.

Filtration at the time of sampling eliminates suspended solids as a sorption medium, but the filter may retain some radiocolloidal radionuclides. Addition of acids, carrier, or other reagents can eliminate observable losses of specific radionuclides by deposition or sorption. Prevention of transfer of radionuclide state from soluble to insoluble or *vice versa* is important if measurements are intended to distinguish oxidation states or degree of complexation in the original sample, or predict radiation exposure to persons drinking the water. Any reagent considered for addition to preserve one radionuclide must be evaluated for adverse effect on preserving other radionuclides.

Sample preservation is also required to prevent bacterial action and changes of form, such as souring of milk and decomposition of vegetation and biological samples, as discussed in Sections 5.5–5.9. These changes can cause sample containers to leak or even explode, make chemical separation of radionuclides more difficult, and change the form of the radionuclide. Sample storage on ice or in a freezer is the common treatment. A preservative such as formaldehyde is added when freezing is not an option.

4.6. Dissolution of Solids

Solids are dissolved so that they can be treated by conventional wet-chemistry processes. The dissolution treatment must be tested to assure that no portion of the radionuclide of interest is lost. The initial steps for organic solids are drying and ashing. Radionuclides in a chemical form with high vapor pressure may be partially lost by volatilization well below their boiling points. Radionuclides such as gaseous tritium and ^{14}C forms can be collected during the drying and ashing steps. Solids then are processed by either wet dissolution with acid or fusion with solid reagents.

A sample prepared for radionuclide analysis generally should be completely dissolved, without residual solids that could retain radionuclides. In contrast, stable-element analysis permits residual solids if the dissolution process has been demonstrated to dissolve completely the substance of interest.

Processing solids by leaching—i.e., incomplete dissolution of the matrix— is acceptable for radionuclide analysis only if knowledge of the retention process supported by leaching tests assures that the radionuclides of interest are leached to a reproducible extent. Without positive knowledge that leaching recovers large and consistent fractions of the radionuclides of interest, results are uncertain.

4.6.1. Acid Dissolution

As indicated in Table 4.1, most samples dissolve at least partially in strong mineral acids such as HF, $HClO_4$, HNO_3, and HCl, used individually and in combinations.

Conventional aqua regia digestion consists of treating a geological sample with a 3:1 mixture of hydrochloric and nitric acids. Nitric acid destroys organic matter and oxidizes sulfide material. It reacts with concentrated hydrochloric acid to generate aqua regia:

$$3\ HCl + HNO_3 = 2\ H_2O + NOCl + Cl_2 \tag{4.1}$$

Aqua regia is an effective solvent for most base metal sulfates, sulfides, oxides, and carbonates. Some elements, however, form very stable diatomic oxides, referred to as refractory species. Aqua regia provides only a partial digestion for most rock-forming and refractory elements. Hydrofluoric acid can destroy silicate matrices completely to liberate trapped trace constituents. Basic solutions can dissolve tissue and many anionic forms of inorganic ions. Complexing solutions such as EDTA are used under conditions that dissolve specific ions (Perrin 1964).

Pressurized vessels in microwave ovens improve dissolution at elevated temperatures and pressures (Kingston and Haswell 1997). The advantages compared to boiling solutions in beakers on hot plates are a shorter dissolution period under controlled conditions of temperature and pressure with lesser quantities of reagents, and—possibly more importantly—avoiding corrosion in hoods and ducts. One disadvantage is a limitation on sample size due to vessel capacity for sample plus reagents. Limited amounts, 1–3 g, of soil, biota, or vegetation are completely

TABLE 4.1. Acid dissolution

Acid	Character	Dissolution behavior (metals)	Dissolution behavior (salts)	Dissolution behavior (organics)
HCl	Strong acid, nonoxidizing	Metals above H in the electrochemical series	Oxides, sulfides, carbonates, phosphates	Organic salts
HNO_3	Strong acid, strong oxidizing agent	Metals, nonferrous alloys	Metal sulfides	Most organics
H_2SO_4	Strong acid, moderate oxidizing agent	Metals	Oxides, carbonates, phosphates	Most organics
HF	Weak acid, strong oxidizing agent	Metals	Silicates	Most organics
HNO_3/HCl (1/3) (aqua-regia)	Strong acid, very strong oxidizing agent	Metals including gold and platinum	Most salts	Most organics
HNO_3/HF	Strong acid, strong oxidizing agent	Metals	Carbides, nitrides and borides of Ti, Zr, Ta, and W.	Most organics

(>99%) dissolved by HF and HNO_3 in 90-ml vessels in a few hours under prescribed conditions (Garcia and Kahn 2001).

4.6.2. Fusion

Fusion is employed to decompose solids that are difficult to dissolve in acids but react at high temperatures to form soluble compounds. To promote the fusion process, a substance called "flux" is employed. Mixing the flux with the insoluble matrix and applying heat can produce complete decomposition. A favorable reagent mix and amount can be selected from published reports (Sulcek and Povondra 1989, Bock 1979), but the sample matrix must be tested to confirm the effectiveness of reagent choice and amount. Fusion can decompose environmental and biological matrices (Sill *et al.* 1974, Williams and Grothaus 1984), even with high silica and alumina content.

Fusion processes can be grouped into acid–base reactions (carbonates, borates, hydroxides, disulfates, fluorides, and boron oxide) and redox reactions (alkaline fusion agent plus oxidant or reductant). Common fluxes are listed in Table 4.2. Fluoride-pyrosulfate and carbonate-bisulfate fusions are used to decompose soil and fecal samples.

Disadvantages of the process are loss of volatile radionuclides at the elevated temperature and the large amount of the fusion reagent added to the sample. Handling the crucibles at the specified high temperatures is a safety concern. Platinum crucibles are expensive, while other crucibles (e.g., nickel and iron) are attacked by some reagents and contribute a contaminant to the sample.

TABLE 4.2. Common fluxes (Dean 1992)

Flux	Melting point (°C)	Types of crucible used for fusion	Type of substances decomposed
Na_2CO_3	851	Pt	For silicates, and silica-containing samples; alumina-containing samples; insoluble phosphates and sulfates
Na_2CO_3 plus an oxidizing agent such as KNO_3, $KClO_3$, or Na_2O_2	—	Pt (do not use with Na_2O_2) or Ni	For samples needing an oxidizing agent
NaOH or KOH	320—380	Au, Ag, Ni	For silicates, silicon carbide, certain minerals
Na_2O_2	Decomposes	Fe, Ni	For sulfides, acid-insoluble alloys of Fe, Ni, Cr, Mo, W, and Li; Pt alloys; Cr, Sn, Zn minerals
$K_2S_2O_7$	300	Pt, porcelain	Acid flux for insoluble oxides and oxide-containing samples
B_2O_3	577	Pt	For silicates and oxides when alkalis are to be determined
$CaCO_3$ plus NH_4Cl	—	Ni	For decomposing silicates in the determination of alkali element

4.7. Carrier or Tracer Addition

The low concentration of radionuclides has stimulated near-universal application of an isotope dilution technique ("reverse isotope dilution") that permits measurement of chemical recovery, termed "yield." As commonly applied outside the radioanalytical chemistry laboratory, isotope dilution consists of the addition of a known amount of radionuclide tracer to a sample that contains its stable element, i.e., the natural mixture of stable isotopes of the element. The tracer is added at the beginning of the procedure and then measured in the separated and purified sample. If the added tracer and the stable isotopes of interest are assured at the beginning to be in the same chemical and physical form, then the fraction of recovered tracer represents the fraction of recovered stable isotope.

Application of this technique avoids the need for quantitative recovery, which enormously simplifies the analytical process. Description of the isotope dilution technique in Section 4.7.1 is used to frame the discussion of reverse isotope dilution with carriers and tracers in Sections 4.7.2 and 4.7.3, respectively.

4.7.1. Isotope Dilution Technique

Isotope dilution begins with the addition of a known amount of calibrated radioactive tracer solution R_i to a sample of mass m_i in solution. By measuring the mass recovered m_f and the radionuclide amount at the end of the procedure, R_f, one calculates the initial mass. Designating S_i and S_f as the specific activity (i.e., the ratio of radionuclide amount to the mass of the same element, at the beginning and end, respectively) and m_{Ri} as the mass associated with the radionuclide tracer initially gives the following relationships:

$$S_i = \frac{R_i}{m_{Ri}} \tag{4.2}$$

and

$$S_f = \frac{R_i}{m_{Ri} + m_i} = \frac{R_f}{m_f} \tag{4.3}$$

The last term in Eq. (4.3) holds because the specific activity is a ratio that remains constant throughout the procedure, if radionuclide and solid have perfectly interchanged when combined. Then

$$m_i = m_{Ri} \left(\frac{S_i}{S_f - 1} \right) \tag{4.4}$$

To calculate m_i if $m_{Ri} \ll m_i$

$$\frac{R_i}{m_i} = \frac{R_f}{m_f} \tag{4.5}$$

In typical isotope dilution, one calculates $m_i = m_f R_i / R_f$.

4.7.2. Carrier Addition

For the purpose of radiochemical analysis, one uses reverse isotope dilution by adding a known mass of stable carrier solution to the radionuclide sample solution, instead of adding a radionuclide tracer to the stable element. The mathematical expression is identical to Eq. (4.5):

$$R_i = \frac{m_i R_f}{m_f} \tag{4.5a}$$

Here, m_f/m_i is the yield.

To assure accuracy, the carrier reagent is calibrated by mass measurement identical to that of the sources prepared for counting. For confirmation, the weight of the precipitate at 100% yield can be predicted from the amount of carrier reagent dissolved in the carrier solution, the amount pipetted into the radionuclide solution, and the ratio of the molecular weight of the precipitate to that of the carrier reagent.

The amount of carrier added to the radionuclide solution conventionally is selected for ease of weighing with an accuracy of 1% or better for yields of 50–100%. Low carrier amounts for processing thin sources for alpha-particle spectrometry (see Section 7.2.1) may require specially selected purification and yield determination techniques.

Under the following circumstances, special efforts are required for yield determination:

- A significant amount of the chemical carrier is in the original sample, that is, m_{Ri} is not far less than m_i. This circumstance becomes obvious when yields exceed 100%. The amount in the sample must be measured before radionuclide analysis. If stable isotopes of the radionuclide of interest are found in the sample, each yield factor must be corrected for the stable isotopes initially in the sample by adding their weight, m_{Ri}, to the carrier weight, m_i. If the amount m_{Ri} is marginal, it may be possible to increase the carrier weight so that the weight originally in the sample becomes negligible.
- The radionuclide and its carrier have more than one chemical state. When multiple oxidation states are present, sufficient oxidation and reduction reactions must to be performed after the carrier is added to the radionuclide solution to ensure that the carrier and radionuclide are in a common state. If a fraction of the radionuclide is solid or gaseous, that fraction must be dissolved. If the radionuclide has formed a complex molecule or ion, sufficient reagent must be added to complex the carrier as well.
- No stable isotopic carrier is available (e.g., for promethium, technetium, radium and its progeny, and the actinides and transactinides). Either nonisotopic carrier or isotopic radioactive tracer can be substituted (see next section). The special problems that arise are discussed in Section 7.2.2.
- Some processing steps do not include the carrier. If the sample initially is a solid, the carrier may be added after dissolution. This will require prevention of radionuclide loss during dissolution. If the carrier is added to the solid, then the problem becomes that the carrier in solution and the radionuclide in the solid

may not respond identically to dissolution. Another gap may occur at the end of the procedure if carrier yield is determined by a method other than precipitation, such as spectrometric measurement of an aliquot, or if the precipitate is dissolved for liquid scintillation counting. In such cases, the last step must be quantitatively controlled.

4.7.3. Tracer Addition

A known activity of a second radionuclide may be added to the solution as a tracer for the radionuclide of interest when no stable carrier is available, or if measuring the tracer is more convenient than measuring the carrier. Preferably, both radionuclides are isotopes of the same element, e.g., ^{85}Sr tracer for ^{90}Sr measurement. The initial count rate of the radionuclide of interest is calculated from its final count rate by Eq. (4.5a), taking m_i and m_f to be the initial and final net count rates, respectively, of the tracer.

As with the carrier, the tracer must be mixed with the radionuclide of interest so that they behave identically in every step of the procedure. If the possibility exists that the chemical form of the added tracer differs from that of the radionuclide of interest, both must undergo sufficient chemical reactions to achieve complete interchange of the two radionuclides.

The tracer and its radiation detector must be selected to avoid cross-talk between their radiations in measuring the two radionuclides. A tracer must be selected that emits different radiations than the radionuclide of interest, or emits the same type of radiation at an energy sufficiently different for resolution by spectral analysis. Use of radioactive tracer is common for actinides that are measured by alpha-particle spectral analysis. A correction factor may be applied if cross-talk cannot be entirely avoided but is small enough to maintain the reliability of the activity calculated for the radionuclide of interest.

4.8. Sample Purification

Carrier addition is widely used, not only for yield determination, but because the added mass avoids problems of radiocolloidal behavior. Other advantages of added carrier are the ability to use precipitation for radionuclide separation and purification, and avoidance of unintended coprecipitation during scavenging. When carrier is not added because no stable isotope is available or the counting source must be very thin, radionuclide deposition on container walls or on suspended solids must be avoided by applying the techniques discussed in Section 4.4, notably use of nonisotopic carrier or low pH.

The need for carrier addition to achieve radionuclide precipitation or prevent radiocolloidal effects is shown by applying the discussion in Section 3.2 and Eq. (3.6) to the precipitation of ^{129}I. The amount of 1 Bq carrier-free ^{129}I is 2.1×10^{-15} mmol, and its concentration in 0.035 l of solution is 5.9×10^{-14} mmol l^{-1}. No precipitate is expected in a solution with silver nitrate added to a concentration of

28 mmol l^{-1} because the product of the concentrations of the two reagents of 1.7×10^{-12} (mmol l^{-1})^2 is less than the solubility product for silver iodide of 8.49×10^{-11} (mmol l^{-1})^2. The precipitate may form, however, if even small amounts of additional iodide were in the water from natural or man-made sources, or if the thermodynamic activities exceeded component concentrations because of other salts in solution.

Although chemical reactions of radionuclides are assumed, for practical purposes, to be identical to those of their natural stable isotope mixtures, minor differences in equilibrium distribution exists, notably in the case of tritium relative to stable hydrogen, but also to a small extent for others, such as ^{14}C relative to stable carbon. During water distillation to purify it for tritium analysis, the tritium concentration of the residue is enriched by about 10% relative to the vapor. Only when all of the water is distilled will the specific activity of tritium (and the concentration in water, in units of Bq l^{-1}) be the same in the distillate and the sample (Baumgartner and Kim 1990). Moreover, because of such differences in equilibrium constants, the tritium concentration can be enriched approximately 10-fold by electrolysis of a water sample from which the produced hydrogen gas escapes into the air (NCRP 1976a).

Chemical procedures for purifying radionuclides that have no stable isotopes are based on studies of radionuclide behavior performed over the years. A crucial aspect of analytical procedures for these elements is addition of reagents that maintain the radionuclide in the redox state appropriate for the separation step.

The essential tools of radioanalytical chemistry practice are the 50-ml glass centrifuge tube and the stirring rod that are used for precipitation separations. Carriers are added with pipettes, and reagents, from dropper bottles. Heat is applied by a Bunsen or Meeker burner, and solutions are cooled in an ice bath. Plastic is substituted for glass for convenience if no direct heat needs be applied, and by necessity when hydrofluoric acid is used. The precipitate is separated from its supernatant solution during the purification stage by centrifuging, and in the final stage, on a 2.5-cm-diameter filter in a filtration apparatus.

4.9. Counting Source Preparation

Conventional source preparation is by carrier precipitation. In the absence of isotopic carrier, a nonisotopic carrier can carry the radionuclide when both are insoluble under identical conditions. Correction for the yield of the radionuclide may be necessary to adjust for the different solubility product of the nonisotopic carrier. The nonisotopic carrier for source preparation may be a different carrier than for purification, i.e., different nonisotopic carriers may be used in a sequence of precipitation separations.

As discussed in Chapter 7, a source may also be deposited by volatilization or electrodeposition for alpha or beta particle counting by Si diode or proportional counter, or added as a solution or suspension to scintillation cocktail for liquid scintillation counting. For electrodeposition, as indicated by Eq. (3.20), the half-electrode equilibrium potential depends on the tabulated standard reduction potential, the equilibrium concentration of the radionuclide of interest, other factors associated with the reactions of water and other ions in solution, and on the characteristics of the electrode surface.

For example, the standard potential of Ag^+ is 0.799 V. If the concentration of carrier-free ^{110m}Ag in solution is about 2 Bq l^{-1}, the concentration of silver is about 10^{-16} M initially and possibly 10^{-17} M at equilibrium. For the value of $\log (Ag^+)^{-1}$ of 17, the added potential 0.059×17 is 1.00 V. In fact, the concentration of Ag^+ in the solution may be orders of magnitude higher because silver in nature, in the laboratory, and accompanying the radionuclide will reduce the concentration-related voltage proportionately. The effect on the voltage by these factors must be measured for the solution processed and the electrode in use.

Radionuclides at very low concentrations that have a large value of E^0 deposit spontaneously on metals that are readily oxidized in the solution. For example, polonium is routinely purified and prepared for counting by spontaneous deposition on a nickel planchet. Spontaneous deposition has been demonstrated with other metals (Blanchard *et al.* 1960). Polarographic separation has been achieved by reducing a radionuclide to an amalgam in a drop of mercury (Kolthoff and Lingane 1939).

An alternative electrochemical source separation and preparation technique is electrophoresis. A drop of the radionuclide solution is pipetted at one end of a moistened paper strip that has an electric potential difference along its length. Each ionic radionuclide moves at its own rate down the strip. Locations of individual radionuclides are identified by their radiation or by dyes affected by the chemical form (Deyl 1979).

Loss of a fraction of a radioactive daughter from a radioactive source can be expected due to the recoil discussed in Section 4.2. For example, the radioactive daughter ^{224}Ra from a very thin source of ^{228}Th can transfer to a facing surface by recoil. This process can be applied beneficially to prepare a pure source of the radioactive daughter or can be a problem in contaminating the detector during alpha-particle spectrometry.

A similar situation is presented by a sample of soil or rock from which the radioactive progeny ^{222}Rn of ^{226}Ra leaves as a gas. The gas may be collected for counting as a purified radionuclide or its loss can cause an erroneous result in counting the sample by gamma-ray spectral analysis.

A different mechanism is attributed to the observed loss of ^{210}Po deposited on a metal backing. The ^{210}Po spreads over an extended area (Sill and Olson 1970), presumably because it has sufficient vapor pressure in the deposited metallic or hydride form.

5
Sample Collection and Preparation

ROBERT ROSSON

5.1. Introduction

Sample collection and preparation are inextricably linked to the practice of radio-analytical chemistry, despite the fact that two separate groups will perform these activities. Ideally, the radioanalytical chemist is part of the team that plans the sampling effort and prepares the quality assurance project plan (see Section 11.1). Such participation in the planning phase benefits the entire analytical process. The radioanalytical chemist can tailor the analytical approach to both the sample media and the radionuclides in the media. Information on the purposes for which the laboratory produces data can enable the radioanalytical chemist to devise laboratory practices to match required detection sensitivity, sample submission rate, and reporting style. On the other hand, the analyst can describe the capabilities and limits of the laboratory to the planning group. Such participation in the planning process can prepare a cohesive sampling and laboratory effort that produces defensible results for the client.

To serve as an effective member of the team, the radioanalytical chemist must have some knowledge of sampling methods pertaining to various matrices. The planning dialogue can influence specifications of sample collection by location, frequency, size, and techniques, particularly when the suite of radionuclide samples is from a complex environment. Sections 5.2 to 5.9 also briefly describe protocols for sample preservation between collection and analysis.

5.2. Sample Information

Sampling and sample treatment should be considered in terms of the data quality objectives (DQO) approach (Section 11.3.1), which integrates the monitoring program, sample collection program, radioanalytical chemistry program, and reporting of results. The magnitude of such initial planning and iterative improvements must be matched to the size of the monitoring program, with more extensive efforts for a larger program.

Environmental Radiation Branch, Georgia Tech Research Institute, Georgia Institute of Technology, Atlanta, GA 30332

Those planning the sampling effort should initially consider the following:

- The radionuclides produced or used in the facility or project
- Radionuclide activity levels
- Variability in radionuclide activity with time and location
- Chemical forms of the radionuclides
- Sampling circumstances, e.g., routine operation or specific incidents
- Radionuclide transport by various media.

These initial items of information guide sampling location, volume, and initial preservation steps. Implicitly, they also guide laboratory selection; incident response will likely require a quick turnaround time and the capability of handling higher radioactivity levels (see Section 13.1), which narrows the choice of appropriate laboratories.

Once the sample has been taken, and a laboratory chosen, this information goes with the sample as part of the chain of custody documentation. When the laboratory accepts the sample, the sample information is reviewed to extract those facts directly pertinent to the analytical process:

- Sample form and matrix
- Analytical requirements
- Format for results
- Sample load

The sample form and matrix type control the initial sample treatment in the laboratory. The analytical requirements dictate the chemical and instrumental procedures. The format specifies units for reporting values, uncertainty, and detection limits, and the context in which this information will be reported. The sample load controls laboratory staffing, scheduling, and turnaround time.

The DQO checklist for the laboratory is a comprehensive compilation of all these pieces of information. It will include descriptions of the following:

- Sample type
- Description of site parameters (e.g., flow rate of the sampled medium)
- Sample collection, storage, and preservation methods
- Sample matrix, including the radionuclides expected and the radionuclides of interest
- Predicted radionuclide concentrations
- Required detection limits
- Available radiation detection instruments
- Available radioanalytical chemistry methods
- Estimated detection limits for selected method and instruments
- Requirements of sample size and counting times to meet specifications
- Personnel skill, time, and cost estimate

The usual types of samples are described in Table 5.1 and the following sections; their analysis is discussed in Section 6.2. Some of these sample types are obtained with sampling equipment such as air and water collectors placed at selected

TABLE 5.1. Sample storage and preservation

Sample type	Collection	Storage container	Preservation
Air filter	5-cm-diameter Filter	Petri dish	None
Air cartridge	Charcoal cartridge	Ziplock bag	None
Air tritium	Desiccant column	Air tight container	None
Water	Bottle	Gallon cubitainers	Add acid, base, or complexing agent
Milk	Grab	Gallon cubitainers	Refrigerate or add formaldehyde
Vegetation	Grab	Ziplock bag	Refrigerate or freeze dry
Biota	Grab	Ziplock bag	Refrigerate or freeze dry
Soil, sediment	Grab	Wide-mouth jar	Refrigerate or freeze dry
Urine	Grab	Gallon cubitainers	Refrigerate, or add acid or complexing agent

locations (Budnitz *et al.* 1983), while other samples are collected by hand, as in the case of soil and vegetation. Particularly important for hand-collected samples is the reported description of sampling site and process in sufficient detail to permit comparisons among sampling periods and locations. All information regarding the sample should be reported on the chain of custody form (see Section 11.2.5) that accompanies the sample to the laboratory.

Identifying the type of sample matrix and radionuclide of interest helps to determine a pretreatment protocol. Some form of pretreatment is usually required for environmental samples. The half-lives of both the radionuclides of interest and interfering radionuclides must be considered to decide how quickly to perform the analysis. The radiation type and energy must be known to select the radiation detector, and possibly the radioanalytical chemistry method to prepare the source for counting.

The concentration of the radionuclides of interest may be inferred from the sampling location. Natural and man-made background radiation values are summarized by the National Council on Radiation Protection and Measurements (NCRP 1987a) and the United Nations Scientific Committee on the effects of atomic radiation (UNSCEAR 2000a,b).

Estimation of the detection limit of the measurement instrument in the time period available for counting, linked with the required detectable concentration and the expected concentration of the radionuclide to be determined, guides selection of sample size for the analysis. Calculation of the minimum detectable activity is discussed in Section 10.4.2. Additional documents of interest are Altshuler and Pasternack (1963), Pasternack and Harley (1971), and Currie (1968). The terms minimum detectable concentration and lower limit of detection also have widespread use. A document that addresses these and other topics pertinent to radiation monitoring and measurement was developed by a committee of the Health Physics Society (EPA 1980a).

All sample information must be taken into account to develop an estimate of the time and resources necessary to process the sample to the satisfaction of the client, as discussed in Section 13.9. For example, an estimate of needed sample volume to obtain valid result with the detector at hand may show that the proposed sample is too large to accommodate the separation procedure unless pretreated to reduce the volume and measured for a longer period in a low-background system.

5.3. Atmospheric Samples

Air sampling is required around nuclear facilities and in populated areas that may be exposed to elevated radionuclide levels to evaluate radiation exposure from external or inhaled radionuclides. Airborne radionuclides may be in the form of a gas, vapor, or particles. Different sampling techniques are employed depending on the radionuclide of interest, its form, and the sample volume required to reach the detection limit. Samples may be collected at fixed stations or from vehicles moving on the ground or in the air.

Air sampling equipment consists of a framework or housing, a sample collector, a collector holder, an air pump, and control and recording equipment, including a flow control device, flow-rate meter, timer, and data recording or transmitting system, as shown in Fig. 5.1 (see Sections 15.2. and 15.4 for examples). The system should provide structural support, shelter, and a tight connection for the collector. The air pump must provide the required airflow. The system must be maintained for continuous operation and protected from vandalism. The number and location of stations must be selected to provide the information for which they are operated, e.g., to detect and quantify airborne radionuclides that are released at a site or distributed across an area.

5.3.1. Particulate Radionuclides

Airborne particles commonly are collected from air on filter media such as organic membranes, paper (cellulose fiber), or glass fibers. Membrane filters are fragile and

FIGURE 5.1. Air sample collector.

can be used only with low-flow-rate systems, while paper and glass fiber filters are sturdier and more easily handled. These filters have high collection efficiency for particles with sizes in the respirable range, i.e., those with an aerodynamic median diameter of around 0.3 μm (Lockhart *et al*. 1964). All except membrane filters have low-pressure drops that permit high flow rates for collecting large volumes of air to achieve increased measurement sensitivity. The filters can meet standard practice requirements of 99% removal of respirable particles at the operating air velocity and pressure drop (NCRP 1976b, Corley *et al*. 1977, EPA 2004).

In view of the influence of particle size on inhalation radiation exposure, some collectors of discrete particle sizes are deployed, typically for research rather than routine monitoring. Impactors with metal plates or Mylar foils that are sprayed with silicone to reduce particle bounce (ACGIH 2001) have proven useful, but can challenge radioanalytical chemistry preparations. An aerosol spectrometer with a series of metal screens (EPA 2004) may present less of a chemical preparation problem.

Collector location and collection period must be selected to avoid excess mass loading. Excessive deposition of matter on the filter will reduce the air flow-rate, change retention efficiency, and result in some collected solids falling off when the filter is handled.

The duration of sampling (start and end times) and the flow rate through the filter must be recorded. The type of pump, the kind of filter substrate (cellulose, glass fiber, polystyrene, etc.), and even the housing for the sampling station should be recorded because these factors can affect data application. Interruption in sample collection or the air-flow record should be avoided because the former causes a gap in the monitoring program and the latter makes the radionuclide measurement pointless because the air volume is unknown.

Radionuclides in particulate samples collected on filter substrates usually are measured first by gamma-ray spectral analysis of the filter enclosed in a thin-walled envelope. Gross alpha- and beta-particle analysis by low-background proportional counter follows (see Section 7.2.4). The particles collected from the air stream include radioactive progeny of natural ^{222}Rn and ^{220}Rn. These radionuclides always present a radiation background to any targeted measurement of radionuclides. To eliminate interference from these progeny in counting, the filters are held for a time between collection and counting to allow for their decay; the short-lived daughters of ^{222}Rn decay mostly in 2 h and those of ^{220}Rn decay in 4 days. The long-lived progeny of ^{222}Rn—^{210}Pb, ^{210}Bi, and ^{210}Po—and cosmic-ray-produced ^{7}Be remain on the filters after 4 days.

A specified fraction of the filter is taken for radiochemical analysis. The first step is filter dissolution, as discussed in Section 6.2.1. The remainder of the filter is archived for possible future analyses. The archive should also contain unexposed (blank) filters.

Airborne particulate radionuclides are also collected in deposition trays. An upward-facing adhesive "gummed film" in the tray has been used to retain the deposited solids during the collection period (NCRP 1976b). The tray collects solids during dry periods and precipitation events unless a sliding cover is activated to close the tray during precipitation. The cover can be moved by a sensor of increased conductivity that causes the cover to close the tray at the beginning of precipitation and opens it after the end (Chieco 1997). The measured deposition during dry periods (for a collector with a cover) or during the entire period (for a collector without cover) can be compared to the results measured for a rainwater collector filter discussed in Section 5.4.3. The particles collected during precipitation may include some dry deposition that the rainwater washed out of the collector and onto the filter.

5.3.2. Gaseous Radionuclides

The usual gaseous radionuclides are tritium, ^{14}C, radioiodine, and noble gases. Tritium (written below as T) can exist as water vapor, hydrogen gas, and organically bound compounds such as CH_3T. Carbon-14 can be found in CO_2, CO, and organic gases. Gaseous radioiodine is believed to exist in three forms: as I_2, HIO, and organic-bound such as CH_3I. The noble gases include neutron-activated ^{41}Ar, several fission products of krypton and xenon, and naturally occurring ^{222}Rn and ^{220}Rn.

Tritium collection. Tritium in air is usually in the form of water vapor and less commonly in the elemental or organic-bound forms. It is generated in nature by cosmic-ray interactions, and at nuclear reactors and tritium-production facilities by ternary fission and neutron activation. Tritium as HT tends to oxidize to water vapor in air. Conversion to and from organic-bound tritium occurs in biota (NCRP 1979).

Tritiated water vapor is collected by condensing or freezing it, by bubbling air through water to exchange tritiated water with nontritiated water, and by sorption on silica gel or molecular sieve (Corley *et al.* 1977). A standard method calls for drawing filtered air through a silica gel column at a rate of 0.10–0.15 l min^{-1}, as shown in Fig. 5.2 (APHA 1972). Sufficient silica gel must be in the column

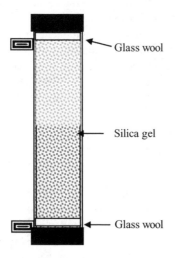

Glass wool

Silica gel

Glass wool

FIGURE 5.2. Silica gel column for collecting tritiated water vapor.

to sorb the maximum humidity expected for the volume of sampled air. Column dimensions must accommodate this silica gel volume and the selected air flow rate. The collected water or used silica gel is taken to the laboratory for tritium analysis (see Section 6.4.1).

Special systems are used to collect separately the tritiated hydrogen gas (HT) and organic-bound tritium from air. A two-component collection train consists of a molecular sieve column, followed by addition of a stream of H_2 carrier gas to the sampled air before it passes through a palladium-coated molecular sieve column. The first column collects tritiated water and the second column collects the two other forms (Ostlund and Mason 1985). The two columns are treated separately at the laboratory, as discussed in Section 6.2.2, to provide the three forms for separate analyses.

Carbon-14 collection. Carbon-14 is formed in nature by cosmic-ray interactions. It is in all carbon-containing compounds that are in equilibrium with ^{14}C in air at a specific activity of 0.23 Bq g^{-1} carbon. Concentration measurements in carbon-containing compounds that are no longer at equilibrium with air, such as dead trees, are used to determine their "age"—the time period since the end of equilibrium with airborne carbon—in terms of the fractional radioactive decay. Fluctuations of cosmic-ray production of ^{14}C in air over the centuries must be considered in this determination (NCRP 1985a). Carbon-14 is also produced at low rates in nuclear reactors, mostly by the (n,p) reaction with ^{14}N and the (n,α) reaction with ^{17}O.

Collection of ^{14}C in air generally is a research effort. A large volume of air is pumped through a collection train that consists of a bubbler with barium hydroxide solution and a tube that contains aluminum–platinum (0.5%) catalyst, where it is heated to 550°C, and then pumped through a second bubbler with barium hydroxide solution. Air passes through the first solution, in which ^{14}C in the form of CO_2 precipitates as barium carbonate and then passes with added CO carrier gas through the oxidation tube in which ^{14}C as CO or CH_4 is converted to CO_2. The gas finally passes through the second solution in which the newly formed CO_2 precipitates as barium carbonate. The two ^{14}C samples are taken to the laboratory for measuring the emitted beta particles (Kahn *et al.* 1971).

Radioiodine collection. Monitoring for radioiodine is associated with fallout, nuclear facilities, and medical use. The most common radioiodine is ^{131}I, but shorter-lived ^{132}I, ^{133}I, and ^{135}I are also fission products, and others, such as ^{125}I, are also used in nuclear medicine. Some long-lived ^{129}I is produced in nature and some is formed in fission and released when processing spent reactor fuel. Iodine in air can be both gaseous and particulate. The existence of radioiodine in a number of chemical states complicates sample collection and analysis.

The typical environmental airborne radionuclide collection train consists of a filter followed by an activated-charcoal cartridge; particulate iodine is retained on the filter and gaseous iodine on the cartridge (Corley *et al.* 1977). A more elaborate collector train distinguishes the various forms of radioiodine. The collectors, in order, are an air filter for particles, cadmium iodide on Chromosorb-P for I_2, 4-iodophenol on alumina for HIO, and activated charcoal for organically bound iodine, such as CH_3I (Keller *et al.* 1973).

Collection efficiency for radioiodine from the air stream by activated charcoal is subject to type of charcoal, size of the granules, depth of the sampling bed, and flow rate. The grain size should be in the range 12–30 mesh and the sampling rate, 0.03–0.09 m^3 m^{-1}, for optimum collection (APHA 1972.). Air streams can also be monitored for radioiodine with silver zeolite and molecular sieve, but activated charcoal is the commonly used sorbent largely due to lower cost.

Noble gas collection. The radioactive argon, krypton, and xenon radioisotopes are monitored near nuclear power plants and related facilities to determine the magnitude of releases of gases generated by fission or neutron activation. Radon radioisotopes and their particulate progeny are measured in homes and mines to determine whether their airborne concentrations are below radiation protection limits.

Grab sampling techniques for noble gas radionuclides include collection in a previously folded large plastic bag or a previously evacuated sampling flask. Low-volume samples can be taken with a hand pump; larger volume samples are collected in metal containers under pressure of 10–30 atm (Corley *et al.*1977). Flowing air is sampled for noble gases by sorption on cooled activated charcoal or by cryogenic condensation.

Radon in air can be collected and measured with an evacuated Lucas cell (see radium analysis in Section 6.4.1). The cell is taken to the sampling site and filled by opening the stopcock. After closing the stopcock, the cell is returned to the laboratory and counted between 2 h and a few days after collection. Radon is also collected on charcoal cartridges that are measured directly by gamma-ray spectral analysis. Radon progeny are collected on filters. They can be measured to estimate radon parent concentrations. Numerous *in situ* detectors of radon and radon progeny, including ionization detectors, proportional counters, and scintillation detectors, are in use (EPA 2004).

5.4. Water

Types of samples include rain, surface water, ground water, drinking water, seawater, and wastewater. The various types have different chemical constituents, solids content, and pH that may require distinctive collection and analysis techniques. Information should be obtained on the expected radionuclide content of the water to guide initial processing, sample preservation, and any requirement for rapid shipping and prompt analysis.

Practical experience suggests that water is the easiest matrix to process. Problems with water samples include the distinction between dissolved and suspended radionuclides and the preservation of dissolved radionuclides in the sample in their original form. The sampling location and process can modify the sample; for example, radionuclides in water may deposit on pipe and valve surfaces and later may dissolve again.

Filtration of water samples at collection removes from the water the suspended and colloidal fractions (for separate analysis, if desired) but some dissolved radionuclides may be retained on the filter and associated particulate matter. If filtration is postponed until the sample reaches the laboratory, some radionuclides may in the interval transfer between the water and suspended materials. This problem does not apply to samples from public water supply systems for water that already has been treated to remove suspended solids.

Sorption of radionuclides onto the container walls is a common problem. The problem does not arise for all radionuclides and types of water, but its occurrence must be checked and may be prevented as discussed in Section 4.5. The usual sample preservation technique is mild acidification, but stronger acid may be necessary to preserve transuranium radionuclides. A basic solution must be used for radioiodine.

5.4.1. Grab Sample Collection

Grab samples provide initial information and periodic checks for surface water or water supply systems. The simplest collection from surface water is bottle or dipper submersion. These methods are effective for collecting water near the top of the water column. To collect grab samples far below the surface, one needs an extendable tube sampler with a check valve to fill the tube with the sample.

A bailer can be used for surface water collection but is primarily intended for collecting ground water. When the bailer is lowered into the water, liquid flows into the bailer through the bottom check valve; when the bailer is retrieved, the check valve closes to retain the water. Other types of collectors are also available (Brown *et al.* 1970).

5.4.2. Continuous Surface Water Sampler

Unattended continuous samplers are used for ongoing radionuclide measurements. A small fraction of the surface flow is pumped into the sampler. One type of automated sampler (see Fig. 5.3) collects samples at regular intervals of minutes to days in a set of sample containers. A single container can be placed at the collector to receive a composite sample.

5.4.3. Rainwater Collector

A rain pan collector accumulates a rainwater sample deposited on a known area. The collector funnels rainwater into a sample bottle. A filter followed by cation- and anion-exchange resin columns can be inserted in the line between the sampling pan and collection bottle, as shown in Fig. 5.4. The filter collects particulate radionuclides, the resins collect cationic and anionic radionuclides, and nonionic radionuclides remain in the water. The pan collects both dry and wet depositions. Collectors with movable shields have been developed that open during rainfall, as discussed in Section 5.3.1. The pans are washed after each collection period to remove the radionuclides associated with dry deposition and clean the pans for

FIGURE 5.3. Automated surface water sample collector.

the next collection period. Acid or another preservative is added to newly placed collection bottles.

Depending on the program, rain samples are collected weekly, monthly, or quarterly. An adjacent rain gage is desirable to record precipitation data, or the collector may be placed near a meteorological station so that radionuclide measurements can be related to weather data. Radionuclide data are reported relative to the amount of water, area of deposition, and rainfall for the period.

5.4.4. Groundwater Collection

Radionuclides in groundwater are identified and quantified to determine sources of discharges or leaks into the ground. They can indicate the presence of radionuclides in subsurface soil, and predict radionuclide concentrations in surface water reached by groundwater. Groundwater usually is collected from wells that have been drilled

FIGURE 5.4. Rainfall collector.

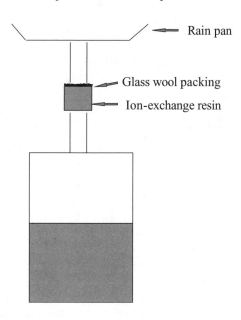

Rain pan

Glass wool packing

Ion-exchange resin

at locations and depths selected on the basis of groundwater flow models. Well drilling and water sampling must follow established guidelines to yield acceptable samples (Wood 1976, EPA 1986, Shuter and Teasdale 1989). Several wells may be bundled at a given site to reach various depths. The number of wells drilled is limited by their cost. A different sampling system consists of collecting water from existing supply wells. The frequency of sampling is related to the estimated subsurface flow rate; for example, if the estimated flow rate is a few meters per year, only an annual sample is needed.

Samples that represent groundwater are collected after several well volumes have been discarded unless little water is available. The sample is acidified promptly for preservation. If one of the radionuclides is not stable in acid, a second sample is collected and preserved as appropriate.

5.5. Milk

Milk is frequently analyzed for radionuclides in a monitoring program because it is one of the few foods that reaches the market soon after collection. Milk commonly is consumed within 7–11 days after milking and occasionally reaches the consumer within 2 days. It may contain relatively short-lived radionuclides and be a dietary source of fission-produced ^{89}Sr, ^{90}Sr, ^{131}I, ^{137}Cs, and ^{140}Ba, as well as naturally occurring ^{40}K. A 4-l sample usually is collected for analysis. Cow's milk samples can be collected to represent a specific herd in the form of raw milk, or a regional pool of pasteurized milk. Goat's milk is collected when this medium may be a significant source of radionuclide intake by humans. Powdered milk may be analyzed for the longer-lived radionuclides ^{90}Sr and ^{137}Cs.

Milk samples should be refrigerated and delivered to the laboratory within 1–2 days. For longer storage, the milk sample should be frozen or a preservative added such as methiolate, formaldehyde, or thimerosal (DOE 1987, EPA 1984), as indicated in Table 5.1. Preservatives may interfere with radioiodine separation (Murthy and Campbell 1960), but can be added if radioiodine is measured directly by gamma-ray spectral analysis.

5.6. Vegetation and Crops

Decisions concerning sampling vegetation and crops include selection of sites, type, and collection technique (e.g., height above ground for cutting grass). Food may be obtained from a private garden, a store, or a distributor. Collection protocols are mainly site specific; crop selection depends on local growth and consumption, guided by availability, location, season, and farming practice. Radionuclide pathways to vegetation and crops are surface deposition and root uptake.

Radionuclides in grasses and leaves are analyzed as indicators of radionuclides in the environment and links in the food chain to animals that feed on them. Food (crop) samples are analyzed to relate radionuclide concentrations to human radiation doses due to consumption. Studies of radionuclide pathways from source to consumer can identify the category of crops—i.e., leafy vegetables, grains, tubers, or fruits—to be sampled.

Sample masses of several kilograms may be needed to achieve specified analytical sensitivity. Sample size and preparation typically should provide 20–40 g of ash from the edible portion. As a rule of thumb, ash weight is approximately 1% of moist weight. The sample should be weighed at the time of collection and analysis. For many sample types, refrigeration and analysis without undue delay are necessary to avoid decomposition.

5.7. Meat and Aquatic Species

Edible tissues from terrestrial animals, aquatic animals, and fowl are measured for radionuclide content to estimate the radiation dose to humans from food consumption. Organs and tissue that concentrate radionuclides, such as radioiodine in thyroids or radiostrontium in bones and teeth, are analyzed as indicators of such radionuclides in the environment. Domestic animals are sampled at the normal time of slaughter. Wildlife samples can be obtained from hunters or by special collections. Because radionuclide concentrations in animal tissues vary widely, depending on species, feed or inhalation intake, location, and individual metabolism, multiple animals must be sampled to establish a normal range of radionuclide concentrations. Radionuclides that emit gamma rays can be measured in live animals by a technique similar to whole-body counting in humans.

Samples from aquatic species, i.e., fish and shellfish, may be the edible portion to focus on human consumption, or the whole fish as indicator of radionuclide levels in the system. Fish can be considered in several categories such as bottom feeders (e.g., catfish), pan fish (blue gill), and predators (bass). Often, more than one fish per species must be collected to accumulate the sample mass needed for analysis. Fish are not good indicators of local radionuclide distribution because of their mobility: i.e., fish caught at one location may have ingested the radionuclide elsewhere, within the limits suggested by their normal movement or imposed by dams.

Fish are sampled by electroshocking or with hoop nets, trout lines, or rod and reel. Certain fish will float to the water surface after shocking (pan fish and predators), but bottom feeders such as catfish are more easily caught with hoop nets or trout lines. Collection with rod and reel simulates sports angler catches.

Samples of meat and fish must be refrigerated for prompt analysis or frozen for extended storage. Fish and other small animals may be dissected promptly after collection or just before analysis. Samples should be weighed when fresh, dried, and ashed as needed for the specified purpose in reporting radionuclide concentrations.

5.8. Soil and Sediment

5.8.1. Soil

Soil usually is an easy matrix to collect in bulk quantity, but one that is difficult to process. This means that the collection regimen must be highly selective in allowing all *pertinent*, but no *extraneous*, soil to be sampled (see also Section 11.3.3). The site must be selected with regard to the radionuclide information required of the program, which may include the consideration of all or some of the following factors:

• amount of area deposition and accumulation
• deposition pattern
• transfer to vegetation or crops
• transfer to groundwater or surface runoff
• resuspension into air
• external radiation exposure

Soil can accumulate radionuclides over long time periods but may not be useful for observing short-term trends. Measurements can provide retrospective information on radionuclide levels. Site description, location, collection date and time, sample area and depth, and sampling method (core, template, trench) should be recorded along with the sample identification number and name of sample collector.

Surface soil can be collected with a scoop or cookie cutter. The scoop sampler is convenient for collecting a grab sample and transferring it directly into a collection container. A cookie cutter is a rigidly framed apparatus that can be sunk into the soil to a selected depth to select the material within the frame for removal and analysis (FRMAC 2002). Vegetation, debris, roots, and stones should be removed at the time of collection to obtain a sample pertinent only to soil and assure that the soil mass is sufficient for analysis. Sandy soil usually retains only a small fraction of radionuclides to which it is exposed, while clay can be highly retentive. In agricultural land, surface soil with elevated radionuclide levels may have been turned over and had chemicals added that affect soil retention capability.

If the interest in radionuclide distribution extends beneath the soil surface, coring is the preferred sampling technique. Cores are taken to a known depth at the sampling site. Depth profiles are separated at selected vertical intervals to provide information on area deposition and downward movement of a radionuclide. Numerous cores may have to be taken because radionuclide retention can vary with changes in soil constituents in a relatively small area. Samples at greater depths can be obtained from well-drilling cores.

5.8.2. Sediment

Radionuclides in sediment are indicators of facility releases to bodies of water and runoff from surfaces in the drainage region. Samples represent the fraction of radionuclides precipitated or sorbed on suspended matter that has settled to the bottom. If such radionuclides are in sediment, elevated levels usually can be found where suspended material tends to settle as the water flow rate decreases, e.g., at widening channels, on the inner side of bends, and behind dams.

Variables that affect radionuclide concentrations in sediment are water movement, water quality, the presence of aquatic vegetation, and sediment characteristics such as clay and organic fractions. Sediment samples collected at time intervals should not be expected to have a consistent pattern of radionuclide concentrations because the location sampled previously is difficult to find precisely, sediment may have moved downstream, and older sediment may be covered by more recent deposits.

A striking effect of the change in water quality is the increased deposition found for ^{137}Cs in estuaries where fresh water contacts saltwater (Corley *et al.* 1977). Impounded water that stratifies, with anoxic (oxygen deficient) conditions near the bottom during winter and mixing in spring, can affect radionuclide solubility (Carlton 1992). In "blackwater" streams, humic acids can complex radionuclides to keep them soluble. Changes in water level may periodically cover soil, vegetation, and debris that take up radionuclides, notably in swamps.

The dredge sampler for sediment acts like a scoop in collecting sediment from the stream bed surface. The dredge has drawbacks because the sampling location on the riverbed is uncertain, it collects only the top 15 cm of sediment, and fine material can escape along with the water on return to the surface. A core sample collected by the slide hammer method can be used to collect a 0.7-m-long sediment sample in water as deep as 5 m (Byrnes 2001).

Collected samples are frozen if preservation is necessary. Both moist weight at the time of collection and dried weight before analysis should be measured.

5.9. Bioassay Samples

Radiobioassay is the determination of the kind, quantity, and location of radionuclides in the body by direct measurement (*in vivo*) or by *in vitro* analysis of material excreted from the body. Spectral analysis permits rapid analysis of radionuclides that emit gamma rays. Computer systems with data analysis in terms of metabolic models support routine use of bioassay procedures for assessing internally deposited radionuclides. The whole-body counter is an example of an *in vivo* procedure.

Need for bioassay in humans depends on the exposure potential and the radiotoxicity of a radionuclide and the degree of radionuclide control and radiation monitoring at the location. Exposure assessment takes into account the physical properties, the quantity, and the concentration of the radionuclide in the process, and monitoring data on surface and airborne contamination. Monitoring levels that exceed 10% of guidelines such as the derived air concentration for each radionuclide are a signal for instituting a bioassay program (NCRP 1987).

Radiotoxicity depends on energy deposition in tissue or organs by the radionuclide, the specific tissue exposed to the radionuclide, and the tissue radiation sensitivity. Energy deposition by a radionuclide is a function of its emitted radiations and half-life. Biokinetic studies have identified for most radionuclides of interest the pattern of movement through the body and the effective turnover rate (the sum of the biological and radioactive turnover rates). Biokinetic information also identifies the appropriate type of sample to be collected among blood, urine, feces, saliva, breath, hair, teeth, nasal swipes, and tissue obtained incidental to unrelated operations, and collection frequency. The measured radionuclide concentrations are combined with biokinetic information to calculate the committed dose equivalent, the indicator of radiation impact on the subject (NCRP 1987b).

Urine is the most common bioassay sample type because it is nonintrusive and readily available, and provides some insight into radionuclide intake, retention, and excretion, but is not ideal for all radionuclides. A preservative can be added to urine samples immediately after collection. Bioassay samples are refrigerated for prompt analysis and kept frozen for extended storage.

5.10. Quality Control in Sample Collection

Quality control (QC) is as important when collecting samples in the field as it is in the laboratory. A valid and defensible laboratory result is meaningless if the sample has been collected improperly or the extent of its reliability is not known (see Section 11.3.3). Many of the pitfalls associated with collection and preservation of environmental samples are enumerated in MARSSIM (EPA 2000c) and HASL-300 (Chieco 1997). The steps taken for field QC vary with the matrix being sampled. The following QC samples that pertain to the sample collection process should be integrated with the QC samples appropriate to laboratory practices that are listed in Section 11.2.9:

- Collocated samples to indicate the variability in the results of sampling at a given location.
- Field trip blanks to identify radionuclide contamination in the collection material (e.g., filter or sorbent) and container.
- Background samples that are from surveillance sites at a direction or distance where no radionuclides from the monitored facility or incident are expected.

The program DQO specifies the types, locations, and collection frequency of these QC samples.

Many types of environmental samples, notably vegetation and soil, are almost inevitably inhomogeneous in ways that may prevent uniform distribution of a radionuclide throughout the matrix. An inhomogeneous radionuclide sample may be due to the nature of the collected material, inadequate preparation, or "hot particles," i.e., radionuclides that are not distributed uniformly in a solid, but associated only with a subset (see Section 10.3.8). Some providers of reference materials state the minimum sample aliquot to be analyzed to obtain a valid result. A larger sample than needed can be taken to eliminate extraordinary efforts in taking reliable aliquots for radionuclide analysis. Analysis of multiple replicate samples indicates the extent of reproducibility, and the blank samples assure that contamination has been avoided before or during transport. Background collection prevents attribution of radionuclides found in these samples to the facility or incident that is being monitored.

6
Applied Radioanalytical Chemistry*

BERND KAHN[1], ROBERT ROSSON[1], and LIZ THOMPSON[1]

6.1. Introduction

A radioanalytical chemistry procedure is a series of steps that leads to the measurement of radiation (or sometimes mass), with the goal of unambiguously identifying the radionuclide of interest and determining its amount in a sample. The chemistry component of the analysis begins with the sample collection and handling described in Chapter 5 and ends with the source preparation discussed in Chapter 7. The radiation emitted by the radionuclide is then measured (i.e., the sample is "counted"), as discussed in Chapter 8. This chapter addresses the practical aspects of the chemical and radiochemical separation processes surveyed in Chapters 3 and 4, which are central to separating interfering radionuclides and solids from a radionuclide in preparation for counting.

Numerous separation methods of the types cited in Chapter 3 were developed and applied in radioanalytical chemistry during the past century. The first 30 years were devoted mostly to nuclear chemistry applications for identifying and characterizing the naturally occurring radionuclides. In the following years, attention shifted to the man-made ones; these activities continue, as exemplified by the work described in Chapter 16. Currently, many methods are devoted to monitoring radionuclides in the environment, facility effluent, process streams, and workers.

The radioanalytical chemist who is responsible for such monitoring selects or designs procedures that meet the client's specifications of sample type, list of radionuclides, measurement reliability and sensitivity, and response time. The analyst also considers the limits imposed by prescribed sample size, solution volume appropriate for chemical separation, and radiation detection instruments at hand. Potentially applicable procedures are selected for these criteria by literature review and then evaluated in a methods development and testing process. A chemical separation procedure can be devised either by selecting the most applicable published method and introducing any needed modifications or by combining pertinent separation steps.

One of the main efforts in measuring a radionuclide of interest is removing interference due to other radionuclides by chemical separation, detector selection,

*This text owes the genesis of its content to Dr. Isabel Fisenne, Department of Homeland Security.
[1]Environmental Radiation Branch, Georgia Tech Research Institute, Georgia Institute of Technology, Atlanta, GA 30332

or spectral analysis. Some sample characterization is always applied to limit the number of radionuclides and their intensity to be expected in a sample. One approach is the following breakdown by origin, pathway, and half-lives (note that the indicated half-lives and number of radionuclides are approximate):

- Samples that contain a few long-lived ($t_{1/2} > 1$ year) radionuclides. This mixture is encountered in monitoring fallout from long-ago tests and in effluent, radioactive waste, and decommissioning actions at nuclear facilities. Some subcategories can simplify separation method selection: (1) Very long-lived (10^4–10^{11} years) radionuclides may be considered when planning protection for thousands of years. (2) A single radionuclide may have to be distinguished from the natural radiation background in contamination incidents, including possible terrorist action. (3) Long-lived naturally occurring radionuclides may have to be measured.
- Samples that contain approximately 10 radionuclides with intermediate to long (1 day–1 year) half-lives. This mix is found when monitoring distant locations for recent fallout from atmospheric nuclear tests and effluent from waste tanks and stacks at operating nuclear power plants. A subcategory is samples that contain natural radionuclides with half-lives in this range as progeny in terrestrial decay chains and cosmic-ray-produced radionuclides.
- Samples that contain numerous radionuclides from certain portions of the periodic table with short (1 h–1 day) and intermediate half-lives. This mixture is found in monitoring process streams at operating nuclear facilities, fallout from nuclear tests, or accidents within hours to days after occurrence. Subcategories include samples from activation analysis, nuclear chemistry studies, and nature.
- Samples that contain a few very short-lived (<1 h) or extremely long-lived ($>10^{11}$ years) radionuclides. These are encountered in nuclear chemistry studies conducted in an academic setting, and include newly prepared and naturally occurring radionuclides.

Circumstances will modify the contents of these categories with regard to radionuclides and half-lives; hence each sample set should be defined in its own terms for planning the analysis. Occasional unusual radionuclide mixtures must be anticipated during a project. The first three categories are discussed in Section 6.4. Examples of very short-lived radionuclides in the fourth category are given in Chapter 16; very long-lived radionuclides are best measured by mass spectrometry (see Chapter 17).

Once the radionuclide content of a sample is specified, available detectors must be selected to measure the emitted radiation. The choice of detection technique influences the need for and extent of the radiochemical separation, as discussed in Chapter 7.

Chemical separation and radiation detection can work together in discriminating against interfering radiation. Chemical separation may not be necessary if the radiation from contaminant radionuclides is not detected by the selected detection instrument or does not interfere with the measurement, e.g., by spectral analysis.

In some instances, purification may not be necessary but can ease detection and improve measurement.

Two or more radioisotopes of the same element that cannot be measured by spectral analysis require integration of effective separation of impurities and radiation detection of the selected distinguishing decay characteristics. To determine the amounts of ^{89}Sr and ^{90}Sr in a sample, for example, interfering radionuclides such as ^{226}Ra and ^{140}Ba must be removed; only then can the two strontium radioisotopes be distinguished in terms of the radioactive decay of ^{89}Sr, ingrowth of the ^{90}Y daughter, and detector response to beta-particle energies, as discussed in Section 6.4.1.

A testing program is needed to confirm the reliability of the selected procedures, especially to achieve high chemical yield and a sufficient decontamination factor for every interfering radionuclide. The analyst must estimate the allowable concentrations of radionuclide impurities and matrix components; below this concentration, interference in measuring the radionuclide of interest is not of concern.

An ongoing quality assurance program provides feedback for maintaining or improving each procedure during routine application. Changes are needed when quality control results fall outside the limits of acceptability discussed in Section 10.5. Ambiguous results concerning the presence or absence of radionuclides of interest require methods diagnostics, the topic of Chapter 12. Reported methods that appear to be better should stimulate efforts to replace existing methods.

6.2. Initial Sample Treatment

The samples that are received for analysis are described in Chapter 5, and the receiving actions in Section 13.3. Incoming samples can be divided into three process streams: one for direct measurement, another for direct chemical analysis of aqueous solutions, and the third for conversion to forms that are soluble in aqueous solution. All solid samples are weighed; some will be dried and possibly ashed, and weighed again after each process. Samples to be counted directly are placed in tared counting container and weighed. Volumetric aliquots for gross activity measurements and for chemical analysis are taken from liquid samples. Samples scheduled for dissolution may be processed with suitable acids or bases or by fusion. Special treatment is given to samples that must be removed from a collector, require concentration, or are in gaseous form.

6.2.1. Aerosol Collectors

Air filters and gummed deposition collection films first are measured by gross alpha- and beta-particle counting and gamma-ray spectral analysis. Usually, 1/2 of the sample then is dissolved to perform radiochemical analysis of the deposited radionuclides. The filter can be dry ashed, and then totally dissolved with an HNO_3–HF treatment.

Whether some insoluble residue remains depends on the type of mass loading. For example, soil on the filter may constitute a partially insoluble matrix that retains some radionuclides, including naturally occurring uranium and thorium. Insoluble residues of ashing and subsequent acid dissolution require fusion, as discussed in Section 6.2.7. The two fractions are combined after the melt is dissolved in dilute HNO_3, and the solution is evaporated almost to dryness and is then taken up in dilute HNO_3.

Some readily soluble filter materials are available (see Section 5.3.1), but the most favorable material may not meet the specifications for large filter area, rapid airflow, and long collection period. Paper filters generally are sufficiently rugged to meet these requirements.

Removing the accumulated particles by washing is unlikely to be quantitative or even consistent. Acid leaching can leave highly insoluble forms of several radionuclides, notably the actinides, on the filter. As an exception, some fallout radionuclides (such as ^{90}Sr) are completely dissolved by acid treatment enhanced by refluxing. The undissolved material (principally silica) is removed by filtration. This process can be applied only to separations that have been verified experimentally to show no radionuclide loss.

6.2.2. Gaseous Radionuclide Collectors

Elements that have intermediate or long-lived radionuclides of interest in the gaseous state are discussed in Section 5.3.2 with the collection techniques for each. The gases are either brought into the laboratory in an airtight and possibly pressurized container or collected on a sorbent such as charcoal, molecular sieve, or silica gel. The samples are measured by gamma-ray spectral analysis either directly in the collector or transferred to another container that has been calibrated for measurement. If separation and purification are required, the gases are flushed into a closed system for processing and subsequent measurement.

Tritium Samples

Airborne ^3H may be collected in any of the forms discussed in Section 5.3.2. Samples may arrive with several forms in the same collector for separation at the laboratory, or the forms may already be separated at collection. Most commonly, tritiated water vapor is brought to the laboratory as a water sample. When collected on silica gel or molecular sieve, tritiated water vapor is removed by heating the collector material in a distillation flask above the boiling point of water. The vapor is transferred to a separate flask and condensed. A measured amount of the condensed water is analyzed for tritium in a liquid scintillation (LS) counter in terms of becquerel per gram water. If the collector is silica gel, the result must be corrected for dilution by a small amount of water originally in the silica gel (Rosson et al. 2000). In addition, a small isotope effect, mentioned in Section 4.8, results in distilling a greater fraction of H_2O than HTO unless the entire sample is distilled.

The other two forms of tritium gas usually are submitted on a palladium-coated molecular sieve collector. Any HT is converted to HTO by palladium-catalyzed oxidation in the collector; the water is removed by heating it above its boiling point and condensing the vapor. The sorbent with the remaining organical-bound tritium is mixed with a Hopcalite catalyst and heated to 550°C to oxidize organic gases to water, which is distilled and condensed (Ostlund and Mason 1985).

C-14 Samples

The sample for ^{14}C analysis, usually collected as part of a research project, has been processed at the time of collection. The ^{14}C is in the form of barium carbonate, precipitated from a barium hydroxide solution. Two samples may be submitted if the CO_2 form has been separated from the CO and organic-bound ^{14}C forms. The barium carbonate precipitate can be measured directly in an LS or proportional counter or further purified.

Radioiodine Samples

Gaseous forms of iodine generally are collected undifferentiated on an activated charcoal cartridge after passing through a filter on which particulate iodine is collected. In special studies, the different gaseous forms may be retained on separate collectors that are brought to the laboratory for analysis of ^{131}I. Direct gamma-ray spectral analysis of filter and cartridge for the characteristic 0.364-MeV (85%) gamma ray of ^{131}I is the preferred method of measurement.

A fresh fission-product mixture also contains ^{132}I (2.3 h), ^{133}I (20.8 h), and ^{135}I (6.6 h) that can be measured by gamma-ray spectral analysis. Measurement of ^{129}I is discussed in Section 6.4.1.

Noble Gas Samples

The noble gases can all be collected at liquid nitrogen temperatures on charcoal, or can be sorbed selectively by controlling the collector temperature near the individual condensation temperatures. They can be further purified in the laboratory to separate the noble gases from each other and from other gases collected with them.

The common radioisotopes of krypton and xenon are relatively short-lived, fission-produced radionuclides; these are measured directly by gamma-ray spectral analysis and require no sample treatment. One exception is ^{85}Kr (10.7 years), a radionuclide that emits mostly beta particles (0.687 MeV maximum energy), with a gamma ray of 0.514 MeV (0.43%) that is too low in abundance for measurement except at high ^{85}Kr levels. A simple technique for ^{85}Kr analysis is to concentrate it by collection at a low temperature, dissolving it in an LS cocktail, and then measuring it in an LS counter (Shuping et al. 1970). The method is used to measure the ^{85}Kr background in ambient air. A 7-week decay interval generally reduces all the shorter lived krypton and xenon fission products to concentrations that no longer interfere with this measurement. A gas-separation system was constructed to separate krypton from xenon by sequential condensation and heating on sorption columns (Momyer 1960).

Airborne ^{222}Rn collected in a scintillator-coated Lucas cell is placed on a photomultiplier tube in the laboratory and counted (see discussion for ^{226}Ra analysis by radon measurement in Section 6.4.1). Radon collected on a charcoal cartridge or other sorbent is measured by gamma-ray spectral analysis. In both cases, a 4-h period of ingrowth is needed to accumulate the radon progeny that contribute to the count rate.

6.2.3. Water

The sample may arrive unfiltered or separated as filtered water and the filter that contains the solids. The water sample is preserved with dilute acid or a preservative suitable for a radionuclide such as ^{131}I that may be lost from an acid solution. Water without suspended solids is ready for evaporation to measure the gross alpha- and beta-particle activity, measure gamma rays by spectral analysis, and perform radiochemical analysis. The solids usually are counted similarly and then processed for dissolution as described in Section 6.2.1 for subsequent radionuclide analysis.

When a liquid volume of several liters is needed to achieve a specified sensitivity for radionuclide detection, the volume usually is reduced for ease of analysis. Evaporation is a simple approach. A faster process—and one that is necessary for samples of seawater or other high-salt-content solution—is precipitation of the radionuclides of interest from the large sample volume with a carrier in bulky insoluble forms such as phosphate, hydroxide, or carbonate. The precipitate bearing the radionuclides of interest is separated by decanting most of the solution and filtering the rest, and is then dissolved for further radionuclide purification.

Certain radionuclides may be collected from even larger volumes (i.e., hundreds of liters) by passing a stream of water through a large filter system to retain insoluble forms and then through large ion-exchange columns to retain cations and anions. The filter and columns can be analyzed with a gamma-ray spectrometer. Radionuclides that do not emit gamma rays can be eluted from the ion-exchange columns with a relatively small volume of a high-salt solution, a complexing agent, or a redox reagent. The filter is processed as discussed in Section 6.2.1.

6.2.4. Milk

Many persons consume milk soon after collection, and the pathway for a few radionuclides from origin to milk includes accumulation steps due to large-area grazing by cows and their lactation process. Before the advent of gamma-ray spectral analysis, the radionuclides ^{89}Sr, ^{90}Sr, ^{131}I, ^{137}Cs, and ^{140}Ba were measured after chemical separation. Today, only the $^{89/90}$Sr pair requires chemical separation.

The brute-force approach to sample preparation was evaporating and ashing a liter of milk. The process requires care to avoid splattering and loss of ash from the large mass of organic material. A major improvement was precipitation of the organic phase of the milk with trichloroacetic acid, followed by radiochemical analysis of radiostrontium in the aqueous phase (Murthy *et al.* 1960). Currently, whole milk is passed through ion-exchange resin to sorb the radionuclide of interest, notably radiostrontium, on cation-exchange resin and radioyttrium on anion-exchange resin, at controlled pH. The radionuclide is then eluted from the resin for measurement (Porter and Kahn 1964).

6.2.5. Vegetation and Crops

Vegetation and crop samples are treated for subsequent purification by removing the bulk of the matrix—typically 99%—in oven drying and ashing. After dissolving the ash with mineral acid, a minor residue of insoluble silica can be volatilized by boiling with HNO_3–HF solution.

One common distinction between processing food and vegetation samples is that the edible portion in food is analyzed separately to provide information on radionuclide consumption. Additional treatment of food before analysis may include washing and other preparation steps.

6.2.6. Meat and Fish

Samples of animal products may be processed whole to determine their radionuclide content, dissected to distinguish radionuclide concentrations in various organs, or separated to analyze only edible portions. Questions may arise about what constitutes the edible sample portion, which differs among persons and population groups. These distinctions affect radiological health protection monitoring because metabolic parameters result in accumulation of radionuclides in specific organs or tissue, such as radiostrontium in bone and radioiodine in the thyroid (Cember 1996).

Unless silica is present (e.g., due to soil or sediment in intestines), the ashed material usually is soluble in dilute HNO_3; minor residual silica is removed by HNO_3–HF treatment (see Section 6.2.5). Soft tissue may also be dissolved in basic solution.

6.2.7. Soil and Sediment

Soil and sediment often are considered the most difficult matrix to dissolve. They may be leached under restricted conditions, dissolved completely in small amounts, or dissolved after fusion. The method is selected on the basis of which radionuclide is to be analyzed and in what form, sample origin, and the difficulty and cost of analysis.

Any leaching process must be tested to determine the leached fraction of a radionuclide as a function of type and strength of acid, type of soil or sediment, leachate/sample ratio, leach period, and temperature. The leached fraction varies for different soils and radionuclides.

Total dissolution is needed for less soluble radionuclide forms. This goal may be achieved by repeated treatment with solutions of the same acid or sequential treatment with different acids. Total dissolution of gram amounts is achieved relatively rapidly in a microwave oven with programmed temperatures and pressures (Garcia and Kahn 2001). The factors discussed above also control total dissolution and must be determined experimentally. Sample mass is limited to a few grams, compared to order-of-magnitude larger amounts that can be treated by leaching.

Refractory species (such as zirconium oxide) are incompletely atomized at the temperatures of a flame or a muffle furnace. If the radionuclide of interest may be in a refractory form, the material must be mixed with fusion reagents and melted at a high temperature. After cooling, the solidified melt is dissolved in a HNO_3 solution. Fusion mixtures that have been tested for many types of soil are listed in Table 4.2. Sample masses are limited by practical consideration of the final sample size, taking into consideration the large amounts of fusion reagents added to the sample.

6.2.8. Bioassay Samples

Urine samples are measured directly by spectral analysis for radionuclides that emit gamma rays. To measure radionuclides that emit only alpha or beta particles, a general approach for urine pretreatment is boiling to dryness and then ashing. The ash is dissolved in mineral acid to prepare the aqueous solution for analysis (Chieco 1997). Other bioassay samples are pretreated similarly.

More convenient procedures for urine analysis include separation of the radionuclide of interest from urine by precipitation or by sorption on a cation-exchange resin column. Ferric hydroxide is a precipitate that carries many radionuclides from a neutral or basic solution. The precipitate is washed with water and then dissolved in dilute mineral acid for further purification of the radionuclide of interest. For ion-exchange sorption, a complexing agent such as ethylene dinitrilotetraacetic acid (EDTA) is added to the urine at a selected pH value to complex interfering ions, such as calcium for ^{90}Sr analysis. After passage of the urine through the column and washing the column with the complexing agent and water, the radionuclide of interest is eluted from the column with a suitable reagent (Sunderman and Townley 1960).

6.2.9. Smears

Paper or cloth smears are used, often in response to regulations, to wipe surfaces of specified area (e.g., $100 \, cm^2$) to check for removable radionuclides. The smears are counted directly by gamma-ray spectral analysis. For gross alpha- or beta-particle measurements, thin smears are counted in a proportional counter or immersed in a cocktail for LS counting. Further analysis for a radionuclide of interest that emits only alpha or beta particles requires ashing the smear and subsequent dissolution with dilute mineral acid.

6.3. Carriers and Tracers

Carriers are added to radionuclide solutions, as discussed in Section 4.7, to perform precipitation separations, avoid radiocolloidal behavior, and to eliminate the need for quantitative recovery by allowing the analyst to calculate the fractional chemical recovery ("yield"). A radioactive tracer (see Section 4.7.3) can be added to the sample in addition to or instead of the carrier to measure yield. Tracer addition may make yield determination easier than carrier addition, or may be necessary if no carrier is available.

6.3.1. Carriers

Cationic carriers usually are obtained as chloride or nitrate forms, and anions, in their sodium form. The reagents must be sufficiently pure so that any impurity does not significantly affect its mass determination or radionuclide measurement. Reagent blanks must be counted after carrier preparation to check purity, and any significant source of radiation must be removed from the carrier.

The carrier solution should be maintained in a stable form, which may require addition of a reagent such as a dilute acid to a carrier that may otherwise hydrolyze over the period of use. The carrier should be filtered after preparation, checked periodically, and discarded if solids have formed or other changes are apparent. Care must be taken not to introduce contamination into a carrier; typically, a master solution is stored separately from frequently used fractions.

The carrier concentration in the solution should be determined in replicate with sufficient precision so that the uncertainty of yield measurement does not exceed 1%. The precipitate for carrier mass determination should in all ways be identical to that obtained at the end of the procedure to avoid chemical form differences such as water content.

The carrier is pipetted into the sample solution at the beginning of the analytical process and mixed thoroughly with the solution. The chemical state of the carrier must be identical to that of the radionuclide. If any difference is possible, steps must be inserted into the procedure at this point to achieve identical form for carrier and tracer. A common process is to oxidize and reduce the carrier and radionuclide through all possible oxidation states. For example, if the radionuclide ^{131}I is not known to be in the form of iodide and the carrier is added as iodide, the mixture is oxidized through molecular iodine, iodate, and periodiate, and then reduced back to iodide.

Nonisotopic carriers can be used when no isotopic carrier exists. Rhenium is a carrier for technetium, lanthanum for promethium and plutonium, and barium for radium. Such chemical homologs do not have identical solubility products, hence their yields may differ and must be compared by testing. Group separations can be performed (see Table 3.2), such as iron hydroxide precipitation to carry tetravalent uranium and rare earth radionuclides.

So-called "holdback" carriers are added to improve separation of one radionuclide from another at very low concentration. A holdback carrier differs slightly from a regular carrier in its function: a regular carrier functions to "carry" the radionuclide of interest out of solution, while a holdback carrier's function is to keep an interfering radionuclide in solution while the radionuclide of interest is precipitated with its regular carrier. Holdback carrier addition prevents radiocolloidal behavior exhibited in sorption of a contaminant radionuclide on the precipitate of the radionuclide of interest. The holdback carrier effectively maintains a consistent decontamination factor.

6.3.2. Tracers

Tracers are used as radiometric means of determining the yield of the radionuclide of interest. Tracers are selected for convenience in availability and counting the emitted radiation. The two radiations may be distinguished from each other by spectral analysis, e.g., ^{242}Pu tracer for ^{239}Pu by counting alpha particles, or by use of different detectors, as in counting the gamma rays of the ^{85}Sr tracer for ^{90}Sr that is determined by counting beta particles emitted by its ^{90}Y daughter.

Although isotopic tracers are preferred, application of nonisotopic tracers provides more choices for selecting a radionuclide with suitable radiation, e.g., ^{133}Ba tracer by gamma-ray spectral analysis for ^{226}Ra by counting alpha particles. As stated in the preceding section, tests must be performed to determine whether yields for the two nonisotopic radionuclides are identical or have a constant ratio.

The radioactive tracer must not be contaminated with other radionuclides that interfere with the yield measurement. Some actinide-series tracers have radioactive progeny with slow ingrowth that should be identified from their decay scheme chain and require periodic purification to remove these progeny.

The amount of tracer to be added needs to be balanced; too much may result in some radiation cross-talk between the radionuclide of interest and the tracer, while too little reduces the precision of yield determination. Typically, the tracer concentration is selected to be a few times the expected concentration of the radionuclide of interest.

6.4. Radionuclide Separation and Purification

Radiochemical analyses are presented in this section according to the categories in Section 6.1 for selected radionuclides that have been measured throughout the world for many years. These methods are discussed briefly to suggest approaches for selecting and improving these and other methods.

Chemical separations were important contributors to early efforts for radionuclide identification and measurement when only gas ionization and solid scintillation detectors were available. Chemical separation identified the radionuclide by element. Subsequent measurements with absorber foils (see Section 2.4.2) identified the type of emitted radiation and estimated its energy. The half-life was determined by repeated measurements under the same conditions. This approach was difficult for a radionuclide that emits radiation of several types and energies, for more than one radioisotope per element, or for incompletely purified radionuclides.

These measurement techniques are reflected in the early compilations of radio-analytical chemistry methods (Coryell and Sugarman 1951, NAS-NRC 1960). As spectral analysis achieved better resolution and higher counting efficiency, chemical separations were avoided when possible in favor of spectral analysis.

Currently, only a few routinely measured radionuclides such as ^{3}H, ^{14}C, ^{55}Fe, ^{90}Sr, ^{99}Tc, ^{129}I, ^{147}Pm, and many of the elements heavier than bismuth remain dependent on chemical separations. Chemical separation still is needed for a few reasons; a radionuclide emits no gamma rays, emits too low a fraction of gamma rays to meet required detection levels, requires removal of solids for counting (e.g., alpha particles), or must be detected with greater sensitivity.

In the future, mass spectrometry (see Chapter 17) may supersede radiochemical analysis for long-lived radionuclides and require a different set of chemical separations. This trend is opposed to a certain extent by chemical separation processes introduced to achieve ever lower minimum detectable activity requirements and by the continued interest in identifying newly created radioelements.

6.4.1. Long-Lived Radionuclides

The long-lived radionuclides associated with nuclear fission and concomitant neutron activation are listed in Table 6.1, together with the long-lived terrestrial radionuclides and the major cosmic-ray-induced radionuclides in nature. Of the fission and activation products in Table 6.1, all but the minor ones are commonly measured; of the naturally occurring radionuclides, ^{40}K, uranium, thorium, and radium are frequently encountered in samples, together with radioactive progeny of the latter three.

H-3 Analysis

Tritium is produced in nuclear reactors and devices by ternary fission and neutron activation of deuterium, boron, and lithium, and is produced naturally by cosmic rays. It is analyzed in air, water, and biota samples and for bioassay. Water vapor in air is collected by condensation on a cold surface, by bubbling through water, or on a sorbent material such as silica gel from which it is then flushed above the boiling point of water. Water from biological material is collected as vapor by heating samples just above the boiling point of water and then condensing the vapor. The NCRP has published a survey of tritium measurements, which provides detailed information (NCRP 1976a).

The sample—typically about 20 ml water—is distilled for purification. Reagents and holdback carriers may be added to the distilling flask to prevent volatile forms of contaminant radionuclides from being distilled. Initial distillation may be

TABLE 6.1. Commonly analyzed long-lived to extremely long-lived radionuclides

Category	Radionuclide	Half-life	Radionuclide	Half-life
Major fission products	^{3}H	12.3 years	^{125}Sb	2.7 years
	^{85}Kr	10.7 years	^{129}I	16,000,000 years
	^{90}Sr	28.8 years	^{137}Cs	30.2 years
	^{99}Tc	214,000 years	^{147}Pm	2.62 years
	^{106}Ru	367 days		
Minor fission products	^{79}Se	65,000 years	^{135}Cs	2,300,000 years
	^{93}Zr	1,500,000 years	^{151}Sm	90 years
	^{107}Pd	6,500,000 years	^{155}Eu	4.9 years
Neutron activation in	^{134}Cs	2.06 years	^{239}Pu	24,100 years
Fission	^{233}U	159,000 years	^{240}Pu	65,700 years
	^{236}U	23,400,000 years	^{241}Pu	14.4 years
	^{237}Np	214,000 years	^{241}Am	433 years
Common neutron	^{55}Fe	2.7 years	^{109}Cd	453 years
Activation products	^{60}Co	5.27 years	^{152}Eu	13 years
	^{59}Ni	75,000 years	^{154}Eu	8.5 years
	^{63}Ni	100 years		
Naturally occurring	^{3}H	12.3 years	^{227}Ac	21.8 years
radionuclides	^{10}Be	2,700,000 years	^{228}Ra	5.76 years
	^{14}C	5730 years	^{228}Th	1.91 years
	^{26}Al	740,000 years	^{230}Th	7300 years
	^{32}Si	280 years	^{231}Pa	32,800 years
	^{39}Ar	269 years	^{232}Th	1.41×10^{10} years
	^{40}K	1.28×10^9 years	^{234}U	245,000 years
	^{87}Rb	4.8×10^{10} years	^{235}U	7.04×10^8 years
	^{210}Pb	22.3 years	^{238}U	4.47×10^9 years
	^{226}Ra	1,600 years		

performed to separate the volatile forms, and the remaining water is then distilled for tritium analysis. For extracting water from solid samples or tritium sorbents, azeotropic distillation in a reflux system conveniently maintains a constant boiling point and liquid cover. The organic solvent is refluxed while the heavier distilled phase is 99.9% water, as described in Section 3.6.1.

The distilled water is counted in an LS system at a selected water-to-cocktail ratio (usually 1:1). Samples may be counted without purification if other radionuclides are known to be absent or can be differentiated clearly from tritium by pulse-height discrimination in the detection system, and if chemicals that cause excessive quenching or fluorescence are known to be absent. Section 15.4.3 illustrates an extension of this measurement technique, where ^{3}H is measured in flowing water with a scintillation counter.

Collection and analysis complexity introduced by forms other than tritiated water is discussed in Section 6.2.2. Tritium is released in these forms from nuclear facilities or converted to them in the environment. These forms generally occur in lesser magnitude than tritiated water; they should be differentiated because of their different pathways and radiation impacts after they enter the body. Biological material can be dried to collect tritium as water vapor and then ashed to collect organic tritium as HTO vapor.

Tritium has also been counted as a gas or vapor mixed with the counting gas in a flow-through ionization or proportional counter. Flow-through detectors are suitable as effluent monitors for tritium when tritium concentrations in air are high relative to concentrations of other radionuclides. Discrimination between tritium and radionuclides with more energetic beta-particle groups is achieved by pulse-height control for the proportional counter.

Fe-55 Analysis

^{55}Fe is formed by the (n,γ) reaction with ^{54}Fe. At the same time, ^{59}Fe is formed by the (n,γ) reaction with ^{58}Fe. Both radioisotopes are produced in iron and steel castings, vessels, or supports for nuclear weapons and reactors. ^{55}Fe ($t_{1/2} = 2.73$ years) decays by electron capture with a K_α X-ray energy of 5.89 keV (24.5%) and a K Auger electron energy of 5.2 keV (61%). In contrast, ^{59}Fe (44.5 d) emits beta particles with maximum energies of 0.466 (53%) and 0.274 MeV (45%) and gamma rays of 1.29 (43%) and 1.10 MeV (57%). The following procedure (Chieco 1997) determines both radionuclides.

Iron chloride or nitrate carrier is added and precipitated to carry the radionuclides as hydroxide, dissolved in strong HCl and collected as the chloro-complex on an anion-exchange column. After washing to remove contaminants, iron is eluted from the column with dilute HCl. Cobalt, manganese, and zinc holdback carriers are added to the solution and iron is precipitated as the cupferate at 10°C. The cupferate complex is destroyed by wet ashing and iron oxide is converted to the chloride by boiling in HCl. The iron ion is complexed in an ammonium phosphate–ammonium carbonate electrolyte and electroplated on a tared copper disc. The disk is weighed to determine the iron yield and then is sprayed with a thin acrylic coating to prevent oxidation of the iron.

^{59}Fe beta particles are counted with a proportional detector or its gamma rays are analyzed with a Ge detector and spectrometer. The sample is then measured for ^{55}Fe content with a thin Ge detector and spectrometer or xenon-filled X-ray proportional detector with a thin (e.g., 140 mg cm^{-2}) beryllium absorber. The ^{55}Fe count rate is adjusted for background, the ^{59}Fe contribution, self-absorption in the plated sample, and the chemical yield, and converted to the disintegration rate. The activity of both radioisotopes is corrected for radioactive decay from the sampling date.

If no ^{59}Fe is in the sample, as confirmed by sample analyses, it can be added as yield tracer for ^{55}Fe analysis. After ion-exchange purification, the elutriant is measured by LS counting ^{55}Fe and ^{59}Fe in different energy regions. Cross-talk from the ^{59}Fe beta particles in the ^{55}Fe Auger electron region must be corrected, as well as color quenching from chemicals.

Sr-90 Analysis

Analysis of ^{90}Sr is a classical example of a precipitation separation. Precipitation in concentrated (around 60%) to fuming (>86%) nitric acid separates strontium and barium from most other elements (Sunderman and Townley 1960), and is particularly useful because so many other fission-produced radionuclides are insoluble in basic solution but soluble in acid.

After strontium carrier is added to a small volume (<10 ml) of ^{90}Sr solution, sufficient fuming nitric acid is added to attain a nitric acid concentration of 14–16 N. The solution with strontium nitrate precipitate is cooled in an ice bath and then centrifuged. The supernatant solution is thoroughly decanted and the strontium nitrate precipitate is dissolved in water. Barium and yttrium carriers are added. Precipitation of barium chromate at pH 5.5 removes ^{140}Ba and natural radium from the supernatant strontium solution (for counting, if needed, of these two separated radioelements). Precipitation of yttrium hydroxide in basic solution then removes the ^{90}Y daughter that has grown into the ^{90}Sr parent. Ammonium oxalate is immediately added to the supernatant solution to precipitate strontium oxalate. The precipitate is washed and dried in the filter holder with alcohol and ether, promptly weighed for yield determination, and counted with a beta-particle detector (Chieco 1997) such as a proportional detector.

The measured count rate must be corrected for contribution by ^{90}Y daughter ingrowth that begins immediately after the hydroxide scavenging precipitation. If time is not of the essence, the measurement can be delayed for 20 days until ^{90}Y ingrowth is essentially complete, to reach equal disintegration rates for ^{90}Sr and ^{90}Y. Individual counting efficiencies must be applied for ^{90}Sr and ^{90}Y because the two are not identical in the proportional counter.

An alternative is to store the supernatant solution after the barium chromate scavenging precipitation for 20 days until ^{90}Y approaches equilibrium with ^{90}Sr, and then precipitate yttrium hydroxide with the ^{90}Y. The hydroxide is dissolved in dilute acid, and oxalic acid is added to precipitate yttrium oxalate. The precipitate is washed and dried, promptly weighed for yield determination, and counted with a proportional detector (Sunderman and Townley 1960). To the supernatant solution from the yttrium hydroxide precipitation, ammonium oxalate is added to precipitate strontium oxalate for weighing to determine the strontium yield. The ^{90}Sr activity is calculated from the ^{90}Y count rate adjusted for ^{90}Y ingrowth into the ^{90}Sr parent, ^{90}Y decay since its separation from ^{90}Sr, yttrium and strontium yields, and ^{90}Y counting efficiency.

One variant is that the radionuclides are dissolved and then measured with an LS counter. Another variant is that ^{85}Sr tracer is used for strontium yield measurement when ^{90}Y is measured to determine the ^{90}Sr activity.

Samples with high calcium content can be problematic because some calcium nitrate may also precipitate in nitric acid at the indicated strength. The additional calcium will cause overestimation in gravimetric strontium yield. The calcium nitrate precipitate (after thorough draining to remove the explosion potential of the nitric acid–alcohol mixture) can be dissolved in a few milliliters of absolute alcohol with little loss of strontium nitrate precipitate. A possibly safer removal of calcium contamination is by reducing the nitric acid strength to 60% or dissolving the strontium and calcium nitrate precipitates in a small amount of water and then precipitating strontium nitrate with concentrated nitric acid for a second or even third time. One interference is a silica precipitate in the strong nitric acid solution from samples such as dissolved soil; this precipitate can be removed by filtration after dissolving the strontium nitrate in water.

As indicated in Section 6.4.2 below, the presence of ^{89}Sr requires calculation by simultaneous equations to determine the disintegration rates of ^{89}Sr and ^{90}Sr in the same sample. Both radionuclides can be expected in a mixture of shorter-lived fission products within a year after formation.

Changes in ^{90}Sr purification were developed mainly to eliminate use of large volumes of fuming or concentrated nitric acid that corrode hoods, ducts, and laboratory equipment. Early modifications applied selective ion-exchange sorption to separate strontium from calcium. For example, calcium in the sample is complexed by EDTA at pH 4.6 so that strontium, barium, and radium can be retained on a strong-base cation-exchange resin while calcium passes through the column (Porter *et al.* 1967). The strontium is then selectively eluted with the same solution at pH 5.1 and precipitated as the carbonate for yield determination and counting.

A more recent technique uses the Eichrom column discussed in Section 3.5. After initial precipitation and dissolution to reduce the solution volume, the solution is adjusted to the specified acidity and passed through the column, which retains strontium. Strontium is eluted and then precipitated as the carbonate.

Tc-99 Analysis

Technetium does not have a stable isotope. Of its long-lived isotopes, ^{95}Tc, ^{97}Tc, ^{98}Tc, and ^{99}Tc, only the latter is a fission product. Short-lived ^{99m}Tc (6.0 h) precedes ^{99}Tc. It is the daughter of ^{99}Mo (66 h), a radionuclide widely used in nuclear medicine. The half-lives of all other technetium fission products are less than 1 h.

The most stable oxidation states of technetium are +4 and +7. Hydrazine and hydroxylamine reduce technetium to +4. Atmospheric oxygen, hydrogen peroxide, and strong nitric acid oxidize technetium to +7. Some reactions can be slow. Other, less stable, oxidation states exist. In the lower oxidation state, technetium as TcO_2 is carried on ferric hydroxide precipitate. In the upper oxidation state, technetium is the anion TcO_4^-. It is sorbed on anion-exchange resins, extracted into oxygen-containing solvents such as hexone (methyl isobutyl ketone) and tributylphosphate, and coprecipitated with rhenium as the phenylarsonium pertechnetate, $(C_6H_5)_4AsTcO_4$. It can be distilled as Tc_2O_7 from sulfuric and perchloric acids (Anders 1960).

The above-cited reactions are used to separate ^{99}Tc for measurement. Rhenium is a common nonisotopic carrier for precipitation or yield measurement. The shorter-lived ^{95}Tc (60 days) has been used as isotopic tracer for yield measurement (Anders 1960). Technetium can be separated from rhenium by coprecipitating the +4 state with ferric hydroxide while rhenium remains in solution. Technetium is separated from the fission-produced radionuclides ^{103}Ru and ^{106}Ru by distillation with Re_2O_7 from sulfuric acid while ruthenium remains behind because it is volatile only in its highest oxidation state (RuO_4). An automated separation procedure is described in Section 15.3.4.

The beta-particle group emitted by ^{99}Tc with a maximum energy of 0.294 MeV can be measured by proportional or LS counter. The latter is more suitable for the low energy, but the former may provide more sensitive detection. ^{99}Tc does not emit gamma rays.

I-129 Analysis

If a thermal neutron source is available, ^{129}I can be determined at low concentration by measuring activation-produced ^{130}I (12.4 h); at the same time, the specific activity can be measured by activation of stable ^{127}I to ^{128}I (25 min). The cross-sections for neutron activation are 6.1 barn for ^{127}I and 9 barn for ^{129}I. Although ^{129}I is a fission product, it is also produced in the environment by cosmic ray neutrons. Attribution to an anthropomorphic source must be checked by background measurements of comparable matrices in terms of the specific activity. Data from such measurements are available (NCRP 1983).

At a magnitude of 0.1 Bq per sample, ^{129}I can be measured directly with an LS counter or a Ge gamma-ray spectrometer in its low-energy region. The radionuclide emits 0.153 MeV (100%) maximum energy beta particles, 0.040 MeV (7.5%) gamma rays, K X-rays in the 0.029–0.034 MeV (70%) region, and conversion and Auger electrons in the 0.025–0.039 MeV (22%) region.

The radionuclide with added carrier can be separated from most other long-lived radionuclides by precipitating silver or palladium iodide from dilute nitric acid solution. As indicated in Section 6.3.1, the oxidation state of radioiodine, of which there are several, must be well defined before a separation step can be trusted. Other purification techniques are solvent extraction of iodine oxidized to I_2 into carbon tetrachloride followed by back extraction of the reduced iodide form into water, or sorption of I^- on anion-exchange resin followed by elution with a strong chloride solution (Kleinberg and Cowan 1960). These processes also lend themselves to concentrating the radionuclide from larger solution volumes to attain a lower detection limit.

Cs-137 Analysis

Initially, ^{137}Cs was analyzed by adding cesium carrier and precipitating a compound that is specific for cesium in dilute acid solution in the presence of sodium, such as the cobaltinitrite, phosphomolybdate, silicotungstate, or tetraphenylborate. Known contaminants, such as ^{32}P or ^{99}Mo in phosphomolybdate, are then removed after dissolving this precipitate by a second precipitation with a different reagent or by a scavenging precipitation for the contaminant. The purified sample is weighed for yield and then counted with a proportional or G-M counter (Finston and Kinsley 1961).

^{137}Cs is collected from larger volumes of water on large cation-exchange resin columns. Anion-exchange resin columns loaded with one of the above-cited precipitation reagents make these columns specific for cesium. A column loaded with ammonium hexacyanocobalt ferrate effectively retains radiocesium (Boni 1966) from liquids with high salt content.

Once gamma-ray spectral analysis became available, the 0.660-MeV gamma ray of the 2.55-min ^{137m}Ba daughter was counted for identification, distinction from other radionuclides, and measurement. Little interference from other gamma rays is encountered with a Ge detector. ^{134}Cs may accompany ^{137}Cs in reactor-produced mixtures. The 2.55-years ^{134}Cs emits several gamma rays, notably at 0.605 (98%) and 0.795 MeV (88%).

Pm-147 Analysis

^{147}Pm is one of three relatively long-lived rare earth fission products; the other two are ^{144}Ce (284 days), which is discussed in the following section, and ^{155}Eu, with a low fission yield. ^{147}Pm emits a 0.284-MeV maximum beta-particle energy group and no gamma rays. In contrast, ^{144}Ce emits gamma rays of 0.134 MeV (10.8%) and ^{155}Eu emits gamma rays of 0.087 MeV (31%) and 0.105 MeV (20%). Contamination by ^{152}Eu and ^{154}Eu is possible if a sample that contains trace amounts of europium, such as soil, was irradiated with neutrons. Both of these radionuclides emit gamma rays.

The typical purification method for rare earths is coprecipitation with ferric hydroxide, dissolution in dilute acid, precipitation as fluoride in strong mineral acid solution, dissolution in strong nitric acid with boric acid to complex fluoride, and precipitation for counting as the oxalate in dilute acid solution (Stevenson and Nervik 1961). Because ^{147}Pm has no stable isotope, another rare earth (such as lanthanum) is added as carrier. The ^{147}Pm precipitate can be counted with a proportional counter, or can be dissolved and measured with an LS counter because of the low beta-particle energy. If small amounts of the other rare earth radionuclides are detected by gamma-ray spectrometric analysis, the beta-particle count rate of ^{147}Pm can be calculated by difference.

^{147}Pm can be separated from cerium and europium because it remains in the +3 oxidation state when cerium is oxidized to +4 or europium is reduced to +2. ^{144}Ce can be scavenged in dilute acid solution by adding cerium carrier, oxidizing to the +4 form with sodium bromate, and then precipitating it with HIO_3 as $Ce(IO_3)_4$. Eu-155 can be separated by adding holdback europium carrier, reducing europium to the +2 state with zinc powder, and then carrying ^{147}Pm on ferric hydroxide precipitate (Stevenson and Nervik 1961).

Rare earths can be separated from each other on ion-exchange columns, as illustrated in Fig. 3.2. Once separation conditions have been defined, the method can be simplified for purifying one radionuclide such as ^{147}Pm from interfering rare earth radionuclides.

Ra-226 and Ra-228 Analysis

Radium analysis in water supplies is required under the Safe Drinking Water Act. The US EPA specifies that ^{226}Ra and ^{228}Ra must be detected at a concentration of 1 pCi/l (0.037 Bq/l). Radium analysis may also be needed in soils, ores, manufactured materials such as concrete, and various contaminated media.

The conventional analysis of ^{226}Ra in water is coprecipitation with barium sulfate carrier from a 1-l volume. After decanting the supernatant solution, barium sulfate is dissolved in EDTA solution and stored in a sealed container for 1–4 weeks to allow for ingrowth of its ^{222}Rn (3.82 days) daughter. The ^{222}Rn gas is flushed with air (sufficiently aged in a tank to assure the absence of radon) into an evacuated 100- to 200-cc ZnS(Ag)-coated "Lucas" cell. A photomultiplier tube views scintillations at the ZnS(Ag) surface through a light pipe at one end of the cell (Lucas 1957). The measured alpha particles are emitted by ^{222}Rn and its progeny ^{218}Po and ^{214}Po. The ingrowth of ^{214}Po, controlled by its precursors ^{214}Pb and ^{214}Bi, reaches equilibrium with that of ^{222}Rn within 4 h.

The activity of ^{226}Ra is calculated from the count rate of the three alpha particles of its progeny. Adjustments are made for the ingrowth and decay of ^{222}Rn, the detector counting efficiency, and any losses during the precipitation of barium sulfate and the transfer of the gas into the cell.

The analysis of ^{228}Ra conventionally followed that of ^{226}Ra in precipitating with barium sulfate and dissolving with EDTA solution. After ingrowth of ^{228}Ac (6.13 h) for 2 days, the ^{228}Ac is precipitated with yttrium carrier as yttrium oxalate. The precipitate is washed and dried with alcohol and ether, weighed, and then counted with a proportional detector. The barium that remained in the EDTA solution again is precipitated as sulfate to determine its yield. The ^{228}Ra activity is calculated from the ^{228}Ac count rate and counting efficiency, the yields of barium and yttrium carriers, and the ingrowth and decay of ^{228}Ac (Percival and Martin 1974).

Various other methods of radium purification by coprecipitation, ion exchange, and radon emanation are available (Kirby and Salutsky 1964). In a recent method, both ^{226}Ra and ^{228}Ra are collected by a barium sulfate precipitate which is weighed to determine the barium yield, and then counted by Ge gamma-ray spectral analysis. The counted gamma rays are emitted by the ^{214}Pb and ^{214}Bi progeny of ^{226}Ra and the ^{228}Ac progeny of ^{228}Ra. The measurement for ^{228}Ra can be performed immediately because ^{228}Ac coprecipitates with barium sulfate, but delay by 1–4 weeks is needed for ingrowth of the ^{222}Rn daughter of ^{226}Ra in the barium sulfate (Kahn et al. 1990). Corrections are required for yield and incomplete ingrowth to calculate the activity.

Uranium Analysis

The uranium isotopes often encountered in the radioanalytical chemistry laboratory are listed in Table 6.2. In natural uranium, the relative decay rates at equilibrium are 1.0 Bq ^{238}U, 1.0 Bq ^{234}U, and 0.045 Bq ^{235}U. Enriched (containing relatively more ^{235}U and ^{234}U) and depleted (relatively more ^{238}U) combinations are also encountered, as are ^{236}U in neutron-irradiated mixtures and ^{233}U from some processes. These uranium isotopes emit alpha particles, characteristic 13-keV L X rays, and generally several weak gamma rays. Several isotopes have numerous minor alpha-particle or gamma-ray transitions that are not listed.

TABLE 6.2. Common uranium radionuclides

Isotope	Alpha particles (MeV)	Percent	Gamma rays (MeV)	Percent	Comments
^{232}U	5.139	0.28	0.058	0.202	
	5.264	31.2	0.129	0.067	
	5.320	68.6			
^{233}U	4.729	1.6	0.054	0.019	
	4.783	13.2	0.317	0.0080	
	4.796	0.28			
	4.824	84.4			
^{234}U	4.605	0.24	0.053	0.12	
	4.724	27.4	0.121	0.041	
	4.776	72.4			
^{235}U	4.209	5.7	0.144	10.5	
	4.322	4.7	0.163	4.8	
	4.358	17	0.186	54	
	4.392	54	0.205	4.7	
	4.411	2.1			
	4.501	1.7			
	4.555	4.5			
	4.597	5.4			
^{236}U	4.445	26	0.045	0.079	
	4.494	76	0.113	0.019	
^{238}U	4.042	0.26	0.063	4.9	(from ^{234}Th)
	4.150	23	0.092	5.5	(from ^{234}Th)
	4.200	73	0.766	0.29	(from ^{234}Pa)
			1.001	0.89	(from ^{234}Pa)

One approach to measuring natural uranium is chemical analysis. For calculating the decay rate, the specific activity of ^{238}U is 12.4 Bq/mg (0.33 pCi/μg). A common method for detection of microgram amounts is fluorimetry of a fused aliquot. Laser fluoroscopy has been developed as a more sensitive method. If a nuclear reactor is available, neutron activation or the fission reaction can detect nanogram amounts (Gindler 1962). Measurement by mass spectrometer has been applied, as discussed in Section 17.8.1.

The stable oxidation states of uranium are 0 (as metal), +4, and +6. Uranium has also +3 and +5 states. In nature, uranium exists as relatively insoluble oxides such as UO_2, U_3O_8, and UO_3 and as the relatively soluble cation UO_2^{2+}. Both +4 and +6 states form soluble complexes with numerous inorganic and organic ions.

Uranium has been extracted from aqueous solutions with many organic solvents and sorbed on either cation- or anion-exchange resins under various conditions. A commonly used procedure that also applies to the actinides is sorption on an anion exchange resin of the chloride complex from strong hydrochloric acid, followed by selective elution at lesser acidity. If precipitation is desired, ammonium uranate, $(NH_4)_2U_2O_7$, is formed with NH_4OH, but many other radionuclides will accompany the uranium salt under the basic conditions. For radiation measurement, the uranium source typically is prepared by electrodeposition (Talvitie 1972), notably by the American Society for Testing and Materials (ASTM) standard method

C1284-94 listed in Appendix A-1. Other actinides that emit alpha particles are also prepared with this method. After purification, uranium isotopes usually are identified and measured by alpha-particle spectrometry with silicon diodes. Tracer ^{232}U can be added to the initial solution and counted with the uranium isotopes of interest.

Table 6.2 indicates some uranium isotopes that emit gamma rays in a significant fraction of decays; these can be measured directly by gamma-ray spectral analysis. In measuring ^{235}U by its most intense gamma ray of 0.186 MeV, interference from ^{226}Ra is possible. For ^{238}U, the listed gamma rays are emitted by the daughter ^{234}Th (24.1 days) and its daughter ^{234}Pa (1.2 min) if they are at equilibrium.

Plutonium Analysis

Plutonium isotopes of interest are listed in Table 6.3. All except ^{241}Pu emit alpha particles, 14-keV L X rays, and multiple weak gamma rays; ^{241}Pu emits beta particles that decay to ^{241}Am, which emits alpha particles and gamma rays. Single or multiple neutron reactions with ^{238}U and ^{235}U form all of the isotopes. The isotopes that emit alpha particles have numerous unlisted minor gamma-ray transitions.

Among plutonium oxidation states from +3 to +6, the most stable forms are +3 and +4. Conversion between oxidation states is used for purification from other radionuclides. Plutonium is oxidized to the +4 state by hydrogen peroxide, permanganate, and nitrite, and reduced to the +3 state by bisulfite and ascorbic acid. A strongly acidic or complexing solution is needed to maintain the selected state and avoid hydrolysis with polymerization (Coleman 1965).

An early conventional method for plutonium analysis was coprecipitation in acid solution with a rare earth fluoride, dissolution in aluminum nitrate solution with sodium nitrite to maintain the +4 oxidation state, and extraction into thenoyltrifluoroacetone (TTA) in benzene. Plutonium was back-extracted into dilute HCl, the acid was evaporated, and plutonium was taken up in HCl–NH$_4$Cl solution

TABLE 6.3. Common plutonium radionuclides

Isotope	Alpha particles (MeV)	Percent	Gamma rays (MeV)	Percent	Beta particles (E_{max} in MeV)
^{238}Pu	5.359	0.013	0.043	0.039	NA
	5.454	28.7			
	5.498	71.1			
^{239}Pu	5.105	11.5	0.052	0.027	NA
	5.129	15.1	0.129	0.0062	
	5.161	73.3			
^{240}Pu	5.123	26.5	0.045	0.045	NA
	5.168	73.4	0.104	0.0070	
^{241}Pu	No alpha (Measure ^{241}Am daughter)	NA	No gamma	NA	0.0208
^{242}Pu	4.856	21	0.045	0.042	NA
	4.900	78.9	0.103	0.0088	

for transfer to an electrodeposition cell and deposition on a stainless steel disk (Coleman 1965).

Currently, anion-exchange resins are used to purify and separate transuranium elements. Plutonium and other radionuclides are coprecipitated with iron hydroxide. The precipitate is dissolved in strong nitric acid and adjusted to 8 N, with sodium nitrite added to control the plutonium oxidation state at +4. The solution is passed through an anion-exchange resin column to retain plutonium. After washing the resin with more 8 N HNO_3, plutonium is eluted with 10 N HCl and dilute NH_4I, repeatedly evaporated with nitric acid to convert it to the nitrate, and then taken up in H_2SO_4–$(NH_4)_2SO_4$ solution at pH 2.0–2.3 (IAEA 1989). The plutonium is electrodeposited as described above for uranium. An alternative is coprecipitation with 0.1 mg of a rare earth such as lanthanum fluoride and collection on a smooth filter (EPA 1984).

Plutonium, uranium, and the other actinide radionuclides can be separated on crown-ether extractant columns (Eichrom columns are discussed in Section 3.5) by selective sorption and elution. A sequential process collects uranium on the first column and the other elements on the second column. Plutonium and other actinides are then eluted with solutions specific for each element.

For tracing plutonium yield, ^{242}Pu is available. The plutonium isotope peaks are measured with a Si diode and alpha-particle spectral analysis. The ^{239}Pu activity is calculated from the product of the ^{242}Pu activity and the ^{239}Pu /^{242}Pu peak area ratio. The ^{239}Pu activity includes any contribution from ^{240}Pu because the energies of their peaks are almost identical. Other plutonium isotopes (except ^{241}Pu) are measured at the same time in terms of characteristic peak areas at the energies listed in Table 6.3. A mass spectrometer is used to separate ^{239}Pu from ^{240}Pu if they must be reported separately. As indicated in Table 6.3, the emitted gamma rays can be used for intense sources but are too weak for sensitive measurements.

Am-241 Analysis

The long-lived isotopes of americium are ^{241}Am (458 years), ^{242}Am (152 years), and ^{243}Am (7400 years). All are formed by multiple neutron interactions with uranium and plutonium. Of particular interest is ^{241}Am because it identifies the erstwhile presence of its ^{241}Pu parent. ^{241}Am emits alpha particles but can also be measured by its 0.059-MeV (36%) gamma ray. ^{242}Am has several isomers that mostly emit beta particles, and ^{243}Am emits alpha particles.

Americium in aqueous solution is in the +3, +5, and +6 states, respectively, as Am^{+3}, AmO_2^{+1}, and AmO_2^{+2}. The trivalent state is most common, but higher oxidation states are achieved by strong oxidation. The highest oxidation states can be used for separating americium from curium and rare earth radionuclides with ammonium persulfate, $(NH_4)_2S_2O_8$ (Penneman and Keenan 1960).

The trivalent actinides such as ^{241}Am follow the same precipitation reactions as the trivalent rare earth radionuclides, notably with insoluble hydroxides, fluorides, and oxalates. Numerous solvent extraction and ion-exchange separations from other trivalent radionuclides are reported. Americium radionuclides can be separated from other rare earth and actinide radionuclides by meticulous control of ion-exchange column elution.

A convenient tracer for ^{241}Am is ^{243}Am (if the latter is not in the sample). Both conventionally are measured with a silicon diode by alpha-particle spectral analysis. The disintegration rate is calculated as discussed above for plutonium.

6.4.2. Radionuclides with Intermediate Half-Lives

A mixture in this category may include the radionuclides listed in Table 6.4 as well as some longer-lived radionuclides in Table 6.1. With few exceptions (^{89}Sr among the major fission products in this list), the radionuclides are detected by gamma-ray spectral analysis. Because in many cases the shorter-lived radionuclides have a much higher decay rate, sequential gamma-ray spectral analysis, obtained promptly and then after intervals such as 1 day, 1 week, and 1 month, allows more precise analysis in separate measurements for the shorter-lived and longer-lived radionuclides.

The radionuclide content in a sample can be predicted for a known source, i.e., a nuclear reactor or device, or a hospital. Conversely, the source may be identified by the radionuclide content of the sample. This is particularly true when, for example,

TABLE 6.4. Common radionuclides with intermediate half-lives

Category	Radionuclide	Half-life	Radionuclide	Half-life
Major fission products	^{89}Sr	50.5 days	^{133m}Xe	2.19 days
	^{90}Y	64.1 h	^{140}Ba	12.8 days
	^{91}Y	58.5 days	^{140}La	40.3 h
	^{95}Zr	64.0 days	^{141}Ce	32.5 days
	^{95}Nb	35.0 days	^{143}Ce	33.0 h
	^{99}Mo	66.0 h	^{143}Pr	13.6 days
	^{103}Ru	39.4 days	^{144}Ce	284 days
	^{105}Rh	35.4 h	^{147}Nd	11.0 days
	^{131}I	8.04 days	^{149}Pm	53.1 h
	^{132}Te	78.2 h	^{151}Pm	28.4 h
	^{133}Xe	5.25 days	^{153}Sm	46.8 h
Minor fission products	^{121}Sn	27.1 h	^{127m}Te	109 days
	^{123}Sn	129 days	^{129m}Te	33.5 h
	^{125}Sn	9.62 days	^{131m}Te	30 h
	^{126}Sb	12.4 days	^{156}Eu	15.2 days
Neutron activation	^{82}Br	35.3 h	^{136}Cs	13.1 days
in fission	^{124}Sb	60.2 days	^{239}Np	2.35 days
Common neutron	^{51}Cr	27.7 days	^{59}Fe	44.6 days
Activation products	^{54}Mn	312 days	^{65}Zn	244 days
	^{58}Co	70.8 days	^{110m}Ag	252 days
Naturally occurring	^{7}Be	53.3 days	^{224}Ra	3.64 days
radionuclides	^{210}Po	138 days	^{227}Th	16.7 days
	^{210}Bi	5.01 days	^{231}Th	25.5 h
	^{222}Rn	3.82 days	^{234m}Th	24.1 days
	^{223}Ra	11.4 days		

the mix is purely natural, or consists of fission products; certain activation products may also be identified. In principle, when the relative amounts of various fission products resemble the distribution for one of the fissionable isotopes (see Fig. 2.2), instantaneous fission can be inferred. The absence of shorter-lived fission product suggests that a considerable interval between production and measurement has occurred.

In practice, the origin of a radionuclide mixture usually is more difficult to identify because of ongoing radioactive decay between origin and sample, differing degrees of retention at the origin, and partial retention along the pathway. At a nuclear reactor, for example, fuel element cladding retains most, but not all, of the fission products, and reactor coolant controls and waste processing retain more. Many of activation products that may originate from nuclear reactor or nuclear medicine facilities can be predicted from experience, but not necessarily associated with a specific facility.

Radioisotope multiples, e.g., $^{89/90}$Sr, $^{103/106}$Ru, $^{141/143/144}$Ce, $^{134/136/137}$Cs, $^{133/135}$Xe, $^{235/238}$U, or $^{238/239}$Pu, are particularly useful for attributing origin and inferring the time from formation to sampling. With the $^{89/90}$Sr pair, for example, if the generation rate of ^{89}Sr atoms relative to ^{90}Sr atoms were the same, the ^{89}Sr:^{90}Sr activity ratio is inversely proportional to the half-lives, i.e., 208. The ratio is about 170 because of the higher fission yield (the relative atom generation rate) of ^{90}Sr compared to ^{89}Sr. After accumulation for 500 days, the ratio is reduced to about 25 because ^{89}Sr no longer accumulates while ^{90}Sr still is far from equilibrium. Even here, conditions usually are complex because various fuel elements may be exposed for different periods and the two radioisotopes may have been released at different times.

More confident attribution usually is possible when radionuclides in a sample can have only limited origin from, say, atmospheric fallout and one or two nuclear facilities. For tritium, ^{14}C, ^{129}I, or uranium in the environment, for example, the specific activity (radioisotope/stable isotope ratio) can indicate the origin. Certain activation products can be attributed to specific nuclear medicine or reactor facilities.

Screening measurements for gross alpha and beta particles may not be particularly useful for fission-product mixtures because generally alpha particles are at extremely low concentration and beta particles are contributed by numerous radionuclides. A gross screening measurement is useful for suggesting additional analyses if it clearly exceeds the sum of the radionuclides measured individually.

^{89}Sr-^{90}Sr Analysis

Determination of ^{89}Sr together with the long-lived ^{90}Sr (see Section 6.4.1) is a widespread analytical endeavor for radiation protection because of the similarity of strontium to calcium in their absorption into bone mass. Strontium separation is performed as described in Section 6.4.1 whether or not ^{89}Sr is in the sample. The distinction between treating a sample that is known to have no ^{89}Sr and one

TABLE 6.5. Calculation of ^{89}Sr and ^{90}Sr /^{90}Y disintegration rate

Description	Formula
Initial measurement	$R_1 = A_{90_{Sr}} \, \varepsilon_{90_{Sr}} + A_{90_{Sr}} \, \varepsilon_{90_Y} \, D_{1 90_Y} + A_{89_{Sr}} \, \varepsilon_{89_{Sr}}$
Second measurement	$R_2 = A_{90_{Sr}} \, \varepsilon_{90_{Sr}} + A_{90_{Sr}} \, \varepsilon_{90_Y} \, D_{2 90_Y} + A_{89_{Sr}} \, \varepsilon_{89_{Sr}} \, D_{2 89_{Sr}}$
Solution for $A_{90_{Sr}}$ at the initial count	$A_{90_{Sr}} = \dfrac{(R_1 D_{2 89_{Sr}} - R_2)}{[\varepsilon_{90_{Sr}}(D_{2 89_{Sr}} - 1) + \varepsilon_{90_Y}(D_{2 89_{Sr}} D_{1 90_Y} - D_{2 90_Y})]}$
Solution for $A_{89_{Sr}}$ at the initial count	$A_{89_{Sr}} = \dfrac{[R_1 - A_{90_{Sr}}(\varepsilon_{90_{Sr}} + \varepsilon_{90_Y} \, D_{1 90_Y})]}{\varepsilon_{89_{Sr}}}$

in which it may be present relates to the radionuclide measurements intended to measure ^{89}Sr and ^{90}Sr separately.

The decay characteristics of ^{89}Sr, ^{90}Sr, and ^{90}Y (see Figs. 9.3 and 9.4) suggest three approaches to distinguishing between the two radiostrontium isotopes: by half-life, the parent–daughter relation of ^{90}Sr and ^{90}Y, and beta-particle energy. The simple approach, described below as Alternative 1, is to perform measurements in a proportional counter twice with an interval of 2–4 weeks to obtain the net count rates (background subtraced), R_1 and R_2, shown in Table 6.5. The different counting efficiencies ε for ^{89}Sr, ^{90}Sr, and ^{90}Y are determined for the detector in use. The decay constants λ are 0.0137 days^{-1} for ^{89}Sr and 0.260 days^{-1} for ^{90}Y. The decay factor fraction $D_{89_{Sr}}$ for ^{89}Sr and the ingrowth fraction D_{90_Y} for ^{90}Y are calculated. For practical purposes, the decay fraction for ^{90}Sr (28.5 years) of 0.998 in a 4-week period can be considered to be 1.0. With these factors, the disintegration rates A of ^{89}Sr and ^{90}Sr may be determined.

Alternative 1 consists of measuring the strontium precipitate as soon as possible (within a few hours) after yttrium separation, and repeating the measurement after 2–4 weeks. As shown by the equations in Table 6.5, the first value consists of beta particles from ^{89}Sr and ^{90}Sr, with a small contribution from the recently separated ^{90}Y. The latter is the calculable fraction $D_{1 90_Y}$ relative to the ^{90}Sr beta-particle activity. The second value consists of the same contributors, except that the ^{89}Sr beta-particle emission rate has decreased according to its known half-life, while the ^{90}Y emission rate has increased to become almost the same as that for ^{90}Sr, i.e., D_2 for ^{90}Y is almost 1.0.

Alternative 2 consists of counting the ^{90}Sr precipitate as before, then dissolving the precipitate in dilute acid and storing it for 2–4 weeks so that the ^{90}Y daughter grows to near equilibrium. The ^{90}Y then is precipitated initially as hydroxide and secondly as oxalate (see ^{90}Sr analysis in Section 6.4.1). The yttrium precipitate is measured as soon as possible (within a few hours) after separation. This measurement provides the ^{90}Y activity. The ^{90}Sr activity is equal to the ^{90}Y activity when corrected for radioactive ingrowth and decay of ^{90}Y. The ^{89}Sr activity is calculated from the initial measurement shown in Table 6.5 by subtracting the contribution by the activity of ^{90}Sr with a small activity of ingrown ^{90}Y.

Alternative 1 is more precise when ^{90}Sr is the major component and Alternative 2, when ^{89}Sr is the major component. An interval in Alternative 1 of a few days instead of 2–4 weeks is feasible because ^{90}Y can be calculated for any time period, but reduces the count rate, and hence the precision. In distinction, if results need not be reported promptly, then the first measurements can be performed 3 weeks after yttrium separation, when the ^{90}Sr/^{90}Y pair is in equilibrium, and the second measurement, after 7 weeks when ^{89}Sr has decayed by about one half-life. The difference in the count rate then is almost entirely due to ^{89}Sr decay.

The two radioisotopes of strontium and ^{90}Y can also be distinguished by energy discrimination with absorber foils in a proportional counter or by spectrometer in an LS counter. These methods usually are less precise than the two above-cited alternatives.

The other cited radioisotope pairs among fission products are measured simultaneously by gamma-ray spectral analysis. Before such analysis became available in the 1950s, these pairs were distinguished after chemical separation by dual measurements such as those described above for $^{89/90}$Sr before and after an interval during which the shorter-lived radionuclides had decayed sufficiently to measure the difference with the needed precision.

6.4.3. Short-Lived Radionuclides

Prompt submission to the laboratory and rapid handling for chemical separation and counting are crucial for measuring a fission-product or activation mixture with half-lives between 1 h and 1 day. The numerous short-lived fission products have been tabulated with their fission yields, precursors, and progeny. Sequential counting, promptly at first and then at increasing intervals as the shorter-lived radionuclides disappear, is the common approach. Rapid separation may be needed to remove radionuclides that are so intense that they obscure gamma rays from other radionuclides. ^{24}Na (15.0 h) and the short-lived noble gases are examples of such obscuring radionuclides. As examples of a rapid separation, ^{24}Na can be carried on a precipitate of sodium zinc uranyl acetate, $NaZn(UO_2)_3(C_2H_3O_2)_9.6H_2O$, by combining solutions of a sodium and a zinc salt with uranyl acetate in acetic acid. The noble gases can be removed from solution by agitation or solvent extraction.

The radionuclides in this category that emit beta particles also emit gamma rays that can be detected by spectral analysis. Short-lived radionuclides that emit alpha particles occur in the natural decay chains and usually are identified by other members of the decay chain that emit gamma rays. One caution to consider is that air filters and other surfaces in the environment collect particulate progeny of ^{220}Rn and ^{222}Rn that emit alpha particles, beta particles, and gamma rays with half-lives of minutes to hours. Observation of such emissions and decays has misled unprepared observers into attributing these radiations to man-made radionuclides.

6.4.4. Very Short- and Extremely Long-Lived Radionuclides

Measurement of radionuclides that have half-lives of minutes or even seconds is a research activity that requires careful preparation and testing (see Chapter 16).

Numerous techniques have been reported (Kusaka and Meinke 1961). Fundamentally, all processes for sample preparation, purification, and measurement must be brief and highly effective. Sample purification before production is helpful. Miniature ion exchange, solvent extraction, and precipitation systems have been developed for processing small volumes. Mechanical transfer from purification to the counting system enhances prompt measurement. Measurement results must take into account the radioactive decay of the radionuclide during measurement shown in Eq. (10.2).

Radionuclides with intermediate, short, and very short half-lives include progeny of long-lived terrestrial radionuclides and products formed by cosmic-ray irradiation of gases in the upper stratosphere. All three terrestrial decay chains—starting from ^{238}U, ^{235}U, and ^{232}Th—include radionuclides with half-lives of seconds and minutes that are readily detected because they are supported by their long-lived precursors. Prompt analysis after collection is needed to analyze very short-lived cosmic-ray-produced radionuclides such as the isotopes of chlorine, ^{34m}Cl, ^{38}Cl, and ^{39}Cl, with half-lives of minutes.

For radionuclides in the very long-lived to extremely long-lived category, the primary consideration is that a longer half-life is associated with the lower probability of decay per atom and lesser decay energy. As a consequence, analysis for such radionuclides requires large samples, thorough purification, and minimal radiation background. Moreover, energetic gamma rays are not emitted. Mass spectrometric separation and analysis (see Chapter 17) generally is the preferred alternative.

6.5. Standard Methods

As indicated in Section 11.2.8, several organizations publish standard methods for radionuclide testing. Prominent among these are the ASTM, Standard Methods, American National Standards Institute (ANSI), and International Organization for Standardization (ISO). Some of these methods are radioanalytical chemistry procedures while others are guides for actions such as processing samples, using standards and controlling detectors.

ASTM is a consensus standards organization with committees that develop and approve standard test methods. Committee members represent industry, universities, and government. Several committees publish methods for radioactivity measurement; the main ones are D19, water (of which subcommittee D19.04 deals specifically with radiochemical analysis of water), C26, Nuclear Fuel Cycle and E10, Nuclear Technology and Applications.

Method development and publication requirements differ slightly among committees. All D19.04 methods must undergo round robin testing prior to approval; C26.05.01 methods must undergo single operator testing. Appendix A-1 lists some current ASTM standards and methods.

Standard Methods for the Examination of Water and Wastewater is a text, first published in 1905, that has remained current through the efforts of its advisory committee. This committee has over 500 professionals from the ranks of the American Water Works Association, the American Public Health Association (APHA), and the Water Environment Federation. Their combined effort produces the *Standard Methods* text; part 7000 of the text deals with the testing of water samples for radionuclides. Appendix A-2 lists the *Standard Methods*, part 7000 procedures. The APHA intersociety committee published *Methods of Air Sampling and Analysis* (2nd ed.) in 1977.

ANSI is the U.S. representative of ISO. Appendices A-3 and A-4 list ANSI and ISO procedures, respectively.

The benefit of a standard method is the scrutiny and evaluation that it has received by fellow professionals. For that reason, a regulatory agency or customer may require application of specified standard methods in the radioanalytical chemistry laboratory. If another method is preferred because of its greater flexibility, rapidity, or ease in application, it may be compared initially or periodically with a standard method to confirm its reliability. Development and approval of these standards takes much time, effort, and money. For that reason, the standards are available from these agencies for a fee, generally ranging from $50 to $75.

6.6. Methods Development or Modification

A method development and testing function is a prerequisite to initiating laboratory operation. Each selected method must be tested for reliability in recovering the radionuclide of interest and removing impurities that interfere with measurement. Method modifications are inevitable if the sample matrices in the laboratory differ from those for which the selected method was designed.

In a functioning laboratory, development and testing is required for trouble shooting (see Chapter 12) and introducing improved methods. Problematic processes must be examined to determine whether the problem is due to method formulation, analyst error, or sample matrix complications. The method as written may be insufficiently stable to handle the wide range of radioactive or stable impurities in the samples, and may require more detailed specifications, added steps, or replacement.

In contrast, the effort in performing a selected method may be excessive if the set of samples has fewer interfering materials, either stable or radioactive, than the matrix for which the method was designed. Simplification may subtract one or more steps that were designed to remove interfering materials that are not in the sample to the extent foreseen. Such changes must be tested to confirm that the deleted steps indeed were unnecessary and that their deletion does not induce other problems.

Even if a method is successful, a better method may become available with greater ease in handling, better reproducibility of results, and lower cost. Improvements may be feasible for the chemical procedure, radiation detection instruments, and computer hardware and software.

One type of improvement is sequential analysis to use a single sample for analyzing several radionuclides. This approach is beneficial if the sample amount is limited by collection, shipping, or handling restrictions, or the largest possible amount is required for adequate detection sensitivity, or the initial analytical steps are identical for several radionuclides. The method must be tested to assure that cross-contamination and other possible interference among the analyses for the several radionuclides are minimal and/or acceptable.

The development and testing function may be part of the analyst's responsibility or may reside with a separate group. The former can be beneficial because it adds to the challenge of the analyst's work and draws on the analyst's knowledge of the specific process, but requires that management allot sufficient time to perform both routine and research functions. The latter approach may benefit from the broad knowledge of specialists devoted entirely to development and testing.

7
Preparation for Sample Measurement

BERND KAHN

7.1. Introduction

The transition from radiochemical separation described in Chapter 6 to instrumental radiation detection in Chapter 8 is source preparation for counting. The analyst wants to prepare a source that represents the radionuclide in the collected sample, can be measured reliably by its radiation, and is stable. The analyst selects a detector that is sensitive to the radiation that characterizes the radionuclide, stable as defined by its QA program, and calibrated for efficiency and—if needed—energy. Source preparation concerns are addressed here for the four types of detectors that are described in Chapters 2 and 8, but these considerations can apply to sources prepared for measurement by other detectors.

7.2. Counting Alpha and Beta Particles with a Gas Proportional Detector

7.2.1. Source Preparation

Sources conventionally are prepared by precipitation to permit gravimetric measurement of stable isotopic carrier yield, as discussed in Section 6.3. The isotopic carrier must be in the same chemical state as the radionuclide of interest, or the sample must be processed to achieve this requirement when carrier is added. The precipitating agent is selected to obtain a pure precipitate of the radioelement with a large, but not necessarily quantitative, yield and reproducible weight. If this precipitate does not completely purify the sample, as is often the case, then previous separation steps should have done so. The various purification steps must eliminate extraneous solids that will add to the carrier yield and contaminant radionuclides that will add to the count rate. The purification steps must reduce such contaminants to a small fraction of the amount to be weighed and counted. Occurrence in nature of significant amounts of the isotopic carrier in the sample must be determined in control samples to correct the yield value.

Environmental Radiation Branch, Georgia Tech Research Institute, Georgia Institute of Technology, Atlanta, GA 30332

The carrier weight generally is in the range of 10–30 mg and the yield is specified to be sufficiently high—typically above 50%—to reduce the uncertainty associated with yield determination to about 1%. To maintain constant weight and counting efficiency, the precipitate should not be volatile, hygroscopic, flaky, or powdery. Any drying or heating during source preparation must not affect comparison with initial carrier weight due to a change in chemical form or differences in degree of hydration.

Sources are also prepared by pouring a liquid or slurry onto a planchet and then evaporating it to dryness. Care must be taken to avoid nonuniform distribution because solids tend to dry as rings and to accumulate near the lip of the planchet. Adding a drop of a solvent such as alcohol that reduces surface tension in water before the onset of dryness can reduce such uneven deposition (Chieco 1997). Planchets have been designed with concentric ridges to distribute evaporated solids more evenly. Evaporation on a level surface prevents lopsided accumulation of solids. Evaporation must be sufficiently slow by controlling heat lamp distance to avoid sample loss by spattering.

Electrodeposition on metal disks is used to prepare very thin sources of the radionuclides discussed in Section 3.7. At very low concentrations in water, certain radionuclides may be deposited on metal disks by sorption or spontaneous electrodeposition (Blanchard *et al.* 1960). The former yields a relatively low percent deposition on more noble metals, as in the case of ^{144}Ce deposited on gold. In contrast, spontaneous electrodepostion results in near 100% yield on a less noble metal, for example, in depositing ^{110m}Ag or divalent ^{59}Fe on magnesium. Yield must be determined separately for this source preparation step and then combined with the yield observed for the chemical purification steps.

7.2.2. Other Yield Determinations

Alternatives for yield determination in the absence of stable isotopic carrier are nonisotopic stable carrier and radioactive tracer, as discussed in Sections 4.5 and 6.3. Technetium, promethium, and the elements heavier than bismuth have radioactive but no stable isotopes.

The yield for a low-mass sample, e.g., 1 mg or less for alpha-particle measurement, can be determined with nonisotopic carrier in an aliquot taken before preparing the counting source. The analytical technique can be instrumental, such as colorimetry or atomic absorption spectrometry. Subsequent source preparation, by precipitation, evaporation, or electrodeposition, must be quantitative or highly reproducible so that a reliable yield value for this final step can be included in the total yield.

If no carrier or distinctive tracer is available, recovery must be quantitative or estimated by separate yield determinations with a solution to which has been added a known amount of the radionuclide of interest. This yield is measured with each batch of samples. Separate determinations require consistent and high yield.

TABLE 7.1. Source thickness relations

Dimension	Ring and disk filter	Planchet
Diameter (cm)	1.6	5.0
Area (cm^2)	2.0	20
Area mass (mg/cm^2) (20 mg sample)	10	1.0
Thickness at density 3.0 g/cm^3 (cm)	0.0033	0.00033
Energy of β particle stopped from sample bottom (keV)	80	15
Energy of α particle stopped from sample bottom (MeV)	7	1

7.2.3. Source Form

Thin sources are prepared for counting alpha and low-energy beta particles that are readily absorbed in solids because counting efficiency depends on sample thickness. Counting efficiency is determined as a function of source thickness by counting several radionuclide sources in the range of source thickness expected for the usual fluctuations in carrier yield. Sources are prepared by mixing a radionuclide standard with carrier and preparing filter sources or evaporated slurries on planchets for various carrier masses. A smooth curve is drawn through the measured points to plot counting efficiency (i.e., the net count rate/disintegration rate) *vs.* sample mass.

The counting efficiency calibration must match the dimensions, uniformity, and electron density of the routinely analyzed sources. Filtering a precipitate with suction usually prepares a sample of uniform thickness unless the sample is too small to cover the filter or so thick that it piles up at the edge.

A source is sufficiently thin if it has an area mass (density × thickness) of a few milligrams per square centimeter. The relation of area mass, mass, thickness, and area is shown in Table 7.1 for sources on a ring and disk or on a planchet. An example of counting configuration is shown in Fig. 8.4. Area masses of 10 and 1 mg/cm^2 will stop beta particles with energies below 80 and 15 keV, respectively, in an aluminum-like material (see Fig. 2.4). A source with an electron density similar to aluminum and area mass of 10 mg/cm^2 will stop alpha particles below about 7 MeV energy, as shown in Fig. 2.3.

Although the electron density of the actual source affects attenuation, the listed values suggest the magnitude of the energies of the alpha particles and weak beta particles from the bottom of the 10-mg/cm^2 source will be stopped. Alpha particles of that energy and a progressively higher fraction of beta particles that originate nearer the surface that faces the detector will reach this surface. Other barriers will combine to further reduce the energy of the emerging particles; these barriers include any source covering, the air between source and detector, and the detector window. If the sum of attenuating media exceeds the range of the alpha and beta particles, none of the particles are detected.

Thinner sources are measured with higher counting efficiency per unit mass, but the highest count rate per sample is attained with an "infinitely thick" source, the thickness of which equals the maximum particle range. This configuration can be useful for detecting the presence of very low levels of radioactivity, although it is of questionable value for quantification. The mass of an "infinitely thick" source is irrelevant for efficiency calculations (except that the surface of a thicker sample is nearer the detector window above it) because radiation emitted only near the surface of such sources is detected.

In a very thin source (<0.1 mg/cm^2), the alpha particle flux at the surface that faces the detector, R_s in particles/s, is one-half of the disintegration rate, as shown in Eq. (7.1a) from Finney and Evans (1935):

$$R_s = \frac{A \Re a}{4m} \left[\frac{2x}{\Re} - \left(\frac{x}{\Re} \right)^2 \right] \quad \text{for} \quad x < \Re \tag{7.1a}$$

In the equation, A is the source activity in Bq, $\Re$ is the alpha particle range in mg/cm^2, a is the source area in cm^2, m is the source mass in mg, and x is the source thickness in mg/cm^2. The equation indicates that, for samples with a constant activity per unit mass, the surface flux decreases linearly with thickness to one-fourth the disintegration rate when the source thickness equals the range, which is about 7 mg/cm^2 for the 5.5-MeV alpha particle emitted by ^{241}Am. Beyond this thickness, the surface flux is predicted to remain constant:

$$R_s = \frac{A \Re a}{4m} \quad \text{for} \quad x \geq \Re \tag{7.1b}$$

Equation (7.1b) shows that the counting efficiency varies inversely with the sample mass when the thickness exceeds the alpha-particle range, i.e., at "infinite thickness."

The counting efficiency, discussed in Section 8.2.1, is related to the flux at the surface in terms of the source-detector geometry and attenuation of alpha particles between source surface and the sensitive-detector volume. Table 8.3 predicts a 31% counting efficiency for the detector-source configuration shown in Fig. 8.4 (but with the ring-and-disk replaced by the 20-cm^2 planchet source). In practice, the counting efficiency is considerably lower than 31%, as indicated by the measured alpha-particle counting efficiency curve shown in Fig. 7.1. The reduction in efficiency even at 0 mg/cm^2 is due to alpha particles that pass through the source, air, and window at angles not perpendicular to the detector for distances that exceed their range. The curve in Fig. 7.1 is approximately a straight line from 0 to 60 mg (i.e., 3 mg/cm^2) and then curves, as the count rate becomes constant at about 130 mg, the alpha-particle range in the source.

The decrease in counting efficiency with thickness in Fig. 7.1 emphasizes the benefit of a thin source for counting alpha particles at low levels. Thin sources can be prepared by using only a small amount of carrier and a light precipitating reagent, with purification processes other than precipitation such as electrodeposition, ion exchange, and solvent extraction, and by flaming the source to remove organic compounds.

For beta particles, reasonable estimates of the curve $\varepsilon / \varepsilon_0$ for counting efficiency as function of mass have been based on their approximately exponential attenuation

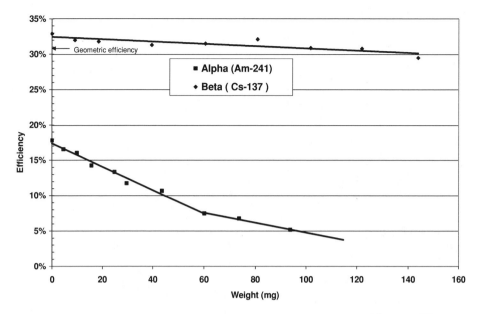

FIGURE 7.1. Alpha- and beta-particle efficiency self-absorption curves for ^{241}Am and ^{137}Cs, respectively, on 5-cm-diameter planchet measured with proportional counter.

over part of the attenuation curve (itself an empirical finding), so that

$$\varepsilon = \frac{\varepsilon_0(1 - e^{-\mu x})}{\mu x} \tag{7.2}$$

In Eq. (7.2), ε is the counting efficiency, subscript 0 refers to the efficiency at zero source mass, x is the area mass of the sample in mg/cm^2, and μ is an empirically derived attenuation coefficient in cm^2/mg. This attenuation coefficient has been related to maximum or average beta particle energy for specific source-to-detector configurations and source backing material (Evans 1955). Typical curves of counting efficiency vs. sample thickness are shown in Fig. 7.1 for ^{137}Cs on a planchet and in Fig. 7.2 for ^{89}Sr,^{90}Sr, ^{90}Y, and ^{14}C on a ring and disk. The value of μ is purely empirical because Eq. (7.2) does not consider beta particles that are emitted at an angle, scattered back, or not in conformance with exponential attenuation (notably near the maximum energy). Back scattering significantly increases the beta-particle counting efficiency above its geometrical efficiency.

More recently, reliable Monte Carlo simulation of electron interactions in the source, its environment, and the detector has been used to calculate the detector efficiency curve. The efficiency curves in Fig. 7.2, calculated with the Monte Carlo n-particle code, version 4, from available beta spectral data, agree with measured efficiency values within the uncertainty of the measurements (Nichols 2006).

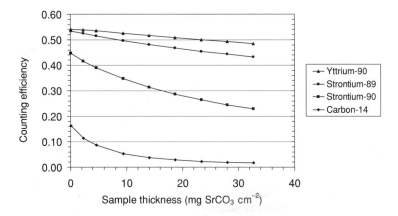

FIGURE 7.2. Beta-particle efficiency self-absorption curves for ^{89}Sr, ^{90}Sr, ^{90}Y, and ^{14}C as 1.6-cm-diameter ring-and-disk source, calculated by Monte Carlo simulation. (from Nichols 2006).

7.2.4. Gross Counting

Determination of gross alpha- or beta-particle activity is a measurement for which the source is prepared for counting without chemical separation. The measurements are performed to:

- assure that the maximum permissible concentration of even the most hazardous radionuclide in the sample is not exceeded;
- eliminate need for further analysis if only a single radionuclide is known to be present;
- compare to the sum of individually analyzed radionuclides for checking whether major radionuclides have been missed;
- indicate the magnitude of total alpha- or beta-particle activity in an unfamiliar sample.

For the first three applications, a radionuclide- and mass-specific counting efficiency must be selected. For the fourth application, a thin sample—below 2.5 mg/cm^2 for alpha-particle counting—should be prepared so that efficiency values are similar at commonly encountered energies. For counting beta particles, the sample should not exceed 10 mg/cm^2. An intermediate-energy (e.g., 0.6–0.8 MeV β_{max}) radionuclide standard provides reasonable efficiency estimates except that the activity of a radionuclide that emits only low-energy beta particles will be underestimated.

An aliquot of the sample as liquid or slurry usually is dried on a planchet and then counted for gross activity measurement. In some instances, the radionuclides are coprecipitated by a carrier such as ferric hydroxide or manganese dioxide that is expected to collect many of them (ICRU 1972), and this solid is filtered or poured onto the planchet as slurry and then dried. Under favorable circumstance, the radionuclide concentration is high enough that a thin sample can be prepared and attenuation by the sample itself is not a major concern. A thicker sample may yield detectable beta-particle counts in the available time period from a less radioactive sample, but the self-absorption fraction is in doubt if the radionuclides are not identified.

Gross alpha- and beta-activity counting can be a useful screening process. It should not be used to delineate radioactivity levels without associated specific radionuclide measurements because of uncertainty about what and how much is being measured. Moreover, some radionuclides may be lost in processing due to volatility during evaporation or incomplete carrying during precipitation.

7.3. Counting Alpha and Beta Particles with a Liquid Scintillation Detector

7.3.1. Source Preparation

Radionuclides typically are measured in a liquid scintillation (LS) counter as an aqueous solution that is pipetted into a vial, mixed with a scintillation cocktail, and counted within a radiation shield by a dual photomultiplier tube (PMT) system. As described in Section 8.3.2, the cocktail consists of a scintillating substance in an organic solvent such as benzene. The mixture may be a cloudy emulsion, but this lack of visible-light transparency does not interfere with the effective transmission of the generated 3-eV photons to the PMT.

Commercial LS systems are designed to use 20-ml screw-top vials available in glass or plastic. The sample volume that results in the most sensitive detection per unit volume is usually 10 ml aqueous solution per 10 ml cocktail, but the optimum volume ratio should be determined empirically. Smaller or larger sample vials have been used in systems designed to meet specific needs, such as limited sample volume or greater sensitivity, respectively. Very large solid scintillator systems have been designed for whole-body counting (Horrocks 1974).

An aqueous sample may be added to the cocktail directly, after minor prior processing, or at the end of a radiochemical separation procedure. Direct addition is the equivalent of gross activity counting discussed in Section 7.2.4 except that some spectral analysis may be possible. Alpha particles can be differentiated from beta particles by deposited energy, pulse shape, and decay time. Self-absorption is of no concern. Quenching and luminescence, discussed in Section 8.3.2, often occur. Identification by maximum beta-particle energy is approximate, and requires comparison to radionuclide standards.

A clearer understanding of what is being measured results from separating and purifying the sample radiochemically before counting. Sample volume should be sufficient for mixing with the cocktail and for taking an aliquot to measure yield. Quenching and luminescent impurities should be removed to the extent necessary for reliable measurement from samples destined for LS counting. Some quenching of the final sample can be anticipated—water is a quenching agent—but a consistent level of quenching leads to more reliable results than wildly variable levels.

The main advantage of the LS counter over the gas proportional counter is the intimate mixing of the radionuclide with the scintillating detection medium to eliminate loss of alpha and beta particles by absorption outside the detector. Moreover, the LS geometrical efficiency is almost 100%. The practical counting efficiency is above 90% for alpha and more energetic beta particles. Energy loss by quenching is minor except for low-energy beta particles from radionuclides such as tritium.

One disadvantage of the sample–cocktail mix is impermanence. The stability of the prepared vial contents should be tested over time. Typically, a sample can be recounted after a few days or weeks, but should not be considered reliable after longer periods.

7.3.2. Beta Particles

LS counters are suitable for measuring radionuclides that emit only very low-energy beta particles or electrons, notably tritium ($E_{max} = 18.6$ keV). When tritium is measured as tritiated water and its activity is reported relative to water weight or volume, no yield measurement is needed. Liquid samples, e.g., water from the environment, process streams, urine, or dissolved solids, can be counted directly or purified by distillation. Results for purified samples are more reliable to the extent that the radioelement can be identified, quenching is stabilized, and luminescent contaminants are removed. Reagents may have to be added to the distillation flask to hold back other potentially volatile radionuclides, such as radioactive iodine, carbon, ruthenium, or technetium.

Examples of other radionuclides counted with an LS system because they emit very low-energy beta particles or electrons are ^{63}Ni ($E_{max} = 66$ keV) and ^{55}Fe ($E_e = 5.2$ keV, 61%), respectively. Each radionuclide normally is purified to remove interfering substances, including other radionuclides, and then prepared as a solution with suitable volume for LS counting and yield determination. Radionuclides that emit energetic beta particles are measured by LS counting because their counting efficiency is near 100% and sample processing is convenient.

7.3.3. Alpha Particles

Measuring alpha particles with an LS counter is an attractive option because the counting efficiency is near 100% and no self-absorption problem exists. After the usual sequence of separations for radionuclides such as thorium, uranium, and transuranium isotopes, the radionuclide is prepared in the final solution for counting and yield determination. A tracer that emits alpha particles at a sufficiently distinct energy is added initially to measure yield. The factor that controls detection sensitivity is the background, typically of 1–2 c/m in the alpha-particle energy region of the LS counter.

Alpha particles emitted by a mixture of several radionuclides can be analyzed simultaneously by energy discrimination with an LS spectrometer. Such separation is better than for beta-particle continua because alpha particles are emitted with discrete energies. Resolution of alpha-particle energy peaks in conventional LS systems is about 0.50 MeV. It is reduced to about 0.40 MeV by deoxygenating the sample to reduce quenching. This resolution is far poorer than for solid state detectors (see Section 7.4), but spectral analysis can be useful for radionuclide mixtures in which only a few alpha-particle groups occur at widely separated energies. A photon/electron rejecting alpha liquid scintillation (PERALS) system has been developed that discriminates against beta particle scintillations by their longer decay time, improves the resolution to about 0.25 MeV, and reduces the background to about 0.001 c/m (McDowell 1992).

7.3.4. Special Applications

Ingenious permutations of the conventional combination of aqueous solution with organic cocktail have helped the analyst. Some samples are extracted into an organic solvent that is soluble in the scintillation cocktail to eliminate water quenching. Sample purification and counting can be integrated by collecting a dissolved or airborne radionuclide on a sorption or ion-exchange matrix that contains scintillating material and is transparent to its scintillations. This material is washed and placed directly into the counting vial for measurement.

Collected particles and prepared precipitates have been counted by dispersing and stabilizing them in a cocktail with silica gelling agent (ICRU 1972). Air filters and smears have been counted by placing them directly into a scintillation cocktail. The counting efficiency for each type of sample must be calibrated for specific configurations, radionuclides, and interfering substances.

Radon-222 gas in water is counted after pipetting the water sample without agitation (to avoid losing the gas) into a counting vial that contains an organic cocktail solvent not miscible with water. Almost no air space should remain in the vial. The vial is closed tightly, shaken to transfer the radon gas to the organic phase in which it is far more soluble (EPA 1978), and counted after a 4-h interval to permit ^{222}Rn progeny ingrowth. The system can be calibrated by pipetting the same volume of a ^{226}Ra standard aqueous solution into a vial that contains the usual volume of immiscible cocktail and sealing the vial. After 4 weeks to accumulate the ^{222}Rn progeny (three alpha particles from ^{214}Po, ^{218}Po, and ^{222}Rn and two beta particles from ^{214}Pb and ^{214}Bi) in the standard to near equilibrium, the mixture is shaken to transfer the ^{222}Rn to the organic phase, and the standard is counted. Alpha and beta particles are counted together at a composite counting efficiency of about 420%, with a background of about 20 c/m.

Flow-through systems have been developed for monitoring radionuclides, especially tritium, in facility effluent, as described in Section 15.4.3. Typically, the sample accumulated in each intermittent collection period is mixed automatically with the scintillation cocktail and transferred to a counting chamber. While the sample is counted, the next sample is collected. After counting, the sample is discharged and the counting chamber is flushed for cleaning.

Solid scintillators such as anthracene have been tested as alternatives for continuous measurements (ICRU 1972). The scintillation material and the radionuclide collector are combined, as discussed above for scintillation beads (Winn 1993). The operating characteristics of these devices, such as the sample volume, count accumulation period, energy resolution, detection efficiency, and background must be arranged to satisfy radionuclide concentration limit specifications.

The LS counter also measures Cherenkov radiation (see Section 2.4.3) in water without addition of scintillation cocktail. The radiation is generated in water by electrons at energies above 0.265 MeV. Hence, about one-half of beta particles in a group with E_{max} of 0.8 MeV generate this light, and the fraction is larger for groups with higher maximum beta-particle energies. The advantages of this process for detecting beta particles are that the water sample volume can be the full 20 ml and the sample is stable. The drawbacks are that quenching still can occur, few photons are produced per beta particle, and the PMT usually is not optimal for detecting the wavelength of the Cherenkov radiation. A 1-MeV beta particle produces about 200 Cherenkov scintillations in water compared to 10,000 scintillations in an LS cocktail (Knoll 1989).

7.4. Counting Alpha Particles with a Silicon Detector and Spectrometer

The main radionuclides that are measured in this system are isotopes of thorium, uranium, and plutonium. Others are the longer-lived isotopes of heavy elements such as radium, protactinium, neptunium, americium, and curium. Conventionally, an isotopic tracer of known activity that emits alpha particles is added at the beginning of the radiochemical procedure (see Section 6.3.2). The solution from which the sample for counting will be deposited is thoroughly purified by the radiochemical procedure to remove interfering radionuclides and solids.

The critical requirement for alpha-particle spectrometry to achieve good characteristic peak resolution, shown in Fig. 9.1 and described in Section 8.3.3, is the preparation of a very thin and uniform source. Less rigorous criteria for resolution can be met if the several radioisotopes to be measured and the tracer have widely different alpha-particle energies, only a single radionuclide is known to be present either initially or after chemical separation, or the measurement is for screening purposes. A sample may be prepared for screening with limited prior separation or none, and without tracer addition, if previous tests have demonstrated that the process causes no losses for all radionuclides of interest.

7.4.1. Tracer Addition

Isotopes available as tracers include ^{229}Th, ^{232}U, ^{242}Pu, and ^{243}Am. Only one or two tracers usually are available per element because of the requirements of distinctive alpha-particle energy, long half-life, and absence of contamination by the radioisotopes of interest. Some radioactive tracers must be purified at intervals of a few percent of the progeny half-life to remove ingrown radioactive progeny that emit alpha particles. Prior tracer purification is not necessary if the impurity is removed as part of the chemical separation before counting sample preparation, or if the ingrown alpha-particle peaks do not interfere with the peaks of the radioisotope of interest and the tracer.

7.4.2. Source Preparation

Electrodeposition is the preferred method for preparing a thin, smooth, uniform, and stable source for counting, as discussed in Section 3.7. Empirically developed procedures are applied to the heavy elements in aqueous solution because they cannot be reduced to the metal; the source is deposited at the cathode in what is believed to be hydroxy forms after partial reduction. Deposition is in acid solution, under various conditions that yield near-quantitative recovery from the solution. Test results specify the volume, reagent content, applied volts and amperes, and optimum time period (see Section 6.4.1). The cathode can be a platinum disk, but a polished stainless steel disk is satisfactory.

In an application of the spontaneous deposition discussed in Section 7.2.1, sources of ^{210}Po are prepared on a nickel planchet. This process separates this radionuclide from other radon progeny or heavy elements that emit alpha particles.

Sources can also be prepared by coprecipitating actinides with 1 mg or less of rare earth—e.g., lanthanum—carrier. Reasonable alpha-particle peak resolution results when the precipitate is counted on a filter that has been prepared with additional rare earth substrate (Chieco 1997).

Very thin and uniform sources have been prepared on targets in a mass spectrometer or a thermal volatilization system (see Section 3.6.1). Such preparations require a relatively intense radioactive source.

7.5. Counting Gamma Rays with a Germanium Detector and Spectrometer

Samples with radionuclides that emit gamma rays are counted conveniently with an intrinsic, high purity, germanium (Ge) detector with spectrometer. Gamma rays from the sample can be counted with minimal processing—often the sample is counted directly in a container that has been calibrated for counting efficiency. The container is tared for weight and calibrated for volume; when filled, it is weighed to calculate the activity per this mass and determine source density for calculating the self-absorption factor. Samples counted to report activity relative to the flow rate, such as filters and sorption media for air and water samples, must conform to a calibrated source volume even if volume and mass do not enter into the calculation of activity.

7.5.1. Source Preparation

Many sources can be prepared directly for gamma-ray spectral analysis. Exceptions are sources that require concentration, stabilization, or radionuclide purification. Concentration improves detection sensitivity, stabilization should prevent changes in source configuration or content, and purification can eliminate interference with counting characteristic peaks.

Liquid samples are concentrated by evaporation or by separating the radionuclide of interest with processes such as filtration, precipitation, solvent extraction, ion exchange, sorption, and distillation. Solid samples are dried, ashed, compressed, or physically sorted. Gases can be sorbed, condensed, or compressed. Some steps may be performed at collection, others, in the laboratory. Recovery of the radionuclide of interest must be quantitative or at a previously determined fraction.

For measurement reproducibility, the source must be stable and of uniform shape. Sources should be prepared so that solids do not shift or decompose. Radionuclides in liquids should not settle, float to the surface, or adhere to container walls. Gaseous radionuclides should not escape from the solid or liquid matrix to collect in vapor spaces or leak from the container. The applied counting efficiency must take into consideration any known nonuniformity of radionuclide distribution in the source.

The good energy resolution a Ge detector system reduces the possibility that a second gamma-ray peak resulting from the presence of other radionuclides in the sample will obscure the gamma-ray peak of the radionuclide of interest. If this does occur, the interfering radionuclide must be removed.

7.5.2. Direct Measurement

Direct measurement of relatively large samples is feasible because energetic gamma rays—those above about 0.1 MeV—will reach the detector with only minor attenuation in large, multiliter, samples. The Ge counting efficiency, although less than for a NaI(Tl) detector of the same dimensions, can be optimized by placing a large Ge detector in close proximity to a large sample. Sensitivity is improved with re-entrant beakers that surround the detector, with well-type detectors, and with multiple detectors connected in parallel to view a large sample. Larger samples may be of irregular shape for which the geometry can only be approximated. Extreme examples of the latter are animals and humans monitored in whole body counters.

7.5.3. Calibration

Source containers must be sufficiently rigid to maintain their shape. They must resist attack and decomposition from contents such as organic solvents, acids, bases, and biological material. They must be closed to prevent spillage, permit storage, and protect the detector and its environment from contaminants. In some instances, samples are sealed to prevent inleakage of air or escape of gases. One technique is to place a solid or liquid sample in a metal can and then seal its lid (Chieco 1997). Gaseous samples are transferred to an evacuated container connected by tubing, valves, and a pump with the sample container.

Efficiency calibration requires a sample of defined shape and volume placed at a defined location relative to the detector. A sample frame may be needed for reproducible placement. Calibration can be performed with a radioactivity standard or by Monte Carlo simulation (see Sections 8.2 and 10.5).

Although gamma rays are much less subject to attenuation than alpha and beta particles, a density correction is needed if the density of the sample deviates significantly from the density of the calibration standards. The effect of density on self-absorption for both the standard and the sample is estimated by Eq. (7.2); μ for this purpose is the photon attenuation coefficient in cm^2/g and x is the sample area density in g/cm^2. Values for μ in some common materials are listed in Table 2.2 and in its cited reference. If a large set of samples with consistent density is analyzed, it may be possible to prepare radioactivity standards at the same density to avoid the need for correction. Interpolating efficiency values as a function of density is feasible at energies above 0.1 MeV because the effect of minor density difference on counting efficiency is small.

When initial nonuniformity cannot be eliminated by mixing the sample, e.g., in whole-body counting, the radionuclide distribution can be estimated by calibrating with standards in various configurations and conditions. The results can establish a range of possible counting efficiencies. Changing sample location relative to the detector for separate counts may yield further information concerning radionuclide distribution. Monte Carlo simulations can also provide needed efficiencies.

A simple example of sample nonuniformity is an air cartridge that may have retained a radionuclide either on the front surface or with decreasing concentration in depth, or distributed uniformly throughout. Counting the cartridge from both sides and comparing the results to a simulation model will permit calculating the activity as a function of depth distribution.

8
Applied Radiation Measurements

JOHN M. KELLER

8.1. Introduction

The radiation detection systems employed in radioanalytical chemistry laboratories have changed considerably over the past sixty years, with significant improvement realized since the early 1980s. Advancements in the areas of material science, electronics, and computer technology have contributed to the development of more sensitive, reliable, and user-friendly laboratory instruments. The four primary radiation measurement systems considered to be necessary for the modern radionuclide measurement laboratory are gas-flow proportional counters, liquid scintillation (LS) counters, Si alpha-particle spectrometer systems, and Ge gamma-ray spectrometer systems. These four systems are the tools used to identify and measure most forms of nuclear radiation.

Some operating parameters can be considered in common for all these detectors, notably detection efficiency and the radiation background. Additional parameters pertain only to detectors with associated spectrometers. These parameters concern the radiation energy peaks that identify and quantify radionuclides, and include energy calibration, energy resolution, peak-to-Compton ratio, and peak shape.

Each of the detection instruments has a specialized field of application. The gas-flow proportional counter is the laboratory workhorse for measuring radionuclides that emit alpha and beta particles in samples before radionuclide separation and then again in the purified fractions. The LS counter also measures radionuclides that emit alpha and beta particles and is particularly useful for measuring low-energy beta particles. The alpha-particle spectrometer is applied for purified samples to measure radioisotopes, notably of the heavy elements, that emit alpha particles. The gamma-ray spectrometer is the laborsaving device for measuring photon-emitting radionuclides, in most cases without chemical separation.

Many ingenious applications of these detectors go beyond such specialization so that their use can be a matter of matching sample characteristics, radiation types, the radionuclide of interest and associated impurities, required sensitivity, and convenience in sample preparation, or even analyst preference. Various other detection systems and instrumental techniques that are mentioned at the end of

Oak Ridge National Laboratory, Oak Ridge, TN 37831

134

this chapter should be considered when available. Some of these other detectors may be more appropriate than the four conventional ones for analyzing samples with very high or low activity and for radionuclides that otherwise would require complex radiochemical separations prior to analysis.

8.2. Common Operating Parameters

8.2.1. Counting Efficiency

Dependable radiation measurement requires a professional understanding of radiation detector calibration, as it applies to each detector and sample. Several calibration techniques are applied:

• Prepare and count a radionuclide standard for each radionuclide and source–detector configuration.
• Prepare and count selected radionuclide standards to determine curves of counting efficiency *vs.* energy for each source–detector configuration.
• Determine separately and then combine all factors that constitute the counting efficiency.
• Apply Monte Carlo simulation to determine curves of counting efficiency *vs.* energy for each source–detector configuration.

A careful analyst will apply more than one of these techniques to assure reliable calibration. The first three will be discussed here to review the basic elements of counting. The fourth approach is a numerical analysis technique that should be given proper grounding in an appropriate text (see Briesmeister 1990) and proper introduction to the student after the basics of counting are understood.

The crucial characteristic that applies to all detector calibration is the practical counting efficiency—the ratio of the observed measurement to the activity of the sample in Bq (or pCi) per sample that may be converted to activity per unit mass (or volume or flow rate). In the simplest case, a measure of the practical counting efficiency ε is given by the relationship

$$R_{\mathrm{G}} = A\varepsilon + R_{\mathrm{B}} \qquad (8.1)$$

where A is the disintegration rate (or activity), R_{G} is the observed count rate of the sample, and ε is the counting efficiency. A measured background count rate R_{B} must be subtracted from the observed count rate to calculate the actual count rate of a sample. The observed count rate is also referred to as the gross count rate and the value after subtracting the background, the net count rate. Most measurements of the count rate of a sample are performed to determine its disintegration rate, but some measurements are performed only to compare count rates, e.g., in half-life measurements or for samples taken to evaluate a purification step.

In the first two calibration approaches cited above, a certified radionuclide standard is measured to obtain the practical counting efficiency of the detector, according to Eq. (8.2):

$$\varepsilon = \frac{R_{\text{G-std}} - R_{\text{B}}}{A_{\text{std}}} \tag{8.2}$$

Here, $R_{\text{G-std}}$ is the count rate observed for the standard and A_{std} is its reported activity. The calibration standard must be traceable to a national standardization agency—in the United States, the National Institute of Standards and Technology. Calibration may be by direct counting of a source in the form prepared by a standardization group, or it may involve processing a standard solution to prepare a counting source. In an indirect approach to calibration, a source may be counted that was previously counted in the laboratory with a calibrated detector.

The radionuclide standard must be accompanied by a certificate (see Section 11.2.6) with detailed descriptions of its chemical, physical, and radio-logical characteristics and the uncertainty of the reported disintegration rate. The uncertainty of calibration depends on the reported uncertainty of the standard, compounded by the uncertainty due to source preparation and measurement.

The calibration value can be accepted if it agrees, within the calculated uncertainty, with the value obtained from another standard or by other means, such as a calibration curve, calculations of individual parameters described below, and by Monte Carlo simulation. The reliability of these approaches depends on meticulous description of the source, the detector, and their environment, and must be defined by the uncertainty associated with these factors.

The third listed calibration approach involves the separate determination and subsequent combination of all factors that constitute the counting efficiency. These factors depend on detector volume, source-to-detector geometry, and the extent of radiation attenuation and scattering. They have differing degrees of impact in different detectors. Some approximate counting efficiency and background values for the detectors discussed here are listed in Table 8.1. The listed PERALS (Photon

TABLE 8.1. Typical values of radiation detector counting efficiency and background

Detector type	Sensitive volume (cm^3)	Radiation type	Energy (keV)	Efficiency (%)	Background (c/m)	Conditions
Proportional	28	alpha	4000–9000	20	0.05	Anticoincidence
		beta	400–3500	45	1	
LS	20	alpha	4000–9000	>95		
	20	beta	18	25	2	10 ml H$_2$O
	20	beta	250–3500	>90	5	20
LS	1	alpha	4000–9000	99.7	0.001	PERALS
Si spec	0.06	alpha	4000–9000	25	0.0001	Electrodeposited source
Ge spec	85	gamma	100–103	0.15	0.5	Flat source
			500–505	0.063	0.3	Flat source
			1000–1006	0.035	0.15	Flat source
			2-000–2007	0.011	0.011	Flat source

Note: maximum beta-particle energy is given

electron rejecting alpha LS) is an LS counter specially designed for improving alpha-particle spectral analysis by increasing counting efficiency and decreasing background.

For calculating the practical counting efficiency from all contributing factors, parameters must be considered for the source–detector system, the detector itself, and the radioactive sample. The source–detector system contributes the following:

- Intrinsic efficiency of the detector, ε_i
- The geometric relation of sample and detector, or solid angle Ω
- Factor for attenuation loss in the sample, detector window, and intervening material l
- Factor for scattering into the detector by sample support and shield s

The detector electronic system controls the following:

- Factor for pulse losses and discrimination in the pulse processing system p

The radionuclide and the sample affect the following:

- The decay fraction of the measured radiation, f
- The fractional extent of radioactive decay or ingrowth, D
- The chemical recovery fraction or yield Y
- The sample mass or volume, m or v

The radionuclide concentration, a, the "massic activity," is calculated from the count rate and the combined factors, which constitute the overall counting efficiency by:

$$a = \frac{R_G - R_B}{\varepsilon_i \Omega l s p f D Y m} = \frac{A}{m} \tag{8.3}$$

If the count rate is in c/s, then the units of A are Bq and of a are Bq per unit mass. Discussion of these individual parameters follows.

The intrinsic efficiency of a detector is a function of its dimensions and the electron density of its detection volume. For alpha- and beta-particle detection in a typical 1-cm deep, gas-filled proportional counter, creation of only a few electron–ion pairs is needed to achieve high probability of detection. The Poisson distribution (see Section 10.3.5), with probability $\Pr(x)$ for the number of events x with a mean number of events μ (see Eq. 10.14), suggests that the probability is only 0.01 when the number of events is zero at a mean number of 4.6:

$$\Pr(0) = e^{-\mu} \tag{8.4}$$

Therefore, the probability of detecting an event is 1–0.01, or 99%. The specific ionization in air is more than 50 ion pair per cm for beta particles and about 400 times as great a value for alpha particles. The specific ionization in the filling gas is similar. Hence, a travel distance of 0.1 cm will exceed 99% intrinsic efficiency for counting both alpha and beta particles in the typical proportional counter, although most particles deposit only a fraction of their energy in the gas.

TABLE 8.2. Relation of intrinsic efficiency to
gamma-ray energy in 6-cm-thick Ge detector

Energy (MeV)	Efficiency (%)
0.1	>99
0.5	94
1.0	87
3.0	83

In an LS counter, the cocktail is about a thousand times as dense as counter gas, hence most beta particles and all alpha particles deposit their full energy in it. This energy deposition produces enough scintillations for all but near-zero-energy beta particles to be recorded.

The sensitive region of the Si diode typically is about 200 µm thick, compared to the range of 20–60 µm for alpha particles with energies of 4–9 MeV, respectively. Hence, detectors are constructed to absorb the entire energy and produce sufficient electrons for good resolution at essentially 100% intrinsic efficiency.

The intrinsic efficiency for detecting gamma rays in the Ge detector, estimated in terms of the attenuation coefficient, usually is less than 100% except for low energies. Table 8.2 shows the percent efficiency in a typical 6-cm-thick detector as a function of gamma-ray energy. The figures are based on Eq. (2.12).

The efficiency is somewhat lower for the noncollimated gamma rays emitted by a sample because some would not traverse the entire thickness of the detector. Larger detectors can be expected to have larger intrinsic efficiency values for energetic gamma rays.

To optimize the geometry factor, a source is placed as close to the detector as feasible without causing damage or contamination. A second requirement is preparing the sample so that its diameter is appreciably less than the diameter of the detector. For the simple case of a point source and a detector with a circular aperture, the solid angle (Ω) can be calculated as follows:

$$\Omega = \frac{1}{2}\left(1 - \frac{d}{\sqrt{d^2 + \rho_d^2}}\right) \tag{8.5}$$

Here d is the distance from the point source to the detector face and ρ_d is the radius of the detector window. The solid angle approaches 2π when d approaches 0, i.e., Ω approaches 0.5 for a point source near the detector surface. Such optimized geometry is desirable in low-level measurements, but not necessary under other circumstances. Some geometry factors for a circular source with radius b, calculated as a function of d/ρ_d and b/ρ_d (Bland 1984), are given in Table 8.3.

Reducing the sample mass and the amount of materials between the sample and the detector reduces radiation attenuation l. This effort is most important for alpha particles and least important for energetic gamma rays, as discussed in Chapter 7. The fractional self-absorption within the sample can be estimated for beta particles and gamma rays by using Eq. (7.2). The fractional attenuation of gamma rays in

TABLE 8.3. Geometry factors for extended source

b/ρ_d	d/ρ_d						
	0.10	0.15	0.20	0.25	0.30	0.35	0.40
0.05	0.4502	0.4257	0.4018	0.3786	0.3562	0.3346	0.3141
0.10	0.4500	0.4255	0.4015	0.3783	0.3558	0.3342	0.3137
0.15	0.4498	0.4252	0.4011	0.3777	0.3552	0.3335	0.3129
0.20	0.4495	0.4247	0.4004	0.3769	0.3543	0.3325	0.3118
0.25	0.4490	0.4240	0.3996	0.3759	0.3531	0.3312	0.3104
0.30	0.4484	0.4232	0.3986	0.3747	0.3517	0.3297	0.3087
0.35	0.4478	0.4222	0.3973	0.3731	0.3499	0.3277	0.3066
0.40	0.4470	0.4210	0.3958	0.3713	0.3478	0.3254	0.3042
0.45	0.4460	0.4196	0.3940	0.3691	0.3453	0.3227	0.3013
0.50	0.4449	0.4179	0.3917	0.3665	0.3425	0.3196	0.2981
0.55	0.4434	0.4159	0.3892	0.3636	0.3392	0.3161	0.2943
0.60	0.4418	0.4135	0.3862	0.3601	0.3354	0.3120	0.2901
0.65	0.4398	0.4108	0.3826	0.3560	0.3309	0.3074	0.2854
0.70	0.4373	0.4072	0.3784	0.3513	0.3258	0.3021	0.2800
0.75	0.4344	0.4031	0.3735	0.3457	0.3199	0.2960	0.2740
0.80	0.4304	0.3977	0.3672	0.3390	0.3131	0.2892	0.2673
0.85	0.4253	0.3910	0.3596	0.3312	0.3052	0.2814	0.2598
0.90	0.4184	0.3823	0.3504	0.3218	0.2961	0.2729	0.2514
0.95	0.406	0.3706	0.3387	0.3106	0.2854	0.2648	0.2420
1.0	0.394	0.355	0.3239	0.2970	0.2729	0.255	0.2316

the sample covering, detector window, and intervening space is calculated by Eq. (2.12). The fractional attenuation of beta particles can be measured for these materials, or estimated from the energy of beta particles required to pass through the material, as suggested by Fig. 2.4. Alpha particles of a given energy are entirely absorbed by intervening materials unless the sum of the material thickness is less than the range for the energy of the particle.

Scattering into the detector, s, depends on the environment of the source and the detector. One subcategory is back scattering from the sample support, which significantly increases the count rate for beta particles when the sample is placed near the detector. It is about 20–30% of the direct radiation count rate for a ring-and-disk source for the efficiency values shown in Fig. 7.2 with an end-window proportional counter. Back scattering increases with beta-particle energy and the Z value of the backing material.

Backscattering is minimal for alpha particles because they are mostly stopped in the backing material (NCRP 1985b). A typical factor is 1.03. Backscattered gamma rays add to the Compton continuum a broad peak at an energy of about 0.2 MeV.

Material around the source and detector, notably the detector housing, cause scattering into the detector. The opportunity for scattering into the detector increases when the source is more distant. This scattering adds a few percent to the count rate for end-window Geiger–Mueller (G-M) detectors when the sample is 2 cm or more distant (Zumwalt 1950), but little for gas-flow proportional counters with the sample only about 0.3 cm from a relatively large window. Scattering, attenuation, and self-absorption effect in G-M counters also differ from those in end-window proportional counters because of the typically larger sample-to-detector distances and smaller detector window (Steinberg 1962). Scattered energetic gamma rays do not affect the efficiency in gamma ray spectrometers because they only increase the Compton-scattering region. Material between source and detector also scatters direct radiation away from the detector, but this effect is small when, as is common, the amount of intervening material is small.

Source radiation that produces relatively weak pulses is lost (to the extent of fraction $1 - p$) when the pulse processing system of a detector discriminates against the low-energy pulses that constitute noise; discriminators are, however, set to minimize count loss. Such discrimination prevents large Ge detectors from measuring low-energy (<30 keV or <3 keV, depending on construction) X rays, reduces the counting efficiency of LS counters in measuring low-energy radionuclides such as tritium, and causes a loss of $<1\%$ in proportional counters. Measurements with a pulse generator or recording low-energy electron or X-ray pulses can identify the energy cutoff to evaluate whether the discriminator setting is acceptable.

The factor p also refers to loss of counts at all energies if radiation arrives at the detector more rapidly than individual pulses can be processed. The dead time (τ) is the detector recovery period before it is capable of recording a second event. The dead time depends on the period and shape of the pulse and is an inherent property of the detecting systems (e.g., gas, liquid, or solid state detectors) and their electronic support. If the count rate is sufficiently low, the average interval between pulses is sufficiently long that the dead time is insignificant. For higher count rates, the following "nonparalyzable" model for correction can be applied:

$$R_{\text{cor}} = \frac{R_{\text{obs}}}{(1 - R_{\text{obs}}\tau)} \tag{8.6}$$

R_{cor} is the corrected count rate and R_{obs} is the observed rate. Count rates should be limited so that the correction does not add more than about 30%.

The dead time is long in a G-M counter because of its large and wide pulses, but short in a proportional counter. For a G-M detector with a dead time of 0.5 μs, a measured count rate of 200 c/s implies a loss of 11% according to Eq. (8.6). In contrast, a proportional counter with a dead time of 5 μs has a loss of only 0.1% at the same measured count rate.

The loss in count rate can be determined by counting a set of samples with increasing known activity and plotting the relation of count rate to activity. The extent of deviation from a straight line at higher count rates indicates the loss. Another technique for determining the dead time compares the count from a split source when combined with the sum of the separately counted components (Knoll 1989).

The distinction between a second event that only adds to the pulse height of the first event or is counted separately is a function of pulse height discrimination and random pulse arrival time, as shown in Fig. 8.1. Further distinctive behavior, also shown in the figure, is a second event that is counted although the first pulse has not completely died away, i.e., the second pulse has some contribution from

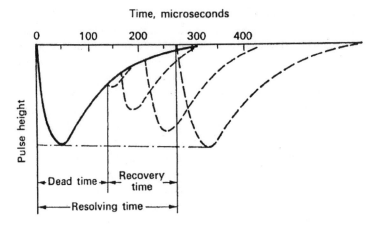

FIGURE 8.1. Resolving time (from Cember 1996, p. 350).

both events. The time required for the first pulse to die away completely is termed the resolving time. The terms "dead time" and "resolving time" are often used interchangeably.

The impact of such loss depends on pulse shape and pulse application. In integral counting, all distinct pulses with height above an acceptance value are counted.

In spectral analysis, the pulse height matters. A clock records the live time for spectrometers as the time interval to use in calculating the count rate. A pulse that results from two radiations detected within the resolving time is known as a "coincidence"; the coincidence is recorded at the summed energy and results in two lost pulses at the energies corresponding to the individual radiations. Coincidences occur randomly, but can be expected with much greater frequency when counting radionuclides that emit two radiations simultaneously.

Radiations that are emitted simultaneously and measured for standardization purposes (see Section 9.3.5) include two gamma rays and beta particles with gamma rays. Occurrence of coincidences can be inferred from information in the compilations of radionuclide decay characteristics cited in Chapter 9. The counting efficiency ε for such coincidences is the product of the efficiency for each radiation. In simplified terms, the fractional detection of a coincident gamma ray relative to either of the two gamma rays is $\varepsilon_1\varepsilon_2/\varepsilon_1 = \varepsilon_2$, which is significant only for large values of ε_1 and ε_2.

In recent years, calibration by cascade summing has become a more recognized measurement technique with the introduction of the larger high-purity germanium (HPGe) detectors which have a higher intrinsic efficiency. Cascade summing or true coincidence summing results when two gamma rays are emitted from the nucleus at nearly the same time and are then detected simultaneously at an energy that is the sum of the two incident gamma rays. True coincidence summing is geometry-dependent and should not be confused with the count rate-dependent random summing mentioned above. The probability of detecting both gamma rays is proportional to the square of the solid angle subtended by the detector, hence the sum peak count rate will increase as the source to detector distance decreases.

Coincidence summing can introduce the problem of misattributing summed gamma rays as single gamma rays when counting sources close to large HPGe detectors. Use of n-type HPGe detectors and Be windows can result in cascade summing problems at lower energy X rays. Gamma rays can even sum with Bremsstrahlung radiation from associated beta particles (Gilmore and Hemingway 1995). Cascade summing problems are observed with radionuclides commonly used as calibration standards such as ^{60}Co, ^{134}Cs, and ^{152}Eu.

The decay fraction f of an emitted radiation and its half-life that are used for measuring a radionuclide and calculating the extent of decay or ingrowth are tabulated in tables of isotopes and on the Internet at the BNL nuclear data bases (http://www.nndc.bnl.gov/ index.jsp). The decay scheme should be examined, as discussed in Section 9.3, before selecting the decay-fraction value. When measuring gamma rays, for example, the gamma-ray fraction must be used instead of a decay fraction from an excited state that includes conversion electrons. Any applicable conversion-electron fraction must be included when converting the beta-particle count rate to activity. Characteristic X rays may be detected when counting beta particles or gamma rays.

Processing a radionuclide with minimal decay between collection and measurement—that is, D is almost 1.0—optimizes detection sensitivity and minimizes the uncertainty of decay correction. Some radioactive decay is inevitable and may be useful for reducing interference from shorter-lived radionuclides.

A separation procedure with the yield Y below unity is expected in radioanalytical chemistry. For quality assurance purposes, the range of acceptable yields may be set between 0.5 and 1.0. Lower or higher yield fractions suggest occurrence of analytical process problems. Carrier or tracer addition to determine yield is described in Section 6.3.

Appearance of the mass (or volume) term, m, in the denominator of Eq. (8.2) suggests that large samples permit more sensitive measurements. Analyzing ever larger samples is a tempting solution for attaining lower detection levels, but larger samples require ever more complex processing and may cause reduction in factors for the other variables in Eq. (8.3), notably those related to detector-sample geometry, radiation attenuation, and chemical yield.

8.2.2. Radiation Background

Radiation detectors should have a low and stable radiation background to reduce interference and uncertainty in sample measurements (see Section 10.3). Contributors to the radiation background for a detector are the following:

- Radionuclides in the detector and nearby components of the detector system
- Radioactive samples placed near the detector
- Radioactive contamination on the detector, sample holder, and other components
- Radionuclides in the counting room walls, floor, and air

- The radiation environment near the counting room
- Other radionuclides in the source that is being counted.

Specifications in purchasing detectors and tests of the received instrument systems can limit the contamination in detector materials and associated nearby components such as the sample holder, preamplifier, and radiation shield to acceptable levels. The background due to cross-contamination from other samples and placing highly radioactive sources near the detector can be prevented by careful laboratory practices. For example, solid and liquid radionuclide sources should be enclosed as thoroughly as is feasible when brought to the counting room.

A reasonable criterion for controlling radiation background from the cited origins is to limit the fraction that any component contributes to the total background. Controls should be more restrictive if great efforts are being made to reduce the radiation background or if the background interferes with measuring a radionuclide of interest at its specified detection sensitivity.

Understanding, stabilizing, and minimizing the radiation background in a counting system maintain confidence in observed net count rates and the calculated detection sensitivity. Stability is important because the net count rate of a sample is calculated as the difference between two measurements—the sample gross count rate and the background count rate—that are performed with the detector at different times. The implicit assumption is that the background has not changed between the measurements. Fluctuations in the background count rate have an especially important impact on the uncertainty of the measured sample content when the background count rate is equal to or larger than the net count rate.

Of primary importance in achieving a low and stable background is locating the counting room in an environment of relatively low terrestrial radionuclide background. Direct radiation from terrestrial radionuclides can be from the ground and from structural materials such as concrete, brick, glass, and wallboard. In addition, airborne radon and its particulate progeny emanate from the ground and walls. Radon and its progeny are major causes of radiation background fluctuation because their indoor concentrations depend on ambient conditions such as soil moisture, air pressure, and inversion conditions. The progeny are sources of both external radiation and surface contamination.

Radiation from radon and its progeny has been minimized by site selection and controlled by sealing floor and walls to reduce emanation and by passing ventilating air through a sorbent that removes the gas with its particulate progeny. The air volume within the detector shield can be kept to a minimum or filled with aged, radon-free, air. Samples are stored within containers to avoid deposition of radon progeny immediately before counting and to await the decay of progeny deposited earlier, in view of the short half-lives of ^{214}Pb ($t_{1/2} = 27$ min) and ^{214}Bi ($t_{1/2} = 20$ min). Air filtration can also reduce the concentration of other particulate airborne radionuclides, such as fission products from nuclear tests or incidents.

The counting room should be located at a distance from sources of penetrating radiation, i.e., gamma rays and neutrons. Avoidance of sources that contribute to background fluctuation by being turned off and on periodically is especially desirable.

Shielding certain detectors contributes significantly to reducing the radiation background. Alpha-particle detectors need no shielding because the detector walls are sufficiently thick to stop external alpha particles. Beta-particle detector containers usually are shielded by 5-cm-thick lead bricks that are more than enough to stop all external beta particles and also stop a large fraction of low-energy gamma rays. Gamma-ray detectors are shielded with 15-cm-thick lead shields or the equivalent in steel.

The low-energy components of cosmic rays can be controlled by shielding with heavy metal (Pb, stainless steel, and in some cases Hg). The energetic ("hard") component of cosmic radiation is not substantially reduced by shielding with 15 cm of lead or the equivalent in other metals. Cosmic-ray components are further reduced with thick walls of low-Z material such as water or borated polyethylene or by placing the detection rooms far underground.

Typical radiation backgrounds for the detectors discussed here are listed in Table 8.1 as criteria for background control, but experience suggests that the background depends strongly on detector and shielding material, as well as radiation in the immediate environment. In gamma-ray spectral analysis, the background decreases with increasing energy, as shown in Fig. 8.2, with peaks due to gamma rays from radon progeny, ^{40}K, and annihilation radiation.

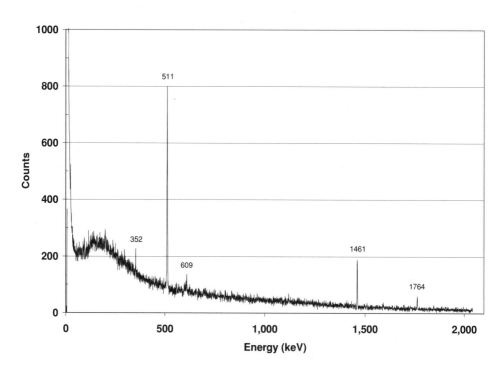

FIGURE 8.2. Gamma ray spectral background for an HpGe detector.

The extremely low background count rates given in Table 8.1 for two types of alpha-particle detectors suggest that few, if any, counts will be accumulated during a background measurement. In that case, statistical considerations in Sections 10.3.1 and 10.4.1 for calculating the uncertainty due to the background and the minimum detectable activity do not apply. To decide whether a measured value near zero is due to the background may depend on the presence of a recognizable peak or on inferences based on experience concerning background fluctuation in the energy region of interest.

Gamma rays from the thermal neutron capture of cosmic ray neutrons within the detector and shielding materials are emitted by cadmium in graded shield liners for gamma-ray spectrometry systems. One of the naturally occurring isotopes of cadmium (^{113}Cd) has an extremely high cross section ($\sim$20,000 barns) for thermal-neutron capture and emits a 558-keV prompt gamma ray. Background due to such prompt gamma rays should be significantly lower than the 511-keV annihilation peak and the 1461-keV ^{40}K peak in the detector background.

The detector, its neighboring structures, and the counted source itself also contribute to the background. Reducing the detector volume can decrease the background count rate, but may also reduce the counting efficiency. This interaction can be evaluated in terms of maximizing the figure of merit ε^2/R_B, where ε is the counting efficiency and R_B is the radiation background count rate. The figure of merit also permits comparison of sensitivity of two detector systems. For example, if ^{89}Sr is counted with an efficiency of 0.96 and a background of 20 c/m in an LS counter and with an efficiency of 0.40 and a background of 1 c/m in an anti-coincidence-shielded gas proportional counter, the respective figures of merit are 0.046 and 0.16. In this comparison, the gas proportional counter is more sensitive when other factors, such as sample volume and counting period, are equal.

Radioactive contaminants in filters, planchets, detectors, and shields may be primordial radionuclides and their progeny in aluminum (e.g., thorium), lead (e.g., uranium progeny), and filters (^{40}K). Materials with low radionuclide content should be selected from carefully screened supplies. Man-made radionuclides have contaminated steel and other metals during processing (from airborne fallout radionuclides) or reprocessing (from radioactive tracers or medical irradiation sources such as ^{60}Co, ^{137}Cs, or ^{226}Ra).

A common and effective way to reduce the external background is use of an anticoincidence system (see Figs. 8.3 and 8.4). A separate detector shields the sample detector. When the shield counter detects a pulse, the electronics system momentarily turns off the detector to reduce the background count rate.

Research that requires extreme gamma-ray detection sensitivity, such as neutrino and double beta decay measurements, has led to enormous reduction in gamma-ray spectrometer background. Contributing to this improvement are extremely low radionuclide contents in materials for the germanium detector, container and shield, and reduction of the cosmic-ray background by underground location, electronic anticoincidence, muon shielding, and shield material with low photon production in muon interactions (Heusser 2003).

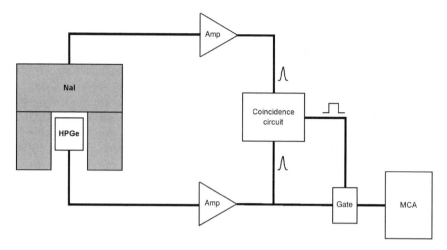

FIGURE 8.3. Schematic drawing of an anticoincidence system for Compton suppression.

The radionuclide that is being counted is another source of background, both by contaminating the detector and its immediate environment, and by emitting other radiations that interfere with counting the radionuclide of interest.

A specific cause of potential contamination is counting radionuclides that emit alpha particles and decay to progeny radionuclides; such sources cannot be covered, and the progeny may leave the source due to recoil (see Section 2.2.1). One possible means of control is to keep the recoiling radionuclide on the source mount by an applied negative charge.

A background of sorts in gamma-ray spectral analysis is the Compton continuum (see Section 8.3.4), which may underlie characteristic photon peaks, as shown in Fig. 2.14. The continuum is due to scattering interactions in the detector and the subsequent exit of scattered gamma rays. This situation is recognized in the description at purchase of a germanium detector by its peak-to-Compton ratio. Larger detectors have better, higher, ratios because the increase in multiple Compton scattering with a final photoelectric interaction results in fewer counts in the continuum and more in the peak. A second detector surrounding the primary detector and operated in the anticoincidence mode (see above) can count photons scattered from the primary detector and subtract them from the continuum.

Photoelectric interaction in materials surrounding the detector can result in characteristic X rays in the lower energy region of the gamma-ray spectrum. For example, the K_α (72 keV) and K_β (85 keV) X rays are almost always part of the background in a spectrum of a detector shielded with lead. Commercially available lead shields for gamma-ray-spectrometer detectors are lined with thin cadmium and copper layers to attenuate these lead X rays.

Beta particles that pass through the can that holds the gamma-ray detector produce another sample-related background. If the detector is sufficiently shielded to stop all beta particles, it will still record the lower-intensity Bremsstrahlung produced in the shield, as described in Section 2.4.3.

The reverse—gamma-ray background in beta-particle detectors—also occurs due to electron-producing interactions of gamma rays in the walls of the gas-filled proportional and G-M detectors. The magnitude of the detection efficiency for energetic gamma rays, i.e., above 0.1 MeV, typically is about 1% of that for beta particles (Knoll 1989).

An effective but expensive technique for monitoring background radiation is operation of one of each set of detectors in the background measurement mode at all times to alert the operator to fluctuations due to line noise, radon in air, or external radiation sources. Another approach is to monitor the count rates recorded in anticoincidence detectors for identifying fluctuations in the external background. Detector contamination between regular background tests may be inferred from a sudden increase to unusually high sample count rates.

8.3. Detectors

8.3.1. Gas-filled Detectors

Different types of radiation detectors that use a counting gas are based on a similar design (see Fig. 2.10). Detectors are either sealed or with a continuous flow of gas. A high voltage applied across the volume of gas in a conducting container forms an electrical field between two electrodes—the detector wall and a central electrode. As radiation passes through this electrical field with enough energy (~25–35 eV per ion pair) to ionize the counting gas, a flow of electrons to the wire anode creates a current pulse between the electrodes. Based upon detector design, this electrical signal is measured as a current, accumulated charge, or pulse which can be related to the incident radiation. Examples of operating parameters for the three basic types of gas-filled detectors are listed in Table 8.4. As described in Section 2.4.1, detector design and operating voltage determine detection performance.

The ionization chamber generally integrates the current over a brief period and measures it with a sensitive electrometer. An early version was an electroscope

TABLE 8.4. Characteristics of common gaseous detectors

	Ionization chamber	Proportional counter	G-M counter
Gas	Air or other gas	Argon	Helium or argon
Quench gas	—	Methane (10%)	Halogen organic (3%)
Dimensions (cm)	—	5.7 dia. × 1.0	3.5 dia. × 10
Window diameter (cm)	—	5.7	3.0
Window area density (mg/cm^2)	—	0.08 or 0.5	0.50 (β) or shield (γ)
Applied volts	approx. 100	800 (α) or 1600 ($\alpha + \beta$)	1000–2000
Amplification	Yes	Pre-amp + linear amp	No
Measurement	Electrometer	Pulse count	Pulse count

charged by friction, in which the current was measured by the rate of separation of two light metal foils that was a measure of their accumulation of electric charge.

The relatively large pulses from alpha particles are measured with a Frisch grid chamber (Knoll 1989) with a resolution of about 40 keV. Its advantage is the capability to measure and perform spectral analysis for samples with relatively large areas, e.g., 500 cm^2.

A pressurized and sealed ionization chamber is used as calibrator for individual radionuclides that emit gamma rays. Their activity is measured in terms of the radiation dose, as discussed in Section 8.3.5.

The G-M counter is a simple and relatively inexpensive gas-filled tube with a count-rate meter; an amplifier may also be present. The G-M detector counts alpha particles, beta particles, and gamma rays with a very thin window, counts beta particles and gamma rays with a thicker window, and counts gamma rays only with a thick shield. Alpha and beta particles interact in the gas; gamma rays interact mostly in the walls, from which electrons enter the gas. The intrinsic efficiency for counting gamma rays relative to beta particles depends on the amount and type of solids surrounding the detection gas.

The fill gas differs among types of detector. Almost any gas is suitable for an ionization chamber. The counting gas for a proportional counter must have an added organic quenching agent. The most commonly used gas is a mixture of argon fill gas (90%) and methane quench gas (10%) readily available as P-10 gas. In the G-M counter, if the tubes are sealed, the quenching agent usually is a small amount of Br_2 or Cl_2; if the detectors are flow type, the quenching agent may be organic compounds such as ethyl alcohol and ethyl formate. The fill gas is usually a noble gas; for example, Q-gas is 97% He and 3% an organic compound.

The most common detector is the gas-flow proportional counter, although ionization chambers and G-M counters can be usefully applied in the laboratory. Several types of gas-flow proportional counters are used in modern counting laboratories.

The first type is a manual detector, either with a thin (typically 80 $\mu g/cm^2$) window beneath which the sample is placed, or without a window but with a sample tray that slides under the detector and becomes the detector bottom. Use of the internal proportional counter eliminates external attenuation of low-energy beta particles but introducing the source into the detector can cause contamination or reduce the potential difference to a value below the applied voltage. The detection volume may be cylindrical (typically about 1 cm high and 3 cm in radius), or may be a hemisphere in the classical 2π groportional counter. For either a thin window or no window, continuous gas flow is maintained. A thicker window (typically 0.5 mg/cm^2) can seal the counting gas in the detector.

The second commonly available system is a detector coupled to an automatic sample changer. In this arrangement, planchets or disks from a stack of 30 or 50 are fed individually by tray to the location beneath the detector window. The controlling computer sets the counting time of typically 10 to 100 min, moves the tray, and records the time and accumulated counts. The computer also controls whether alpha or beta particles or both are measured and whether samples are to

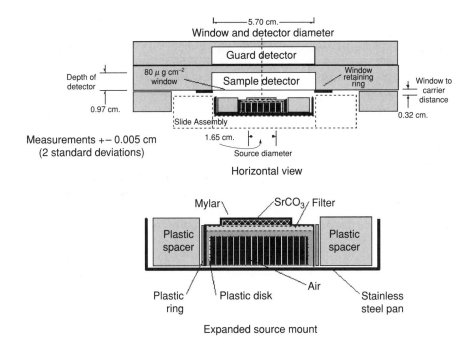

FIGURE 8.4. Configuration of the source and detector for the Tennelec LB5100, an end-window, gas-flow, anticoincidence proportional counter (from Nichols 2006).

be counted repeatedly. A typical pancake-shaped detector is shown in Fig. 8.4. It includes a guard detector connected in anticoincidence to reduce the external radiation background. The sample is inserted immediately beneath the detector. A ring-and-disk sample is shown, but a planchet may be substituted by removing the spacer.

A third, usually more expensive, system has multiple detectors to count multiple samples simultaneously. Configurations available commercially are modular with groups of four detectors arranged over a slide-out drawer. Modules can be added for 8, 16, or even 64 detectors. This counting system is useful for low-level samples that require long counting time (i.e., 24–96 h). A high sample load can justify the cost.

Two internal counters facing each other and electronically combined to produce single pulses constitute a 4π detector. A thin source deposited on a thin backing is placed between the two detectors. Measurements at near 100% counting efficiency approach the absolute disintegration rate of a radionuclide that emits beta particles.

A gas-flow proportional counter is operated at a low bias voltage (~600–800 V with P-10 gas) if alpha particles are to be detected in the "alpha only" mode. By increasing the bias voltage to ~1500 V with P-10 gas, the counter will respond to both alpha and beta particles. Operation at this higher voltage is referred to as "simultaneous" or "$\alpha + \beta$" mode. These voltages can be decreased by increasing the linear amplifier gain and *vice versa*. The optimal operating voltage for each mode is selected after finding the plateau by plotting count rate *vs.* bias voltage with an alpha-particle and a beta-particle source (see Section 2.4.1). The pulses require a preamplifier and an amplifier for counting. Pulses are proportional to energy deposition, hence are hundreds of times larger for alpha particles than for beta particles.

Plateau curves are generated with a pure alpha-particle emitter (^{238}Pu, ^{244}Cm) and a pure beta-particle emitter (^{63}Ni, ^{90}Sr/^{90}Y, ^{99}Tc) unless a combination of radionuclides is selected to represent an actual sample. The analyst chooses, on the basis of the two curves, a bias voltage for "alpha only" mode that is below the point where beta particles just begin to be detected (see Fig. 2.11). At this voltage, and also by application of pulse-shape discrimination, beta-particle "cross talk" is minimized. The voltage for $\alpha + \beta$ counting is selected near the midpoint of the beta-particle plateau. This plateau is not quite horizontal because an increase in the applied voltage adds a few counts that previously had just been excluded by the lower-energy discriminator.

The next step is to determine the counting efficiency for alpha- and beta-particle measurements. The alpha-particle counting efficiency for thin samples with a proportional counter is almost constant over the energy range of 4–9 MeV. This energy range includes the naturally occurring isotopes of thorium, uranium and their progeny, and also the actinides in global fallout from nuclear weapon testing. A few naturally occurring rare earth elements (e.g., ^{147}Sm, ^{144}Nd, and ^{152}Gd) emit alpha particles in the 2–3 MeV range, but these radionuclides have very low decay rates associated with their long half-lives and are not of interest at most radioanalytical chemistry laboratories.

Alpha particles that enter the gas proportional detector deposit energies of the order of 1 MeV, far more energy than is needed for detection, hence alpha-particle counting efficiency is a function of the source–detector geometry minus losses from alpha-particle attenuation in the sample mass, intervening air, and detector window. The ratio of count rate to disintegration rate (cpm/dpm) for alpha particles from a very thin source mounted on stainless steel and counted within a windowless 2π gas-flow proportional counter is ~51.5%. The additional 1.5% over the expected 50% from the 2π geometry is due to backscattering of the alpha particles from a stainless steel source mount.

End-window proportional counters with thin windows that separate sample from detector have the alpha-particle counting efficiency shown in Table 8.1 for thin samples on planchets. The lesser counter efficiency relative to the internal detector is due to less than 2π geometry and attenuation in sample, air space, and window.

Beta particles deposit typically about 2 keV of energy, in the range of 0.5–10 keV, in the gas proportional counter. Pulse height discrimination readily distinguishes these beta-particle pulses from the much more energetic ones generated by alpha particles to provide a "beta only" mode. Cross-talk with alpha particles can occur because the various angles at which the particles enter the detector lead to considerably lower energy deposition for a small fraction of the particles. Beta particles emitted with near-zero energy are not detected when their energy is deposited in the sample, air, and detector window.

Unlike alpha particles, a significant fraction of beta particles that enter the detector has been backscattered. Due to the energy distribution of beta particles, the overall beta-particle counting efficiency for a proportional counter depends on the fraction of low-energy beta particles absorbed in the sample, air, and window. To quantify the activity for ^{63}Ni (66.9 keV) or ^{99}Tc (292 keV), for example, a separate standard is needed for each radionuclide. The beta counting efficiency is in the 65–75% range for radionuclides with maximum energies of 0.3–3.5 MeV on steel planchets in a windowless gas-flow proportional counter. The efficiency is less in an end-window counter (see Table 8.1), and can reach zero for beta-particle groups with maximum energy below about 30 keV.

Other versions of the gas proportional counter system permit measurements of special sources or radionuclides. Radioactive gas samples can be counted in a detector that has valves for drawing a vacuum or admitting gases. A measured volume of the gas sample is mixed with counter gas and counted in the detector (ICRU 1972). Alternatively, a flowing gas stream can be mixed with flowing counting gas. Radioactive liquid samples can be placed in a thin-walled container beneath an end-window counter, or a flowing stream can be passed through this container.

Modified versions of the gas proportional counter are used for spectral analysis of X rays in the energy region of a few kiloelectron volts. Detectors are constructed with a greater depth and a window with a low-Z material such as Be, and with a high-Z counting gas such as Xe at a pressure of several atmospheres. The resolution is about 0.6 keV at 5 keV (Knoll 1989).

8.3.2. Liquid Scintillation Counters

Counting with LS systems is the most widely used technique for measuring radionuclides that emit only low-energy beta particles. LS counting can be used for measuring any radionuclide in any liquid sample, but it is the preferred method for low-energy beta emitters, especially ^{3}H, and volatile radionuclides, e.g., ^{99}Tc as TcO_4^-, that could be lost during evaporation or drying before counting in a gas-flow proportional detector. Examples of radionuclides routinely measured by LS are listed in Table 8.5.

TABLE 8.5. Low energy and potentially volatile b-particle-emitting radionuclides

Radionuclide	E_{max} (keV)	$t_{1/2}$	Volatile Forms
^{3}H	18.6	12.3 years	H_2O, H_2
^{14}C	156.5	5730 years	CO, CH_4, CO_2
^{32}P	1709.0	14.3 days	—
^{35}S	167.4	87.2 days	SO_3, H_2S
^{36}Cl	709.0	301,000 years	Cl_2, HCl
^{45}Ca	258.0	163 days	—
^{63}Ni	66.9	100 years	$NiCl_2$, $Ni(NO_3)_2$
^{99}Tc	292.0	213,000 years	TcO_4^-

Traditional LS applications in radioanalytical chemistry are minor compared to the wide use of LS techniques in the life sciences (biological and medical) for measuring labeled proteins, genetic materials (RNA and DNA), and other types of biological tracer work. Even nonvolatile forms of ^{32}P and ^{45}Ca that are frequently used to label biological and medical samples are counted by LS, as indicated by Table 8.5. Cherenkov radiation from ^{32}P beta particles may be counted directly in water or other biological fluids without adding scintillation cocktail. The LS technique is also useful in counting radionuclides smear samples collected for health physics (radiation protection) applications.

The basic design of a commercial LS counter was developed more than fifty years ago. A charged particle that moves through the mix of sample and scintillation cocktail transfers energy in ~5 nanoseconds with a yield of ~10 photons per keV to generate many photons at energies of a few electron volts that distribute isotropically. The set of simultaneously generated photons per charged particle creates a pulse in each of the two photomultiplier tubes (PMTs) that face the sample vial. The signal from each PMT is fed into a coincidence circuit which produces an output pulse only when the two PMT signals occur within the resolving time of the circuit (~20 nanoseconds). The coincidence pulse drives a gated circuit that passes only a valid signal and excludes randomly produced electronic noise from the PMTs.

A significant improvement to LS instrumentation was summing the pulses from each PMT for an output proportional to the total intensity of the scintillation event. This design for an LS system with pulse coincidence detection and summation (Fig. 8.5) has remained basically unchanged over the last forty years. More recently, instrument vendors have replaced a traditional three-channel system with a multichannel analyzer (MCA) and software for spectral analysis.

The spectrometer permits confirmation of the observed radionuclide by its spectrum, within the limitations of spectral distortion by energy losses and the radiation background associated with the measurement. It also permits setting the lower and upper energy discriminators for optimum ε^2/R_B values for each radionuclide of interest. The energy ranges beyond the region of interest usually are measured at the same time to confirm the absence of contaminant radionuclides. Two or three radionuclides may be detected simultaneously by such discriminator control

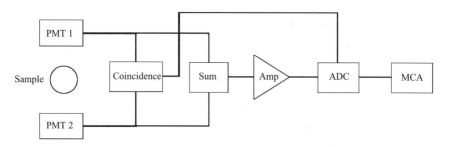

FIGURE 8.5. Functional diagram of a modern liquid scintillation counter.

if their maximum energies are considerably different so that their beta-particle groups can be clearly distinguished.

Most commercially available LS systems have automatic sample changers to measure multiple sources. A computer controls the system to specify counting schedule, counting period, energy regions, and data recording, calculations, and storage. Concentrations for specified radionuclides, the counting uncertainty, and corrections for luminescence and quenching are calculated as discussed below.

For LS counting, a radioactive sample is dispersed in a scintillation counting solution referred to as a *scintillation cocktail* (see Section 2.4.2). In the radioanalytical chemistry laboratory, typically 10 ml of aqueous sample is shaken with 10 ml of cocktail in a 20-ml glass or plastic counting vial. The solvent absorbs the energy deposited by the charged particle and transfers it to the scintillator compound. Additional considerations for the solvent are high transmission of the photons emitted by scintillator compounds and a low ^{14}C content (to ensure low background for counting).

Solvent molecules that can effectively transfer the nuclear radiation energy have a high density of π electrons; they include aromatic compounds such as toluene, xylene, and pseudocumene (1,2,4-trimethylbenzene). These solvents were used for many years, but now have been replaced in most commercial scintillation cocktails with more environmentally friendly solvents such as diisopropylnaphthalene (DIN), dodecylbenzene (LAB), and phenylxylylethane (PXE).

The primary scintillator compound becomes excited (electrons are elevated to a higher energy state) when the energy deposited by radiation is transferred from the solvent. The excited scintillator compound promptly returns to the ground state while releasing energy in the form of a photon that can be detected by a PMT. One of the more widely used scintillator compounds is PPO (2,5-diphenyloxazole). Characteristics that define an effective scintillator compound include a fluorescence wavelength (the energy of the emitted photon) that is effectively detected by the PMT, a brief decay time (interval for return to ground state), and a good quantum yield (number of photons emitted per number of molecules excited).

The peak wavelength emitted by the primary scintillator should be within the sensitive range (300–425 nm, or 4–3 eV) of the PMT used in the LS system. Some older scintillation cocktail recipes used a secondary scintillator to shift the initial photon emitted to a slightly longer wavelength (400–420 nm range). Most modern LS systems do not require this secondary scintillator because new PMTs are sensitive to the photons emitted by the primary scintillator.

Because the solvents used for most scintillation cocktails are not good solvents for aqueous samples or ionic species, detergents or surfactants are included in the cocktail to help emulsify aqueous samples. The formulation of a scintillation cocktail can be complex, but most laboratories purchase prepared scintillation cocktails.

The counting efficiency for charged particles with an LS counter can approach 4π geometry, as shown in Table 8.1. It decreases to about 25% for ^{3}H because of low beta-particle energies.

Counting efficiency for LS is reduced or degraded by several factors, notably chemical and color quenching. Chemical quenching occurs when an impurity in the

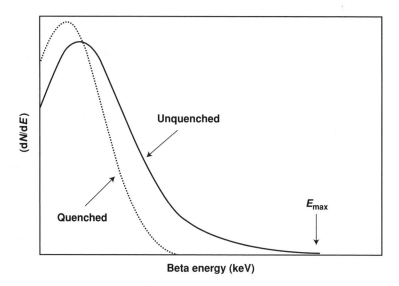

FIGURE 8.6. Example of quenching in a beta-particle energy deposition spectrum in an LS counter.

sample or scintillation cocktail interferes with energy transfer between the solvent and the scintillator. Color quenching is the attenuation of scintillator photons in the solution due to light absorption. The difference between the two types of quenching lies in the type of energy absorbed; chemical quenching absorbs deposited charged-particle energy while color quenching absorbs emitted-photon energy.

Loss of energy, i.e., a reduction in the pulse height recorded at the PMT, is the result in both cases as quenching shifts the energy spectrum to lower energies (see Fig. 8.6). Any pulses shifted to below the low-energy discrimination are lost and reduce the counting efficiency. Factors that can influence the degree of quenching include the sample matrix, sample preparation, and choice of scintillation cocktails. Several techniques have been developed to correct the effect of quenching.

The traditional method for quench correction is to add an internal standard to each sample to determine the counting efficiency for each sample matrix. This method continues to be one of the most accurate but is labor intensive and expensive. A set of samples is counted without the internal standard. A duplicate set, with a small volume (e.g., 0.1 ml or less), of known activity of the internal standard spiking solution is added to each sample, and is then counted. The counting efficiency for each sample is as follows:

$$\text{Efficiency} = (\text{cpm}_{s+i} - \text{cpm}_s)/\text{dpm}_i \qquad (8.7)$$

The values of cpm_s and cpm_{s+i} are the net count rates for each sample without and with the added internal standard, respectively, and dpm_i is the activity of the internal standard added. The activity, dpm_s, for each sample is calculated by dividing the cpm_s for each sample by the efficiency for each sample. This method assumes that the chemical form of the radionuclide added as the internal standard is the same as in the sample and that the added internal standard spiking solution does not affect sample quenching.

Instrument vendors developed various techniques to measure the degree of quenching in each sample with quench correction curves. The quench correction curve is prepared from a set of increasingly quenched standards with the same level of activity. Then a plot of efficiency *vs.* the degree of quenching is prepared for each radionuclide of interest. The success of quench correction curves depends on skill in identifying the quenching agent and measuring the degree of quenching, which is typically represented by a quench indicating parameter (QIP). Techniques developed to define various QIPs are discussed in LS instrument manuals, vendor literature, and other LS application descriptions (Horrocks 1974). The main types of QIP used for quench correction curves are either based on the sample spectrum or an external standard spectrum. The QIPs based on sample spectra are derived from mathematical transformations of spectral data that provide a measurement related to either the average or maximum spectral energy.

An external standard spectrum to determine the QIP is popular with users and instrument vendors. The external gamma-ray source (e.g., ^{137}Cs, ^{133}Ba, or ^{152}Eu) that is part of the detector system induces a Compton-electron spectrum in the scintillation cocktail. Each sample is automatically counted with and without the external standard. The Compton-electron spectrum produced in each sample vial is applied with mathematical techniques to derive a QIP for a quench correction curve.

One method offered by instrument vendors to determine the activity of a radionuclide is *efficiency tracing* (ET) (Takiue and Ishikawa 1978) for most beta-particle radionuclides except tritium. The ET method uses a single unquenched ^{14}C standard to calibrate the LS system by collecting the spectrum and determining the efficiency in several energy regions. Then, an unknown sample is counted in the same energy regions. The cpm measured for the unknown in each counting region is plotted *vs.* the efficiency determined for the ^{14}C standard in the same regions. The generated curve is extrapolated to 100% efficiency to obtain the dpm for the unknown activity.

As an example, the information in Table 8.6 summarizes the data collected for a ^{14}C standard (134,900 dpm) in six energy regions of the beta-particle spectrum. The table also includes the data from a simulated unknown sample with ^{90}Sr and ^{90}Y in secular equilibrium with a total beta-particle-activity of 1037 dpm.

TABLE 8.6. Counting data collected to demonstrate ET method.

Energy Region (keV)	^{14}C (cpm)	^{14}C Efficiency (%)	^{90}Sr + ^{90}Y (cpm)
100–2000	13,550	10.04	777
80–2000	26,788	19.86	825
60–2000	45,897	34.02	877
40–2000	71,228	52.80	918
20–2000	100,951	74.83	971
0–2000	130,157	96.48	1032

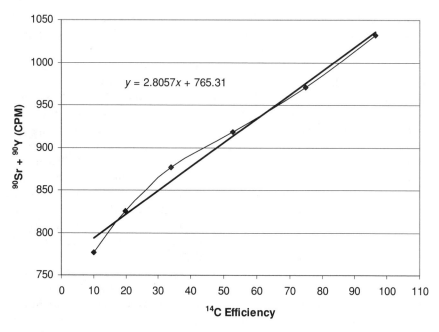

FIGURE 8.7. Example of efficiency tracing for direct measurement of activity.

The cpm of the "unknown" is plotted *vs*. the ^{14}C efficiency from each energy region as shown in Fig. 8.7. The equation in Fig. 8.7 is the best fit to the data as determined by linear regression. This linear equation is then used to extrapolate the count rate of the "unknown" at 100% [14]C efficiency, which is equivalent to the activity (dpm) for the ^{90}Sr $+ ^{90}$Y in the sample. The calculated activity for this example of 1045 dpm is within −0.8% of the known value.

Chemiluminescence and photoluminescence are other forms of interference that can reduce the accuracy of LS techniques. "Chemiluminescence" describes the emission within the scintillation cocktail of photons that result from a chemical reaction; common initiators are samples with an alkaline pH or the presence of peroxides. Photoluminescence can occur when the scintillation cocktail is exposed to ultraviolet light. Some substances in the cocktail, notably the scintillator, are excited and then emit light when the species return to ground state. The effect of photoluminescence is reduced in LS systems by decay when the sample train is held in a dark environment for a few minutes prior to counting. On the other hand, chemiluminescence may have a slow decay rate that requires a change in sample preparation to eliminate the chemical that causes it. Some LS systems identify luminescence by pulse shape and indicate its relative extent.

8.3.3. Alpha-Particle Spectrometers

Most modern alpha-particle spectral analysis is performed with a semiconductor such as a surface barrier detector with a very thin dead region in front of the active

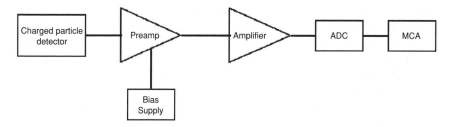

FIGURE 8.8. Block diagram of an alpha-particle spectrometer system.

(depletion) zone of the detector. Only a very thin dead region is acceptable because of the low penetrating power of alpha particles. The sample is counted in a small vacuum chamber that provides a nearly air-free path between the sample support plate and the detector. Without the vacuum, the alpha-particle spectrum is degraded by interactions with air molecules.

Commercially available systems integrate in a single unit the high-voltage power supply, amplifier, analog-to-digital converter (ADC) and multichannel analyzer with the counting chamber, detector, and preamplifier, as shown in Fig. 8.8.

In general, larger detectors have worse resolution. A good compromise for most laboratory applications is a Si diode detector with an active area of 3–5 cm^2, a depletion zone of about 0.02 cm, and a resolution of 15–20 keV. This resolution is sufficient for baseline separation of ^{239}Pu (5.15 MeV) from ^{238}Pu (5.50 MeV) despite complications due to multiple alpha-particle groups per radionuclide. The counting efficiency for surface barrier detectors depends on the geometrical factor and can be expected to be constant at fixed geometry for very thin sources that emit alpha particles in the typical energy range of 4–9 MeV. As described Section 7.2.3, preparation of a uniform and very thin source, usually by electrodeposition, is essential for obtaining good peak resolution. This requirement may be relaxed to the extent that low-energy tailing at a peak does not interfere with calculating the activity, identifying the peak energy for the alpha particle, or measuring lower-energy alpha particles.

Spectrometer energy is calibrated for these energies with available thin sources of alpha particles at energies known within 0.1 keV. The energy peaks in an acceptable system are almost symmetrical. Many radionuclides emit additional alpha-particle groups at energies slightly lower by 30–50 keV than the main characteristic peak. These are not clearly resolved from the characteristic peak, but do not interfere seriously with calibration or detection because they are at much lesser intensity.

Detectors within this range of dimension and resolution are suitable for most environmental radioanalytical chemistry applications. The alpha-particle emitters of interest include isotopes of thorium, uranium, neptunium, plutonium, americium, and sometimes curium (^{244}Cm at 5.80 MeV). Most of the major alpha particles on this list can be resolved. Exceptions are ^{233}U from ^{234}U and ^{239}Pu from ^{240}Pu, which require mass spectrometric separation to resolve them (see Table 6.3). A pair that can be distinguished only after chemical separation is ^{238}Pu and ^{241}Am at 5.50 MeV.

Computation of activity is simplified because the same counting efficiency applies to all alpha particles in the usual energy range. The activity of a radionuclide is calculated simply from the activity of the added tracer multiplied by the net accumulated counts for the peak of the radionuclide of interest and divided by the net counts for the tracer peak. Separate values of the counting efficiency and the yield are not needed, although they may be of interest to monitor tracer activity and process yield, respectively.

Alpha particles are also measured by spectral analysis in Frisch grid chambers (see Section 8.3.1) and specially designed LS counting systems (see Section 7.3.3). Although solid-state detectors are most commonly used, the other detection systems, if available, can simplify the processing of certain samples.

8.3.4. Gamma-Ray Spectrometers

That nuclear radiation stimulates light emission in certain substances was discovered at the beginning of the nuclear age. A flash of light soon was associated with passage of single radiation. Some materials were found to be more sensitive and in some forms to be particularly responsive to certain radiations. As technology improved, these scintillation materials were coupled to a PMT to record the flashes of light. The PMT detects the photons that constitute the flash of light from the scintillator and converts each set of photons to an electron at the photocathode. The number of electrons is then amplified for a factor of $10^7 -10^{10}$ by being accelerated through a series of dynodes. Commercial PMTs are available that have appropriate dimensions, provide suitable amplification, optimize responses at a range of photon energy, and minimize thermionic noise.

The most common scintillation crystals used in these systems were sodium iodide doped with thallium, referred to as NaI(Tl) detectors (see Section 2.4.3). The advantages of the NaI(Tl) crystal for gamma-ray detection are high detection efficiency and ability to machine the detectors in various dimensions and shapes. A common configuration is a cylinder sealed in an aluminum can, where the sample is placed on one end while the other end is coupled by light pipe to a PMT. Some cylinders are manufactured with a central well into which a small sample may be placed for better counting efficiency. Cylindrical crystal dimensions are reported by diameter and height; the $3'' \times 3''$ crystal used as comparison standard for reporting relative counting efficiency in terms of ^{60}Co has a diameter of 7.6 cm and a height of 7.6 cm.

In the late 1940s, the pulses generated in a PMT were amplified and sorted by pulse height with a single-channel analyzer that was moved incrementally through the entire gamma-ray energy region. This analysis depicted a spectrum such as the one in Fig. 2.14 for each radionuclide that emits a single gamma ray, or a multiple gamma-ray spectrum for two or more gamma rays. In time, ease of analysis was improved, first by placing 10 or 20 channels in parallel and ultimately with systems that had 1024 channels or more, by multiples of 2^x.

Lithium-drifted semiconductor detectors were the next generation of detectors for gamma-ray spectrometers. Based on lithium-drifted germanium and silicon, they were designated as Ge(Li) and Si(Li), respectively. The Ge(Li) detector was applied to the entire gamma-ray energy range while the Si(Li) detectors were applied for low-energy gamma rays and X rays. The Si(Li) detectors, because of their lower Z value and smaller size, were more suitable for lower gamma-ray energies and continue to be good choices for X-ray spectrometers in modern counting laboratories.

The lithium drifted detectors provided greater resolution but at a higher cost and lesser counting efficiency relative to the NaI(Tl) detectors. Additionally, the Ge(Li) detector must be kept at liquid nitrogen temperature at all times to prevent lithium ions from drifting in the germanium to destroy the detector.

The intrinsic HPGe detector now has replaced the Ge(Li) detector. This detector is made of sufficiently pure germanium that does not require lithium drifting for its radiation-sensitive properties. The HPGe is not subject to adverse effects if allowed to come to room temperature, but must be cooled to liquid-nitrogen temperature during use to achieve good resolution.

The most common type of detector in current use for gamma-ray spectrometry is the coaxial HPGe detector in an aluminum can. These detectors are available with either a p-type or n-type crystal. The efficiency of a p-type detector typically decreases quickly below $\sim$100 keV whereas the n-type detectors can be used at energies as low as 3 keV with a beryllium window.

Selection of detector type and size should be controlled by its application, both immediate and long-term. In addition to cost, important performance characteristics of a semiconductor detector for gamma-ray spectrometry are the resolution, efficiency and peak-to-Compton ratio.

Resolution (see Section 2.5.4) is reported as the peak full width at half maximum height (FWHM) for solid state detectors, and as FWHM divided by the peak centroid energy for NaI(Tl) detectors. The FWHM increases slowly with peak energy.

A standard method is described in IEEE standard test procedures (IEEE 1996) to define the relative detection efficiency for a coaxial detector. The relative efficiency ε_R for a coaxial detector is defined as follows:

$$\varepsilon_R = \frac{\varepsilon_{HPGe}}{\varepsilon_{NaI(Tl)}} \tag{8.8}$$

In Eq. (8.8),

ε_{HPGe} = absolute efficiency at 1332 keV for a coaxial detector measured 25 cm from the source.

$\varepsilon_{NaI(Tl)}$ = absolute efficiency at 1332 keV for a 3″ × 3″ NaI(Tl) scintillation detector measured 25 cm from the source.

The absolute efficiency in this calculation is the ratio of the net number of counts measured in the 1332-keV full-energy peak divided by the number of gamma rays emitted by the ^{60}Co source at this energy during the same time interval.

TABLE 8.7. Resolution vs. efficiency

Energy (keV)		NaI(Tl) (30 mm × 30 mm)		HPGe (~30% Relative Eff.)	
		ε_{abs} (%)	FWHM (keV)	ε_{abs} (%)	FWHM (keV)
60	(^{241}Am)	50	8	22.4	1.2
			(14%)		
662	(^{137}Cs)	32	50	5.0	1.5
			(7.5%)		
1332	(^{60}Co)	25	95	2.5	1.9
			(7.1%)		

Application of this relation is shown in Table 8.7 for comparing the efficiency and the resolution of two types of detectors with different dimensions. The comparison has been extended to three photon energies to demonstrate that comparison at 1332 keV may not be sufficient for the purposes of the user because both relative efficiency and resolution change with energy. The comparison for practical application also depends on sample dimensions and source-to-detector distance. Table 8.7 data support the observations of better resolution but worse efficiency in the HPGe detector than the NaI(Tl) detector.

The peak-to-Compton ratio is an important indication of detector performance for measuring lower-energy gamma rays in the presence of higher energies. The Compton continuum that is from 0 to ~200 keV below the full-energy peak (see Eq. 2.16) of ^{60}Co increases the spectral background across the entire low energy region below the Compton edges. Most detector manufacturers report the peak-to-Compton ratio as the full-energy peak count rate divided by the net count rate for the same number of channels in the middle of the Compton continuum. This ratio typically ranges from 40 to 70 for ^{60}Co measured with coaxial HPGe detectors currently produced. The peak-to-Compton ratio is higher for detectors with better efficiency and resolution.

The spectral response of a detector is more complex than described in Section 2.4.4 because of the bulk of the detector. The observed Compton continuum consists of single plus multiple successive scattering interactions. When such multiple Compton scattering interactions are terminated by a photoelectric interaction, the pulse is added to the full-energy peak. Most of the counts in a full-energy peak for gamma rays above 100 keV are due to such multiple scattering plus a final photoelectric interaction.

Photoelectric interactions are recorded at the full energy peak despite the binding energy loss (see Eq. 2.17) because the emitted X ray that follows electron emission is detected simultaneously with the gamma ray. In small detectors, where the X ray has a high probability of escape, a small second ("escape") peak is seen at the energy predicted by Eq. (2.17), which is below the gamma-ray energy by 28 keV in the NaI(Tl) detector and by 10 keV in the Ge detector. The dual peaks are readily apparent at low gamma-ray energies.

Interaction by pair production results in a spectrum that includes escape peaks at the full energy minus 511 keV and minus 1022 keV, when either one or both of the positron annihilation photons do not interact with the detector. Compton scattering of these photons adds to the continuum.

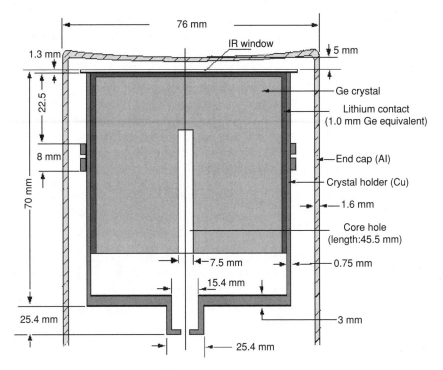

FIGURE 8.9. Detailed description of a Ge detector (from Wang *et al.* 2002).

The previously cited sum peaks occur for two or more coincident gamma rays, for example, at 2505 keV for ^{60}Co. Interactions outside the detector commonly are detected as a peak at 511 keV due to annihilation radiation and at about 200 keV due to Compton scattering at 180°. Gamma rays produced by cosmic-ray interactions in or near the detector are observed as discussed in Section 8.2.2.

All of these interactions by gamma rays with the detector can be modeled by a gamma-ray simulation program such as the previously cited Monte Carlo n-particle code, version 4. Modeling requires precise information on the location and material of the source, detector, and surroundings. A description of the detector such as that shown in Fig. 8.9 must be obtained from the supplier because the detector container is sealed.

Selection of the type of detector for gamma-ray counting should be informed by the analyst's intention for the measurement. For example, NaI(Tl) detectors have excellent efficiency and are useful for counting low-activity samples with few radionuclides that emit gamma rays. Germanium detectors have superior energy resolution, as shown in Fig. 2.14, for identifying numerous radionuclides that emit gamma rays in a sample. Both systems must be evaluated in terms of their background characteristics. Sodium iodide crystals of the same size will have higher backgrounds inherently in spectral analysis because the peaks cover a wider energy range; larger detectors have higher backgrounds because of their size.

Current efforts to select better detectors for spectral analysis have identified several materials, notably cerium-doped lanthanum chloride or bromide for scintillation systems and cadmium zinc telluride for solid-state systems. The energy resolution of the new crystalline scintillators is better than two-fold that of thallium-doped sodium iodide and the detectors are less responsive to temperature change. One drawback is the ^{138}La radionuclide that constitutes 0.09% of lanthanum in nature; moreover, thorough purification from thorium is necessary. The new solid-state material has much better counting efficiency (due to its higher Z) than the same volume of germanium and can be used at room temperature, but resolution comparable to germanium has not yet been achieved.

8.3.5. Other Radiation Detectors

Some other detectors in common use either measure radiation in terms of exposure or by tracks. Among the former are ion chambers, dosimeters of various kinds, and photographic film. Tracks can be observed in other types of photographic film, cloud chambers, and track-etch films.

Radiation exposure is defined as the energy deposited by radiation per unit mass. For a gamma ray emitted by a point source, the radiation exposure rate i is related to the gamma-ray decay fraction f, the gamma-ray energy E (in MeV), the activity of the radionuclide a (in Bq), the linear energy absorption coefficient μ_a, the density of air ρ, and the distance x:

$$I = \frac{(4.7 \times 10^{-15}) f E a \mu_a}{4\pi \rho \, x^2} \tag{8.9}$$

If a radionuclide emits several gamma rays, their values on the right side in Eq. (8.9) are summed. The unit of exposure rate is coulomb/kg/s. Typical exposure rates are in microcoulomb/kg/s (3881 Roentgen equals 1 coulomb/kg$^-$). Thus, the activity of a gamma ray can be calculated from its decay characteristics f and E and a measurement of the exposure rate. Exposure rates are read directly with ionization chambers and exposures, as light emission with thermoluminescent dosimeters and as darkening of developed photographic film. These readings are calibrated by exposure to a radiation standard. A common laboratory application is measurement of single radionuclides that emit gamma rays by insertion into a calibrated ionization chamber (Knoll 1989).

A sensitive measurement of alpha-particle tracks consists of placing a very weak source on suitable photographic emulsion for a long period and then developing the film and counting the number of tracks (Knoll 1989). Although track length depends on the angle at which the alpha particle penetrated the film, tracks of alpha particles that almost parallel the surface can indicate the alpha-particle energy. Tracks of alpha particles or heavier ions can also be counted by etching material, such as certain plastics, with a basic solution because the tracks dissolve faster than the nonirradiated material. These tracks can be viewed with a microscope or can be enlarged by an electrical discharge to view by the lesser magnification of a card reader (Knoll 1989). Cloud chambers provide visual observation of alpha particles and electrons that pass through a gas saturated with a vapor that condenses on the ions formed along their tracks (Ehmann and Vance 1991).

9
Radionuclide Identification

BERND KAHN

9.1. Introduction

Systematic identification of potential radionuclides of interest usually is considered at the beginning of a project, with further needs developing if results are questioned in the course of the project. In practice, one identifies a radionuclide by finding a match of measured decay characteristics to listed values. This comparison may not be a simple matter. The effort entails selecting appropriate radiation detectors, correctly interpreting the resulting data, and being aware of distinctive formation and decay characteristics that can distinguish otherwise similar radionuclides. Correct radionuclide identification can be crucial to planning protective measures, especially in emergency situations, by defining the type of radiation source and its radiological hazard. Discussed here are the information used for radionuclide identification, the sources of this information, and the application of the information in the radioanalytical chemistry laboratory.

A radionuclide is defined by its half-life and type, fraction, and energy of emitted radiation. Also of interest are its atomic number (element) Z, its mass number (isotope) A, whether it exists in an excited or ground state, and any parent–progeny relation, as discussed in Chapter 2. The radioanalytical chemist needs this information initially to plan the analytical program for measuring expected radionuclides appropriately, and later to identify and quantify any unexpected radionuclides. The professional should be familiar with the decay schemes of commonly encountered radionuclides and with sources of information for all decay schemes.

Each radionuclide among the more than one thousand that are known has a unique decay scheme by which it is identified. For this reason, among others, researchers have studied decay schemes over the years and their reported information has been compiled and periodically updated. The compiler surveys the reported information for each radionuclide and attempts to select the most reliable information for constructing a self-consistent decay scheme. The fraction of beta particles that feed an excited state must match the fraction of gamma rays plus conversion electrons emitted by the excited state. The energy difference between any two states must be consistent with the energies of the transition radiations plus the recoil energy of the atom that emitted the radiations.

Environmental Radiation Branch, Georgia Tech Research Institute, Georgia Institute of Technology, Atlanta, GA 30332

Compilations of decay scheme information are issued to maintain this information reasonably current, notably in the Table of Isotopes (Firestone 1996) and in supporting nuclear databases, such as that maintained by the Brookhaven National Laboratory at http://www.nndc.bnl.gov (Jan. 2006). These resources organize information by Z and A. Compilations prepared in the last two decades can be considered reliable except for the few radionuclides that are still being created (see Chapter 16), rare and incompletely described radionuclides, recent improvements in precision, and occasional errors. Such problems may be identified and then resolved by referring to two or more independent sources or to the original decay scheme research cited in the compilation.

Because the compilations are used for various nuclear physics purposes, much of the recently published information is far more extensive than needed by the radioanalytical chemist. Notably, information on numerous radiations emitted by radionuclides at less than 1% wastes the time of the analyst who wishes to identify a radionuclide by one or two major radiations. Detailed information on radiations of minor intensity becomes useful only when these radiations affect measurement results for the characteristic radiations or lead to misidentification.

Simplified decay information has been extracted from detailed compilations for practical use in identifying and measuring radionuclides (Kocher 1977, NCRP 1985b), and also is given in wall charts that usually are arranged by proton *vs.* neutron numbers, as shown in Fig. 16.2. These charts provide a useful overview of stable and radioactive nuclides to indicate patterns of radionuclide formation, decay, and parent–progeny relations. Tabulations of radionuclides ordered by radiation type and energy, separated for longer and shorter half-lives, are given in Appendix D of the Table of Isotopes and other compilation (Wakat 1971, Shleien 1992) for quickly matching measured with known values to identify a radionuclide.

The main tools of the radioanalytical chemist for identifying and quantifying a radionuclide are chemical separation and radiation measurement, especially by spectral analysis. The former separates radionuclides by element, i.e., Z value, and removes potentially confusing contaminants. The latter determines the type of radiation, its energy, and relative intensity at various energies. These measurements are repeated during the period of study to determine the half-life of radionuclides that are sufficiently short-lived to observe a decrease in the count rate.

9.2. Decay-Scheme Knowledge for Planning and Analysis

The radioanalytical chemistry laboratory is designed, staffed, and operated to consider radionuclide characteristics for the following purposes:

- Perform measurements within time periods consistent with half-lives
- Determine radionuclide activity by measuring emitted radiations that have sufficiently large and accurately known decay fractions
- Recognize detection limitations and uncertainty imposed by the measured radiation

- Utilize spectral analysis for radionuclides with appropriate radiations
- Select radiation detectors that optimize detecting the radionuclides of interest
- Chemically separate radionuclides for element identification
- Chemically separate radionuclides for contaminant removal before radiation measurement

Some of the above-listed purposes may be addressed simultaneously.

9.2.1. Half-Life Measurements

The half-life $t_{1/2}$ of a radionuclide is calculated by performing two successive measurements of the purified sample with a time interval sufficient to observe significant decay:

$$t_{1/2} = 0.693(t_2 - t_1)/\ln C_1/C_2 \qquad (9.1)$$

In Eq. (9.1), t is the time of measurements on the first and second occasion, C is the net count rate at those times, ln represents the natural logarithm, and 0.693 is the natural logarithm of 2. The equation is another form of Eq. (2.6a). A more precise determination of the half-life is obtained with more measurements plotted on a graph of ln count rate $vs.$ time of measurement. If several radionuclides are in the counted sample, the curve of best fit is not a straight line, but may be resolved into component straight lines when only a few radionuclides are present (see Fig. 2.1).

The half-life must be known to bring radionuclides with relatively short half-lives to the laboratory promptly before they have decayed below the level needed for precise measurement. The measured activity is then compensated for radioactive decay in terms of the half-life. The accuracy of the reported concentration depends on the reliability of both the half-life and the decay interval values.

Half-life values can be used to distinguish among radionuclides that emit similar radiations. For a sample with a mixture of radionuclides, repeated measurements over intervals similar to the shorter half-life reduce interference by the shorter-lived radionuclide when measuring the longer-lived ones.

Half-lives of relatively long-lived radionuclides may not be measurable if the count rate does not decrease significantly in the time period available for analysis. A count rate that actually increase in this interval indicates the presence of parent–progeny pairs or chains, as discussed in Section 9.3.1.

9.2.2. Type and Energy of Radiation

The radiation type guides selection of the detection system and sample size, which in turn affect the magnitude of the radiation background and the detection limit, as discussed in Section 8.2. The gas-filled proportional and liquid scintillation (LS) counter respond efficiently to alpha and beta particles. Low-energy X rays can also be detected in them, while energetic gamma rays pass through with little interaction. Thin silicon diodes respond efficiently to alpha particles and low-energy beta particles and electrons. A large Ge detector responds efficiently to gamma rays and beta particles.

Application of shielding can make a detector more selective. A relatively thin window or shield prevents alpha particles and low-energy beta particles from entering an end-window proportional counter. A thicker shield prevents beta particles from entering the Ge detector.

Pulse selection or rejection can distinguish by type and energy of radiation and also optimize the ratio of counting efficiency to the background count rate. Alpha and beta particles can be distinguished from each other by pulse height or pulse shape selection in the proportional and LS counters. Two or three beta-particle groups with distinctly different maximum energies can be counted simultaneously in LS counters. Pulse-height rejection at low energy is designed to reduce background noise, but also rejects radiation at very low energy. A typical Ge spectral analysis system designed for measuring energetic gamma rays has a low-energy cutoff near 30 keV, but systems designed to count lower-energy gamma rays and X rays reduce the cutoff to 1 keV or less. The low-energy cutoff for the proportional counter typically rejects only electrons below 0.1 keV.

Detection of energetic gamma rays is optimized by a thick Ge detector, and of low-energy X rays by high-Z gas in a proportional counter. Increasing sample size improves gamma-ray measurement efficiency per unit sample mass until a maximum value is reached, beyond which the increasing distance of sample portions from the detector begins to reduce the average counting efficiency. For beta-particle measurement, backscattering by a thick, high-Z material increase counting efficiency, but increasing the bulk of the sample soon decreases efficiency because of self-absorption. Alpha particles are much more sensitive to count-rate loss by self-absorption.

9.2.3. Radiation Decay Fraction

The radiation fraction per radionuclide decay is one of the factors for calculating radionuclide activity from measurement that are described in Section 8.2. The intensity of a characteristic peak measured by spectral analysis or of a count rate measured by proportional or LS counter must be divided by the decay fraction (among other factors) to obtain the activity of the radionuclide. Decay fraction values can be obtained from the above-cited decay scheme compilations.

Some interpretation may be required to obtain parameters of interest from the values reported in compilations. Gamma-ray intensities may be given in fractions that can be used directly, or relative to a selected—usually the most intense—reference gamma-ray intensity set at a value of 100. In the latter case, all other gamma-ray intensities must be calculated by multiplying the given intensity of the reference gamma ray by the appropriate fraction.

The fractional emission of conversion electrons may be reported directly, or relative to the associated gamma ray as ce_k/γ and ce_k/ce_{lmn}, where γ represents the decay fraction of the gamma ray and ce represents the decay fraction by conversion electrons from the k, l, m, and n shells. The decay fraction for the entire internal transition can be summed from this information on the assumption that contributions are minimal from electron shells more distant from the nucleus. The energies and fractions of characteristic X rays that accompany conversion electron transitions may be reported in compilations or the energies and relative intensities from the various K and L subshells are tabulated separately in compilations such as Table 7 in Appendix F of the Table of Isotopes (Firestone 1996). The fractions of characteristic X rays per conversion electron from that shell can be calculated by multiplying the relative intensities from the shell by the K or L fluorescence yield (see Table 3 in the same Appendix F).

The uncertainty information for these values that is needed to estimate the uncertainty in the calculated radionuclide activity may be given in the compilation or in the cited original studies. Any consistent deviations beyond this uncertainty in quality assurance test comparisons from the reported values may suggest, among other causes, an erroneous decay fraction in current use by the laboratory.

9.2.4. Application of Radiation Detection and Spectral Analysis

The end-window proportional counter detects beta particles with energies above approximately 25 keV because lower-energy beta particles are absorbed in the thin window, a narrow air gap, a thin film covering the sample, and the sample itself. Lower- energy beta particles can be detected in very thin samples measured in an internal proportional counter. An end-window proportional or Geiger–Mueler (G-M) counter with sufficient space between window and source for inserting aluminum foils and sheets can be used to obtain a curve of count rate *vs.* absorber thickness for determining the maximum beta-particle energy (see Section 2.4.2) from the beta-particle range.

All alpha particles in the conventional energy range are detected by the proportional counter unless the attenuation between source and detector exceeds 4 mg/cm^2. They can be distinguished from beta particles by operating the counter at an applied voltage too low to detect beta particles, or by pulse-height discrimination to measure only the more energetic alpha-particle pulses if the detector is operated at the higher applied voltage for detecting both alpha and beta particles. Very thin foils (about 1–12 mg/cm^2) can be used to determine alpha-particle energies by their range (see Section 2.4.1).

LS counters are useful for detecting radionuclides that emit beta particles with energies of only a few kiloelectron volts such as ^3H (18.6 keV maximum energy). For counting ^3H, an energy range from 1 to 19 keV can be selected to count most tritium beta particles with a reasonably low background count rate. Considerably more energetic beta particles (in the range of several hundred to several thousand kiloelectron volts) can be counted separately at the same time in the energy range from 20 keV to their maximum energy. If two beta-particle groups with sufficiently different maximum energies are present, the upper range can be further subdivided.

Use of the LS counter for alpha-particle spectral analysis is discussed in Section 8.3.2. Source preparation is simpler, but energy resolution is worse than with the solid-state detector. Special source preparation and electronic pulse-shape selection can improve resolution.

The Ge detector with gamma-ray spectrometer is the measurement instrument of choice for radionuclides that emit gamma rays. Its high resolution minimizes interference from other gamma rays in the sample so that many radionuclides can be identified and quantified unambiguously in a single sample. The gamma ray is displayed as a full-energy peak, together with a Compton continuum, that usually is not used for identification or quantification but interferes with lower-energy peaks. At energies above 1022 keV, the annihilation interaction peaks discussed below are also produced.

The radionuclide is identified by the energy at the midpoint of the characteristic full-energy peak. It is quantified in terms of the count rate in the channels that define the full-energy peak. Subtracted from this count rate is the count rate in these channels due to other gamma rays discussed in Section 10.3.7; in simple cases, the background count rate per channel is the average of the count rates in one or more channels on each side of the peak.

If the radionuclide has multiple gamma rays, a set of several of the more intense ones can be considered characteristic to provide more convincing identification and more counts for reduced uncertainty. Weaker gamma rays are excluded from this set if they have a less reliable decay fraction values and more interference by gamma rays from other radionuclides.

Gamma-ray spectral analysis of a radionuclide that emits only low-intensity gamma rays provides results of marginal reliability unless the radionuclide has been purified so that little interference is expected, or is in large amount. Examples of radionuclides measured with gamma rays of low decay fraction are plutonium isotopes and ^{85}Kr.

Radionuclides are also measured by gamma-ray spectrometer if they emit positrons that are then annihilated with the creation of 511-keV gamma rays. The efficiency of counting annihilation radiation is slightly less than that of 511-keV gamma rays emitted by a source because some annihilation occurs in solids beyond the source. The 511-keV annihilation radiation is also created by gamma rays more energetic than 1022 keV that are emitted by the source. These interactions produce not only a 511-keV peak in the detector, but also escape peaks that correspond to the full gamma ray energy minus 511 and 1022 keV.

The characteristic X rays of the atomic electron energy levels of an excited daughter isomer are also measured conveniently by the Ge detector with gamma-ray spectrometer. These X rays are associated with conversion electrons emitted by the isomer, and their energies and emission fractions are listed in radionuclide compilations (see Section 9.2.3). Conventional detectors that measure gamma rays above about 30 keV are used to identify K X rays from elements heavier than Xenon.

Thin Ge detectors with spectrometer systems are designed to detect gamma rays with energies as low as a few kiloelectron volts to measure K X rays of lighter elements and L X rays of heavier elements. Peaks are resolved for energy sublevels such as Kα1, Kα2, and Kβ that differ by several kiloelectron volts for the more energetic K X rays. Low-energy X rays are also spectrally analyzed with gas proportional counters designed for that purpose, as described at the end of Section 8.3.1.

Care should be taken to avoid misidentifying a radionuclide by peaks that are artifacts of the detector. Most important are the escape peaks from annihilation radiation cited above, and from low-energy radiation at the full energy minus 10 keV for the K X ray and 1.2 keV for the L X ray in Ge. Sum peaks are displayed for a radionuclide that emits two or more gamma rays simultaneously ("in coincidence"), as discussed in Section 9.3.5 for ^{60}Co, i.e., a 2505-keV peak for two gamma rays with energies of 1173 and 1332 keV. Although not a peak, the Compton edge, approximately 200 keV less than the full-energy peak (see Eq. 216) in the usual gamma-ray energy region, may be misinterpreted as such when seeking barely detectable radionuclides. Similarly, a slight increase in the Compton continuum due to the backscattered electrons is associated with this edge at about 200 keV. The radiation background in this energy region consists of peaks for annihilation radiation, progeny of ^{222}Ra and ^{220}Rn gas, the terrestrial radionuclides ^{40}K, uranium, and thorium, and cosmic-ray-produced radionuclides, as discussed in Section 8.2.2.

The Si detector with spectrometer is used with thin sources to identify and quantify radionuclides that emit alpha particles. All alpha particles are in the appropriate energy range for detection unless attenuated in a thick source. Chemical separation of the element of interest and meticulous preparation of the source usually are needed to obtain well-resolved peaks. Figure 9.1 shows the spectrum of a

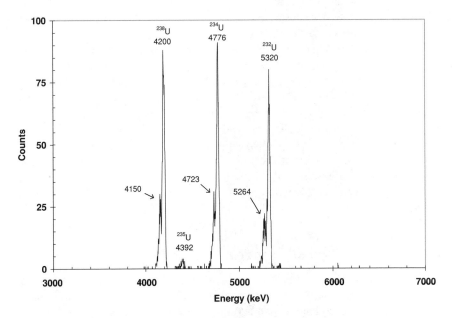

FIGURE 9.1. Alpha-partial spectrum of uranium with ^{232}U tracer, counted for 60,000 s.

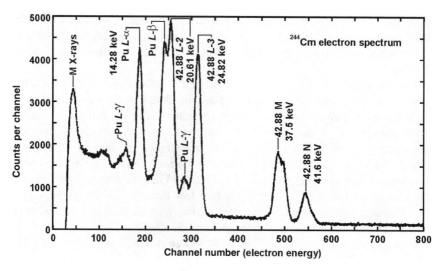

FIGURE 9.2. Conversion electron s measured whith a Si(Li) detector (from Knoll 1989, p. 464).

separated uranium source, with the large peaks for ^{238}U and ^{234}U, the usual small peak for ^{235}U, and the peak for the radioanalytical tracer ^{232}U. Many radionuclides emit several alpha-particle groups at energies that—because of the small energy difference and the lesser quality of the detector and the source cannot be resolved. Some smaller peaks of lower-energy alpha particles are shown in Fig. 9.1. The combined count rates of the unresolved peaks must be matched to the same composite group in a known source for calibration, or the combined decay fractions of all unresolved peaks must be used for calculating activity.

For some radionuclide mixtures, a group separation, e.g., for actinides, is satisfactory for measuring its component radionuclides by alpha-particle spectral analysis. As discussed in Section 6.4.1, further chemical separation is needed for radionuclides that emit alpha particles of almost the same energies, or even a mass spectrometer for radioisotopes of the same element with almost identical alpha-particle energies such as ^{239}Pu and ^{240}Pu.

A thin solid-state detector with spectrometer is also useful for identifying and quantifying conversion electrons. Figure 9.2 shows the spectrum of conversion electrons from a thin source of ^{244}Cm. For radionuclides that also emit beta particles, the conversion electron spectrum may be underlain by the beta-particle continuum.

9.2.5. Application of Chemical Separation

Chemical separation to characterize a radionuclide by element and purify it for quantification by radiation measurement is needed when direct spectral analysis does not achieve identification and quantification. Separation procedures are selected on the basis of the chemical behavior of the radioelement of interest and of possible interfering radioelements. When little is known about the presence of various radionuclides in the sample, an attempt is made to separate the radioelement of interest from all conceivable radioactive contaminants. At best, the identity and approximate amounts of contaminant radionuclides are known, and a purification scheme can be developed as discussed in Chapter 6.

A purification procedure can be considered appropriate if separation from other radionuclides is so effective that, in the absence of the radioelement of interest, no radioactivity is detected in the purified sample. That is, the measured radioactivity is zero within the measurement uncertainty. If the radioelement of interest is present, then contaminants should not observably increase its measured activity. To plan the procedure, the amount of each contaminant radionuclide that remains after applying the separation methods is calculated in terms of its decontamination factor (DF) (see Section 3.1). Meeting the stated criteria depends on the initial concentrations of the radioelement of interest and of each contaminant radionuclide in the sample, and on the achieved DF. Because these values generally are not known initially, reasonable concentrations must be assumed and subsequently revised when initial measurement results become available.

The other important purpose of chemical separation is to remove nonradioactive substances that interfere with the selected separation procedures and measurement of the purified source. In many instances, dissolved solids must be removed from the sample solution to measure carrier yield without interference and to obtain thin sources for counting alpha and beta particles, as indicated in Sections 7.2 and 7.4.

9.2.6. Radionuclide Confirmation

Under usual conditions, the above-described processes of detector selection, chemical separation, and spectral analysis are sufficient for identifying a radionuclide. Often, only one or two of these processes must be applied. A question of identity may arise under special circumstances:

- Two radionuclides emit the same type of radiation with almost identical energies
- Count rates are too low for clear peak energy identification
- The emitted radiation cannot be examined by spectral analysis
- Chemical separation is not possible for radioisotopes of the same element
- Chemical separation is imperfect for detecting a minor constituent

Resolving uncertainty is especially important if the radionuclide is relatively hazardous or characterizes a source or circumstance of particular concern.

Radionuclides are confirmed by applying redundant processes. A radionuclide is identified by gamma-ray spectral analysis and checked by chemical separation followed by a second spectral analysis measurement. Different analysts are assigned to analyzing the same sample by different chemical separations. The expected absence of gamma rays is confirmed by gamma-ray spectral analysis after the usual measurement for alpha- or beta-particle activity. Measurements are repeated to determine the half-life or a parent–daughter relation. Tabulations of energy and half-life are scanned to look for other radionuclides with the measured decay scheme.

9.3. Decay Scheme Information and Application

Decay schemes for seven sets of radionuclides—^{89}Sr, ^{90}Sr/^{90}Y, ^{137}Cs/^{137m}Ba, ^{40}K, ^{60}Co, ^{22}Na, and ^{241}Am—that exhibit the various aspects of spontaneous nuclear transformations are shown in Figs. 9.3–9.9 as a guide to their interpretation and application. Each decay is discussed separately in Sections 9.3.2–9.3.7. These figures were selected from earlier publications—the first six from Lederer *et al.* (1967), and the seventh from Martin and Blichert-Tolf (1970)—to show the conventional formalism for radioanalytical chemistry purposes, without the details that are useful to some practitioners, but excessive and confusing for radionuclide identification and quantitation. The data cited in these sections are more recent (2005), from the Brookhaven National Laboratory nuclear data tables, than the ones shown in the figures. Uncertainty in the reported values generally is associated with the last numeral on the right. As the minor differences between Figs. 9.3–9.8 and Fig. 9.9 suggest various compilers prepare decay schemes in similar but not identical formats.

The following conventions apply to the information of primary interest to the radioanalytical chemist presented in these decay schemes:

- element symbol below energy level line, with value of A to upper left and value of Z to lower left
- half-life at energy level line in years, days, min, or s
- the energy at each level relative to the ground state, in MeV
- the decay energy (Q) in MeV
- alpha-particle decay, α, an arrow (sometimes with a double line) to lower left
- beta-particle decay, β^-, an arrow to lower right
- positron decay, β^+, a vertical line followed by an arrow to lower left

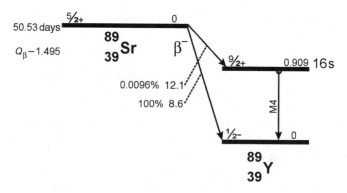

FIGURE 9.3. Decay scheme for ^{89}Sr (modified from Lederer *et al.* 1967, p. 392).

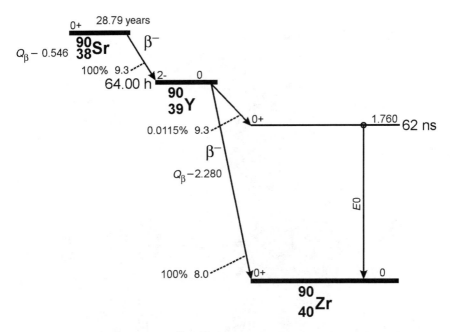

FIGURE 9.4. Decay scheme for ^{90}Sr/^{90}Y (modified from Lederer *et al.* 1967, p. 392).

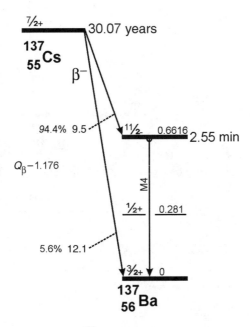

FIGURE 9.5. Decay scheme for ^{137}Cs (modified from Lederer *et al.* 1967, p. 399).

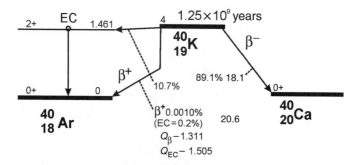

FIGURE 9.6. Decay scheme for ^{40}K (modified from Lederer *et al.* 1967, p. 384).

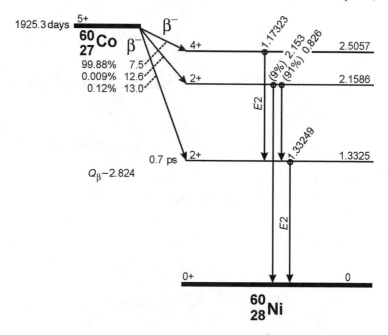

FIGURE 9.7. Decay scheme for ^{60}Co (modified from Lederer *et al.* 1967, p. 389).

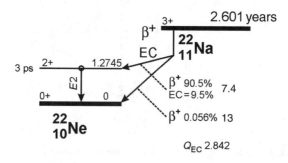

FIGURE 9.8. Decay scheme for ^{22}Na (modified from Lederer *et al.* 1967, p. 382).

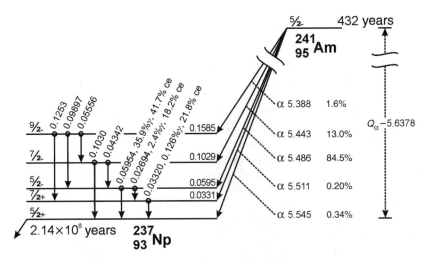

FIGURE 9.9. Decay scheme for ^{241}Am (modified from Martin and Blichert-Tolf, 1970, p. 144).

- electron capture (EC) an arrow to lower left
- internal transition by gamma ray and conversion electrons with characteristic X rays and Auger electrons, a vertical arrow downward
- alpha-particle energy, next to appropriate arrow, in MeV
- gamma-ray energy, the energy of the excited state to right on energy level line, or above appropriate arrow for multiple gamma rays (see Figs. 9.7 and 9.9), in MeV
- the decay fraction, next to appropriate arrow, in percent

Additional information necessary for radionuclide identification and analysis but not shown in Figs. 9.3–9.9 is given in Sections 9.3.2–9.3.7; it includes the fraction of conversion electrons and X rays in an internal transition and the beta-particle maximum energy.

Also in these figures is information primarily of interest to the physicist that can provide some guidance to the radioanalytical chemist:

- nuclear spin, e.g., 5/2−, 2+, shown at left on energy level line, that relates to the intensity of transition
- gamma ray transition polarity, e.g. M4, E0, shown sideways on internal transition vertical arrow
- log ft value, e.g., 12.1, 8.6, listed after the beta-particle energy, that defines the spectral distribution of beta particles

Applications of these items are discussed in nuclear physics texts and summarized in the nuclear and radiochemistry texts cited in Chapter 1.

9.3.1. Radioactive Decay and Ingrowth Rate

First consideration usually is given to the half-life shown on or next to the line in the decay scheme that represents the radionuclide energy level. For example, the 2.55-min half-life of ^{137m}Ba in Fig. 9.5 suggests that, once ^{137m}Ba is separated from its ^{137}Cs parent, rapid methods must be applied to separate and count it. The ^{90}Y 64-h (Fig. 9.4) half-life requires prompt attention and priority for analysis after separation from its parent ^{90}Sr. The longer half-lives of all other examples do not require special scheduling efforts for analysis.

Short half-lives enable identification by decay measurement. They also require precise records of the time of sample collection, separation from the longer-lived parent, and the beginning and end of counting to calculate ingrowth and decay. Decay correction for reporting the activity at the time of collection is needed for all radionuclides, but is significant under usual sample analysis scheduling only for radionuclides with short and intermediate half-lives; ^{89}Sr is an example of the latter.

Equations (2.6) and (9.1) describe radioactive decay. Figure 2.1 shows two straight lines on a semilogarithmic scale that represent decay by two radionuclides in the sample, observed by measurements repeated over a period of time. The precision of each calculated slope, characterized by the decay constant λ, is optimized by obtaining large net count rates, numerous repeated measurements, reliable background count rates for subtraction, and meticulously reproduced source–detector configurations for the sequential measurements. The value of the decay constant can be obtained with a least-squares fit for each of the component lines. Because the important aspect of these measurements is their relative values, the counting efficiency of the detector is not needed.

The curve toward a horizontal asymptote of the measured line in Fig. 2.1 suggests the presence of at least a second radionuclide with a smaller decay constant, i.e., a longer half-life. Because the curved line represents the sum of two or more straight lines, the contributing straight lines with their decay constants may be determined by fitting first the line for the longest-lived radionuclide, and then, in turn, lines for the shorter-lived components. Fitting multiple lines reduces the precision compared to a single straight line. Measurements at times later than shown in Fig. 2.1 may identify a third component if the line continues to curve toward the horizontal, or at least suggest an upper limit for a third component if the difference between the measured count rate and the background count rate is near zero and hence uncertain.

Equation (2.8) describes the decay of a radioactive parent with ingrowth of its radioactive daughter. The three types of ingrowth relations—secular, transient, and no equilibrium—are discussed in Section 2.2.4 and shown in Figs. 9.10–9.12.

The secular equilibrium example of the ^{226}Ra – ^{222}Rn pair in Fig. 9.10 is analogous to the ^{90}Sr – ^{90}Y and ^{137}Cs – ^{137m}Ba pairs discussed below. The slope of the line that represents the disintegration rate of the parent with its half-life appears horizontal because of negligible decay on the scale of 10 daughter half-lives. It is approached by the disintegration rate of the daughter after about six half-lives. The two disintegration rates become equal beyond that time.

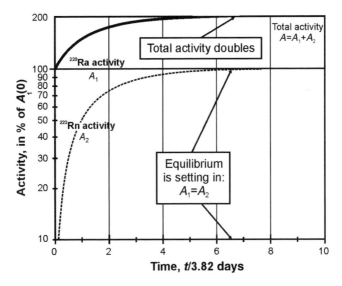

FIGURE 9.10. Radioactive daughter ingrowth—secular case (from Vertes *et al.* 2003, p. 275).

Transient equilibrium occurs when the half-life of the daughter is only somewhat shorter than that of the parent (Fig. 9.11). The slopes of the disintegration rate of the daughter approaches that of the parent after about six half-lives. The important distinction from secular equilibrium is that the disintegration rate of the daughter

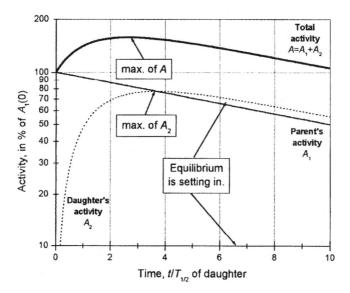

FIGURE 9.11. Radioactive daughter ingrowth—transient case (from Vertes *et al.* 2003, p. 273).

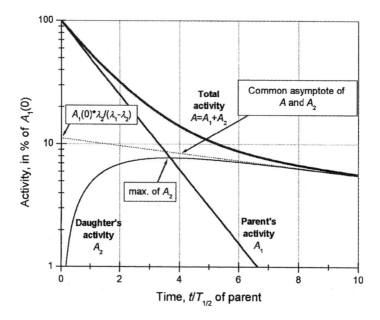

FIGURE 9.12. Radioactive daughter ingrowth—no equilibrium case (from Vertes *et al.* 2003, p. 276).

ultimately exceeds that of the parent by the factor $t_1/(t_1 - t_2)$, where t is the half-life and the subscripts 1 and 2 refer to the parent and daughter, respectively, as inferred from Eq. (2.8).

No equilibrium is reached when the daughter half-life exceeds that of the parent. As shown in Fig. 9.12, the daughter first accumulates as the parent decays, and will remain for some time after the parent no longer can be detected. The daughter will decay with its own half-life when, for practical purposes, the parent no longer remains.

The decay constant of a daughter in equilibrium with its parent may be inferred from repeated measurements of the count rate during ingrowth by use of Eq. (2.8), or directly determined by chemically separating the daughter and then performing repeated measurements of the count rate as the daughter decays.

9.3.2. Beta-Particle Decay: ^{89}Sr and ^{90}Sr/^{90}Y

The ^{89}Sr decay scheme in Fig. 9.3 is simple. For practical purposes, the radionuclide emits a beta particle group with a maximum energy of 1495 keV (plus the associated neutrino group discussed in Section 2.2.2). The half-life is 50.5 days. An obscure beta particle group of 590 keV maximum energy, followed by gamma rays of 909 keV with an intensity of about 0.01%, is of no practical consequence in radioanalytical chemistry.

The ^{90}Sr/^{90}Y decay scheme in Fig. 9.4 shows a parent and daughter (progeny) pair. The parent has a long 28.8-years half-life and the daughter has a short 64.0-h half-life. Each emits a pure beta-particle group in 100% of the decay, which is directly to the ground state. The maximum beta-particle energies are 546 keV for ^{90}Sr and 2280 keV for ^{90}Y. The minor (0.0115%) radiation emitted by ^{90}Y of a 519-keV maximum-energy beta-particle group followed by 1760-keV gamma rays is usually of no practical consequence. If the ^{90}Y is separated from ^{90}Sr, the subsequent daughter ingrowth is described as secular equilibrium discussed in Section 9.3.1.

9.3.3. Beta-Particle Decay to Excited State: $^{137}Cs/^{137m}Ba$

The ^{137}Cs/^{137m}Ba decay scheme in Fig. 9.5 shows two beta-particle groups emitted by ^{137}Cs. The 1176-keV maximum-energy beta-particle group decays to the ground state in 5.6% of the decays and the 514-keV maximum-energy beta-particle group decays in 94.4% of the decays to an excited state in ^{137}Ba. The ^{137m}Ba is in secular equilibrium with ^{137}Cs. This metastable state decays with a 2.55-min half-life. The decay of ^{137m}Ba is 85.1% by 661-keV gamma rays and 9.4% by conversion electrons.

The sum of conversion electron and gamma-ray decay of 94.4% equals the percent decay of the lower energy beta-particle group. The conversion electrons consist of K electrons at 624 keV (7.66%), L electrons at 656 keV (1.39%), and M, N, and O electrons at about 660 keV (total of 0.3%).

The X rays that accompany conversion electrons are X(K) at about 32 keV and X(L) at about 4.5 keV. With a Ge detector, several groups of X(K) energies can be distinguished, such as Kα1 at 32.2 keV (3.6%), Kα2 at 31.8 keV(2.0%), and Kβ at 36.4 keV (1.3%). The sum of these X-ray transitions is less than the fraction for conversion electrons because other transitions at lesser intensities occur and because Auger electrons are also emitted. The Auger electrons have average energies of 26 keV (0.8%) for the K shell and 3.7 keV (7.3%) for the L shell.

An important point in measurements is that conversion electrons are counted with the usual end-window beta-particle detectors and not with gamma-ray spectrometers. The decay fraction counted under these conditions with a beta particle detector is 1.094. The K Auger electrons may also be counted with beta particles and conversion electrons in an internal proportional counter. In a Ge detector and spectrometer, in addition to the gamma ray, the various K X rays are counted.

9.3.4. Electron Capture and Beta-Particle Decay: ^{40}K

The decay scheme of the naturally occurring radionuclide ^{40}K (see Fig. 9.6) is 89.1% by beta-particle emission, 10.9% by electron capture, and 0.001% by positron decay. The 1311-keV maximum-energy beta-particle decay is to the ground state of ^{40}Ca. Electron-capture decay is followed instantaneously by emission of a 1461-keV gamma ray from the excited state of ^{40}Ar to its ground state. Electron capture decay to the ground state is 0.2% per disintegration.

Electron-capture decay is associated with emission of 3-keV (1%) argon K X rays and of K Auger electrons at an average energy of 2.7 keV and an intensity of 7.3%. Positron decay produces two 511-keV gamma rays in the course of annihilation, or 0.002%.

The 1461-keV gamma ray at 0.107 per disintegration and the beta-particle group at 0.891 per disintegration are commonly measured. The other radiations—notably positron decay—are at too low intensity for practical purposes.

9.3.5. Beta Particles and Gamma Rays in Coincidence: ^{60}Co

The decay scheme of ^{60}Co shown in Fig. 9.7 consists, for measurement purposes, of a 318-keV maximum-energy beta-particle group (99.88%) followed instantaneously by a 1173-keV gamma ray and a 1332-keV gamma ray. Minor beta-particle groups that constitute 0.12 and 0.009% of the decay go to the indicated lower excited state of ^{60}Ni. The major beta-particle group and the two following gamma rays constitute 99.9% per disintegration. Conversion electrons are about 0.03% of total decay, and the characteristic K X rays at about 7.5 keV are 0.01% of total decay.

The major beta-particle group is in coincidence with the two gamma rays that follow it, and one gamma ray is in coincidence with the other. These relations can be applied to detect ^{60}Co in the presence of numerous other radionuclides by counting only radiations in coincidence with each other. For example, a beta-particle detector and a gamma-ray detector can be coupled in a system to count each types of radiation separately and also to count in a third recorder those beta particles and gamma rays that are detected simultaneously, as defined by the time constant of the system. In the case of ^{60}Co, either one or both gamma rays may be recorded for beta-gamma coincidences.

A special benefit of counting radiations in coincidence is that the process permits the absolute measurement of the activity of the radionuclide. For the beta-particle counting efficiency ε_β and the gamma-ray counting efficiency ε_γ (in count/disintegration), the activity A is related to the three count rates. The count rate R_β of beta particles in the beta-particle counter, the count rate R_γ of gamma rays in the gamma-ray counter, and the coincidence count rate $R_{\beta\gamma}$ (in count/s) yield A (in Bq) by:

$$R_\beta = \varepsilon_\beta A \tag{9.2}$$

$$R_\gamma = \varepsilon_\gamma A \tag{9.3}$$

$$R_{\beta\gamma} = \varepsilon_\beta \varepsilon_\gamma A \tag{9.4}$$

The three equations can be solved for A while eliminating the counting efficiencies:

$$A = \frac{R_\beta R_\gamma}{R_{\beta\gamma}} \tag{9.5}$$

Eq. (9.5) becomes complicated (NCRP 1985b) when some beta particles are counted in the gamma-ray detector or *vice versa*, or the decay scheme is more complex than for a beta particle followed by a single gamma ray. For example, conversion electrons may be in coincidence with beta particles and X rays, or several beta particle groups or gamma rays that have different counting efficiencies may be emitted, some in coincidence and others not in coincidence.

For ^{60}Co, gamma–gamma coincidence counting can be performed with a single Ge detector and gamma-ray spectrometer by recording separately the peak count rates for the two coincident gamma rays and their coincidences at the summed full-energy peak. Corrections are necessary for the nonpeak count rate and any significant angular correlation between the two gamma rays (NCRP 1985b). Coincidences between other pairs of radiations are also used for absolute measurement.

9.3.6. Positron Emission and Electron Capture: ^{22}Na

The decay scheme of ^{22}Na in Fig. 9.8 typifies positron emission accompanied by some electron capture. The main decay is by 546-keV maximum-energy positron group (90.5%) and electron capture (9.5%) to the ^{22}Ne-excited state. The latter decay is followed essentially instantaneously by the emission of a 1275-keV gamma ray (99.95%). A very small fraction of positron emission has 1820-keV maximum energy (0.056%) and goes to the ground state. The fraction of 0.85-keV K X ray is minimal (0.2%). The fraction of Auger K electrons at an average energy of 0.82 keV is 9.2%.

The major positron group and the gamma ray can be measured by coincidence counting (see above for ^{60}Co). In addition to the noted gamma ray, 511-keV annihilation radiation is detected by the gamma-ray detector; it is generated at the rate of two per positron, i.e., 1.81 per disintegration.

9.3.7. Alpha-Particle Decay and Gamma Rays: ^{241}Am

The ^{241}Am daughter of ^{241}Pu shown in Fig. 9.9 emits mainly a 5486-keV alpha particle (84.5%) to an excited state of ^{237}Np. Emission of two other alpha particles at successively slightly lower energy and considerably less intensity is typical of many alpha-particle emitters. In addition, two alpha particles of less than 1% intensity each have slightly higher energies, and 18 alpha particles of less than 0.02% intensity each have been reported between energies of 4757 and 5469 keV. The energy difference between the decay energy Q and the alpha particle to the ground state is due to the atom recoil energy discussed in Section 2.2.1.

The main gamma ray has an energy of 59.5 keV at an intensity of 35.9%. Two other gamma rays of 26.3 and 3322 keV have intensities of greater than 0.1%, and many gamma rays of extremely low intensity are emitted with energies between 922 and 43 keV. The L X ray of 14 keV has an intensity of 36.9%, while the K X ray of about 100 keV has an intensity of only about 0.003%. Numerous conversion electrons are emitted; the most intense groups are near 37 keV (33%) and 11 keV (17%). Numerous L Auger electrons at about 10 keV are also emitted.

The radionuclide activity usually is determined by measuring the 5486-keV alpha particle or the 59.5-keV gamma ray. Americium-241 is often measured to determine its beta-particle-emitting 14.35-years ^{241}Pu parent.

9.4. Radionuclide Characterization for Special and Routine Occurrences

The first effort in a characterization project usually is devoted to screening the sample for radionuclide content. Screening includes:

- Prediction of the radionuclides and their concentrations from a known or assumed source
- Attribution to source from other information
- Inference of the radionuclide mixture and overall decay rate in submitted samples based on type of sample and prescreening radiation scans of sample
- Initial gamma-ray spectral analysis
- Initial gross alpha- and beta-particle counting
- Observations of decay rates in gamma-ray spectra and gross counts
- Rapid chemical separations and radiation measurements for important radionuclides with little or no gamma-ray emission
- Integration of information developed within laboratory and by others
- Confirmatory and quality control measurements

Subsequent measurements pursue the implications of the screening results for monitoring persons and places, and seek radionuclides that are inferred to be present but would not have been found by such screening.

Sample characterization for radionuclide content under emergency response conditions requires a combination of rapid processing and careful evaluation that is best planned in advance. Included in advance planning must be a prescreening facility for distinguishing between samples with higher and lower radionuclide levels for sending them to separate laboratories with detectors dedicated to measuring these different levels. Protocols must be written to assure availability of skilled analysts, sample tracking and quality assurance for data reliability, and minimal cross-contamination to maintain the viability of detectors and samples. Exercises should be performed at appropriate intervals to test these plans and accustom analysts to emergency response actions.

Radionuclide identification to explain puzzling radiation measurement results or respond to clients concerning a routinely submitted sample requires the same efforts. The main distinction is that usually the sample radionuclide content and level is more narrowly defined and the response can be relatively unhurried.

9.4.1. Screening Measurements

Screening measurements of incident response samples begin as soon as possible after sample receipt and logging. A necessary first step is prescreening to prevent a sample from being taken to a laboratory if its radionuclide content is unsuitably high for the usual controls of personnel radiation exposure and contamination of the analyst, the laboratory, and radiation detectors. External radiation is measured with conventional low-level radiation monitors that include the G-M detector and the NaI(Tl) detector with spectrometer. Surface contamination is monitored by wiping with paper or cloth smears the outside of shipping and sample containers and measuring radionuclides on the smears with a G-M or LS counter.

Once the accepted sample is moved to the appropriate laboratory, aliquots of the sample are transferred to tared containers that provide reproducible measurements. The weight is recorded for solids and the volume for liquids. Samples for gamma-ray spectral analyses are placed in closed containers. Samples for gross alpha- and beta-particle measurements with a proportional (or G-M) counter are stabilized by evaporating liquids or slurrying and drying solids and by covering them with thin film. Samples for measurement with an LS counter are mixed with scintillation cocktail in a counting vial.

The results of brief initial counting—typically for a period of 1–5 min—are reviewed to determine whether they provide sufficient information on radionuclide content. If not, a longer counting period, say 1–2 h, is selected to provide as much reliable data as possible on the presence or absence of specified radionuclides within the necessarily limited period.

Detector quality control records are reviewed to assure that control samples and the radiation background have been measured recently and that the detectors in use are within control limits (Section 11.2.10). Brief control source and back-ground measurements are performed before the screening process begins to assure that detectors continue to operate appropriately and have not been contaminated recently. The detection limit in terms of activity per sample is calculated for all radionuclides of interest to determine whether a null result will meet radionuclide detection requirements for the submitted samples.

Gamma-ray spectral analysis typically is performed with a Ge detector plus spectrometer for radionuclides that emit gamma rays in the energy range of 30–4000 keV. All peaks are noted, and those believed to be full-energy peaks are tentatively identified by comparing them to a table of gamma-ray energies for all radionuclides. Assignment of multiple peaks to a radionuclide increases confidence in the identification. Wide peaks are tested for resolution to consider whether more than one gamma ray exists per peak energy region. Energies of background radiation—both in the sample and at the detector—and escape peaks are listed to eliminate such peaks from consideration. Concentrations of the tentatively identified radionuclides are estimated on the basis of listed gamma-ray decay fractions, sample mass or collector flow rate, and available detector counting efficiencies as function of energy.

If the decay of peaks in the gamma-ray spectrum is observable, the half-lives associated with these peaks are estimated and compared to tabulated values to confirm identification by energy. Gamma-ray peaks that appear constant are considered to be emitted by radionuclides that have half-lives long compared to the observation period.

At the same time, other aliquots are counted with gas proportional and LS counters to determine the presence of alpha and beta particles. The detectors are operated in a mode that distinguishes between alpha and beta particles. The LS counter can suggest maximum beta-particle energies and show alpha-particle energies from spectrometric readings. The activity of total alpha and beta particles per sample is estimated by assuming a typical or intermediate counting efficiency. These efficiency values must be revised once the samples are better characterized because factors such as beta-particle energy, sample self-absorption (in the proportional counter), and quenching (in the LS counter) affect counting efficiency.

If the gross alpha-particle activity is sufficiently high, a thin sample—of the order of 1 mg/cm^2—is prepared for alpha-particle spectral analysis with a Si detector plus spectrometer. The energy range of interest usually is 4–10 MeV.

Measurements are repeated at specified intervals to observe radioactive decay or ingrowth. Counting usually is performed after increasing intervals, e.g., 0.5, 2, 8, and 24 h, as the initially rapid overall decay becomes slower. Decay or ingrowth measurements for screening are ended when no further change is observed in the count rate.

Any change in the net count rate associated with alpha- and beta-particle counting provides a composite decay–rate curve that may be analyzed for a few component half-lives by the procedure discussed in Section 9.3.1. These half-lives are used to identify the radionuclides in the sample by scanning a list of radionuclides tabulated by half-life. If the sample has many radionuclides with short half-lives, only a general conclusion can be drawn concerning the presence of such radionuclides.

Initial measurements may stimulate further analysis with different sample mass or type of measurement. A larger sample is prepared if the count rate was too low for reliable measurement, and a more favorable geometry or longer counting period may be selected. Too intense a sample suggests use of a smaller sample, greater distance of source from detector, and shorter counting period. If the gamma-ray spectrum suggests the presence of gamma rays below 30 keV, the sample can be measured with an X-ray spectrometer to energies as low as 1 keV. A limited table of gamma-ray energies and radionuclide half-lives used for identification—either on paper by computer—may be modified to include additional radionuclides and remove ones inappropriate for the recorded spectra. Other detectors that are considered more useful for the type of sample or expected radionuclides may be selected for further measurements.

When no radiation is observed above the detector background, the estimated detection limit is compared to the regulatory limit or radiation protection guidance to determine whether sample analysis results are sufficiently sensitive. If not, larger samples, longer counting periods, and more efficient detection are required for more sensitive measurements. Any radionuclides that have been tentatively identified and quantified are compared to initial predictions to assure consistency or consider reasons for differences.

Additional samples may have to be collected if an observed radionuclide can be attributed either to the source of interest or to the ambient background due to another source. These samples are analyzed to examine the pattern and concentration range of the radionuclide and establish as reliably as possible the distinction between the two sources.

9.4.2. Information from Others

Prior to or in parallel with the screening measurements, available information is collected to suggest the set of radionuclides that can be anticipated in the submitted samples. A known radionuclide origin, such as a nuclear weapon, nuclear reactor, nuclear fuel cycle facility, accelerator, hospital nuclear medicine group, or irradiation source, will limit the list of potential radionuclides in a sample and possibly identify specific radionuclides. Even better are results of measurements performed at the site of origin to identify the radionuclides to be analyzed.

Various transport mechanisms from source to sample can be expected to modify the relative amounts of the component radionuclides. The type of sample may suggest the influence of the transport mechanism on the relative radionuclide content of the sample. For example, an air filter retains solid particles in preference to gases, and a filtered water sample contains only soluble radionuclides. Radiation monitoring of the sample in its container at collection, during transport, and on arrival at the laboratory may indicate the magnitude or absence of emitted gamma radiation, the extent of radioactive decay during transport, and the presence or absence of specific radionuclides.

Results obtained at other laboratories can suggest the identity and quantity of radionuclides at the origin even if different samples are analyzed. Cooperation with other groups that participate in the response can be decisive in focusing on the appropriate set of radionuclides instead of wasting time following false trails. On the other hand, confirmation of reported radionuclides in the laboratory is necessary because information from others on occasion has been unreliable and misleading.

9.4.3. Rapid Radiochemical Separations

A thin source for alpha spectral analysis should be prepared if screening or outside information indicates the presence of radionuclides that emit alpha particles but few or no gamma rays. The presence of alpha particles is predicted by a positive result for gross alpha particle activity and by gamma rays attributed to some actinide radionuclides. Suitable analytical methods are discussed in Chapter 6. A simple approach for a liquid sample is to purify actinides with a small ion-exchange resin column and prepare a thin source by electrodeposition. For solid samples, purification is preceded by ashing and dissolution. Tests performed in anticipation of an emergency should demonstrate the shortest time period required for these processes to function reliably. Calculations can suggest the sample amount needed to meet the detection limit requirements in a reasonable counting period.

If the gross beta-particle measurement is well in excess of the total beta particle activity inferred from gamma-ray spectral analysis, then radionuclides that emit only beta particles must be measured. The search usually can be narrowed by information on other radionuclides or from screening. The relatively few radionuclides that emit energetic beta particles but not gamma rays include ^{32}P, ^{89}Sr, ^{90}Sr, ^{91}Y, ^{99}Tc, and ^{147}Pm. Methods presented in Chapter 6 can be tested for obtaining results rapidly by reducing the number of steps and simplifying the remaining steps, but the reduced effort must provide sufficient DFs for anticipated impurities. Incomplete decontamination is acceptable only if the impurity is minor and the measurement technique corrects for the impurity.

Radionuclides that emit only low-energy beta particles may not be detected by gross beta-particle analysis with an end-window proportional counter, but can be detected, although at low efficiency, by LS counting. These radionuclides include ^{3}H, ^{14}C, ^{32}Si, ^{33}P, ^{35}S, ^{45}Ca, ^{59}Ni, ^{63}Ni, ^{228}Ra, and ^{241}Pu. The long-lived radionuclides among these have very low decay rates and usually are not of concern as immediate health hazards.

9.4.4. Quality Assurance

Some aspects of conventional quality assurance discussed in Chapter 11 cannot be applied to screening processes that require data reporting soon after sample receipt. Nevertheless, statements based on measuring radionuclides—notably their concentrations relative to exposure limits or their absence—must be carefully checked in emergencies because of their impact on efforts for protecting humans and area control and remediation.

Screening in response to emergencies is particularly subject to problems due to the unexpected situation and the possibly unprepared persons who collect and transport samples. In the laboratory, problems can arise due to large differences in radionuclide contents among samples, unexpected radionuclide mixtures, unusual modes of detector operation, and the pressure to report analytical results immediately. Problematic actions that need to be prevented because they adversely affect results include:

- Unreliable chain of custody descriptions of collection process, sample content
- Sample contamination during collection or delivery
- Laboratory cross-contamination
- Application of unfamiliar rapid methods
- Operation of detectors in unusual modes
- Hurried data calculation, analysis, and reporting

Special preparation is needed to anticipate and prevent these problems.

A sound approach is to prepare written instructions and conduct emergency exercises. A special reception area for radiation scanning of incoming samples, described in Section 13.3 for routine samples, should be prepared for receiving emergency samples. Control-source and background measurements must be interspersed with sample measurements despite the urgency given to measuring incoming samples. Collection and analysis of selected samples in duplicate should be specified. Calculations and data interpretation should be checked by at least a second person. Any results that may lead to major action decisions in the matter of concern must be evaluated for consistency with internal and outside information before they are reported.

The results that guide emergency response decisions must be susceptible to subsequent evaluation. Chain-of-custody information, procedural instructions and instruction changes, analyst comments, radiation measurement records, and QC data must be in writing and preserved despite the confusion that tends to accompany emergency response. Occurrence of contamination, instrument failure, analytical problems, and calculating error must be annotated. Samples and measured sources must be retrievable for subsequent confirmatory or expanded measurements.

9.5. Identification Needs

Radionuclide monitoring requires identification and quantification for the following interrelated purposes:

- Associate radionuclides with their origin
- Relate identified to expected radionuclides
- Determine extent and magnitude of radionuclide contamination
- Compare radionuclide concentrations to concentration limits
- Show absence of specified radionuclides

The essence of any monitoring program is to distinguish between radionuclides contributed by the activity, incident, or process that is being monitored and radionuclides that occur in the sample as "background." The background in the sample conventionally is considered to consist of the terrestrial and cosmic-ray-produced radionuclides. For careful measurements at low concentrations, the background must be considered to include man-made radionuclides in fallout from nuclear devices tested in the environment, a power source from a burned-up satellite, and airborne and waterborne discharges from nuclear power fuel cycle facilities. Man-made radionuclides also reach the environment by discharges from radionuclide production, research, and application, from stored and buried wastes, and from various incidents (Eisenbud 1987).

Analysis is planned for the radionuclides that are expected from the object of monitoring, with emphasis on radionuclides for which relatively large amounts are estimated to occur, be released, and reach the sample. Under some circumstances, radionuclides that are measured most readily or most sensitively may be used as tracers for accompanying radionuclides. Gamma-ray spectral and gross activity analyses are useful for testing predictions of radionuclide amounts, releases, and transfers, and also to seek unexpected radionuclides.

For initially undefined sources of radionuclide origin, such as certain accidental releases or terrorist actions, the wider net that must be cast by screening is described in Section 9.4.1. Measurements obtained near the origin of the radionuclides provide the best guidance for analyzing samples collected at greater distances. The radionuclides selected for analysis and the precision of analysis are guided by the plans to identify the source at the origin and to understand the circumstance that led to the dispersion of the radionuclides.

Once the radionuclides of interest are identified, a sample collection pattern should be developed to delineate the region of contamination and the radionuclide concentrations that affect radiation exposure. The usual concern is to identify locations of external exposure by direct radiation and internal exposure by inhalation and food consumption. The lower limits of detection required for analysis are specified by the customer to assure reliable detection in accord with the regulation or guidance that limits the concentration of each radionuclide. The lower limit of detection usually is required to be 2–10 times below the concentration limit to assure reliable measurement at the limit.

Establishing the absence of a radionuclide becomes important when radioanalytical chemistry is used to demonstrate that a radiological incident, such as a threatened terrorist act, did not occur. Absence of a substance in a sample is a relative conclusion that could be altered by a more sensitive measurement. A radionuclide is reported as measured or as "less than" the lower limit of detection. A more sensitive detection method may replace the "less than" description with an actual value, or continue to report "less than," but at a lower value. Such "less than" values are based on net count rates that may be zero or a positive value sufficiently near zero that is too uncertain, as discussed in Section 10.4, to be reported. This net count rate can be the difference between the gross count rate and (1) the detector background count rate or (2) the count rate in the sample attributed to background from various sources, as defined above.

10
Data Calculation, Analysis, and Reporting

Keith D. McCroan[1] and John M. Keller[2]

10.1. Introduction

The reputation of the radioanalytical chemistry laboratory is based on the extent to which its reported results are judged to be reliable and reported in a form responsive to customer needs. The effort to obtain and provide such data requires the competent execution of every analytical step in the process, as well as adherence to the quality assurance tenets discussed in Chapter 11. The result of these interlocking activities should be an accurate and defensible data set. This chapter addresses the processing and evaluation of those data.

Four types of calculations are typically performed that pertain to measured radionuclide values:

- Conversion of the instrument signal (counts or count rate) and other measured quantities to an activity concentration, typically expressed in becquerels (Bq) or picocuries (pCi) per unit mass (or per unit volume, area, or time);
- Calculation of the uncertainty of the result in terms of an estimated standard deviation or a multiple of the standard deviation;
- Calculation of the *critical value*, which is used to decide whether the radionuclide is "detected"; and
- Calculation of the *minimum detectable activity* (MDA).

These types of calculations are discussed in Sections 10.2 through 10.4.

For the reasons discussed in Chapter 11, quality control (QC) measurements must be performed. Replicate measurements are only one aspect of the overall program of QC. The additional calculations associated with QC measurements fall into three main categories: instrumental, analytical, and methodological. These topics are covered in Section 10.5.

Preparing, compiling, and presenting the information developed by a radioanalytical chemistry laboratory is not merely a clerical exercise; it is fully as important as the analysis and measurement effort. The work must assure that the reported values meet the data quality objectives (see Section 11.3.1) in being reliable, providing

[1]National Air and Radiation Environmental Laboratory, USEPA, Montgomery, AL 36115
[2]Oak Ridge National Laboratory, Oak Ridge, TN 37831

the information required by the initial plan, and conveying to the reader a clear understanding of the radiological impact of the subject. Section 10.6 describes the process of data review, the steps that ensure the viability of the data. Section 10.7 addresses some of the important details of data presentation.

10.2. Calculating the Activity

A typical radioanalytical measurement involves radiation counting, although other measurement techniques, such as ICP-MS, are sometimes used. The initial processing of the raw counting data usually involves two steps. The first step is division of the gross sample count C_G by the length of the counting period t_G to calculate the gross count rate R_G. The second step is subtraction of the detector background count rate R_B from the gross count rate to calculate the net count rate R_N.

Conversion of the net count rate to activity (or massic activity or volumic activity) produces the final result that is reported to the client. The mathematical model for calculating the final result a often has a form similar to that shown in Eq. (8.2), and in Eq. (10.1):

$$a = \frac{(C_G/t_G) - (C_B/t_B)}{\varepsilon Y m D F} = \frac{R_G - R_B}{\varepsilon Y m D F} \tag{10.1}$$

where

a	is the massic activity of the sample (or volumic activity)
C_G	is the gross count observed when the sample is counted
t_G	is the duration (live time) of the counting period for the gross sample measurement
C_B	is the count observed when the background (or blank) is counted
t_B	is the duration of the counting period for the background (or blank)
ε	is the practical counting efficiency
Y	is the chemical yield
m	is the mass of the sample aliquot analyzed (or volume)
D	is the decay or ingrowth factor
F	is the decay fraction (or radiation emission probability)
R_G	is the gross count rate (C_G/t_G)
R_B	is the background (or blank) count rate (C_B/t_B)

If the count times are expressed in seconds and the mass in grams, then the result of the calculation is given in becquerel per gram (Bq/g).

The practical counting efficiency ε represents the probability that any particular photon or particle of radiation emitted by the sample source will be recorded by the detector. As explained in Section 8.2, its value may depend on many factors, including the detector, the type and energy of the radiation, the composition of the source, and the geometry of the source–detector configuration. It includes the loss factor in the pulse analysis system and attenuation and scattering fractions associated with the sample-detector system. All of these factors are discussed further in Section 8.2.

The mass m in Eq. (10.1) may represent mass of material that is moist, dried, or ashed. The concentration may be reported on the basis of moist weight to indicate the radionuclide concentration in the environment, or on the basis of dried or ashed weight for reliable comparison with measurements of other samples or by others. For such comparisons, aliquots of a sample are weighed moist as collected, then after drying at 110°C, and then again after thorough ashing, both at least overnight.

The decay or ingrowth factor D is a correction factor for decay or ingrowth of the radionuclide before and during counting. For the simple situation where the radionuclide of interest is not supported by a parent in the sample, the decay factor is given by

$$D = e^{-\lambda t_D} \times \frac{1 - e^{-\lambda t_R}}{\lambda t_R} \tag{10.2}$$

Here λ denotes the radionuclide decay constant, t_D denotes the time from sample collection to the start of sample counting, and t_R denotes the real time, or clock time, of the sample counting measurement, which may be longer than the live time t_G. When the time t_R is short relative to the half-life of the radionuclide, the product λt_R may be very small, and as λt_R approaches zero, the correction factor for decay during counting, $(1 - e^{-\lambda t_R})/\lambda t_R$, approaches 1. If the value of this factor is sufficiently close to 1, it may be omitted from the expression for D. When it is included, it must be calculated carefully to avoid large round off errors when λt_R is small.

Note: For small values of λt_R, the factor $(1 - e^{-\lambda t_R})/\lambda t_R$ may be approximated either by $e^{-\lambda t_R/2}$ or by the sum of the first few terms of the series $1 - \lambda t_R / 2! + (\lambda t_R)^2 / 3! - (\lambda t_R)^3 / 4! + \cdots$.

The model for a gamma-ray spectral analysis is somewhat more complicated than suggested by Eq. (10.1), and in the worst cases may be much more complicated. In gamma-ray spectrometry, the gross count C_G equals the sum of counts in channels that represent a photopeak. The radionuclide being measured may or may not be present in the instrument background (i.e., as a peak in the background spectrum), but the total background correction always includes a contribution from the Compton baseline of the spectrum, which is usually estimated from the observed counts in several channels on each side of the peak. The shape of the baseline under the peak is often assumed to be a straight line, although stepped functions are sometimes used. If the radionuclide being measured emits measurable gamma rays at more than one energy, the activity is typically calculated as a weighted average of the results obtained for selected peaks. In some cases, gamma rays from two or more radionuclides may have energies so nearly equal that the associated peaks in the spectrum overlap and combine to form multiplets. Such individual peaks often can be resolved by nonlinear regression. For all these reasons, most laboratories rely on software packages to perform the required data analysis for gamma-ray spectrometry.

10.3. Measurement Uncertainty

The uncertainty of each result must be reported to indicate its reliability. This information is necessary when comparing a measured value with a guidance level or regulatory limit to state with confidence that the true value is above or below the level or limit. Measurement uncertainty must also be considered when comparing results within or between laboratories.

An internationally accepted guide to the subject of measurement uncertainty is the *Guide to the Expression of Uncertainty in Measurement*, commonly referred to as "the GUM" (ISO 1995). The GUM defines standard terminology, notation, and methodology for evaluating and expressing measurement uncertainty. This chapter uses the GUMs terms and symbols, some of which are summarized briefly below. For more information, see the GUM itself (ISO 1995), NIST Technical Note 1297 (NIST 1994), or the Multi-Agency Radiological Laboratory Analytical Protocols Manual (MARLAP) (EPA 2004).

The uncertainty of a measurement is defined in the GUM and in the *International Vocabulary of Basic and General Terms in Metrology* (ISO 1993) as a "parameter, associated with the result of a measurement, that characterizes the dispersion of the values that could reasonably be attributed to the measurand." Thus, the uncertainty may be used to find bounds for the likely value of the measurand. The traditional approach to uncertainty in radioanalytical chemistry, which is also the GUMs approach, is to express the uncertainty of a result as either an estimated standard deviation or a multiple thereof. An uncertainty expressed as a standard deviation is called a *standard uncertainty*. The same concept has traditionally been called a "one-sigma uncertainty." A multiple of the standard uncertainty, such as the "two-sigma" or "three-sigma" uncertainty, is called an *expanded uncertainty*. A laboratory may report the uncertainty of a result as either the combined standard uncertainty or an expanded uncertainty, but the type of uncertainty should always be explained.

Radioanalytical measurements can have many causes of uncertainty. One of the best known is the inherent randomness of nuclear decay, radiation emission, and radiation detection, which is often referred to as "counting statistics." Radiation laboratories generally recognize the existence of counting statistics and account for it when they report the uncertainty of a measurement. The component of the total uncertainty due to counting statistics is the *counting uncertainty*, which has traditionally been called the "counting error."

In some situations, other components of the total uncertainty may be as large as or larger than the counting uncertainty. A laboratory that reports only the counting uncertainty may give data users an unrealistic view of the data's reliability. All known sources of uncertainty and all the uncertainty components should be evaluated if they are believed to be significant.

10.3.1. Uncertainty Evaluation and Propagation

In radioanalytical chemistry measurements, as in many analytical measurements, the final result is not obtained by direct observation of a measuring instrument.

Instead, one uses a mathematical model of the measurement to relate the values of *input quantities*, which are observed or previously measured, to the value of the desired *output quantity* (the measurand), which must be calculated. The mathematical model is an equation or a set of equations that describe exactly how one calculates the value of the measurand from the observed values of the input quantities. The model for a radioanalytical measurement often resembles Eq. (10.1), but for the purpose of explaining uncertainty evaluation, a model that follows the GUM is written abstractly in the form

$$Y = f(X_1, X_2, \ldots, X_N) \tag{10.3}$$

where $X_1, X_2, \ldots, X_N$ denote the input quantities, and Y denotes the output quantity (not the yield in this case), or measurand.

A particular observed value of an input quantity X_i is called an *input estimate*. If we denote the input estimates by lowercase variables $x_1, x_2, \ldots, x_N$, then the *output estimate*, y, is calculated as follows:

$$y = f(x_1, x_2, \ldots, x_N) \tag{10.4}$$

Each input estimate x_i has an uncertainty, $u(x_i)$. In principle the standard uncertainty of an input estimate can be evaluated in many ways. For example, if the value of the input quantity can be measured repeatedly, then a series of observations of the quantity, $x_{i,1}, x_{i,2}, \ldots, x_{i,n}$, can be obtained to calculate the arithmetic mean $\bar{x}_i$ and experimental standard deviation $s(x_{i,k})$:

$$\bar{x}_i = \frac{1}{n} \sum_{k=1}^{n} x_{i,k} \quad \text{and} \quad s(x_{i,k}) = \sqrt{\frac{1}{n-1} \sum_{k=1}^{n} (x_{i,k} - \bar{x}_i)^2} \tag{10.5}$$

The experimental standard deviation of the mean $s(\bar{x}_i)$ is obtained by dividing $s(x_{i,k})$ by $\sqrt{n}$:

$$s(\bar{x}_i) = \frac{s(x_{i,k})}{\sqrt{n}} = \sqrt{\frac{1}{n(n-1)} \sum_{k=1}^{n} (x_{i,k} - \bar{x}_i)^2} \tag{10.6}$$

One then lets x_i equal the arithmetic mean of the observations and lets $u(x_i)$ be the experimental standard deviation of the mean:

$$x_i = \bar{x}_i$$
$$u(x_i) = s(\bar{x}_i) \tag{10.7}$$

This type of uncertainty evaluation is an example of what the GUM calls a *Type A evaluation* of uncertainty, which is defined as an evaluation of uncertainty based on the statistical analysis of series of observations (ISO 1995).

Many methods of uncertainty evaluation do not involve the statistical analysis of series of observations; these are called *Type B evaluations*. One of the most common Type B methods of uncertainty evaluation in radioanalytical chemistry is the common practice of estimating the standard uncertainty of an observed count C by its square root $\sqrt{C}$ (see Section 10.3.4).

Another way to perform a Type B evaluation is to estimate the maximum possible absolute error b in an input estimate x_i and to assume that the distribution of the estimator is either rectangular or triangular, with half-width equal to b. To evaluate the standard deviation of the estimator, if the distribution is rectangular with half-width b, the standard deviation (i.e., the standard uncertainty of x_i) equals $b/\sqrt{3}$. If the distribution is triangular, the standard deviation equals $b/\sqrt{6}$.

A rectangular distribution is a more conservative assumption than a triangular distribution. A rectangular distribution is appropriate if one believes that the true value lies in the interval $x_i \pm b$, and all possible values in this interval are equally likely. A triangular distribution is more appropriate if one believes that values near the center of the interval are more likely than those near the edges. For example, a rectangular distribution is ideal for evaluating the uncertainty due to rounding a number. A triangular distribution may be used to evaluate the uncertainty associated with the capacity of volumetric glassware or pipetting devices, where the manufacturer states a tolerance for the true value but the nominal capacity is assumed to be most likely.

In both of the preceding examples, the value of the half-width b is usually given. In other cases, the determination of b may require experience and professional judgment. For example, one would use judgment to estimate the maximum possible deviation of a meniscus from the capacity mark in a volumetric flask.

Sometimes a pair of input estimates x_i and x_j may be correlated, because they are not determined independently of each other or because there is some effect in the measurement process that influences the observed value of each. The estimated covariance of x_i and x_j is denoted by $u(x_i, x_j)$. A Type A evaluation of covariance may be performed in some cases by making a series of paired observations of the two input quantities, $(x_{i,1}, x_{j,1})$, $(x_{i,2}, x_{j,2})$, ..., $(x_{i,n}, x_{j,n})$, and performing the following calculations:

$$x_i = \bar{x}_i = \frac{1}{n} \sum_{k=1}^{n} x_{i,k}$$

$$x_j = \bar{x}_j = \frac{1}{n} \sum_{k=1}^{n} x_{j,k}$$

$$u(x_i, x_j) = \frac{1}{n(n-1)} \sum_{k=1}^{n} (x_{i,k} - \bar{x}_i)(x_{j,k} - \bar{x}_j) \qquad (10.8)$$

Other possible methods for estimating the covariance $u(x_i, x_j)$ are described by MARLAP (EPA 2004).

All the uncertainties $u(x_i)$ and covariances $u(x_i, x_j)$ of the input estimates combine to produce the total uncertainty of the output estimate, y. The mathematical operation of combining the standard uncertainties and covariances of the input estimates x_i to obtain the standard uncertainty of the output estimate y is called *propagation of uncertainty*. The standard uncertainty of y obtained by uncertainty propagation is called the *combined standard uncertainty* of y and is denoted by $u_c(y)$. The following general equation, which the GUM calls the "law of propagation of uncertainty," describes how the combined standard uncertainty is calculated:

$$u_c(y) = \sqrt{\sum_{i=1}^{N}\left(\frac{\partial f}{\partial x_i}\right)^2 u^2(x_i) + 2\sum_{i=1}^{N-1}\sum_{j=i+1}^{N}\frac{\partial f}{\partial x_i}\frac{\partial f}{\partial x_j}u(x_i, x_j)} \qquad (10.9)$$

In this equation, $\partial f/\partial x_i$, called a *sensitivity coefficient*, denotes the partial derivative of the function $f(X_1, X_2, \ldots, X_N)$ with respect to X_i, evaluated at $X_1 = x_1$, $X_2 = x_2, \ldots, X_N = x_n$. A sensitivity coefficient $\partial f/\partial x_i$ represents the ratio of the change in the output estimate y to a small change in one input estimate x_i.

If the input estimates are uncorrelated, then all the "covariance terms," which are those terms that involve $u(x_i, x_j)$, are zero, and the uncertainty equation reduces to the simpler form shown in Eq. (10.10) below:

$$u_c(y) = \sqrt{\sum_{i=1}^{N}\left(\frac{\partial f}{\partial x_i}\right)^2 u^2(x_i)} \qquad (10.10)$$

The following applications show how Eq. (10.10) can be applied to some simple models when the input estimates are uncorrelated. In these equations, the variable x_i denotes input estimates and the variable y denotes the calculated output estimate; a and b are constants.

Application 1: Addition and Subtraction

$$y = x_1 \pm x_2$$

$$u_c(y) = \sqrt{u^2(x_1) + u^2(x_2)}$$

Application 2: Multiplication

$$y = x_1 x_2$$

$$u_c(y) = \sqrt{x_2^2 u^2(x_1) + x_1^2 u^2(x_2)} \qquad \frac{u_c(y)}{y} = \sqrt{\frac{u^2(x_1)}{x_1^2} + \frac{u^2(x_2)}{x_2^2}}$$

Application 3: Division

$$y = \frac{x_1}{x_2}$$

$$u_c(y) = \sqrt{\frac{u^2(x_1)}{x_2^2} + \frac{x_1^2 u^2(x_2)}{x_2^4}} \qquad \frac{u_c(y)}{y} = \sqrt{\frac{u^2(x_1)}{x_1^2} + \frac{u^2(x_2)}{x_2^2}}$$

Application 4: Exponential

$$y = ae^{bx}$$

$$u_c(y) = abe^{bx}u(x) \qquad \frac{u_c(y)}{y} = b\,u(x)$$

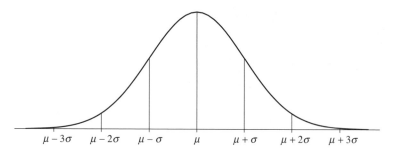

FIGURE 10.1. A normal distribution.

10.3.2. Expanded Uncertainty

Most laboratories multiply the combined standard uncertainty $u_c(y)$ by a factor k, which is typically a value such as 1.96 or 2. The laboratory then reports the product of k and $u_c(y)$ as the uncertainty of the result y. The purpose of expanding the uncertainty in this manner is to obtain an interval about the measured result $y \pm k\, u_c(y)$ that has a high probability of containing the true value of the measurand.

When $k = 1.96$ or 2, this product has traditionally been called the "two-sigma" uncertainty. The GUM calls the product $k \times u_c(y)$ an *expanded uncertainty* and calls the factor k itself a *coverage factor*.

The common choice of $k = 1.96$ or 2 is based on the assumption that the distribution of the measured result is approximately normal, or Gaussian. If the distribution is normal and centered on the true value of the measurand, there is approximately a 95% probability that the measured result will fall within 1.960 standard deviations of the true value.

Figure 10.1 shows a normal distribution with intervals about the mean μ represented in terms of multiples of the standard deviation σ. The area under the curve between $\mu - \sigma$ and $\mu + \sigma$ is approximately 68.3% of the total area, and between $\mu - 2\sigma$ and $\mu + 2\sigma$ it is approximately 95.4%. More than 99.7% of the area falls between $\mu - 3\sigma$ and $\mu + 3\sigma$.

The combined standard uncertainty is only an estimate of the true standard deviation, rather like an experimental standard deviation, and for this reason the probability that the expanded uncertainty interval $y \pm 1.96\, u_c(y)$ will contain the true value of the measurand may actually be less than 95%. In fact it is misleading for a laboratory to use as precise a coverage factor as $k = 1.96$, which implies that the true standard deviation is known and that the coverage probability really is 95%. The use of $k = 2$ better indicates the approximate nature of the uncertainty evaluation and does not imply so strongly that the coverage probability is exactly 95%.

Note: Although it is common to use a fixed coverage factor, such as 2, to calculate the expanded uncertainty, the GUM also describes a method for calculating a coverage factor k_p for each measurement to provide a desired approximate coverage probability p. The method is derived by treating the standard uncertainty of each input estimate $u(x_i)$ as if it were an experimental standard deviation with a specified number of degrees of freedom ν_i. When $u(x_i)$ is determined by a Type A method of evaluation, ν_i is the actual number of degrees of freedom for the evaluation. When $u(x_i)$ is determined by a Type B method of evaluation, the value of ν_i is based on the estimated relative standard deviation of $u(x_i)$ (i.e., the uncertainty of the uncertainty). Then the number of "effective" degrees of freedom ν_{eff} for the combined standard uncertainty $u_c(y)$ is evaluated. Finally, the coverage factor k_p is chosen from a table of percentiles of Student's t-distribution with ν_{eff} degrees of freedom. For example, if the number of effective degrees of freedom is 12, then the coverage factor $k_{0.95}$ for 95% coverage is the 97.5% percentile of the t-distribution with 12 degrees of freedom, which equals 2.179. For more information about this method of calculating coverage factors, see the GUM (ISO 1995) or MARLAP (EPA 2004).

10.3.3. Application to Radioanalytical Chemistry

A typical mathematical model for a simple radioanalytical measurement might have the form of Eq. (10.1), which is shown again below:

$$a = \frac{(C_G/t_G) - (C_B/t_B)}{\varepsilon \times Y \times m \times D \times F} = \frac{R_G - R_B}{\varepsilon \times Y \times m \times D \times F} \tag{10.1}$$

where all the symbols are defined in Section 10.2. If there are no correlations between any of the input estimates, the law of propagation of uncertainty (with some algebraic manipulation) implies that

$$u_c(a) = \left[\frac{u^2(C_G)/t_G^2 + R_G^2\left[u^2(t_G)/t_G^2\right] + u^2(C_B)/t_B^2 + R_B^2\left[u^2(t_B)/t_B^2\right]}{\varepsilon^2 Y^2 m^2 D^2 F^2} \right.$$
$$\left. + a^2 \times \left(\frac{u^2(\varepsilon)}{\varepsilon^2} + \frac{u^2(Y)}{Y^2} + \frac{u^2(m)}{m^2} + \frac{u^2(D)}{D^2} + \frac{u^2(F)}{F^2} \right) \right]^{1/2} \tag{10.11}$$

The uncertainty of the input quantities in this model is discussed below in Section 10.3.8. Some quantities, such as the decay factor D and the yield Y, are not directly observed but are calculated from other observed quantities. Their uncertainties are calculated by uncertainty propagation before being used in the equation above.

The uncertainties of the count times, $u(t_G)$ and $u(t_B)$, and the uncertainty of the decay factor, $u(D)$, usually are negligible and can be omitted from the uncertainty equation. Then the simplified uncertainty equation becomes

$$u_c(a) = \sqrt{\frac{u^2(C_G)/t_G^2 + u^2(C_B)/t_B^2}{\varepsilon^2 Y^2 m^2 D^2 F^2} + a^2 \times \left(\frac{u^2(\varepsilon)}{\varepsilon^2} + \frac{u^2(Y)}{Y^2} + \frac{u^2(m)}{m^2} + \frac{u^2(F)}{F^2} \right)} \tag{10.12}$$

The uncertainty of a count time may be relatively significant if the count time is very short (say less than 1 min), if the number of counts is very large, or if the dead time rate is large. The uncertainty of the decay factor is most significant when the radionuclide is short-lived and the decay time is either uncertain or long.

The preceding uncertainty equations presume that all pairs of input estimates are uncorrelated, which may or may not be true. One of the most common examples of correlated input estimates in radioanalytical chemistry occurs when the chemical yield Y is calculated from an equation involving the counting efficiency ε. This happens in measurements by alpha-particle spectrometry with an isotopic tracer. In this case, the uncertainty equation can be simplified by treating the product $\varepsilon \times Y$ as a single variable. What happens in effect is that the efficiency cancels out of the activity equation and for this reason its uncertainty can be considered to be zero:

$$u_c(a) = \sqrt{\frac{u^2(C_G)/t_G^2 + u^2(C_B)/t_B^2}{\varepsilon^2 Y^2 m^2 D^2 F^2} + a^2 \times \left(\frac{u^2(\varepsilon \times Y)}{(\varepsilon \times Y)^2} + \frac{u^2(m)}{m^2} + \frac{u^2(F)}{F^2}\right)}$$

$$(10.13)$$

Another example of correlated input estimates occurs when both the counting efficiency ε and the yield Y depend on the mass of a precipitate or residue on the prepared sample source. In this case, dealing with the correlation is less simple. It may be necessary to replace the variables ε and Y in the activity equation by the expressions used to calculate them, or to include the covariance term for ε and Y in the uncertainty equation.

10.3.4. Counting Statistics

Nuclear decay, the emission of radiation from a decaying atom, and the detection of emitted radiation by a detector are inherently random phenomena. Their occurrences cannot be predicted with certainty, even in principle, although they can be described probabilistically. The randomness of these processes causes the result of a radiation-counting measurement to vary when the measurement is repeated and thus leads to an uncertainty in the result, called the "counting uncertainty."

Suppose one places a radioactive source in a radiation counter and acquires counts for a period of time t, and the total number of counts observed in this time period is C. A common Type B method of evaluating the standard uncertainty of C is to take its square root. That is, $u(C) = \sqrt{C}$. The rationale for this approach, and some of its limitations, are described below.

One option for evaluating the uncertainty of the number of counts observed is to perform a Type A evaluation described in Section 10.3.1, where one repeats the measurement n times, obtaining the values $C_1, C_2, \ldots, C_n$, and calculates the arithmetic mean $\bar{C}$, the experimental standard deviation of the values $s(C_i)$, and the experimental standard deviation of the mean $s(\bar{C})$. The arithmetic mean $\bar{C}$ is then used as the measured value and the experimental standard deviation of the mean $s(\bar{C})$ is used as its standard uncertainty.

When one performs repeated measurements of this type with a long-lived radioactive source, what is often observed is that the experimental standard deviation $s(C_i)$ is approximately equal to the square root of $\bar{C}$. One can explain this observation by probability theory if one makes the following assumptions about the measurement problem:

- The source in the detector contains a large number, N, of atoms of some long-lived radionuclide.
- Each atom of the radionuclide has the same probability p of decaying during the counting period, emitting the radiation of interest, and producing a count. When the half-life of the radionuclide is long, p is a very small number.
- All atoms of the radionuclide decay and produce counts independently of each other. Decay by one atom that produces a count has no impact on whether another atom decays and produces a count.
- No atom produces more than one count during the counting period.
- Counts are not produced by other sources.

Given these assumptions, the distribution of the observed total number of counts according to probability theory should be *binomial* with parameters N and p. Because p is so small, this binomial distribution is approximated very well by the *Poisson distribution* with parameter Np, which has a mean of Np, and a standard deviation of $\sqrt{Np}$. The mean and variance of a Poisson distribution are numerically equal; so, a single counting measurement provides an estimate of the mean of the distribution Np and its square root is an estimate of the standard deviation $\sqrt{Np}$. When this Poisson approximation is valid, one may estimate the standard uncertainty of the counting measurement without repeating the measurement (a Type B evaluation of uncertainty).

Although all the assumptions stated above are needed to ensure that the distribution of the total count is binomial, not all the assumptions are needed to ensure that the Poisson approximation is valid. In particular, if the source contains several long-lived radionuclides, or if long-lived radionuclides are present in the background, but all atoms decay and produce counts independently of each other, and no atom can produce more than one count, then the Poisson approximation is still useful, and the standard deviation of the total count is approximately the square root of the mean.

Note: One does not evaluate the uncertainty of a count rate, $R = C/t$, by taking the square root of R. If the total count C has a Poisson distribution and t has negligible uncertainty, the standard uncertainty of R is given instead by $\sqrt{C}/t$ or $\sqrt{R/t}$.

The Poisson approximation is not always valid. For example, if the half-life of one or more radionuclides in a source is short relative to the count time t, the Poisson distribution may not be a good approximation of the binomial distribution. Also note that some radiation-counting measurements involve the counting of particles emitted by a radionuclide and its short-lived progeny, where one atom may produce several counts as it decays through a series of short-lived states. A well-known example of this type of measurement is alpha-counting ^{222}Rn and its progeny in an alpha scintillation cell, or "Lucas cell." For such measurements, neither the binomial nor Poisson model is valid. Another example is the use of the Poisson model to estimate the total uncertainties of gross and background counts measured by gamma-ray spectrometry. While the use of the Poisson model is applicable in this case, there are some complications to consider (see Sections 10.2 and 10.3.7).

10.3.5. The Poisson Distribution

If X is a random variable with a Poisson distribution, and μ denotes its mean, then the following equation describes the probability, $\Pr(X = x)$, of observing any particular value x (for non-negative integers x):

$$\Pr(X = x) = \frac{\mu^x e^{-\mu}}{x!} \tag{10.14}$$

When the mean μ is small, the distribution of values about μ is skewed because $\Pr(X = x)$ cannot be negative. When μ is large—say 20 or greater—the Poisson distribution approaches a Gaussian distribution like the one shown in Fig. 10.1 (but with the standard deviation $\sigma = \sqrt{\mu}$). The probability function for the Gaussian approximation of the Poisson distribution is shown in Eq. (10.15):

$$\Pr(X = x) = \frac{1}{\sqrt{2\pi\mu}} e^{-\frac{(x-\mu)^2}{2\mu}} \tag{10.15}$$

Figure 10.2 shows how the Poisson distribution approaches the Gaussian approximation as the mean μ increases. In the figure, the height of a bar over a value x represents the actual Poisson probability $p(x)$ of observing that value, while the height of the smooth curve over the point x represents the Gaussian approximation of this probability. When μ is only 3, it is clear that the Poisson distribution is skewed and the Gaussian approximation does not describe it well. When $\mu = 10$, the distribution is more symmetric and superficially Gaussian in shape. By the time μ reaches 20, the Gaussian approximation is good enough for many purposes, and the approximation only improves as μ increases further.

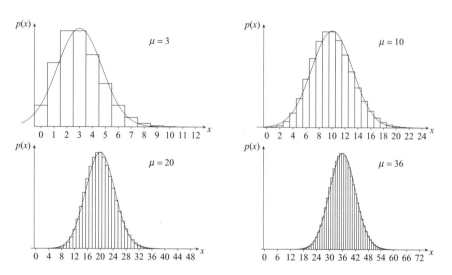

FIGURE 10.2. Poisson probability functions with Gaussian approximations.

10.3.6. Statistical Test for a Poisson Distribution

Many common mathematical formulas used for calculations of uncertainty, detection capability (see Section 10.4), and quality control in the radioanalytical chemistry laboratory are based on the assumption of Poisson counting statistics and are valid only if that assumption is valid. In particular, the formulas depend on the fact that the mean and variance of a Poisson distribution are numerically equal. For this reason it may be a good idea to test the Poisson assumption from time to time.

When a statistical population has a distribution that is approximately normal, a chi-square test can be performed on a random sample from the population to check its variance. Since the Poisson distribution is approximately normal whenever its mean is large enough (e.g., 20 or more), the chi-square test can be adapted to check whether the mean of a set of counting data equals its variance. For example, if a long-lived source is counted n times with the same detector to generate the counts $C_1, C_2, \ldots, C_n$, then one may calculate the chi-square statistic given by Eq. (10.16):

$$\chi^2 = \frac{1}{\bar{C}} \sum_{i=1}^{n} (C_i - \bar{C})^2 \qquad (10.16)$$

Under the hypothesis that the data distribution is Poisson, this statistic has (approximately) the chi-square distribution with $n - 1$ degrees of freedom; so, one can test the hypothesis by comparing χ^2 to appropriate percentiles of that distribution. Chi-square tables are commonly given in books and handbooks about statistics; many are also found on the Internet, though you must be careful to choose a reliable source. Generally, one does not expect the variance of the data to be less than the mean. The question in most instances is whether the variance is larger. So, one usually performs a one-sided test rather than a two-sided test. For example, if $n = 10$, one might compare the calculated value of χ^2 to the 95th percentile of the chi-square distribution with nine degrees of freedom, which equals 16.9. Then, if $\chi^2 > 16.9$, one would conclude that the distribution was not Poisson and that the variance was larger than the mean.

Note: It is appropriate to choose the 95th percentile for a one-sided test at the 5% level of significance, or the 2.5th and 97.5th percentiles for a two-sided test. Other significance levels are possible, but 5% is common.

Although a one-sided test is most common, a two-sided version of the test might be used in some situations. In this case one would compare the value of χ^2 to the 2.5th percentile and the 97.5th percentile of the chi-square distribution with $(n-1)$ degrees of freedom. For example, if $n = 10$, these percentiles are 2.7 and 19.0. Any value of χ^2 less than 2.7 or greater than 19.0 indicates that the distribution is not Poisson.

10.3.7. Spectrometry

Uncertainty evaluation for gamma-ray spectral analysis is more complicated than for most other radiation measurements. First, gamma-ray counts for one radionuclide are observed in several channels of a multichannel analyzer and form a more-or-less Gaussian photopeak that must be identified and integrated. Second, the counts estimated to be associated with the Compton baseline under the peak must be subtracted from the total peak counts, as discussed in Section 10.2. Third, any gamma-ray-emitting radionuclides of interest that are present in the detector background must be subtracted from the peak counts. Fourth, mathematical techniques may have to be applied to resolve peaks from different gamma-ray emitters that overlap to form multiplets. Each of these issues affects the uncertainty evaluation. At times, other complications include summing, escape peaks, and Compton edges. On the other hand, observation of a peak yields additional information by confirming the presence of the specific radiation, compared to the simple use of counts to calculate the uncertainty discussed in Section 10.3.4.

The problem of uncertainty evaluation for background corrections in alpha-particle spectrometry is that the background count rate may be extremely low. Counting for several days may yield only a few, or even zero, counts. At least two issues may arise. First, the Poisson uncertainty estimate is relatively imprecise when the mean count is so low. In the worst case, both the background count C_B and the gross count C_G are zero, so that the uncertainty equation will calculate zero uncertainty for the measured activity, which is inappropriate for a laboratory to report. In any situation where it is possible to observe zero counts ($C = 0$), the Poisson uncertainty estimator $\sqrt{C}$ should be replaced by $\sqrt{C + 1}$. Low-level alpha particle spectrometry is a situation where zero counts may indeed be observed. Second, it becomes important to consider possible variability in the background (e.g., from spurious counts), which may be relatively large in comparison to the Poisson uncertainty.

10.3.8. Other Components of Uncertainty

The many other possible sources of uncertainty in a radioanalytical measurement include:

- Measurements of masses and volumes
- Variable instrument background
- Dead time
- Tracer activity
- Calibration
- Contaminants in tracers and reagents
- Values of physical constants such as radionuclide half-lives and radiation emission probabilities
- Laboratory subsampling of heterogeneous material

MARLAP (EPA 2004) describes methods for evaluating the uncertainties of masses and volumes, but it also points out that these components of uncertainty tend to be negligible. Generally these uncertainties can be neglected if balances and volumetric devices are maintained and used properly.

The uncertainty due to varying instrument background can be significant. If the background varies, the assumption of pure Poisson counting statistics to evaluate the uncertainty of the net count rate for a sample may seriously underestimate the uncertainty. Options include replicate background measurements to determine its uncertainty (Type A evaluation), or to evaluate an additional component of uncertainty to be added to the Poisson counting uncertainty.

Uncertainty due to dead time can be avoided by keeping the dead time rate low (e.g., less than 10%). If the dead time rate is not negligible, the gross count rate must be calculated for the live time rather than the clock time (real time); if the dead time rate is not excessive, no additional uncertainty component for dead time is usually needed.

The uncertainty due to the concentration of a tracer used to measure the chemical yield may be significant. It should be evaluated and propagated when determining the uncertainty of the yield.

Calibration uncertainty should be evaluated and propagated. The causes of calibration uncertainty include uncertainties in the reported values of calibration standards, measurement uncertainty in the laboratory such as counting statistics, and in processing standards.

Significant contamination in tracers and reagents should be measured and corrected for if it cannot be eliminated. Corrections for contaminants may introduce uncertainty that should be propagated when calculating the combined standard uncertainty of the sample activity. Options for assessing such contamination include direct measurement of the contaminant's concentration in particular solutions, or replicate measurements of reagent blank samples.

Uncertainties in physical constants may or may not be significant, depending on the measurement. Half-lives in particular tend to be well known and usually contribute only negligible uncertainty to the result of a radioanalytical measurement. Errors in radiation emission probabilities may contribute additional uncertainty to some measurements. Uncertainties for both categories are available from tables of isotopes and the public Web site of the National Nuclear Data Center at Brookhaven National Laboratory.

Laboratory subsampling of heterogeneous material may contribute significant uncertainty to the result of the measurement, especially if the sample contains tiny high-activity particles, i.e., "hot particles." Currently, the most fully developed theory of sampling errors is that of the French geologist Dr. Pierre Gy. The theory explains that the relative uncertainty due to subsampling decreases if one either takes a larger subsample or grinds the sample to reduce particle sizes before subsampling. MARLAP summarizes the simpler aspects of Gy's sampling theory. It suggests a simple equation that can be used by default to evaluate the relative standard uncertainty due to subsampling particulate solid material, such as soil, when "hot particles" are not suspected and a reasonable effort has been made to homogenize the sample before subsampling (EPA 2004). The equation suggested by MARLAP is

$$u_r = \sqrt{\left(\frac{1}{m_S} - \frac{1}{m_L}\right) \times kd^3} \qquad (10.17)$$

where

u_r denotes the relative standard uncertainty due to subsampling
m_S is the mass of the subsample (aliquot)
m_L is the mass of the entire sample
k is, by default, 0.4 g cm^{-3}
d is the maximum particle diameter (size of a square mesh), in cm

For example, if a very large sample ($m_L = \infty$) is ground until it passes through a sieve with mesh size $d = 0.1$ cm, and a subsample of mass $m_S = 0.25$ g is removed for analysis, this equation estimates the relative subsampling uncertainty to be $u_r = 0.04$, i.e., 4%.

10.4. Radionuclide Detection

One of the most controversial topics in analytical and radioanalytical chemistry involves measures of detection capability and the methods used to decide whether an analyte is present in a laboratory sample. Despite the controversy and confusion, the basic theory is simple. The only difficulties that ought to exist are those associated with the application of the theory.

The term *detection capability* refers to the ability of a measurement process to *detect* the analyte, where to "detect" the analyte means to decide, based on the result of an analysis, that the analyte is present in the sample. Detection capability is often called "sensitivity," although the term *sensitivity* is also commonly used to mean the ratio of the change in an output quantity to the change in an input quantity (e.g., the slope of a calibration curve). The latter meaning of sensitivity is the one implied by the term "sensitivity coefficient" in Section 10.3.1.

The laboratory may often be required to make detection decisions about samples, but when the analyte activity is low enough, the relative uncertainty in the result may make it difficult to distinguish between a small positive activity and zero. The performance characteristic of the measurement process that describes its detection capability is called the *minimum detectable value, minimum detectable activity, minimum detectable concentration* (MDC), or *lower limit of detection* (LLD). These terms have been used to denote the theoretical concept of the smallest true value of the analyte in a sample that gives a specified high probability of detection.

10.4.1. The Critical Value and the MDA

The common approach to detection decisions in radioanalytical chemistry is based on statistical hypothesis testing. In a hypothesis test, one formulates two mutually exclusive hypotheses, called the *null hypothesis* and the *alternative hypothesis*, and uses the data to choose between them. The null hypothesis is presumed to be true unless there is strong evidence to the contrary. When such evidence is present, the null hypothesis is rejected and the alternative hypothesis is accepted.

Typically in radioanalytical chemistry, the null hypothesis is the hypothesis that no analyte is in the sample. Even if no analyte is present, the net result of the measurement has uncertainty, and, if the measurement were repeated a number of times, a distribution of results about zero, including both positive and negative values, should be observed. Although results near zero are most likely, in principle there is no upper or lower bound for what the result might be. Observation of a positive result in a single measurement does not necessarily constitute strong evidence that the analyte is present. The result must exceed some positive threshold value, called the *critical value*, to lead one to conclude that the analyte is really present. The question is: how to determine the critical value?

Before choosing the critical value, one specifies one's tolerance for *Type I errors*, defined as erroneous rejection of the null hypothesis when it is true. Type I errors are sometimes called *false positives* or *false rejections*. Regardless of the choice of the critical value, there is always the probability of a Type I error. For large enough critical values, this probability is small and generally can be tolerated. The tolerable probability that one specifies is called the *significance level* of the test and is usually denoted by α. In radioanalytical chemistry, it is common to set $\alpha = 0.05$. If $\alpha = 0.05$, then analyte-free samples should produce false positive results at a rate of only about one per twenty measurements.

The MDA for a radioanalytical measurement process is defined as an estimate of the smallest true activity (or massic activity or volumic activity) in a sample with a specified probability $1 - \beta$ that the net count rate will exceed the critical value, to conclude correctly that the sample contains the analyte. The probability β of a *Type II error* is defined as the error of failing to reject the null hypothesis when it is actually false. A Type II error is also called a *false negative* or *false acceptance*. It is common to choose $\alpha = \beta = 0.05$, but other values of α and β can be selected.

10.4.2. Calculation of the Critical Value and MDA

In radioanalytical chemistry, the critical value can be calculated by either of two common approaches. One is based on repeated measurements of blank samples, and the other on the assumption of Poisson counting statistics. The former method is generally applicable to any measurement process for which the distribution of measurement results is approximately normal. The latter method is useful when the Poisson assumption is valid, but it may give misleading results in other situations.

Suppose the laboratory performs measurements of n blank samples, where the blanks simulate real analyte-free samples as nearly as possible and the results $B_1, B_2, \ldots, B_n$ are expressed as yield-corrected activity. Then the laboratory may calculate the arithmetic mean $\bar{B}$ and the experimental standard deviation $s(B_i)$ of the values (see Eq. 10.5), and it may blank-correct the absolute activity found in a subsequent measurement of a real sample by subtracting the mean $\bar{B}$ from it. In this case the critical value L_C for the net absolute activity is calculated as shown in Eq. (10.18):

$$L_C = t_{1-\alpha}(n-1) \times s(B_i) \times \sqrt{1 + \frac{1}{n}} \tag{10.18}$$

In this equation $t_{1-\alpha}(n-1)$ denotes the $(1-\alpha)$-quantile of the Student's t-distribution with $n-1$ degrees of freedom. For example, if $\alpha = 0.05$ and $n = 10$, then $t_{1-\alpha}(n-1) = t_{0.95}(9) = 1.833$.

If one performs the series of blank measurements described above, calculates L_C by Eq. (10.18), and subsequently calculates the blank-corrected activity A for an analyte-free sample, the probability of observing a value of A greater than L_C is approximately equal to α. If one makes the detection decision by comparing A to L_C, the false positive rate should be approximately α.

If $\alpha = \beta$ and the number of blank measurements n is not too small (say at least 5), and if the measurement variance does not increase rapidly with activity, the minimum detectable *absolute* activity L_D may be approximated as shown in Eq. (10.19):

$$L_D = 2 \times L_C \tag{10.19}$$

Typically, one expresses the MDA as a massic or volumic activity by dividing L_D by the amount of sample analyzed. If $\alpha \neq \beta$ or if variance increases too rapidly with activity, Eq. (10.19) is inappropriate. For these cases see MARLAP (EPA 2004).

When the radiation counting statistics are essentially Poisson for the background count and there are no interferences, the critical net count rate (critical value of the net count rate) may be calculated as follows:

$$S_C = z_{1-\alpha}\sqrt{\frac{C_B}{t_B}\left(\frac{1}{t_G} + \frac{1}{t_B}\right)} \tag{10.20}$$

where

S_C is the critical net count rate in s^{-1}
C_B is the observed background count
t_B is the background count time in s
t_G is the sample (gross) count time in s
α is the significance level (e.g., $\alpha = 0.05$ by default)
$z_{1-\alpha}$ is the $(1-\alpha)$-quantile of the standard normal distribution (equal to 1.645 when $\alpha = 0.05$).

In this case one makes the detection decision for a sample by comparing the observed net count rate R_N to the critical net count rate S_C.

When Eq. (10.20) is used for the critical value and $\alpha = \beta$, a commonly used approximation formula for the MDA is

$$\mathrm{MDA} = \frac{\dfrac{z^2}{t_G} + 2z\sqrt{R_B\left(\dfrac{1}{t_G} + \dfrac{1}{t_B}\right)}}{K} \tag{10.21}$$

where

z $= z_{1-\alpha} = z_{1-\beta}$

R_B is the background count rate in s

t_G is the sample (gross) count time

t_B is the background count time

K is an appropriate estimate of the denominator by which the net count rate is divided to calculate the final result of the analysis.

The denominator represented by K above might, for example, be the product $\varepsilon \times Y \times m \times D \times F$, which appeared in the previous example of a mathematical model for a radioanalytical measurement (Eq. 10.1). If there is significant variability in this denominator, a somewhat low estimate of its value should be used to avoid underestimating the MDA.

In many cases, the assumption of pure Poisson counting statistics is invalid because other sources of variability affect the distribution of measurement results. In these cases, the laboratory should consider determining the critical value and MDA based on a series of blank samples, as in Eqs. (10.18) and (10.19). When the Poisson model is valid but the background level is so low that the Poisson distribution is not approximately normal, other formulas for the critical value and MDA tend to give the best results, as discussed in MARLAP (EPA 2004).

Note: Radioanalytical chemists should avoid the common mistake of using the MDA as a critical value.

10.5. Calculations for Quality Control

The quality of measurements in the radioanalytical chemistry laboratory is evaluated by QC measures for both instrumental measurements and sample analysis, as outlined in Section 11.2.8. QC ensures that the measurement process remains in a state of statistical control so that uncertainty estimates are valid, and ideally should help to keep measurement uncertainties small. QC may also provide assurances that the quality of measurements does not decrease when there are changes in personnel, instrumentation, or methods.

Initial QC is devoted to assuring that analytical methods and detection devices are suitable for the purpose of the program and includes estimating the uncertainty of analytical results. Subsequent methodological QC efforts during data processing evaluate the acceptability of the results and add information on the degree of reliability.

10.5.1. Instrument Quality Control Calculations

Radiation detection instruments are checked at frequent intervals to assure that they are operating and responding correctly. To estimate analytical uncertainty and sensitivity, information must be obtained about the stability of pertinent instrument parameters, especially background count rate and counting efficiency. Additional parameters for spectrometers are energy response and peak resolution.

At installation, all instrument controls (voltage, gain, bias, discriminator, etc.) must be set according to supplier instructions, and the settings must be recorded for periodic inspection and consideration of need for future adjustments. Initial tests should check whether the count rate for a test source and the radiation background count rate are reasonable and meet purchase specifications. Spectrometers should be tested with sources that emit radiation at known energies to assure that the desired range of energies can be observed and that peak shape and resolution are acceptable relative to purchase specifications.

Once accepted, the instrument must be calibrated for counting efficiency and, if a spectrometer, for energy response. For the former, the radionuclide standards must be prepared by the national calibration facility—NIST in the United States—or by another facility in a manner traceable to NIST (ANSI/IEEE 1995). Standards used for calibration may be supplied as a point source, an extended source of the same geometric configuration as the samples that will be counted, or as a sealed solution which is converted by the user to the desired form (see Section 8.3). A certificate that contains all appropriate information described in Section 11.2.6 must accompany all standards. The typical relative standard uncertainty of radionuclide standards is in the range of 1–2%.

Handling the standard, placing the source near the detector, recording data, and processing results all must be meticulous operations because the reliability of all future calculations of activity from count rate depend on the efficiency calibration. Accumulated calibration counts should be sufficiently large to achieve a relative counting uncertainty of 0.01 or less (at least 10,000 counts). To perform measurements in a brief period (such as 1–10 min), the source usually is prepared to yield a relatively intense count rate, but not so intense that the count rate is affected by dead-time losses (see Section 8.2.2) or system contamination becomes a threat.

For calibrating energy response in terms of kiloelectron volts per channel, any set of radionuclides can be used, as long as their radiation energies are better known than the precision of measurement that will be reported for results from the calibrated instrument. For example, if radiation energies are to be reported to four significant figures, the calibration radiation should be known with uncertainty in the fifth significant figure, e.g., 1.1732 MeV for the lower energy ^{60}Co gamma ray.

Calibrated instruments at specified control settings are checked by measuring an instrument-control source and the radiation background as described in Section 11.2.9. Initially, both measurements are repeated to establish a mean value and calculate the standard deviation and its multiples for defining acceptable limits for operation. A useful device for keeping track of instrument performance is the control chart shown in Fig. 10.3. The control chart consists of parallel lines at the mean value and the upper and lower 2σ and 3σ uncertainty values; its x-axis is the calendar. The lines are horizontal for both the source and the background if the source rate of radioactive decay is negligible during the time period of the chart. For a control source that has observable decay during the period of observation,

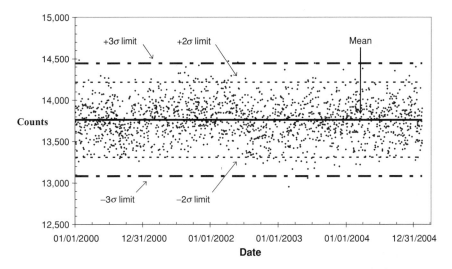

FIGURE 10.3. Typical radiation detection instrument control chart.

the mean and bounding values for the source must curve downward with the slope
defined by the radioactive decay constant.

The instrument control source is not a standard. Its main criteria are stability—
i.e., no observable change in the radionuclide content (other than by decay) or
configuration—despite frequent handling over the period of use, and known radioactive constituents. Its radionuclide content is selected to have a count rate that
is high but not excessive, as discussed above, so that a 1- to 10-min counting period is sufficient to accumulate about 10,000 counts. The background count usually
requires a much longer time period—50 to 1000 min, depending on the type of
detector—but still will have a larger relative standard deviation than the source.
Hence, the upper and lower bounds for the source control chart will be much closer
together, i.e., have a smaller relative standard deviation, than for the background
control chart.

During routine operation of the instrument, the control source and the background are measured at selected intervals and the number of counts is entered on
the control chart. This may be done by hand or by computer, as in Fig. 10.3. A
measurement point that falls within the 3σ control limits is acceptable.

Any value outside the 3σ limits indicates that a problem exists and correction is
required. The instrument should be checked for easily corrected problems such as
control settings, computer instructions, applied voltage, radioactive contamination,
ambient radiation levels, temperature changes, dirt, and noise. Points outside of
the 3σ limit may occur, for example, if small errors in the decay correction cause
an upward or downward trend in the data that results in outlying values over time.
If the problem cannot be corrected at this level of trouble shooting, the instrument
is scheduled for repair. Values that are too low generally result from component
failures or a decrease in the applied voltage. Values that are too high often result

from high ambient radiation levels—notably high radon concentration in counting-room air—or an increase in the applied voltage.

Normal instrument operation should result in control chart values that lie within the 2σ limits about 95% of the time and between the 3σ limits more than 99% of the time if source and background counts are randomly distributed. Nonrandom distribution requires evaluation of instrument operation, whether the distribution is too narrow or too wide. Any obvious trend in the distribution so that the center line no longer represents the mean value should also be evaluated, leading either to redetermining the mean and standard deviation or considering the problems cited above.

Instead of the above-cited control chart to test the hypothesis that the measured values are randomly distributed, a tolerance chart may be established to compare periodic measurements with established acceptance limits (see ANSI N42.2, Appendix A). In this system, an acceptable relative standard deviation—say 2% for the standard source count and 10% for background count—is selected and horizontal lines that represent these values are drawn as upper and lower acceptance limits. These limits should be wider than the statistically based limits; if they were narrower, they would be exceeded frequently. They need not be symmetrical on the positive and negative side. Their purpose is to keep the counting efficiency and background count rate within limits that give acceptable results even if the instrument and its output are subject to nonrandom fluctuations.

10.5.2. Analytical Quality Control Calculations

The usual QC sample program for routinely analyzed sample sets consists of laboratory blanks, laboratory-fortified (or spiked) blanks, and matrix spikes (in EPA terminology), as described in Section 11.2.9. The laboratory blank is a sample prepared with laboratory water. The laboratory-fortified blank is laboratory water with a known amount of radioactive tracer added. This tracer usually is the same radionuclide that is being analyzed. The matrix spike is a replicate of an actual sample to which a known amount of radioactive tracer has been added. At selected intervals, all are processed in the laboratory and counted exactly as are the routinely analyzed samples.

A number of different test statistics are used for evaluating the results of radio-analytical chemistry QC analyses. The test statistic for a QC analysis should take into account the combined standard uncertainty of measurement.

Laboratory blank measurements represent the radiation background count rate of the samples: the sum of the count rates due to detector background and any radioactive contaminants in the prepared counting source. Ideally, the latter contributes little to the background count rate. The cause of any significant increase over the detector background count rate should be investigated and—if possible—eliminated. The recommended test statistic for a reagent blank analysis is computed as follows:

$$Z_{rb} = \frac{A}{u_c(A)} \tag{10.22}$$

Here A is the measured sample activity and $u_c(A)$ is its combined standard uncertainty. Acceptance limits for Z_{rb} may be analogously to 3σ control limits on a control chart or at other values based on the desired false rejection rate; they could, for example, be based on the estimated number of effective degrees of freedom for the combined standard uncertainty $u_c(A)$, described in the note in Section 10.3.2. Thus, the result of the reagent blank analysis A is found to be acceptable if its absolute value does not exceed a specified multiple of its combined standard uncertainty $u_c(A)$.

Laboratory-fortified blanks and matrix spikes both test the analyst's ability to obtain the expected result. The extent to which the net radionuclide concentration of the fortified blank (corrected for yield and radioactive decay) deviates from the expected value for the tracer radionuclide concentration is a measure of analytical bias. Any consistent deviation from the expected value should be investigated to eliminate the cause. Typical causes are the wrong counting efficiency, an analytical problem with interchange between carrier and tracer, unreliable yield determination, or erroneous tracer radionuclide concentration.

The recommended test statistic for a blank spike is computed as follows:

$$Z_{bs} = \frac{A - K}{\sqrt{u_c^2(A) + u_c^2(K)}} \tag{10.23}$$

Here A is the measured sample activity, K is the added spike activity, and $u_c(A)$ and $u_c(K)$ are the combined standard uncertainties of A and K, respectively. Acceptance limits for Z_{bs} may be set at value such as 3.

The percent recovery ($\%R$) for a blank spike is computed as in Eq. (10.24):

$$\%R = \frac{A}{K} \times 100\% \tag{10.24}$$

Here A represents the measured spiked sample activity and K is the actual activity of the spike added.

The difference between the activity of the matrix spike and that of the unspiked matrix should be approximately equal to the activity of the added radioactive tracer. The recommended test statistic for a matrix spike analysis is computed as in Eq. (10.25):

$$Z_{ms} = \frac{a_S - a_U - a_K}{\sqrt{u_c^2(a_S) + u_c^2(a_U) + u_c^2(a_K)}} \tag{10.25}$$

Here a_S is the measured spiked sample result, a_U is the unspiked sample result, a_K is the spike added, and $u_c(a_S)$, $u_c(a_U)$, and $u_c(a_K)$ are the estimated standard deviations in a_S, a_U, and a_K, respectively. All variables are again in the same units. If required, the percent recovery for a matrix spike is computed as in Eq. (10.26):

$$\%R = \frac{a_S - a_U}{a_K} \times 100\% \tag{10.26}$$

where a_S represents the measured spiked sample result, a_U the unspiked sample result, and a_K the actual concentration of spike added.

Any difference in tracer results between the spiked and unspiked blanks and the spiked and unspiked matrix usually can be attributed to the effect of different chemical and physical forms of the radionuclide of interest and its tracer. In that case, the tracer does not represent the radionuclide of interest, and some treatment of the tracer plus radionuclide of interest, such as a redox process, will be needed to place them into identical forms. The cause of such inconsistency needs to be examined and resolved if the difference exceeds the uncertainty of the two values.

Accumulated results from laboratory-fortified blanks and matrix spikes are a measure of both precision and bias. To determine precision, numerous replicate values can be examined and the standard deviation can be calculated. The mean value is compared to the expected tracer concentration to determine bias. For this application, the results collected at different periods must be seen to belong to the same set, i.e., there is no obvious temporal difference among measurements due to analytical or measurement problems. The calculated standard deviation value can be compared to the suitably propagated components for counting and for the rest of the analytical process. Causes of unexpectedly large or small values of the standard deviation should be examined.

To obtain reliable values of the standard deviation, the accumulated tracer-related count must be sufficiently large, e.g., at least several hundred counts. The amount of tracer that is added must be similar to the amount of the radionuclide of interest in order to represent that level of activity but large enough to achieve a relatively small standard deviation of counting in a reasonable period. The latter is especially of concern for the matrix spike, where the result is the difference between counts that are often relatively small for the spiked and unspiked samples.

Combined analytical and measurement bias is also evaluated by periodic analyses of interlaboratory comparison samples described in Section 11.2.10. The radionuclide concentrations in these samples should have been measured with great care. Samples are submitted for blind analysis, i.e., they are not identified as test samples.

10.5.3. Analytical Method Test Quality Control Calculations

The applicability of radioanalytical chemistry methods is tested initially with radioactive tracers and realistic mock samples. A tracer that can be measured reliably and conveniently, such as a radionuclide that emits gamma rays, is preferred. If initial tracer tests are successful, tests are repeated with the media that will be analyzed. These tests must demonstrate that the radionuclide of interest is recovered consistently with good yield and that no interfering radionuclides or solids remain. The extent of reproducibility is determined by analyzing actual samples in replicate for chemical and radionuclide yield. Replicate samples are identical samples from the same batch, processed and counted separately to assess the variability or uncertainty in the analysis. The recommended test statistic for a duplicate analysis is computed using Eq. (10.27):

$$Z_{dup} = \frac{a_1 - a_2}{\sqrt{u_c^2(a_1) + u_c^2\, a_2()}} \qquad (10.27)$$

Here a_1 and a_2 are the two measured sample results, and $u_c(a_1)$ and $u_c(a_2)$ are their combined standard uncertainties. Acceptance limits for Z_{dup} may be set at ± 3 or at other values if desired.

If required, the relative percent difference (RPD) for a pair of duplicate analyses may be computed using Eq. (10.28):

$$\text{RPD} = \frac{|a_1 - a_2|}{(a_1 + a_2)/2} \times 100\% \tag{10.28}$$

where a_1 and a_2 represent the two measured results. If the average of the two results a_1 and a_2 is less than or equal to zero, the RPD is undefined.

Applicability of radioactive tracer tests to actual radionuclide measurements is always in question because the radionuclide in the sample may be in a different chemical or physical form. If the radionuclide has multiple oxidation states in nature, e.g., iodine or plutonium, a tracer study will be applicable only if the initial step in the procedure provides for the interchange of all possible oxidation states, or if tracers have been used in all possible oxidation states. Tests can become even more elaborate if the radionuclide of interest is an integral part of a solid sample matrix, e.g., biological material or soil.

In such cases, it may be desirable to reproduce the actual sample with tracers; for example, radioactive tracer is added to the roots of the growing vegetation that is to be analyzed. If going to such lengths is not feasible, all portions usually discarded during the testing of the procedure should be analyzed for the radionuclide to check for losses not indicated by chemical yield measurements.

10.6. Data Review

Erroneous values have a way of creeping into results despite the control measures in the Quality Assurance Plan (QAP). The data review process must be designed to remove such mistakes. Data review should accompany each step of the analytical process from sample receipt to calculation of activity. Thereafter, weeding out erroneous values becomes a management responsibility during data compilation for presentation and retention. Especially for the reality checks described below, the data compiler must be knowledgeable about radioanalytical chemistry processes as well as the pattern of analytical results related to sources of radionuclides, or must be assisted by specialists in these topics.

The data review process is conducted at the following points in the analytical process:

- Sample collection and shipment data review at sample receipt.
- Sample identification review at initial sample preparation and each subsequent sample transfer.
- Sample amount, analytical method application, sample response, and yield review during analysis.
- Sample counting and computer data processing review at radiation measurement.

- Calculational results review.
- QC results review, including instrumental, analytical, and methodological QC, to maintain statistical control of the measurement process.
- Confirmation of the preceding steps during data compilation.
- Search for missing samples and data during data compilation.
- Reality check of analytical results during data compilation.

The listed items emphasize that each sample must be processed with alertness at every step. At collection, the sample must be accompanied by all of the information required to calculate results and to assure that the correct sample is collected in the specified manner. The sample must be preserved and stored appropriately. Familiarity with the sampling plan and previous shipments is desirable to enable recognition of deviations from the plan. Care in sample identification avoids samples being switched, lost, or analyzed for the wrong constituents. Observation of strange behavior during analysis may indicate reasons for unusual yields or counting results. Application of inappropriate calculations and factors, by hand or computer, are common causes of error.

Problems are best recognized and corrected at the time of occurrence. If found later, during data compilation, they may be difficult or even impossible to remedy. At the very least, however, false results will not be reported.

Findings based on QC results of unacceptable circumstances during a period of operation (e.g., unduly elevated sample radiation background or deviation of reported tracer concentration addition from the expected value) invalidate sample data for the entire period to which the QC data pertain. This period extends back to the previous QC tests, although an investigation may narrow the problem period by dating the cause of error, e.g., onset of contamination or error in reagent preparation. The inverse inference, that correct results in the QC program validate the processes during the period since the previous QC test, is not necessarily true because errors may have occurred between QC tests.

One type of reality check is comparing a radionuclide measurement with values for similar samples collected at the same time or previously, either in the same program or reported by others. It is usually invoked when the result under consideration appears to be unusually low or high. As indicated in Chapter 12, the best response is reanalysis. If the questioned value is confirmed, an unexpected situation has been found. If the second result does not confirm the questioned value, the source of error should be sought in the analysis, radiation detection, sample identification, preservation, or possible contamination.

Reports of false positives or false negatives are a common problem, especially for gamma-ray spectral measurements with data analysis by computer. A false negative value may be suspected because the radionuclide is known to be at the monitoring site. A false positive value may be suspected because the radionuclide is unlikely to be at the location or in the sample—for example, a short-lived fission product where a nuclear reactor has been inactive for years. Results and calculations should be reexamined if the reviewer has doubts about the presence or absence of a reported radionuclide. Repetition of the measurement for a longer period or with a more sensitive detector can be ordered; ultimately, the sample can be reanalyzed if portions are still available.

In a gamma-ray spectrum, the reexamination may consist of considering details of the sample plus detector background. For example, peaks in soil samples at 766 and 835 keV that may be falsely attributed to ^{95}Nb and ^{54}Mn, respectively, actually can be minor gamma rays in the natural uranium and thorium decay chains. For measurements near or below the MDA, the usual peak identification and quantification software can be replaced with one that is more or less responsive to channel-by-channel fluctuations.

A reality check may also focus on scanning the raw data and initial calculations to seek major deviations from the norm. This search typically unearths typographical errors such as decimal-point shifts and digit reversals, or shifts of entire lines or columns. Obvious identifiers of error are improbable weights and volumes, unacceptable yields, and unusual detection efficiencies. More subtle sources of error are wrong decay schemes or interference by the natural background.

The chemical and physical characteristics of a radionuclide that affect its interaction with the sample matrix may offer suggestions whether its presence can be expected. For example, thorium and radiozirconium compounds are insoluble in water except under highly acid conditions and would not be expected in drinking water. Anions generally are not sorbed on soil and remain in groundwater. Generalities such as these, however, can be invalid under local conditions. Radiocesium ions are strongly retained in some soil types but not in others. Radioiodine can be found in airborne particles and also as a gas. Plutonium and the radiolanthanide compounds are insoluble in water near neutral pH values but can be soluble as organic complexes at the same pH values.

Uranium and technetium are soluble in water when oxidized but insoluble in reduced form. Radioisotope pairs, parent–progeny relations, and specific activity may provide guidance in assigning the origin of radioactive material and identify questionable results for a specified location. Certain radioactive materials have "signatures" of uranium or plutonium isotopes at known ratios. Ratios for other radionuclide pairs, such as $^{89/90}$Sr and $^{103/106}$Ru, can suggest the time interval since formation. The time interval can also be suggested by ingrowth of shorter-lived progeny into long-lived parents. The $^{134/137}$Cs ratio provides information on the neutron environment at origin. The specific activity of ^{3}H, ^{14}C, and ^{129}I can distinguish between natural and anthropomorphic creation.

The data reviewer may be guided in expecting a certain radionuclide concentration in a sample matrix by the radionuclide concentration in a related matrix and the reported transfer factor between these matrices. Relations exist among environmental samples of air, rain, vegetation, and milk, and between industrial samples of airborne particles and workers' urine. These relations are complex and often time- and site-related, but calculations ranging from rule of thumb to an elaborate computer program may suggest the magnitude of radionuclide concentration ratios among such matrices. A major inconsistency in the concentration of a radionuclide among related sample matrices suggests reevaluating the analytical data, although the conclusion may be that the data are reliable but the applied transfer coefficients are inappropriate.

A common problem in radioanalytical chemistry laboratories, especially those that analyze environmental samples, is the existence of an overwhelming number of samples without detectable radionuclide activity, which can lead analysts and data reviewers to expect such results as a matter of course. Reviewers who expect undetectable results for samples may tend to interpret small positive values as false positives even when they are not. The tendency to let one's preconceptions influence one's judgment in data interpretation must be resisted. Detection decisions and other evaluations of the data should always be based on objective criteria. If the laboratory's objective criteria indicate the presence of an unexpected radionuclide, an investigation may be needed to confirm that the measurement process is performing properly.

10.7. Data Presentation

The main criteria for presenting radioanalytical chemistry data are correctness, clarity, and usefulness to the intended audience. In some instances, the customer or regulator may prescribe the report format. When the format is left to laboratory management, its design should be directed by the data quality objectives plan (see Section 11.3.1) to respond to the program purposes. Tables, graphs, footnotes, and explanatory text should be combined to present the information unambiguously. Data presentation should be guided by the following items:

- Results that indicate an exceeded limit or guidance value should be highlighted.
- Every result should be accompanied by an explicit measure of uncertainty, such as the combined standard uncertainty or an expanded uncertainty. The measure of uncertainty should be clearly explained, and for an expanded uncertainty, the coverage factor should be stated.
- Results should be rounded appropriately to avoid (1) unnecessary loss of information and (2) misrepresentation of measurement accuracy. One acceptable rounding method is to round the reported uncertainty to two figures and to round the result of the measurement itself to the same number of decimal places as its uncertainty.
- Detectable and undetectable results should be clearly distinguished.
- The basis for detection decisions should be explained.
- All analytical results must be available in a form that permits review and use in further calculations.
- Missing data and questionable results must be listed and explained.
- Subsequently changed results due to corrections or new information must be reported when available, and replace earlier erroneous results.

The data presenter has a wide choice of formats, but should realize that critical concerns arise at two levels—the value associated with a guidance or limiting level and the critical value, which distinguishes between a detected and a nondetected

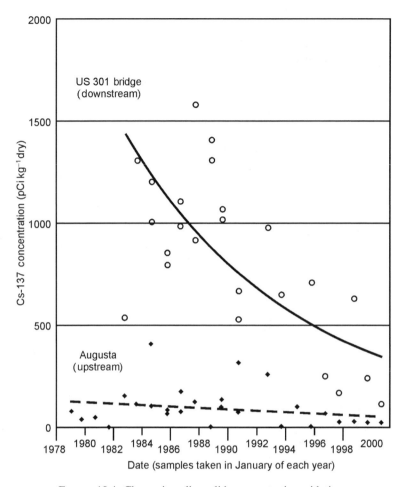

FIGURE 10.4. Change in radionuclide concentration with time.

result. The guidance or limiting level is established by a regulatory agency or by management. It may be direct, as a maximum contaminant level in drinking water, or indirect in relating the concentration in a sample to a radiation dose limit. The critical value (see Section 10.4) is arbitrary to the extent that it is related to a chosen false positive rate; association with 5% false positives is conventional, as discussed in Section 10.4.

An unresolved point of contention is the extent to which undetectable results should be reported. One view is that all data accepted after review must be reported. Values that are statistically indistinguishable from zero, including negative results, should be included to permit subsequent data review, analysis, and numerical treatments, including summing or averaging by collection location or time. The opposing view proposes a reader-friendly table that presents only the detected values, while all other results are stated to be undetectable. One resolution is to

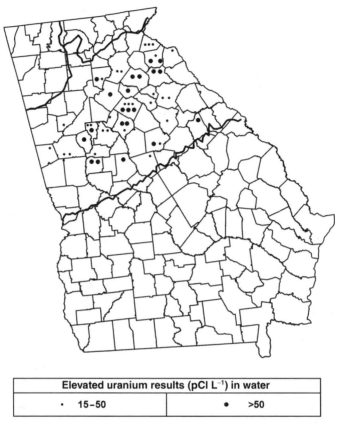

Elevated uranium results (pCl L⁻¹) in water	
• 15–50	● >50

FIGURE 10.5. Pattern of radionuclide concentration in water.

prepare both types of table, the former for the professional reader and the latter for overview by management and the public. Typically, the client of a laboratory should be knowledgeable enough to interpret a full report that includes all data, but tables and data summaries prepared for others may be tailored to the needs of the audience.

Multiple results, such as weekly data in an annual report, can be tabulated as the mean value with its standard deviation or values of the mean, highest, and lowest value by location and sample medium. Additional information should be provided if elevated radionuclide levels occur at specific times or locations. Patterns of changing radionuclide concentrations can be presented as time-line graphs. Spatial patterns of radionuclide concentrations can be presented as maps with concentration isopleths, which indicate the points at which each radionuclide has a specified constant concentration.

Presentations in figures can clarify and emphasize relations of radionuclide concentrations to release levels or release points. Figure 10.4 compares the higher radionuclide concentrations in river water downstream ("US 301 bridge") from a source to the lower concentrations upstream ("Augusta"). The downstream results are highly variable from measurement to measurement, but the gradual decrease with time is apparent in the graph. Figure 10.5 shows that the elevated uranium concentrations in groundwater supplies are clustered in only one of the physiographic regions marked by the heavy lines. Further pictorial information concerning contributions of radionuclides from various sources or the relation of radionuclide concentrations to their limits can be presented in bar graphs or pie charts.

The extent of detail depends on the purpose of the report. Even a brief progress report should characterize samples and their data so that they can be traced to more detailed records. The significance of the results in terms of magnitude, reliability, and pertinence must be understandable by the reader. Detailed reports for long periods or at the completion of the project should also provide all necessary information on sample collection and processing, analytical and radiation detection methods, quality assurance, and data processing, either as part of the report or by references. The names of the persons who participated in the program and their individual contributions are an important part of the record to permit further detailed review.

The units for reporting data are still under discussion at the time of this writing. The universal trend is toward SI units, which include the becquerel (Bq) for activity and the meter, kilogram, and second for length, mass, and time, respectively. However, in the United States at this time, regulatory agencies use the curie (Ci) for activity, defined as 3.7×10^{10} Bq. Moreover, usage such as river mile markers and cubic feet per second (cfs) remains common. A typical response in the United States is to report curies and its submultiples mCi, μ Ci, and pCi (10^{-3}, 10^{-6}, and 10^{-12} Ci, respectively) with SI units for length, volume mass, and time, with the exception that locations, flows, and such may remain in earlier English units.

11
Quality Assurance

LIZ THOMPSON,[1] PAMELA GREENLAW,[2] LINDA SELVIG,[3] and KEN INN[4]

11.1. Introduction

Quality assurance (QA) describes the effort by a laboratory organization to produce trustworthy results. Every laboratory, no matter how small, must maintain a continuing effort to confirm instrument calibration, measurement reproducibility, and applicability of analytical methods. These efforts must be documented so that the results achieved by the laboratory can be used confidently in a decision-making program. Various decision-making programs are supported by radioanalytical chemistry. A partial list of these programs includes

- protection of workers, population, and the environment in accord with regulations and professional guidance;
- determination that contractual requirements have been met, with regard to facility operation or decommissioning;
- verification that a decontamination response is necessary for a nuclear incident with the potential for environmental contamination;
- litigation regarding nuclear attribution or worker compensation;
- demonstration of compliance with guides or regulations for pollution control, e.g., for drinking water or air quality;
- comparison of results among laboratories;
- confirmation of compliance with international trade agreements, e.g., food import;
  verification of compliance with nuclear nonproliferation treaties such as comprehensive test ban monitoring; and
- publication of research.

This list emphasizes the importance of the data generated by a radioanalytical chemistry laboratory. The results produced by radiochemistry procedures are often subject to intense scrutiny and even expression of skepticism by sample submitters,

[1]Environmental Radiation Branch, Georgia Tech Research Institute, Georgia Institute of Technology, Atlanta, GA
[2]Department of Homeland Security, Washington, DC
[3]Centennial High School, Boise, ID
[4]National Institute of Standards and Technology, 100 Bureau Drive, Gaithersburg, MD 20899-8462

regulators, or members of the public. Laboratory results are routinely questioned in court, and must be defensible (see Section 11.5). The subject of contention may be a few positive or nondetected values or the entire contents of a report. The basis for such challenges is inconsistency with results from another laboratory, changes from earlier measurements, differences from predictions, or generalized distrust of the sampling process or laboratory. In some unfortunate instances, skepticism is justified. Bad data are produced by inappropriate analyses, and also by unreliable sampling, failed sample preservation, confusion in sample transfer, errors in data processing and reporting, and—worst of all—dishonesty. The contamination status of important sites (see Section 11.3.3) has been questioned as a result of monitoring failures. To prevent recurrence of these failures, a body of QA measures was developed and instituted.

As these QA measures became more pervasive and intricate, their formal organization as a quality assurance plan (QAP) was deemed necessary. Every laboratory should prepare a QAP that meets all the requirements stated by its operating license, its nuclear materials license (see Section 13.8.2), and all local, state, and federal regulations. In addition, the laboratory must prepare a QAP suitable for each customer who may have different format preferences or requirements.

Although the specific content of each QAP is determined by the individual entity, certain common elements must, by both consensus and regulation, be addressed. Consensus standards are QA formats that have been approved for broad use. They define the elements that must be addressed in a quality system, for that system to be considered comprehensive and acceptable. Consensus standards often have a "point of view" because they apply to a particular field or industry. For instance, ANSI-ASQ E4-1994 (American Society for Quality [ASQ], 1995) was developed for environmental programs by the ASQ and authorized by the American National Standards Institute (ANSI). It serves as the basis for the QA plan of the EPA (2000a), among many other entities. Some standards are even more specific in their scope: ASME NQA-1-1989 (ASME, 1989) sets forth requirements for the establishment and execution of QA programs for siting, designing, constructing, operating, and decommissioning of nuclear facilities. The elements of a QAP discussed in this chapter (Sections 11.2.1–11.2.13) are those of the format suggested by ANSI/ISO/ASQ Q9001-2000 (ISO, 1987), but the goal of this chapter is to indicate the utility of QA measures to the student rather than to insist on a rigid and formalized structure.

The laboratory that participates in large projects also must conform to the data quality objectives (DQO) of the project for which the samples are submitted and the data are used, and the quality assurance project plan (QAPP) for the project. The DQO and QAPP are intended to address every aspect of the project, including justifying the need for a sampling effort, selecting sampling sites and matrices, collecting samples, reporting the entire project results, and interpreting these results. These planning tools are discussed in Section 11.3. The DQO process is mentioned here only as it applies to the overall QA effort.

The fraction of time devoted to QA measures is an important economic consideration that balances the situation with the funds dedicated to the program.

Application of these measures to reduce the frequency of reanalysis required in the radioanalytical chemistry laboratory can be justified by a cost–benefit analysis. The general purpose of supporting acceptance of analytical results has a less determinate economic benefit. The time and effort devoted to QA has increased from about 10% of the total analytical workload to 25–30% at present, as regulatory agencies have required more quality control (QC) measurements and more supporting documents. This increasing cost includes QC measurements, organization and actions instituted to ensure correct and defensible data output. The costs and benefits of QA are considered in Section 11.4.

Acronyms are used pervasively in this chapter. Although they may annoy the reader, these terms are in common parlance when discussing QA, both in the literature and during communication with certain regulatory agencies.

11.2. The Quality Assurance Plan

The *Multi-Agency Radiation Survey and Site Investigation Manual* (*MARSSIM*) describes the quality system as consisting of all efforts devoted to produce quality results that are "authentic, appropriately documented, and technically defensible" (EPA, 2000c, Chapter 9). The purpose of the quality manual or QAP is to describe, in a single document, all elements of the quality system. This includes all laboratory policies regarding quality and all of the QA measures implemented by the laboratory. Some records may be included in the QAP itself, while others are filed as specified in the QAP. The QAP defines every element of laboratory operation and represents the laboratory's commitment to quality. The interconnected elements of QA policy that must be included in the QAP are

- organization,
- personnel training,
- laboratory operating procedures,
- procurement documents,
- chain of custody records,
- standard certificates,
- analytical records,
- standard procedures,
- QC sample analysis program and results,
- instrument testing and maintenance records,
- results of performance demonstration projects,
- results of data assessment,
- audit reports, and
- record retention policies.

These categories are also addressed by other standards and guidance documents, as mentioned in Section 11.1; however, the order, title, description, degree of importance, and amount of detail given each topic differs among publications. For a point-by-point comparison of different standards, see *MARSSIM* (EPA, 2000c, Appendix K) or Appendix B of ASQC (1995).

The categories listed above are described in the following subsections. The degree of detail is intended as guidance for management to formulate an effective QA laboratory organization and to enable an auditor to overview the pertinent components of the organization in a single document.

11.2.1. Organization

The QAP must contain a staffing chart that presents the laboratory organization and describes the function and responsibility of each individual within the organizational hierarchy (see Fig. 13.7). The resume of each staff member should also be copied and filed in the plan. This information defines the chains of authority and communication and records the qualifications for analyzing samples and reporting results so that responsibilities can be assigned to the appropriate analytical and management staff persons. Every staff person is considered responsible for performing duties and reporting in accord with the QAP. The information must be kept current.

A QA officer must be designated to be directly responsible for managing the QAP with respect to preparing the plan, supervising its application, and reviewing its results. The QA officer must be competent in the fields of radioanalytical chemistry and QA. He or she assists in reviewing analytical data reports to address problems with analyses and radiation detection instruments and recommend problem solutions. The QA officer must be free to discuss results and findings directly with supervisors, analysts, and operators. He or she must report directly to upper management, with authority to recommend and initiate corrective action. In a large laboratory, the QA officer supervises a separate and independent QA staff; in a small laboratory, the QA officer may also fulfill other functions.

The QA officer and staff procure, store, and dispense QC samples, report QC results, and evaluate the implication of these results for the analytical program. The QA officer and staff prepare radionuclide sources for counting and solutions for "spiking," i.e., tracing the radionuclide of interest.

11.2.2. Personnel Training

The QAP must document all staff hiring and training. A competent staff is the foundation of reliable analysis. Assurance of competence begins by preparing position descriptions that specify the schooling, experience, skills, and responsibilities of each position, and hiring supervisors, analysts, and operators whose qualifications match those descriptions (see Section 13.7.1). Once hired, the person must be trained to perform the assigned duties. Such training should address both general knowledge that can be conveyed in lectures, and specific analytical and instrumental operations that are taught by practice. Emphasis should be placed on work practices that conform to the QAP and safe operation (see Chapter 14).

Periodic evaluations of employee performance should be recorded, with notes on deficiencies that require greater effort, additional training, or replacement. The QC program is an important means of checking analysts' and operators' skills and reliability.

Continuing education should be available to expand analyst and operator skills and provide opportunity for advancement. Cross-training for other laboratory positions permits flexibility in work assignments. Attention to new methods and improved instrumentation assists the laboratory in maintaining a state-of-the-art program.

11.2.3. Laboratory Operating Procedures

Laboratory procedures must be documented and readily accessible in the QAP manual. Individual procedures must be available for review when resolving an analytical problem, and the entire file for auditing laboratory operation. The file should contain all currently applied radioanalytical chemistry and instrumental procedures, variants of procedures (e.g., for special matrices or contaminants), and ancillary procedures (e.g., for reagent preparation, calibration, and QC). Each procedure and procedural change should be signed and dated. The QA officer is responsible for assuring that actual laboratory operation is reflected exactly in the manual. Periodic updating is necessary to formalize as revisions the insertions, deletions, and additions that can be expected to accumulate in the analyst's or operator's copy of a procedure.

Each radioanalytical chemistry procedure should be written in conventional style, with its title descriptive of radionuclide and medium, introductory discussion and method summary, list of equipment and reagents, numbered analytical steps, numbered measurement steps, and equations for calculating sample activity. Notes and comments should identify the origin of the procedure and address special precautions and optional variants. Appendices should contain information such as reagent preparation, instrument operation, and tabulated values for use in calculations.

Analogously, measurement procedures should include a descriptive title, introduction, numbered procedural steps, and radionuclide activity calculation should describe operation of radiation detection instruments. Appendices should record initial setup instructions, dial settings, maintenance actions and schedule, and instructions on applying computer software for control, information storage, and calculation. The location should be given for files of instrument description and operating instruction manuals and of records of purchase, repair contracts, repair history, and QC data. Modifications in the instrument, computer control, and operating instructions should be listed, dated, and signed.

All QA processes should be described in sufficient detail to perform them without added spoken instructions. These procedures include instrument calibration, QC measurements for radioanalytical chemistry, radiation detection, and ancillary instrument use, data review, and interactions with laboratory staff, both at specified intervals and in response to incidents. The locations should be given of filed records such as certificates, QC results, and reports of actions and audits.

Procedures must be included in the QAP for all other aspects of laboratory operation that can affect data quality, including sample receipt, storage, transfer, and disposal, equipment application, and data collection, processing, reviewing, and reporting. Guidance must be presented for ancillary activities for worker safety, laboratory security, and emergency response.

11.2.4. Procurement Control Documents

Written instructions and procedures must be prepared to control and document the receipt of equipment, supplies, and services that affect the quality of measurement. Examples are reference materials (e.g., radionuclide standards), reagents, supplies (e.g., graduated cylinders, pipettes, planchets), and computer software and hardware. Controls should ensure that only correct items are accepted. If specifications are not met or the material is otherwise unsuitable, the QA officer should be notified and the material returned.

Items with specified storage conditions and shelf life must be controlled by written instructions to prevent storage and use beyond limiting conditions, e.g., temperature, humidity, and expiration date. Equipment is used for its intended period of time and under the correct conditions, and undergoes maintenance work at specified intervals. Similar control measures must be prepared for installing, testing, using, and maintaining computer software and hardware. Test and calibration results must be documented and maintained.

National consensus standards that pertain specifically to computer software are available for reference, such as ANSI/IEEE Standard 730-2002 (IEEE, 2002). The user of equipment driven by software should refer to such documents to understand how the software will react under given conditions. When a user is unfamiliar with the intricacies of a software program, the results and their presentation are open to question.

11.2.5. Chain of Custody and Sample Handling

Each analytical result obviously must apply to the sample for which it is reported. Any mix-up must be avoided. The Chain of Custody system was developed to ensure that the reported result applies to the sample collected at the recorded time and location, and not to another sample inserted by mistake or to deceive. The recorded parameters are available for comparison with those specified in the monitoring protocol. Meticulousness in adhering to protocol is crucial for samples analyzed for legal or regulatory purposes. The format of the Chain of Custody form must match the QAP for all samples.

A sample Chain of Custody and Analysis Request form is shown in Fig. 11.1. The form must be signed by all its custodians with a listing of the periods of custodial control. It reports the sample type, its origin, the times of collection and transfer, and any treatment.

The practical application of the Chain of Custody form in the laboratory is twofold: information for sample analysis and data calculation, and formal evidence of the sample's history. The form must be preserved for both purposes. The information on the form may also serve as criterion for sample rejection at the laboratory because of contents with elevated radioactivity or hazardous material.

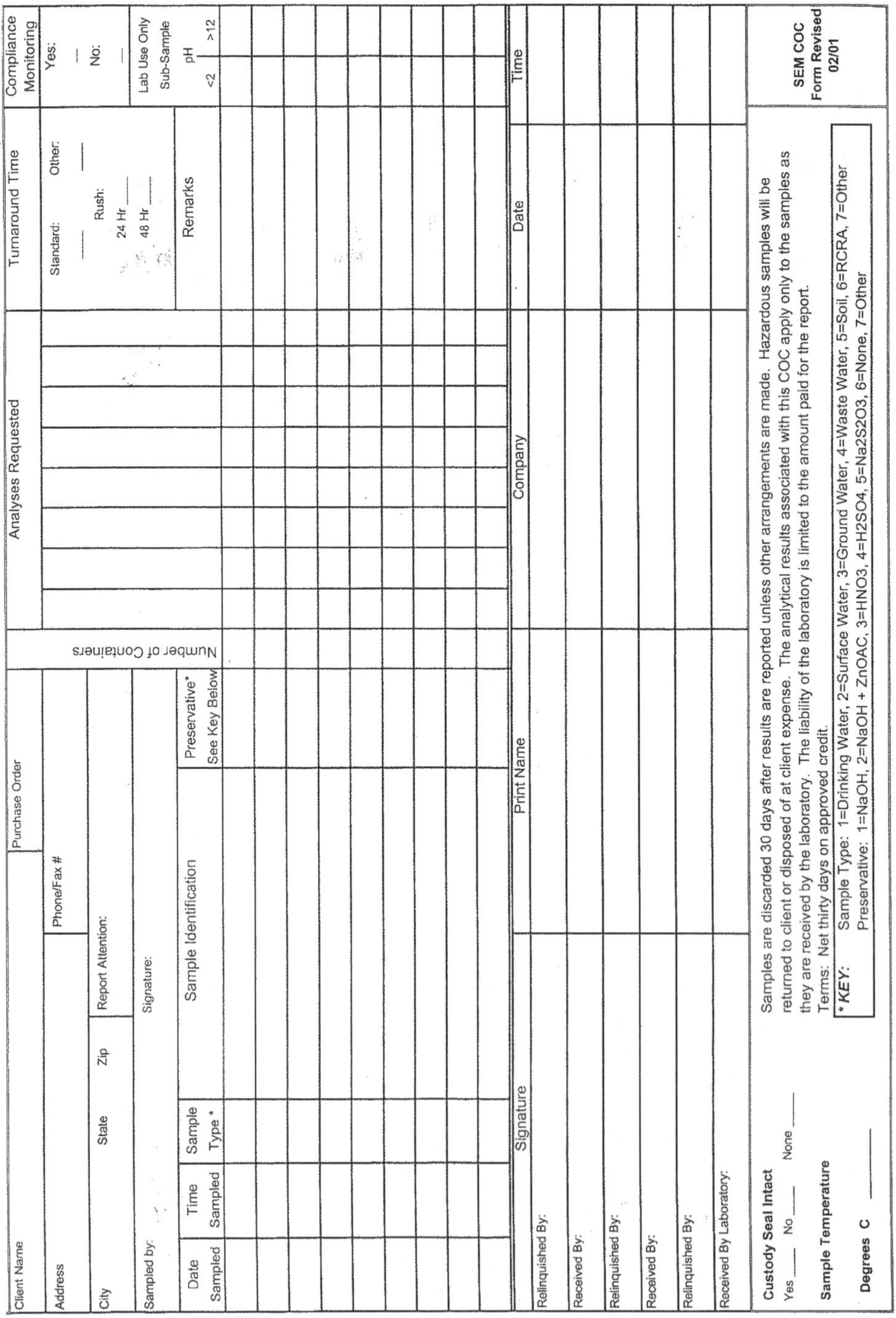

Client Name | Purchase Order | Number of Containers | Analyses Requested | Turnaround Time | Compliance Monitoring

Address | Phone/Fax #

City | State | Zip | Report Attention:

Sampled by: | Signature:

Turnaround Time: Standard: ____ Other: ____ Rush: 24 Hr ____ 48 Hr ____

Compliance Monitoring: Yes: ___ No: ___

Lab Use Only Sub-Sample pH <2 >12

Date Sampled	Time Sampled	Sample Type *	Sample Identification	Preservative* See Key Below	Number of Containers	Analyses Requested								Remarks	pH <2	pH >12

Signature	Print Name	Company	Date	Time
Relinquished By:				
Received By:				
Relinquished By:				
Received By:				
Relinquished By:				
Received By Laboratory:				

Custody Seal Intact
Yes ____ No ____ None ____

Sample Temperature

Degrees C ________

Samples are discarded 30 days after results are reported unless other arrangements are made. Hazardous samples will be returned to client or disposed of at client expense. The analytical results associated with this COC apply only to the samples as they are received by the laboratory. The liability of the laboratory is limited to the amount paid for the report.
Terms: Net thirty days on approved credit.

KEY: Sample Type: 1=Drinking Water, 2=Surface Water, 3=Ground Water, 4=Waste Water, 5=Soil, 6=RCRA, 7=Other
Preservative: 1=NaOH, 2=NaOH + ZnOAC, 3=HNO3, 4=H2SO4, 5=Na2S2O3, 6=None, 7=Other

SEM COC Form Revised 02/01

FIGURE 11.1. Sample Chain of Custody and Analysis Request form.

Once the sample is accepted at the laboratory, an internal chain of custody begins, controlled by a Laboratory Information Management (LIM) system that may be manual or computerized. A unique sequential identification (ID) number is assigned to the sample and affixed to its containers as they move through the laboratory. The ID is recorded by the sample preparer, the analyst, and the radiation detector operator, and is associated with all records of chemical yield, counting data, calculations, and reported values.

11.2.6. Standards Certificates

Standard radioactive Material (SRM) solutions are radioactive materials with accurately known radionuclide content and radioactive decay rate or rate of particle or γ-ray emission. They are used primarily to calibrate radiation detection instruments and to prepare QC samples that test analytical accuracy. The supplier prepares radionuclide standard solutions in flame-sealed glass ampoules. Other standard radionuclides are in the form of point sources (usually on thin backing) or as solids in configurations that represent actual samples.

The QA officer is responsible for appropriate use of SRMs and evaluation of the resulting instrument calibration data, as discussed in Section 8.3.1. The SRM solutions and their dilutions should be stored in containers that are tightly stoppered and meticulously labeled to describe the contents and their origins.

Activity calibration of SRMs must be performed by the National Institutes of Standards and Technology (NIST) or be traceable to NIST. If the radionuclide standard source is purchased from a supplier other than NIST, its accompanying certificate must confirm that the standard is traceable to NIST. Traceability requires reporting an unbroken chain of comparisons to stated references. NIST traceability is discussed at http://ts.nist.gov/traceability/ (December 2005). A terminal date for use of a specific standard is appropriate when the radionuclide decays to a small fraction of its original concentration, an initially minor interference has become major, or chemical decomposition occurs with time.

The SRM must have a calibration certificate with key information; an example is shown in Fig. 11.2. Standards certificates must be filed with the sample ID numbers they accompanied, and the calibrated instrument identifiers.

All certificates must contain information that identifies the physical, chemical, and radiological properties of the SRM. Physical properties include density and mass. Chemical properties identify composition such as chemical form, acidity, and salt content. Radiological properties are the name of the radionuclide, the decay scheme, half-life, daughter radionuclides, reference time for calculating decay, and radionuclide impurity amounts. The most important information is the radionuclide activity or concentration and the uncertainty of this value. Details must be provided about the technique used to determine activity and the basis for the uncertainty value. Technical information such as the production method also may be given on the certificate.

To view or download a NIST certificate in its entirety, go to the following URL: https://srmors.nist.gov/pricerpt.cfm (December 2005). NIST produces a

PROPERTIES OF SRM 4241C
(Certified values are shown in bold type)

Source identification number	NIST SRM 4241C-		
Physical Properties:			
Source description	Dried deposit of barium chloride on a filter-paper disk sealed between two layers of polyester tape that are supported on an aluminum annulus		
Point-source specifications	Aluminum annulus O.D. 5.4 cm Aluminum annulus I.D. 3.8 cm Aluminum annulus thickness 0.05 cm Polyester tape thickness 0.006 cm Filter paper diameter 0.6 cm		
Chemical Properties:			
Point-source composition	Chemical Formula	Mass (g)	
	$BaCl_2$ $^{133}BaCl_2$	7×10^{-4} 3×10^{-8}	
Radiological Properties:			
Radionuclide	**Barium-133**		
Reference time	**1200 EST, 1 January 1999**		
Activity	**kBq**		
Relative expanded uncertainty (k=2)	**0.60%** [b]		
Photon-emitting impurities	None detected [c]		
Half lives used	Barium-133: (10.52 ± 0.13) a [d] Radium-226: (1600 ± 7) a [d]		
Measuring instrument	Pressurized "4π" γ ionization chamber A calibrated using a barium-133 solution whose activity was determined by 4π(e+X)-γ-anticoincidence counting		

FIGURE 11.2. Summary SRM certificate for ^{133}Ba (Certificate available at https://srmors. nist.gov/certificates/view_cert2gif.cfm?certificate=4241C) (December 2005).

great many SRMs, all of which are included in this list; the 4000 series consists of radionuclide SRMs. Click an SRM number—e.g., 4241c for the Barium-133 point source—and then click on the "C" to view the certificate for specific radionuclide or matrix. The full NIST certificate contains more than the information shown in Fig. 11.2 and may be several pages of documentation and instructions on use.

Radionuclides that are used as comparison sources, e.g., for determining instrument count rate stability and as tracers, must be of sufficient radiochemical purity and activity to eliminate interference in counting and permit correction for decay, but in most instances they need not have an accurately known disintegration rate. The absolute count rate also is unimportant for energy calibration because only the energy must be accurately known.

11.2.7. Analytical Records

The basic recording medium in the chemistry laboratory and the counting room has always been the notebook. The numbered notebook pages are dated and signed by the user. Each analyst enters all pertinent information in a personal notebook, including the sample ID, what operations were performed (at least by reference to a method or an instrument), calculations, and any pertinent observations. Ancillary activities, such as QC measurements, reagent preparations, or instrument modifications, are also recorded. Unanticipated occurrences and findings also should be recorded. Changes or corrections should not be erased or whited out, but rather indicated by restating the information and drawing a line through the superseded information.

Separate notebooks are associated with sample log-in and each instrument, such as analytical balance and pH meter in the laboratory and radiation detector in the counting room. Each use is recorded in terms of operator, date, sample ID, and pertinent notes. All completed notebooks are filed permanently.

More recently, storage of information in computers has tended to superseded use of notebooks. The computerized LIM system can include all such information and relate it to sample ID and QC data presentation. Use of notebooks often continues on a limited basis. This dual data record may be a lag in technology development or reasoned record duplication in recognition of computer imperfections. In either case, storage of all analytical records, including raw data printouts, whether hard copy, disk, or tape, is mandatory, and the information must be retrievable, as dictated by the QAP.

11.2.8. Standard Procedures

American national standard procedures for performing tasks such as calibrating and operating instruments are written by specialists under the aegis of professional societies and published for the ANSI. Many standards are available for nuclear procedures; ANSI standards for some of the radiation detection instruments discussed in this text are listed in Appendix A-3. A standard that defines traceability of radionuclides, ANSI N42.22 (1995), is included in the list.

ANSI procedures should be considered for use if applicable. Their correct application ensures that the procedures under consideration are deemed generally acceptable in the profession and require no further demonstration of validity. Some clients require that these standards be used. Standard methods for radiochemical analysis are discussed in Section 6.5.

11.2.9. Quality Control Samples

A radioanalytical chemistry laboratory must have QC procedures to verify that the quality of combined chemical analysis and radiation measurement complies with accuracy requirements. The QC program is the direct-measurement component of the QA program, and is controlled by the QA officer. A detailed description of QC activities is part of the laboratory QAP manual.

Internal QC samples are prepared by the QC officer and inserted into the sample flow at specified intervals, normally to accompany each batch. The conventional QC sample types are as follows:

- *Reagent or laboratory blank*: To verify that method interference caused by contaminants in solvents, reagents, glassware, and other sources during sample processing is known and minimal.
- *Replicate*: To measure the reproducibility of an analysis.
- *Matrix spike* (an actual sample to which a known amount—the spike—of the radionuclide of interest has been added): To check yield by comparing the results of parallel analyses for the spiked and unspiked matrix. The amount of added spike is taken 1–20 times that of the radionuclide of interest to permit precise measurement of both.
- *Traceable reference material*: To test initially the accuracy and reproducibility of a procedure without considering the influence of the matrices. Sufficient spike is added for precisely measuring the counting source and any normally discarded fractions.
- *Blind traceable reference material*: To evaluate the accuracy of the analyst. Sufficient spike is added for precise measurement.

All but the fourth QC sample type are inserted in batches of routine samples. Although inserting these QC samples without the knowledge of the operator ("blind") ensures that they are not treated with special care, this may not be possible in small laboratories, where the analyst may recognize the QC samples. Distinctive features are their radionuclide concentration (i.e., radionuclide concentrations are zero in blank samples and elevated in spiked samples), salt content (blank samples may have none), or the manner of their insertion into the sample flow. The last of the above types also is used for external QC in the form of interlaboratory comparison samples (see Section 11.2.11).

The results of QC sample analyses are evaluated with statistical tests (see Section 10.5) and presented in tandem with the sample batch results to which they correspond. Good QC results give confidence that the laboratory procedures and personnel operated satisfactorily for that batch. Bad QC results show the need for investigation and remeasurement. The QA officer keeps track of the laboratory's QC efforts, and the results are filed as specified by the QAP.

The fraction of QC samples depends on the number of samples per batch. If a batch consists of about 20 samples, then as much as 20% of the sample load is devoted to QC samples. Larger batches not only reduce the fractional cost of QC analysis but also place more sample results at risk of a bad QC result.

11.2.10. Instrument Control Records

The QA effort for radiation detection instruments is directed toward correct installation, accurate calibration, and stable operation. For correct installation, the instrument must be shown to function as designed by the manufacturer and specified by the purchaser. Accurate calibration depends on reliable radionuclide standards handled correctly. Stable operation is demonstrated by comparing control parameters such as the count rate for a stable radiation source and the detector radiation background measured at selected intervals.

The QA manual should have a directory for file locations of instrument manuals and control records. Control records for each instrument should contain the date of the test, the name of the tester, the results, and any pertinent observations. Among these observations may be notes related to deviations and actions taken to correct these deviations. The QAP should require that all instrument control records are retained for the life of the instrument, and longer if subsequent reviews may occur.

Once radiation detection instruments are operational, they must be calibrated with radionuclide standards for response in terms of counting efficiency and, for some detectors, with radionuclides in terms of energy and peak resolution. The counting efficiency is affected by radiation type and energy and by sample dimensions, density, and placement. Counting efficiency results obtained with one standard should be confirmed by replicate measurements and with other standards. Such additional measurements also can be used to estimate the standard deviation of the efficiency values. In many instances, the efficiency values can be checked by calculations based on the geometry and associated factors (see Section 8.2.1) and by computer simulation. Curves or equations that represent trends in the variation of efficiency with factors such as radiation energy and sample thickness (see Figs. 7.1 and 7.2) can provide efficiency values by interpolation where none were measured and call attention to efficiency measurements that are questionable because of their difference from interpolated values.

The results are recorded in permanent form, defined by estimated uncertainty values (see Section 10.3), and applied in the calculations that are used for reporting radioactivity levels. Calibrations are repeated at specified intervals. More recent values are compared with earlier ones to resolve whether the new values are within the uncertainty of the old values, should supersede them, or require further measurements.

Background QC: At specified intervals, in many instances daily or for each batch, the background count rate for each system must be measured. The count rate is recorded and plotted on control charts, either by hand or by computer. The mean value of the background is found by averaging at least 20 measurements. The $\pm 2\sigma$ and 3σ geviations are calculated from the individual and mean values, and these multiple-standard-deviation lines are plotted on the control chart (see Section 10.5.1). Once the control chart is established, each newly measured value is recorded. The measurement should be repeated if it falls outside the 2σ band to distinguish between a random event and an instrumental problem. Remedial action with the detector or its environment is necessary if the repeated measurement is beyond the 3σ band.

Instrumental QC: For a routinely operating radiation detector, a count rate comparison source is measured initially about 20 times to establish the mean value and the standard deviation. A control chart with lines for the mean value and $\pm 2\sigma$ and 3σ values is prepared. The source then is counted at frequent intervals—typically daily or per batch—to demonstrate that the system is functioning properly. The count rate is recorded and plotted on the control chart for the specific system, as discussed above. This value is compared with the 2σ warning) limits and the 3σ (out-of-control) limits, and the procedure is repeated if the 2σ boundary is exceeded. Appropriate action must be taken if the measurement is beyond the out-of-control levels. One option is to continue use of the detector but recalculate the mean value and warning limits for the control chart. This approach is reasonable only if the values have reached another stable level. The other option is instrument repair.

Spectrometers have additional parameters for which response should be tracked. These include the count rate in spectral energy regions of interest, peak channel numbers at selected energies, difference in channel numbers between two specified energy peaks, peak energy resolution at specified energies, and the "peak to Compton" ratio (see Section 8.3.4). A QC chart can be established for each parameter by replicate measurements. This chart will display each data point as well as lines at the mean value and at $\pm 2\sigma$ and 3σ values.

The number of instrument QC checks typically represents about 5% of the total number of measurements made for a set of samples. The QC data and records of problems and responses should be kept accessible for the life of the instrument, and possibly beyond, to indicate the extent of instrument reliability. Changes in values may be examined for periodic patterns related to parameters such as temperature or airborne radon concentrations to consider remedial measures.

11.2.11. Performance Demonstration

Proficiency testing by interlaboratory comparison is the definitive process for demonstrating reliability of a radioanalytical chemistry laboratory at the time of measurement. In contrast to the QC samples described in Section 11.2.9, which are controlled by the QA officer within the laboratory, results of an interlaboratory comparison are evaluated and reported by an external group. The external group distributes the test matrix as a "blind" sample to participating laboratories, i.e., the radionuclide concentration in the distributed matrix is known to the external group but is unknown to the participating laboratories.

Typically, the external group is a testing laboratory such as the Radiological and Environmental Sciences Laboratory (RESL), a DOE laboratory that administers the Mixed-Analyte Performance Evaluation Program and DOE Laboratory Accreditation Program. The RESL obtains NIST-traceable SRM solutions, spikes a matrix with a known aliquot of the standard, and divides the matrix into many identical solution or solid fractions. The testing laboratory measures the radionuclide concentration in several representative fractions to ensure that they agree with the predicted value for the dilution and with each other within the estimated uncertainty of measurement. The interlaboratory comparison fractions are then distributed to participating laboratories at prearranged times.

Several matrices and radionuclides may be prepared for distribution to match the sample matrices and radionuclide concentrations handled by the participating laboratories. The radionuclide activity in these matrices usually is a compromise between the relatively low activity in the samples processed by the participating laboratories and the higher activity that improves precision. As a rule of thumb, the spiked concentration of the unknown sample should be about 20 times the

TABLE 11.1. Interlaboratory comparison testing levels

Measurement	Radionuclide[a]	Typical MDC[b] (per L or per sample)
I BETA particle activity Maximum energy > 100 keV	^{90}Sr	1 Bq (27 pCi)
II ALPHA particle activity isotopic analysis	$^{228/230}$Th or ^{232}Th	0.02 Bq (0.5 pCi)
	$^{234/235}$U or ^{238}U	0.02 Bq (0.5 pCi)
	^{238}Pu or $^{239/240}$Pu	0.01 Bq (0.27 pCi)
	^{241}Am	0.01Bq (0.27 pCi)
III GAMMA ray activity	^{134}Cs, ^{137}Cs, ^{60}Co	0.2 Bq (5 pCi)

[a] A laboratory may elect to be tested for a specific radionuclide or for the category. The testing laboratory will select the test radionuclide if a category is requested.

[b] The upper bound of the testing range should not exceed 20 times the stated MDC.

minimum detectable concentration (MDC) for a radionuclide. Performance on test samples with analyte concentrations much less than 20–200 times the MDC may not meet the statistical requirements for bias and precision testing. Table 11.1 shows the MDC for a few radionuclides.

Upon receipt of results reported by the participating laboratories, the testing laboratory publishes a summary report. The report contains the value and uncertainty obtained by the testing laboratory, the values and uncertainties reported by each participating laboratory (usually with encoded names), and notification that each reported value is acceptable, acceptable with a warning (i.e., barely acceptable), or unacceptable. For Table 11.1 and Fig. 11.3, the testing laboratory is the Environmental Measurements Laboratory (EML). The EML administered the Quality Assessment Program to test the quality of the environmental measurements reported to the Department of Energy by its contractors. Data and reports from recent Quality Assessment Program cycles can be found online at http://www.eml.doe.gov/qap/reports/ (December 2005).

Table 11.2 shows the EML testing results of one laboratory for two different matrices. The "EML Value" listed in the tables is the mean of replicate EML determinations for each nuclide. The contractors' results ("Reported Value") are divided by the "EML Value" to obtain the "Ratio." The column titled "Last Evaluation" gives the laboratory a comparison of its present and preceding performance.

This ratio is expressed graphically in Fig. 11.3 as the distribution of the resulting ratios for the ^{40}K analysis performed by the 109 laboratories participating in the program for that period. The lines on the graph define ranges of acceptability for the ratio value; these control limits are based on the historical distribution of data, are different for every matrix, and are specifically defined for each Quality Assessment Program period.

The upper and lower limits define the total range of acceptable results; a value above the upper (95th percentile) or below the lower (5th percentile) limit is marked Not Acceptable (N) by the testing laboratory. The upper middle (85th–

TABLE 11.2. Individual laboratory results for two matrices.

Radionuclide	Reported value	Reported error	EML value	EML error	Ratio	Evaluation	Last evaluation
Matrix: AI (Air filter) in units of Bq/filter							
[241]Am	0.286	0.017	0.340	0.040	0.841	W	—
[60]Co	31.930	0.530	33.500	0.870	0.953	A	A
[137]Cs	102.500	1.670	99.700	2.300	1.028	A	A
[54]Mn	45.500	0.800	43.800	1.130	1.039	A	W
[238]Pu	0.509	0.022	0.520	0.010	0.979	A	A
[239]Pu	0.323	0.015	0.330	0.010	0.979	A	A
[90]Sr	2.550	0.120	2.800	0.140	0.911	A	A
[234]U	0.244	0.013	0.240	0.003	1.017	A	A
[238]U	0.226	0.012	0.240	0.010	0.942	A	A
Matrix: SO (Soil) in units of Bq/kg							
[241]Am	14.8	1.9	15.6	1.0	0.949	A	—
[212]Bi	50.9	2.1	60.6	4.0	0.840	A	A
[214]Bi	56.2	63.4	67.0	2.3	0.839	W	A
[137]Cs	1,504.0	24.0	1,450.0	73.0	1.037	A	A
[40]K	659.0	11.0	636.0	33.0	1.036	A	A
[212]Pb	56.4	1.1	57.9	2.9	0.974	A	A
[214]Pb	63.4	1.2	71.1	2.3	0.892	A	A
[239]Pu	29.8	1.6	23.4	1.1	1.274	W	A
[234]Th	113.3	3.5	127.0	7.1	0.892	A	—
[234]U	126.0	5.2	120.0	0.5	1.050	A	A
[238]U	128.9	5.3	125.0	0.3	1.031	A	A

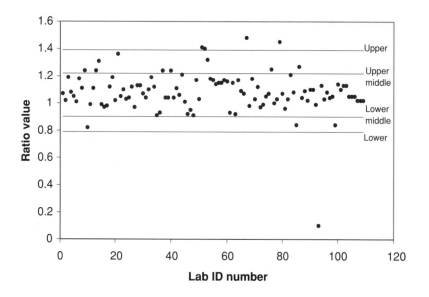

FIGURE 11.3. EML interlaboratory comparison results for [40]K in vegetation.

95th percentile) and lower middle (5th–15th percentile) limits define the warning levels; depending on whether a result lies inside or outside the middle limit lines, it is labeled Acceptable (A) or Acceptable With Warning (W), respectively. Other parameters, such as the mean of all submitted values and the mean of acceptable submitted values (all submitted values except outlying values), may be reported.

Any results in the W or N categories for a participating laboratory should trigger a thorough QA effort within that laboratory to remedy the situation (see Chapter 12). The simplest causes are typographic, data transfer, and calculating error. A second line of consideration is an error in analysis or counting that affects only the intercomparison sample and can be checked by reanalysis. If these causes are not applicable, the bad result reflects an inherent defect in analysis or counting efficiency. This conclusion is supported if previous interlaboratory comparison results were at or near the warning level. The results may hint at what problem to seek by being biased low or high.

In view of the importance placed on agreement of results by a participating laboratory with those reported by the testing laboratory, the testing laboratory must take great care in preparing test samples and calculating its published radionuclide concentration value and uncertainty. Nevertheless, errors have occurred in standard certificates and testing laboratories. A significant difference between the mean of acceptable values by many participating laboratories and the testing laboratory value suggests that such an error may have occurred. For the values in Fig. 11.3, for example, the ratio for the mean of all acceptable values of 1.07 indicates reasonable agreement of results by the testing and tested laboratories.

The existence of a testing laboratory for interlaboratory comparison is so important for assuring reliability of radioanalytical chemistry laboratories that it should be supported by either a Federal agency or a group of laboratories. In the absence of a testing laboratory, a cooperative venture may be arranged among several laboratories to perform round-robin testing of a single procedure or a collaborative study to develop and test standard methods. In these cases, results are compared among several laboratories. Agreement among all participants within specified uncertainty values usually satisfies the participants but disagreements cannot always be resolved by attributing bad results to a given laboratory.

Even acceptable interlaboratory comparison results may require further QC efforts at a laboratory if the test sample matrix is not identical to routine sample matrices. Possible concerns are that the radionuclide in routine samples is in a different chemical form, that the radionuclide is at much lower concentration, or that the matrix has different constituents. The analytical and counting procedure should be examined to determine which factors are important, and whether additional QC tests are indicated.

Acceptable interlaboratory comparison results reinforce acceptable internal QC results by generating confidence in analytical and measurement processes. The information usually is required by the client and for laboratory accreditation. Participation in these comparisons should be at least on an annual basis for every matrix type and radionuclide that is analyzed, and would be beneficial on a more frequent cycle if the workforce turns over frequently.

11.2.12. Documentation of Data Assessment

A significant fraction of laboratory effort must be devoted to weeding out false results by consistent application of the QA techniques discussed in the earlier sections of this chapter, preferably before such results are reported. The QC program and interlaboratory comparisons are designed to identify systematic errors in chemical analysis and radiation measurement due to problematic methods or analysts. By chance, QC results instead may identify an occasional error due to lack of attention in analysis, measurement, recording, or calculation. More commonly, occasional errors are found during data review (see Section 10.6).

The QAP document should specify, in addition to the QC and performance demonstration activities described in Section 11.2.11, the format for data review, including documentation of actions and outcomes. For example, the review can begin with matching the data summary to lists of received samples and requested analyses, individual analytical results, and the applied calculations to look for data gaps and inconsistencies. In the age of computers, the reviewer checks reliability of data entry and use of codes appropriate to the calculation.

The QAP should describe responses and corrective actions when laboratory performance levels are out of compliance. All samples identified in the report as processed by out-of-compliance procedures, systems, or analysts must be considered potentially deficient. Appropriate deficiency report forms need to be issued to inform the pertinent analysts, operators, supervisors, and managers. In some instances, a sample that does not meet general QA criteria nevertheless may be acceptable for the intended application, e.g., a measured value is well below the permissible level although its uncertainty exceeds specifications.

Under favorable circumstances, bad data can be replaced promptly with good data by finding misplaced samples, correcting false entries, reanalysis, remeasurement, or recalculation. Less favorably, extensive studies may be required to correct systematic errors (see Chapter 12). The deficiency forms should indicate whether immediate corrective action is possible.

11.2.13. Audit Reports

Audits are periodic independent surveys, assessments, or site visits to ensure that the QAP is appropriate and that radioanalytical chemistry laboratory operation follows the QAP. Generally, a client or accreditation program sends a crew of audit specialists, but internal audits also are performed.

The audit begins with a meeting with laboratory management, notably the QA manager, to discuss the plan, and a review of the QAP manual. The auditors examine all aspects of the laboratory, notably its physical plant, analyst and operator performance, analyst training, sample handling, record keeping, QA measures, QC results, and data report preparation. Supervisors, analysts, and operators are interviewed. Checks on data validity include comparing summed times recorded for sample analyses to the time available, comparing counting times and background count rates with those required for the estimated data uncertainty, and following selected samples through the entire analysis process.

The audit plan, including the schedule, the personnel involved, procedures, and checklist is the responsibility of the audit organization. A copy of this document should be filed with the QAP to record the extent of the audit. All information, comments, and recommendations provided by the audit team at the exit interview and in the visit report should be added to the filed audit plan.

Every item in the report that requires a response must be addressed. Information must be provided for action items to describe the outcome of the action taken. All responses should be filed with the audit plan, together with resulting documents such as certificates of accreditation or notification that all outstanding items have been resolved.

11.2.14. Record Retention Policies

The QAP should define record storage requirements with regard to location, time period, security, and format. Paper records are commonly stored in lockable, fire-resistant file cabinets. Active computer records are backed up at frequent intervals to disks or tapes that are stored similarly. After the specified period of years, data files are transferred to a secure storage location. The time period to ultimate disposal may be controlled by regulations, management policy, or a specific need for records (or lack thereof) after an extended period of time.

Old records may be needed for data mining, review, or correction by the laboratory, project review by the client or regulator, or legal challenges (see Section 11.5) related to matters such as worker health or environmental degradation. A thorough review ordinarily requires the existence of original records to support both chain-of-custody and reported values. The information can be expected to include the stated method, detector, calibration effort, counting time, calculations, data uncertainty, and analyst names.

Although record storage nominally may be "permanent," consideration must be given to the actual time of preservation with regard to record stability, space needs, and caretaker cost. Many data sets have no conceivable need for permanent storage and can be discarded after several years. Nevertheless, the decision to discard records should be reviewed at the scheduled time to consider any overlooked potential need.

11.3. Quality Assurance in Data-Collection Activities

An agency or a laboratory charged with collection as well as analysis of a sample set must take additional QA measures, as discussed in Section 5.10. A framework for a radioanalytical data collection activity (DCA) must be established before the first analysis is performed to present the purpose of the program and the needed information, even if sample collection and analysis are separate assignments. This integration of effort, imperative to the smooth operation of the project, includes implementation of the DQO process and subsequent preparation of a QAPP. The body of literature describing these activities is large; an introductory description with references is given here.

TABLE 11.3. Individual elements of the DQO planning process

DQO step	Comments
Step 1: State the Problem	Define the problem to be studied, examine budget, and determine time frame of study.
Step 2: Identify the Decision	Decide what specific decisions are to be made on the basis of the results of the study, what questions must be answered to make those decisions, and the order of priority given to each decision.
Step 3: Clarify Inputs to the Decision	Decide what information sources will be needed to answer the defined question(s).
Step 4: Define the Study Boundaries	Define the study area spatially and temporally, as well as the environmental medium of the study.
Step 5: Develop a Decision Rule	Determine how the data will be used to choose among alternative actions, including statistical parameters to be used and logic behind the application of these parameters.
Step 6: Specify Limits on Decision Errors	Define the implications of a decision error, i.e., the cost or health effects that result from a wrong decision.
Step 7: Optimize the Design	Design the DCA in accordance with the needs defined in Steps 1–6 and improve as needed.

11.3.1. The Data Quality Objectives Process

A DCA can be a major undertaking that requires attention to myriad details from many different disciplines. The purpose of the DQO process is to allow planners from different backgrounds to work together, each learning to understand the role of the others in the overall framework of the DCA. Together, the planners can address objectives and possible hurdles at the beginning to obtain a more realistic concept of the scope of the project.

An illustration of the DQO planning process (EPA, 2000b) can be found online at http://www.epa.gov/quality1/qs-docs/g4-final.pdf (December 2005). It consists of six steps, plus the final optimization step, as indicated in Table 11.3. Each step is separate, but all the steps are iterative. The comments pertaining to each step in Table 11.3 are general, and do not begin to address the practical details of data collection, which are more fully covered in Chapter 5.

11.3.2. The Quality Assurance Project Plan

The result of the DQO process is a set of informed planning elements that are expanded into what is often called a QAPP. This document dictates the specific elements of a DCA. The QAPP ensures that the sample collection area is logical and well-defined, the samples collected are representative of the area with respect to distribution, number, and frequency, the samples are collected and documented properly, and the samples reach the radioanalytical chemistry laboratory in an unbroken chain of custody. At the laboratory, the QAPP defines the types of analyses that will be performed, the number of measurements per sample, the required reliability and sensitivity of analytical results, and the response time. The QAPP also determines what will be done with the resulting data after analysis by defining the reporting format and the distribution and/or presentation of reports.

The QAPP is not to be confused with the QAP discussed in Section 11.2. The QAPP is distinct from a laboratory QAP in that it defines an entire project, from start to finish, rather than governing only the handling of samples once they reach the laboratory. Like the QAP, the QAPP should be considered a work in progress because results of analyses are to be used to revise the sampling and analysis program for a more effective response to the information needs of the client.

11.3.3. Example of Insufficient Data-Collection Planning: Love Canal

In July of 1970, the Environmental Protection Agency was established. Its mandate was to repair environmental damage that had already been done and to find ways to change behavior (e.g., waste dumping) that would cause more environmental problems in the future. This mandate required the agency to develop large-scale methodologies for cleaning up contaminated areas and to establish guidelines for future waste generation and removal. This unprecedented multidisciplinary effort unsurprisingly experienced a number of setbacks. One of the most glaring occurred in 1980 in the Love Canal area of upstate New York (Maney, 2002).

The Love Canal region was believed to be highly contaminated by toxic chemicals; the EPA was charged with assessing the level of contamination. The EPA study would serve as the basis for determining whether the area was habitable or, if not, what level of environmental remediation would be required to make it so. The affected area was largely residential, which served to intensify media scrutiny and heighten the furor over what was already a difficult situation.

The result of the EPA's study was that low levels of contamination by hundreds of different chemicals were evident in the area (DHHS, 1982). Almost immediately, the agency was barraged by questions:

- Do the low levels of contamination found by the study accurately reflect reality?
- Will the contamination spread further, or become concentrated in certain areas?
- What are the health effects of the contamination?
- Is it necessary, or even possible, to clean it all up?

The Congressional Office of Technology Assessment eventually was asked to review the EPA's study and its findings (OTA, 1983). The OTA determined that, on the basis of the EPA study, it was simply not possible to determine whether unsafe levels of contamination existed in the Love Canal area. The EPA study was deemed inadequate.

The following problems with the study were identified:

- Documentation was lacking on the handling and transfer of the samples and on the methods used to analyze the samples.
- The sampling effort was inadequate to the task of establishing the level or pattern of contamination. In effect, too few samples were taken and the choice of location

for those samples that were collected was not satisfactory, given the geography of the site.

- Acceptable control sites were not established so that the data collected from the contaminated area could not be compared with those collected from a reliably "clean" area.
- Too few replicate samples were taken to define the statistical uncertainty of contaminant concentration at a given sampling site.
- Insufficient effort was devoted to interpreting the data in terms of its impact on public health. While the effect of many of the *individual* chemicals on humans was well documented, no research was conducted to determine the effect—the synergy—of *many* toxic chemicals in concert.

The EPA clearly needed to develop technical guidelines for environmental monitoring; this included the implementation of sampling and analytical protocols, and the establishment of acceptable techniques for the documentation and presentation of analytical results. The alternative was to continue to produce results that were indefensible and expensive. The EPA realized that protocols were lacking, and published some interim guidelines on QA (EPA, 1980b), but these had not yet been widely implemented by the time the Love Canal situation came to light.

The problems faced by the EPA were symptomatic of the entire field of chemical and radiochemical analysis at that time. Laboratories were increasingly scrutinized and questioned: "How can we (the recipients of the data) be sure that the data are what you (the generators of the data) say they are?" Formal proof of the data and the process that produced it was required to answer this question. This need led directly to the formulation of the DQO and QAPP systems.

11.4. Cost and Benefit of Quality Assurance

Laboratory management initially considered that about 10% of their work should consist of QC measures. However, new regulations have steadily increased that percentage; as mentioned in Section 11.1, QA/QC efforts currently constitute 25–30% of the analytical workload. The expansion of QA activities is certainly a benefit in that it ensures that heightened attention is paid to achieving quality results with a concomitant reduction in error. On the other hand, QA measures are expensive and time-consuming. Moreover, voluminous documentation may give a false impression of reliability, and the time devoted to preparing documents may subtract from time better spent in training the analyst and operator. Estimates of the costs of QA measures and the costs of correcting errors in the absence of QA are useful to determine whether the expenditure on QA measures is justified in light of the benefit derived.

On the basis of the appraisals of the ASQC, Ratliff (2003) has outlined four principal categories for quality costs, as applied to a laboratory environment:

1. *Prevention costs* refer to the management elements that prevent "bad" data from being generated in the first place.

TABLE 11.4. Principal categories of QA active costs (based on Ratliff, 2003)

Cost category	Elements
Prevention	Preparing and maintaining the QAP
	Engaging in the DQO process
	Preparing and maintaining the QAPP
	Document preparation and revision
	Personnel training
	Preventative maintenance: facility and instrumentation
Appraisal	Document upkeep and filing
	QC measures in procurement
	Preparation of QC samples
	Self-assessments
	Participation in interlaboratory testing
	Continuing education
	Instrument calibration
	Data validation
	Data analysis
	Preparation of QA/QC reports to management
Internal failure	Disposal of defective materials and/or equipment
	Extra training costs
	Repeat analysis
	Time spent on internal investigation
	Time/revenue lost because of interruption of sample throughput
External failure	Time spent cooperating with external investigation
	Time/revenue lost because of interruption of sample throughput
	Legal costs
	Cost of corrective action
	Loss of reputation

2. *Appraisal costs* refer to ongoing efforts to maintain and demonstrate quality levels.
3. *Internal failure costs* are incurred to discover and remedy unacceptable data before they leave the laboratory.
4. *External failure costs* are associated with efforts to regain customer satisfaction and confidence if unacceptable data leave the laboratory.

Table 11.4 displays an expanded listing of costs within each category. Most expenditure results from the *time spent* on QA measures, rather than from the cost of materials. Time is of great value, and management should make efforts to quantify the time spent on each of these QA elements. The analyst can help in this effort by communicating with management regarding the allocation of time for each element. For instance, it may be apparent to those in the laboratory that more time should be spent on training or establishing a more efficient document filing system. On the other hand, the staff may believe that some elements are consuming too much time; perhaps too many reports are being prepared or report formats are too elaborate. The dialogue of itself can be informative.

Another aspect of QA made apparent by Table 11.4 is that the elements categorized as Internal and External failure are fewer, but more expensive. If QA efforts in the Prevention and Appraisal categories are managed properly, the cost of failure can be avoided.

The less quantifiable result of successful QA measures is the confidence they awaken in the laboratory output of analytical results. Absent such trust, at best an entire set of measurements may have to be repeated; at worst, the conclusion concerning an entire monitoring program supported by the analyses may be rejected. If these alternatives are unacceptable, then sufficient QA measures are worth the effort.

Beyond reassuring regulators and stakeholders that analytical results reflect reality, a thorough QA program described by a QAP manual has other benefits (Ratliff, 2003):

- Some agencies, such as the NRC, require implementation of a QA system before an operating license is granted.
- Accreditation by various groups requires that a QA system be in place. The QAP manual serves as documentation that this has been done.
- A QAP manual serves as a historical record for the laboratory. It may be referred to in times of investigation, or merely as an aid to memory.
- A QAP manual may be utilized as a promotional tool to emphasize ongoing laboratory commitment to quality.

The QAP manual must, of course, reflect the efforts of effectively trained and responsible analysts and operators, QA staff, and laboratory management committed to producing reliable results.

11.5. Defensibility of Scientific Evidence

As mentioned several times in this textbook, the regulatory environment surrounding radioanalytical chemistry laboratories has become increasingly restrictive. As the level of regulation increases, so too does the pressure on a laboratory to demonstrate that it conforms to these regulations. Therefore, the legal environment has become just as pervasive an influence. This impact is felt in two primary situations, both of which are becoming increasingly common in this litigious society:

1. Laboratories that pollute, renege on contracts or obligations, flagrantly disobey the terms of their license, or fail in some other way will be sued either by their government or by a private citizens' group.
2. A laboratory analyst, hired as an expert witness or simply called to testify, may need to present scientific evidence, either in defense or prosecution of these laboratories, or pursuant to some other complaint.

In both of these situations, it is certain that the reliability of the evidence—i.e., the data—will be called into question to determine its admissibility. The court will ask the laboratory and/or the analyst to defend the data. The most recent, broadly applicable Supreme Court ruling that addresses this issue is related to Daubert v. Merrill Dow (1993). This ruling makes clear that data and testimony must meet a standard that the trial judge deems reasonable to admit the testimony, and indicated the questions that would be used to establish their admissibility. As stated in Daubert:

Many considerations will bear on the inquiry, including whether the theory or technique in question can be (and has been) tested, whether it has been subjected to peer review and publication, its known or potential error rate, and the existence and maintenance of standards controlling its operation, and whether it has attracted widespread acceptance within a relevant scientific community. The inquiry is a flexible one, and its focus must be solely on principles and methodology, not on the conclusions that they generate. (http://www.daubertontheweb.com. Accessed August 27, 2005)

These are some of the same questions that are used to evaluate a standard method (see Sections 11.2.8 and 6.5). Accurate record-keeping in the QAP ensures that standard laboratory procedures are documented, and accurate record-keeping in a laboratory notebook ensures that a written record exists to indicate that said procedures were indeed followed. Although it may seem exaggerated, the importance of instituting QA procedures and keeping QA records cannot be overemphasized.

12
Methods Diagnostics

Bernd Kahn

12.1. Introduction

Ideally, a laboratory processes every sample successfully and then correctly reports its results. In practice, laboratory supervisors and staff devote much time to examining the causes of failed analyses and attempting to avoid such failures in the future. Such efforts in the laboratory to identify and then prevent or remedy problems are usually not considered as a formal topic. This chapter proposes a systematic approach to investigating the cause of a situation, i.e., diagnosis of laboratory methodologies, to solve the current problem and prevent its recurrence. The laboratory management structure must accommodate response to current problems in terms of staff, time, and budget. It must also take measures to prevent future incidents by instituting and adhering to a quality assurance (QA) program, as discussed in Chapter 11.

QA is the institutional system of methods diagnosis. Implementation of a QA program formalizes analyst training, written procedures, and the careful review of results. This formalization demands that each staff member perform the duties of his/her position and take direct responsibility for the result. Within this framework, the analyst takes responsibility for the analytical process, the detector operator for the measurement process, and the supervisor for producing the results. The effect is to create a laboratory environment that pervasively supports reliable analysis and dependable reporting.

The quality control (QC) tests discussed in Sections 10.5 and 11.2.9 are integral parts of QA designed to check results. Some QC measures are prompt indicators that warn of problem occurrence at the time of analysis; others are delayed indicators that require backtracking to find when a problem first arose. Control charts for radiation detector operation are an example of a prompt indicator of reliability. Records of deviations from the norm in an analysis or a measurement may also be prompt indicators if immediately considered. Periodic blank, blind, and replicate analyses, especially interlaboratory comparisons, are delayed indicators for which results may not be available for days or weeks after a problem has arisen. Review and assessment of compiled data are delayed indicators of information quality.

Environmental Radiation Branch, Georgia Tech Research Institute, Georgia Institute of Technology, Atlanta, GA 30332

The magnitude of the current effort devoted to QA (see Sections 10.5 and 11.4) is a measure of the concern about problems that are encountered in radioanalytical chemistry. The goals of this chapter are twofold: first, to emphasize that QA requirements are instituted not just to convince others of the dependability of the laboratory but also to provide information for correcting problems and second, to provide practical insight into laboratory problem solving.

12.2. Problem Origins in the Laboratory

The problems that present themselves in the laboratory are often not immediately obvious. Samples decompose, lose their identity, or disappear entirely. Analytical methods and instruments fail suddenly. Analytical results are mishandled. The analyst may be careless, absentminded, confused, or incompetent. Laboratory supervision may be at fault in issuing ambiguous directions, failing to maintain close control over laboratory operations, or ignoring warning signs. Mistakes may be caught quickly if the analyst recognizes common warning signs and understands the range of problems that they may indicate. Some common problems and their warning signs are listed in Table 12.1. Note that the listed warning signs do not correlate to a single problem; each warning sign may signal the presence of one or several causal problems.

Constant alertness by the supervisor, analyst, and detector operator working together can help to recognize warning signs and respond promptly to a problem. The aim is that few analyses will be wrong and that failed analyses are corrected.

Problem identification and correction, while necessary, are reactive solutions. It is highly desirable that the lessons learned from previous problem occurrences be used to prevent future problems. Future problems can be prevented by identifying the *reason* that the problem occurred. General causes of laboratory problems are the following:

- periods of change in laboratory operation, procedures, or personnel;
- sample variability;
- method instability;
- unreliable analyst behavior;
- deadline pressure;
- unreliable QC samples;
- deficient supervisory interaction.

These issues are discussed in detail in the following sections.

12.2.1. Periods of Change

Most problems tend to occur at the time of change, notably when a laboratory begins to perform its analyses. A competent manager insists on detailed manuals for method performance and instrument operation, thorough staff training, and well-tested instruments with associated computers. Written instructions must

TABLE 12.1. Observed problems and warning signs in radioanalytical chemistry

Problem	Warning signs
Preservation	
Loss during storage	Unexpected sample appearance or pH
Transfer among phases	Inhomogeneous sample
	Changes with time
Preparation	
Loss during sample preparation	Losses
	Incomplete or inconsistent dissolution
Radiochemical analysis	
Inaccurate aliquots of sample	Separation step not fully functional
Inaccurate aliquots of reagent	Strange product appearance (amount, form, color)
Inaccurate reagent preparation	Low or excessively high yields
Reagent instability	
Contamination	
Method not followed	
Inappropriate method	
Incomplete purification	
Counting source preparation	
Sample losses	Source unstable
Wrong sample dimensions	Dimensions inconsistently thick or wide
Source impurity	
Measurement	
Detector peripherals malfunction	Nonreproducible count rate
Detector malfunction	Varying background
Calibration error	Unusual result
Unstable background	Inconsistent result
Interference	
Calculation error	
Data processing and reporting	
Misinterpretation	Unexpected result
Data mishandling	Inconsistent data pattern
Data transfer error	

cover all aspects of laboratory operation, as described in Section 11.1 for the QA plan. Training is given to inculcate laboratory practices and methods application as well as to review underlying chemical and physical principles. "Cold" (nonradioactive) tests are performed in which the trainer first demonstrates the process, then supervises every step to check analyst performance and eliminate faults, and finally challenges the analyst to perform well independently. During this training, the analyst should accept the responsibility to perform reliable work.

Analytical unreliability may arise when new analysts or operators, samples, methods, or instruments are introduced. Loss of an analyst or operator and replacement by another is a common predictor of unreliability in analytical results, as is the shift of an analyst or operator to a new assignment or a temporary one for a vacation period or heavy workload. Thorough education and training can help prevent bad analyses that are caused by unfamiliarity with a process.

In the same category lies the procurement of new instrumentation and the application of new computer software for measurement, data processing, or reporting. If the new instrument or software is radically different than that used previously, a class may have to be scheduled to familiarize the staff with the particulars of the equipment. Less drastic changes, such as the updating of a model or version, probably can be handled by reading the manuals accompanying the new product.

Even subtle changes can cause problems. Examples are preparation of new reagents and interactions by different persons in the chain of sample preparation, analysis, and measurement. Here again, thorough training is the greatest preventive measure in avoiding mistakes. Analysts should be notified in writing when the laboratory makes any changes in reagent concentration or identity, and these changes should be entered in the QA manual. Changes in the procedure for sample transfer or alteration in the chain-of-command in the laboratory should be conveyed to the staff so that everyone knows their new duties within the management framework.

12.2.2. Sample Variability

A competent analyst realizes that a sample matrix can vary widely in kinds and amounts of nonradioactive components and that any method must be tested for stability with the entire range of components. Methods taken from publications or laboratory manuals should specify the limits of applicability with respect to potentially interfering substances. The analyst should expect that constituents beyond the tested range could undermine the analytical method, especially in generic sample media such as soil, vegetation, or tissue. Analysis of samples that have constituent amounts beyond the indicated limits may fail because of interference with chemical separation or carryover with the source prepared for measurement.

The analytical problem can result in appearance of a solid where none should appear, a clear solution instead of a precipitate, or an unusually high or low yield for the radiation detection source. The analyst's notes concerning the procedure may point to the cause. Otherwise, each step of the analytical method must be examined to find the step in which the constituent interferes and the reason for the interference, e.g., competition for reagent, change in pH, or change in the redox potential (pE) that is needed for the reaction.

Interference by an unanticipated radionuclide may be recognized when an unusually elevated value of the radiation measurement is questioned. The sample should be subjected to detailed determination of the radionuclide decay scheme (see Section 9.2) to attribute the measured radiation either to the radionuclide of interest or to interfering radionuclides. Further chemical purification or spectral analysis may confirm the presence or absence of interfering radionuclides.

12.2.3. Method Instability

Even when matrix constituents are within an acceptable range, the method may be written in a way that permits analytical variability to the extent that the method will fail sometimes. Factors such as mixing period, temperature, pH after reagent addition, and amount of reagent may be specified too imprecisely to achieve consistently the conditions for successful analysis.

Such inconsistency may be associated with unexpectedly wide fluctuations in yield or decontamination factor, but may be difficult to assign to the imprecise instruction at fault unless the analyst is sufficiently observant. Without such observation, a step-by-step test of the analysis may be required.

Occasional method failure can occur before or after chemical separation. A variable fraction of the radionuclide may be lost during storage or initial treatment before the carrier or tracer is added, or interchange between the carrier or tracer and the radionuclide of interest may be incomplete. During counting, instrumental effects such as quenching may be inaccurately assessed.

12.2.4. Analyst Behavior

Some analysts are less meticulous than others. Even a careful analyst may have moments of absentmindedness during which a reagent is not added to one sample but is added twice to another, a step is omitted, or the wrong phase is discarded. The incidence of such errors can be reduced by focusing on the work at hand and reducing the opportunity for distraction, but cannot be entirely eliminated.

An odd but apparently common effect is a gradual series of unauthorized changes in the procedure over time. Comparison of the performed procedure with the written procedure after several years may show a subtle change in reagents, order of addition, time of mixing, or sample volume. One can speculate that each individual change was in response to once-only situations, but the effect is that one final modest change disables the method. Periodic internal audits should detect these unapproved changes before ultimate method failure.

Most unfortunate is the employment of an analyst or operator who is uninterested, incompetent, or even untrustworthy. These characteristics usually are signaled by aberrant behavior such as irregular attendance or lack of attention to work in parallel with poor analytical results. Without a thorough change in behavior, only dismissal prevents further analytical problems by such analysts or operators.

12.2.5. Deadline Pressure

Confusion caused by ad hoc changes in the work schedule adversely affects analysis and measurement reliability. Routine work scheduling considers the normal processing time for each type of sample, the number of samples to be processed, and all ancillary responsibilities such as reagent preparation, information recording, and communication with others. Unusual deadline pressures may demand faster analysis, adding samples to be analyzed in parallel, or accepting responsibility for extraneous assignments (see Section 13.9.4). Change orders may be verbal instead of written and may be ambiguous. When such pressures occur, as they inevitably do, supervisors must make special efforts to give clear orders in writing and to avoid giving assignments that confuse the analyst. Preplanning can ameliorate deadline pressure.

A source of both distraction and additional pressure is frequent interruptions by managers with inquiries concerning the status of analyses and suggestions for timesaving method modifications. Such interactions may be inevitable, but they should be minimized. Analysts and operators should be shielded, as much as possible, from sources of unnecessary distraction, which only increase the likelihood of error.

12.2.6. Quality Control Sample Reliability

Preparation of QC samples for internal testing deserves particular care because of their impact on the perceived reliability of the analytical program. The problems listed in Table 12.2 can falsely indicate error in the analytical process that stimulates unwarranted methods diagnostic efforts by reviewing and repeating analyses. The QC staff must take thorough precautions in QC sample preparation to be meticulous, maintain accuracy and traceability to National Institute of Standards and Technology (NIST) standards, prevent radionuclide contamination, and keep reliable and detailed records. These efforts are sufficiently important to require direct overview by the QC supervisor.

The QC program tests and provides information on the aspects of laboratory work listed in Table 12.2. Any suggestion from QC results, e.g., a specific analysis is unreliable, will interrupt routine analyses and cast doubt on their results until the problem is identified. The resulting methods diagnostics should search for causes due to either individual analysts or operators or a more general loss of laboratory control, as categorized in Table 12.3.

12.2.7. Supervisory Interaction Needs

A minor deviation from routine response in the chemistry method or the radiation detector can warn of an incipient problem. To recognize such signals, the analyst and the operator must be observant of and sensitive to the process that they control, maintain careful records, and report this information promptly to the supervisor. For this aspect of QA to be effective, the supervisor has to have an ongoing daily dialogue with the analyst and the operator.

TABLE 12.2. QC Sample purpose and problems

Information provided
   Day-to-day consistency and precision of results
   Data reliability and accuracy of results

Potential errors
   Reported radionuclide concentration, uncertainty, impurities, or chemical form
   Applied values of half-life, decay fractions, or energy
   Calculation of radionuclide dilution or decay
   Mass measurement, losses, or other aspects of source or sample preparation
   Excessive delay period before use
   Calculated or recorded results

TABLE 12.3. Common problems that can be traced to individuals or the general laboratory environment

Individual problems	General laboratory problems
Wrong sample	Sample misidentification
Inaccurate reagent preparation	Reagent decomposition or instrument malfunction
Misinterpreted instructions	Radionuclide cross-contamination
Unwarranted assumptions	

Review of the process prompted by analyst or operator concerns should involve the analyst, the supervisor, and possibly a specialist. This review may be able to distinguish among possible causes in chemical analysis such as matrix difference, method instability, bad reagents, or analyst error. In radiation detection, source problems, detector malfunction, and data analysis must be distinguished. The discussion should focus on what the analyst or operator remembers about the measurement series in question, in contrast to records for similar analyses; this should help determine when the problem was first observed and the differences in the process since then.

12.3. Problem Response

12.3.1. Sample Consistency

The reliability of a sample result is questioned when comparison measurements show different results. Measurements from the following samples are usually compared with measurements of the radionuclide of interest in the laboratory:

- Aliquots of replicate samples analyzed at other laboratories.
- Laboratory QC samples (see Section 11.2.9) such as replicates.
- Samples collected at related locations or times.
- Related radionuclides measured in the sample.
- Samples collected at the point of origin or in related media.

The first two bulleted items provide direct comparisons, and the other three are guides for anticipated radionuclide levels. A measured result that differs greatly from expectations may raise concerns that stimulate interest in reviewing the compared data and perhaps repeating the analysis, but by itself is not an indicator of erroneous measurement.

The initial response to a question of reliability should be an examination of data uncertainty (see Section 12.3.8). Data can be concluded to be inconsistent only if the ranges of uncertainty of the compared values do not overlap. The evaluator must decide whether to accept an overlap by multiples of the standard deviation such as 1σ or 2σ, and realize that occasionally and randomly a measurement lies beyond the specified range.

Consistency in replicate samples depends on the meticulousness of replication as well as the reliability of the compared results, hence both possible causes must be considered. Split samples of a homogeneous medium such as an aqueous solution are readily prepared; their consistency depends on accurate pipetting or weighing in preparing aliquots and the continuing stability of the comparison samples. Ideally, every split sample is measured at the time of preparation to determine the mean radionuclide concentration and the standard deviation for the distributed samples. Measurement of a sufficiently large subset provides these values but leaves in question the radionuclide concentration in unmeasured aliquots.

An important consideration in preparing liquid samples is assurance of radionuclide solubility. Radionuclides subject to hydrolysis and "radiocolloidal" behavior (see Section 4.2) require a suitably acidic medium. Other radionuclides may require specific reagents to prevent reactions that lead to insolubility or volatility, or a stable environment of temperature or darkness.

Preparation of aliquots from heterogeneous material such as dried soil or biota ash requires formal sample splitting (see Section 5.10). A radionuclide may not be uniformly distributed on particles if the solid has various components, e.g., sand and clay in soil, that retain the radionuclide with different affinities, or if the radionuclide is incorporated in specific particles, i.e., "hot particles." Measurements of such solids can be compared only if the radionuclide distribution among aliquots is demonstrably uniform.

12.3.2. Sample Preservation

A common problem in water samples is the "radiocolloidal" behavior observed in extremely low concentration radionuclide samples, which represent the majority of all radionuclide samples. In brief, in deionized water, many radionuclides are sorbed on container walls and suspended material (see Section 4.4). This effect is reduced at higher salt concentration and increased acidity, hence the EPA requires acidification to pH 2 or less with HNO_3 at sample collection for analyzing water from public supplies.

The extent of loss from solution by surface sorption appears to be a complex function of the chemical and physical characteristics of the radionuclide, the solution, and the surface. Prevention of loss can only be inferred from studies that replicate the samples that are being submitted and their containers. Following are some simple tests of radionuclide loss from solution:

- Perform separate gamma-ray spectral analyses of the sample and its container.
- Insert a thin plastic bag in the container to hold the sample and subsequently perform separate analyses of the ashed bag and the liquid sample for radionuclides.
- Measure radionuclide sorption on the surface of a test coupon that represents the container wall that is immersed in the solution.
- Compare radionuclide measurements of a solution to which various reagents, such as acids, complexing agents, or carriers, had been added to prevent sorption on container surface.
- Compare radionuclide measurements of solutions in containers with walls pretreated with various solutions.

The first two tests measure the fraction of radionuclide loss to walls, while the others only show relative retention in solution. A deficiency in these tests for samples with low radionuclide content is that the obtained count rate often is too low for precise determination of loss. A radioactive tracer solution can provide higher count rates but may not represent the conditions in the actual sample.

Another common problem is the separation of samples into two phases: solids and liquids for solids such as sediment, vegetation, and tissue; solution and suspended solids for water. For such separation

- radionuclides distribute disproportionately and nonreproducibly between the phases;
- aliquots taken for analysis do not represent the entire sample;
- sample weight at analysis differs from weight at collection.

Freezing samples or initially filtering liquids can reduce these problems. In the absence of freezing, a preservative should be added to prevent decomposition of biota and accumulation of decomposition products. The analyst should plan to minimize the problem by prompt analysis and either mix the sample before taking an aliquot or sample the liquid and solid phases in proportion.

12.3.3. Sample Preparation

Selective collection of a radionuclide, e.g., gaseous radioiodine in air sorbed on charcoal and particulate radioiodine in air retained on a filter, requires further information to calculate the concentration of the radionuclide in the medium. Both the fraction of the form in the sampled medium and the fraction retained by the collector must be known. Information available from reported studies may not apply precisely to the circumstances under which the samples were collected.

Another concern relates to samples in which the radionuclide is not uniformly distributed, such as biological tissue, vegetation, or soil. The measured radionuclide concentration may represent either the entire sample or a defined fraction, e.g., the soluble portion, a specified particle size range, vegetation without soil (and vice versa), or a dissected animal organ or tissue. Any separation of the defined fraction should be performed as early as possible in the process, possibly at collection. The separation process often is imperfect, either because of losing some of the radionuclide to other fractions or retaining some of the other fractions. Serious error is introduced when the radionuclide concentration in the extraneous fractions far exceeds its concentration in the fraction of interest.

The sample may be prepared by concentrating it in processes such as

- filtering of particles from air or water;
- sorbing of airborne gases;
- sorbing ions;
- evaporating water;
- drying, ashing, and dissolving solids.

One concern is partial or complete loss of the radionuclide before carrier or tracer addition to monitor such loss. Carrier or tracer often are added after these concentration and dissolution processes to achieve complete interchange with the radionuclide in solution (see Section 4.5.2). If, on the other hand, carrier or tracer is added before dissolution, interchange of carrier or tracer with the radionuclide of interest may not occur.

Possible loss may be anticipated and prevented by considering the known behavior of its chemical form. For example, if its form has relatively high vapor pressure, either the processing temperature must be kept well below its boiling point or its redox environment should be controlled to prevent formation of the volatile form.

Sample nonuniformity must be avoided when a source is prepared for direct radionuclide measurement by gamma-ray spectral analysis (see Section 7.5.1) or gross alpha- and beta-particle counting. Some samples that were thoroughly mixed just before measurement may remain mixed during the measurement, while others separate into fractions by solids settling or gas emanation. For a mixture of gas with liquid or solid, e.g., radioactive noble gases in solution or radon in soil, the container must be filled completely to avoid a gas phase at the top and sealed to avoid gas loss. Canning a sample is one way to retain gaseous radionuclides.

12.3.4. Radiochemical Analysis

Problems in purifying a sample for radionuclide measurement can be due to inappropriate procedures or analyst error. The circumstances can be categorized by frequency as follows:

- The procedure suddenly is no longer successful.
- Analyses fail for a few random samples per batch.
- Analysis fails infrequently.

The first case strongly suggests a problem of recent origin: a reagent was badly prepared, a fatal change was inadvertently introduced into the procedure, or a new and incompletely trained analyst is at work. The second case suggests that at least one of the separation steps is unreliable as written when the sample composition varies. Infrequent failure suggests a highly unusual sample matrix or analyst aberration, as discussed in Section 12.2.3.

The simplest cause of failure—and the one most readily resolved—is some temporary aberration or absentmindedness, which may be conceded by the analyst or operator and can be confirmed by repeating the analysis or measurement. For a more serious analyst problem, repeated analyses or measurements can be assigned to a more experienced staff member.

Use of a reagent or carrier that is badly prepared, incorrectly labeled, inappropriate to the method, or no longer effective, can be pinpointed by first repeating the analysis with a completely different set of reagents and, if successful, then checking each original reagent in turn. In tandem with chemical failure, instrumental failure also must be considered. When a detector in the counting room becomes suspect, use a second detector to check the earlier results from the first detector.

Carrier and radionuclide tracer yields are convenient criteria of method efficacy. Low yields suggest the action of interfering substances, but may also be caused by analyst error, as discussed above. Yields that are unexpectedly high, especially in excess of 100%, suggest the presence of the carrier or tracer in the original sample. For example, some samples analyzed for radiostrontium to which strontium carrier has been added may already contain a few milligrams of strontium. Excessively high yields can be caused by the presence of relatively massive amounts of chemicals that behave similarly to the radionuclide of interest, e.g., calcium or barium in a process that purifies strontium in the sulfate or carbonate form, when the calcium or barium is incompletely separated during purification. Imperfect exchange with the radionuclide (see Section 4.5.2) is a condition that makes the yield irregular and yield measurement, unreliable.

Sometimes the sample parameters simply do not mesh with the constraints of a given method. Attempted analysis of a sample in which one or more constituents exceed the amounts in the matrix for which the method was reported to be applicable can result in failure. As discussed in Section 12.2.2, sample matrices such as water, soil, and biota can be so variable that a procedure developed for one set of samples may not be able to control interference from the different constituents of the samples under consideration. In response, analyses can be performed with smaller samples or more reagents.

Analysts should consider the method's written statements of limitations with regard to stable substances and the decontamination factor list for important interfering radionuclides and compare those limitations with the characteristics of the analyzed sample. For methods that do not have sufficiently defined limits, the step at which the method fails will indicate the point of departure for a study of possible interfering substances. This point can be identified by measuring carrier or tracer levels at each step to find where serious loss occurs. Tests of the inferred cause of failure include increasing reagents, inserting new purification steps, or repeating existing steps. The presence of interfering radionuclides is shown by special measurements, e.g., for spectra or radioactive decay.

The evaluation may show that the method is imprecisely written such that the analyst views it differently than the writer. This oversight will become apparent only in discussions with the analyst and must then be rectified by rewriting the procedure.

12.3.5. Preparation of Source for Radionuclide Measurement

Problems in comparing replicates measured for radionuclide content may be related to the type of radiation and its measurement, as discussed in Chapter 7. A sample evaporated on a planchet for counting alpha and beta particles with a proportional counter tends to form a nonuniform deposit with rings or off-center spots, for which measurement results of replicates are insufficiently similar. Moreover, the dimension of the prepared source may not match that of the disk or point calibration source.

A sample precipitated for alpha- and beta-particle counting that is collected on a filter or slurried onto a planchet usually is replicated more consistently than when evaporated. Precipitates should be selected to be stable chemically (for example, not hygroscopic) and physically (no loss by flaking). Solids should be in a defined weight range, as light as is convenient for high chemical recovery and accurate weighing (conventionally within 0.1 mg). Filter materials should be selected for high precipitate retention, uniform surface, flat filter, and low background radioactivity (for example, glass–fiber filters contain radioactive ^{40}K).

For LS counting, conventionally in 20-ml vials with mixed aqueous sample and organic cocktail, common problems are light quenching and luminescence (see Sections 7.3 and 8.3.2). Chemical or physical separations, such as distillation of the aqueous sample under conditions that prevent carrying of interfering substances for tritium analysis, can remove substances that suppress light emitted by nuclear radiation or emit light stimulated by nonnuclear sources. Modern counters have computerized data analysis that can quantify corrections for both problems. Cocktails must be checked periodically to assure uniform counting efficiency. If the cocktail–water mixture separates over time, the counting efficiency will change. If a powder is counted in a scintillation gel, settling can change the detector response.

Samples prepared for alpha-particle spectrometry are very thin and uniform. The purification procedure should have achieved high yield and low contamination by extraneous solids. Failure to prepare a thin sample is shown by poor peak spectral resolution with excessive low-energy tailing. Incomplete radiochemical purification in preceding steps is revealed by the appearance of peaks from radionuclides, which should have been removed.

Prior separation generally is unnecessary for Ge gamma-ray spectral analysis because the high resolution separates the characteristic gamma rays from multiple radionuclides. Separation may be useful to remove radionuclides that obscure the peaks of interest with a Compton continuum from gamma rays at higher energy or with a peak at almost the same energy.

This convenience of gamma-ray spectral analysis with a Ge detector permits counting of samples in their original shape and large volumes if detector calibration and background count rates are available for that geometry. Obtaining a radioactive efficiency calibration source in a complex geometry may be difficult, but calibration may be performed by Monte Carlo simulation. A correction factor must be introduced for gamma-ray attenuation when the calibration source and the samples are large and have significantly different densities. This effect is greatest when counting gamma rays below about 100 keV, but is noticeable even at 1 MeV.

Problems with regard to sample uniformity and stability for gamma-ray spectrometry were discussed in Section 12.3.3. The ease of directly placing a sample into a container and counting should not divert the analyst from maintaining uniformity once a sample is prepared.

12.3.6. Radionuclide Measurements

Some problems encountered with the various detectors for radionuclide measurements have generic aspects related to ambient conditions, detector and ancillary equipment functions, computer operation, and the operator. Other problems are directly associated with the type of detector, its ancillary equipment, and its functions.

Breakdown in control and stability of the immediate detector environment with regard to cleanliness, temperature level, power supply, and radiation background interferes with reliable radiation detector operation. Electronic components function best at a cool, constant temperature in a dust-free environment. Special low-temperature and power-supply-stability controls are needed to stabilize the response of gamma-ray spectrometers and liquid scintillation systems.

Radiation exposure near detectors should be both low and stable (see Section 8.2.2). Major fluctuation in the radiation background that affects a radionuclide measurement must be recorded so that either an appropriate background value is subtracted from the measured count rate or the count rate measurement is repeated during a stable-background period. Periods of elevated background may be confirmed by matching information on the time of occurrence with situations such as

- exposure to external radiation source;
- radionuclides at elevated levels brought into the counting room;
- radionuclide contamination of detector systems within their shields;
- airborne radionuclides in the counting room.

An insufficiently shielded external radiation source can influence background measurements from some distance. Airborne radionuclides typically are gaseous radon and its particulate progeny, but after nuclear tests or major nuclear accidents may include fission and activation products. Efforts should be made to control such exposures for long-term radiation background stability.

Counting efficiency and background values initially are predicted from measurements by the detector supplier or at other laboratories. Once acceptable levels are confirmed, repeated measurements provide the basis for QC control charts for the check source and background count rates (see Section 11.2.10). Problems are indicated by subsequent QC measurements if values are beyond the control limits, drift toward control limits, or change abruptly. The more frequent the measurement, the sooner a problem can be recognized.

The first response to an outlying value is to repeat the measurement. If the problem is confirmed, the next step is checking the instrument and peripheral settings for inadvertent changes that can be corrected. Next, the possibility is considered that the background change is due to the environment or that the efficiency change is due to a test source problem (see below). If none of these noninstrumental causes is responsible, the detector is taken out of operation for detailed testing and repair.

Inaccurate detector calibration is one of the main problems in detector use. Conventionally, detectors are calibrated with radionuclide solution standards traceable to NIST (see Section 8.2.1). These sources are used to calibrate the detection system for counting efficiency and, for a spectrometer, energy per channel and peak resolution. Errors occur if

- a commercial supplier provides the wrong activity value for the standard solution;
- the user prepares the source badly from the solution;
- the source location relative to the detector is not accurately reproduced;
- the measurement is recorded inaccurately;
- a calculation is erroneous.

A correct calibration source may become less reliable with time by incorrect radioactive decay adjustment, increase in the contaminant fraction, damage in handling, or poorer statistical power due to radioactive decay.

In some cases, efficiency and energy calibrations are obtained from a standard source of the radionuclide of interest that has a reported uncertainty value. In other cases, these values are interpolations between data points obtained with several measured radionuclide standards and plotted as function of energy so that the uncertainty of curve fitting must be considered. Monte Carlo simulation has become sufficiently accurate for energy calibration at the usually attained standard deviation of 1–2% for reliability. This approach eliminates the need for interpolation, but depends for accuracy on detailed information on the dimensions of the detector (as in Fig. 8.9), the source, and detector-to-source configuration. The simulation should be tested with at least one measurement each at low and high gamma-ray energy to confirm the utilized information.

Calibration errors can occur if the instrument is incorrectly adjusted, settings are accidentally changed during operation, or components fail. Care must be taken that the computer code is appropriate to the geometric relation of sample and detector, type of sample, radionuclide, and output that is to be calculated. Utilized constants such as type of radiation, energy, half life, and decay or ingrowth fraction must be checked (see Section 9.2). Other pertinent information that must be confirmed relates to the sample dimensions, density, weight, times of measurement, counting period, and radiation background value. Detection system effects must be addressed such as resolving time losses and coincidence summing, as discussed in Section 8.2.1.

The problems listed above for calibration also apply to analysis of routine samples. Consideration must be given in these calculations to the consistency of the periodic test sources, the sample collection and processing dates, and information that relates amount measured to mass or volume collected. Also considered must be numerical adjustments for drying or ashing the sample, concentrating the radionuclide, and procedure yield.

Problems that may affect proportional counters relate to control settings, the flowing gas, the detector window, and the automatic sample changer associated with many detector systems. Wrong control settings can affect the operating voltage, pulse height discrimination, anticoincidence operation, and amplification. The count rate is affected when the wrong gas is supplied, the gas flow rate deviates from specifications, or the gas tank is shut off or empty. The detector window may be the wrong thickness or damaged. The sample changer may have the wrong sample location information or otherwise may malfunction.

For LS counters, successful operation depends on functional dual PMTs and their electronic circuits. Controls must be at appropriate settings. Some systems have spectrometer and pulse shape and timing recognition circuits to restrict background and evaluate the extent of quenching. Luminescence from ambient light sources can be eliminated during a period for dark-adapting the sample before counting it. Luminescence from contaminants must be quantified with associated detector circuits and computer software or by comparison with prepared test samples.

Alpha-particle spectrometer systems often have multiple Si detectors operated in parallel with a single spectrometer. A vacuum must be maintained in each detector cell because attenuation in air causes low-energy tailing in each peak. The adjustable detector-to-source distance affects the counting efficiency and resolution. Power supply stability is required when low-activity sources are measured for long counting periods, e.g., several hundred thousand seconds. Results may be based on only a few counts collected at a peak energy region over this extended time. At higher count rates for which actual peaks can be viewed, knowledge of the energies and intensities of multiple peaks for each radionuclide is required to avoid misattribution of radionuclide activity.

A computer controls gamma-ray detector and spectrometer systems because of their complexity. The Ge detectors require extremely stable voltage and cooling by liquid nitrogen during operation to support high resolution. Failure to replace coolant and freezing of coolant lines are occasional problems. Settings must be determined initially and then maintained as long as QC measurements show acceptable ranges of the background, comparison source count rate at characteristic peaks, energy calibration, and peak resolution. Error in data interpretation can arise when a radionuclide that is counted at high efficiency emits gamma rays in coincidence (see Section 9.4.5). Significant differences in density between the standard and the unknown source can cause error.

12.3.7. Computer Software Application[1]

Modern software is, for the most part, reliable and capable. Problems can occur when software instructions are inappropriate, misinterpreted, or misapplied, and when the system malfunctions. Malfunctions can affect scheduling of counting periods and repetitive counting patterns; processing, storing, and compiling data; accumulating QC measurements; presenting control charts; and flagging questionable data. The most common problems are related to misinterpretation of spectral analysis data.

Spectral analysis software can fail in the following for the indicated causes:

• Proper identification of a radionuclide due to faulty energy calibration or radionuclide energy library.

---

[1] Contribution by Douglas Van Cleef, ORTEC, Oak Ridge, TN 37830.

- Report of correct activity data due to faulty calibration or emission-fraction library.
- Calibration for efficiency and energy due to faulty calibration sample or process.
- Calibration maintenance due to spectrum degradation by drift in applied voltage, electronics, or high count rate.
- Low-count-rate peak detection and integration, due to inappropriate application or nonapplication of alternatives to peak fitting such as region-of-interest or directed fit instructions.
- Background or blank subtraction due to unsuitably high or fluctuating background.

These problems can be prevented or at least minimized by learning to understand both the principles of spectral analysis and the capabilities and weaknesses of the software.

12.3.8. Uncertainty Analysis

Uncertainty estimates are needed for reported data, most importantly to show whether a measured value

- is unambiguously above or below applicable limits or guidelines;
- meets the reliability specified in the Quality Assurance Project Plan (QAPP);
- agrees with replicate measurements performed at other laboratories.

Uncertainty is commonly reported as the standard deviation, σ, or its expanded versions, notably 2σ or 3σ (see Section 10.3.2). The standard deviation or its multiple can be calculated according to Eq. (10.6) by replicate analyses. Although in many instances the reported values are based only on the uncertainty of counting, determined by propagating the uncertainty of the gross count and the background count (see Section 10.3.1), other contributors to the uncertainty can be ignored only if their impact is minor.

The magnitude of the estimated standard deviation is of concern if it is larger than

- dictated by either regulation or convention for the work at hand and/or
- expected from experience with current samples and analyses.

A standard deviation smaller than needed suggests that a briefer counting period, a less sensitive detector, or a smaller sample can be applied. A standard deviation smaller than expected indicates a detector problem.

To improve, i.e., reduce, the standard deviation of an analytical result, the components of the standard deviation value (see Section 10.3.1) must be evaluated to identify for reduction the major contributors to the uncertainty of the sample measurement result. Multiple measurements of the gross count and the background count provide direct measurements of the counting uncertainty. The measurements may indicate that longer counting periods are needed or that the background must be reduced or stabilized. Other contributors to analysis uncertainty, such as sample losses not compensated by carrier yield determination, inaccurate measurement of sample mass or volume, and inaccurate yield determination, should be measured if they may be of the same magnitude as the counting uncertainty. The uncertainty in analyst-related activities often is obtained by "Type B evaluations" (see Section 10.3.1), an estimate of uncertainty based on experience or statements by others, and may have to be checked with repeated measurements by statistical analysis (a "Type A evaluation").

12.4. Response Organization

The preceding discussion suggests that problems and errors in the laboratory can be identified by an effective combination of the following activities:

- analyst and operator alertness and interaction with supervisor;
- QC measures;
- data review;
- lessons learned record.

Management must support these activities to avoid analyst error, method breakdown, and loss of laboratory control. A methods evaluation and development group in a large laboratory, or an individual in a small laboratory, should be responsible for systematically preparing methods for laboratory application, investigating method breakdown, and evaluating replacement methods. A group member should be available for discussing concerns by the analyst and supervisor. The group should review the literature to identify alternative methods of consideration and to consider potential problems discussed by others, such as MARLAP (EPA 2004).

An independent QA group or individual is needed to assure that a QA Plan is prepared to include all methods and training requirements, among other items. The group prepares QC samples, inserts them in the routine sample flow, reports results, and assures that every instance of loss of laboratory control is evaluated and remedied.

A data review group or individual (who may be the supervisor) must consider every value reported by the laboratory in terms of internal consistency and the pattern of past and nearby sample values and the uncertainty of these values. The group or individual must consider missing and questionable values, and investigate analysis and measurement problems to identify their causes.

13
Laboratory Design and Management Principles

CHARLES PORTER[1,3] AND GLENN MURPHY[2]

13.1. Introduction

The primary responsibilities in managing a radioanalytical chemistry laboratory are to perform accurate analytical measurements and report the results in a timely manner. Fine-tuning the design elements and management practices of the laboratory will invariably help a laboratory to meet those responsibilities. This chapter is designed to give students an overview of what a modern radioanalytical laboratory looks like and how it functions. The laboratory features discussed in this chapter apply directly to laboratories processing environmental and bioassay samples with low radionuclide content, but can be extrapolated to laboratory environments where higher level samples are processed.

The early part of this chapter discusses the design and operating practices that support analytical processes in an environment favorable for efficient work. The design incorporates state of the art technologies in sample flow during processing, hood design, ventilation systems, and waste disposal. The latter part of the chapter addresses the staffing, costs, and attitudes appropriate for a reputable laboratory. Management and operating considerations include personnel, operating costs, and service orientation.

13.2. Design and Operational Elements of a Radioanalytical Chemistry Laboratory

In the past, radioanalytical chemistry laboratories processed samples resulting from monitoring nuclear weapons development facilities, fallout from nuclear weapon tests in the atmosphere, and nuclear power stations. At present, monitoring cleanup of former nuclear facilities is a major source of samples, and efforts are

---

[1] ELI Group, Inc., 3619 Wiley Rd., Montgomery, AL 36106
[2] The Matrix Group & Associates, Inc., 118 Hidden Lake Dr., Hull, GA 20646
[3] Charles R. Porter, ELI Group, Inc., 3619 Wiley Rd., Montgomery, AL 36106; email: radlab@charter.net

being devoted to preparation for monitoring radiological accidents and incidents. The laboratories are located at DOE contractor facilities, and other federal, state, and local agencies, or are commercially operated to perform contract analyses for government and private industry. If the demand for processing radioactive samples increases, building new radioanalytical chemistry laboratories or modernizing old ones will become necessary. The design presented here is based on a laboratory recently constructed in association with the authors. The original design includes a radioanalytical chemistry section for performing a variety of radiochemical analyses, as well as a section for the analysis of heavy metals, hazardous chemicals, and volatile and non-volatile organics. The combination characterizes a mixed waste laboratory designed to handle a sample stream of both radioactive materials and hazardous chemical materials. This chapter gives an overview of the salient features of a state-of-the-art radioanalytical chemistry laboratory.

A primary consideration in laboratory design is the magnitude of the radioactivity in the samples that the laboratory will process. The Department of Energy has designated four hazard levels for radiological samples:

1. High Level (Type A)
2. Intermediate Level (Type B)
3. Low Level (Type C)
4. Environmental Level (Type D)

Note: These hazard levels are not to be confused with the DOE classification of nuclear waste into high-level, low-level, mixed low-level, transuranic and 11e(2) byproduct material categories. These nuclear waste categories are established by DOE Order 5820.2A, which can be viewed online at http://www.directives.doe.gov (Dec. 2005). See DOE/EM (1997) for more information on nuclear waste. To reiterate, waste hazard levels are different than laboratory hazard levels, although the defining terminology is similar.

Recent experience at DOE sites has shown that most of the environmental samples collected today are levels C and D. Hence, the laboratory under consideration in this chapter is designed for the analysis of levels C and D samples. These are environmental or bioassay samples that contain radionuclides at low concentrations, i.e., approximating levels of naturally occurring radionuclides. Samples at levels A and B generally will be analyzed in on-site government laboratories for a variety of reasons, i.e. transportation restriction, sample assay limitations, sample security, and national security. In Table 13.1, the authors provide their suggested activity levels to match the four categories identified by the DOE.

Table 13.1 is merely a guide. Each laboratory should develop specific quantity limits. In some cases, the license under which the laboratory operates will specify the quantity limits. For instance, the NRC issues specific radioactive material licenses to facilities, and each license specifies the maximum quantity limit for a given radionuclide. At government owned and operated sites, the DOE facilities do

TABLE 13.1. Correlation of radiological hazard levels to radionuclide concentration levels

Guidance Level	Quantity Level	Quantity Level
High Level Facility	mCi and above	40 MBq and up
Intermediate Level Facility	μCi to mCi	40 kBq to 40 MBq
Low Level Facility	nCi to μCi	40 Bq to 40 kBq
Environmental Level Facility	pCi to nCi	0.04 Bq to 40 Bq

not have quantity limits imposed by any regulatory condition or license, although specific bases for operation may establish operational limits.

Whether of a regulatory or operational genesis, the quantities established are based on several factors. These include the radionuclides to be processed; their physical and chemical form; their decay scheme and half life; the education, skill and training level of the analysts; the physical design of the laboratory; and the radiological monitoring and controls imposed. Other conditions and factors also may justify raising or lowering the quantity limits.

Regardless of the hazard level associated with the laboratory, several physical design features are held in common. All radioanalytical laboratory facilities consist of a receiving and initial processing area (Area A), a set of individual analytical laboratories and radiation detection ("counting room") areas (Area B), and the administrative and support facilities (Area C). These core elements constitute the radioanalytical chemistry arm of the facility shown in Figs. 13.1–13.3. The three figures fit together like a puzzle; they are displayed separately to facilitate easier viewing. This particular facility has an overall footprint of 261 ft. by 161 ft (80 × 49 m), or about 42,000 square feet (3,900 m^2). A few aspects of the complete mixed-waste laboratory are included in these figures while others are omitted.

Numbers have been added to the layout in Figs. 13.1–13.3 to indicate sample flow through the laboratory and help identify specific laboratory components. Samples are received at the loading dock (1) and temporarily held in a storage area (2) until ready for sample acceptance processing. The samples are moved to the sorting area (3) where chain of custody is verified, sample containers are opened, and the samples are inspected for acceptance. They are logged into the Laboratory Information Management System (LIMS) (4) and bar codes are applied. One of the key attributes of LIMS software is the ability to track the status of an individual sample as it moves through the facility. Every sample bar code should be scanned when the sample leaves one area and moves into another area. Each sample thus can be readily located during the analytical process; a technician that moves the sample to the wrong area is notified immediately by the LIMS.

The samples are held in the storage area (5) to await sample processing. Initial preparations begin in area (6). Chemical purification and counting source preparation are performed in the laboratory areas (7) through (11) and (19) through (25); the laboratory spaces have different sizes to handle various procedures. Preparation and counting of tritium samples are isolated in area (12), due to its unique

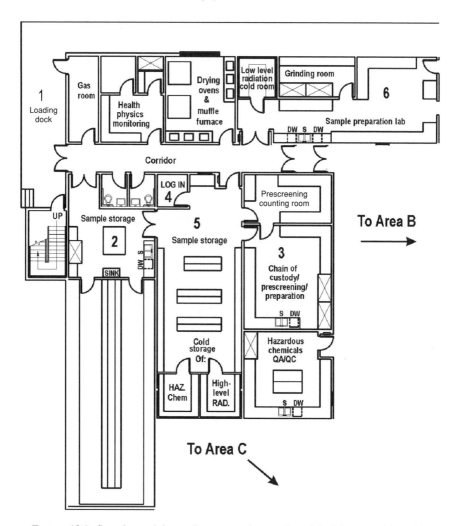

FIGURE 13.1. Sample receiving and pre-screening portion of the laboratory (Area A).

preparation and counting requirements. The prepared samples are moved into area (13) for counting. Areas (14) through (18) contain laboratory space to conduct analysis by various instrumental techniques, such as mass spectrometers (see Chapter 17) and atomic absorption spectrometers. Areas (26) through (29) identify administrative and support facilities.

The results of these analyses are transferred electronically (via the LIMS) to workstations, where the data are validated and verified, and reports are prepared. After the data are reported and accepted by the client, sample residues are moved to the radioactive waste storage facility for disposal (see Section 13.4.5) or returned to the client. The design features of individual laboratory areas are described in Sections 13.3 to 13.6.

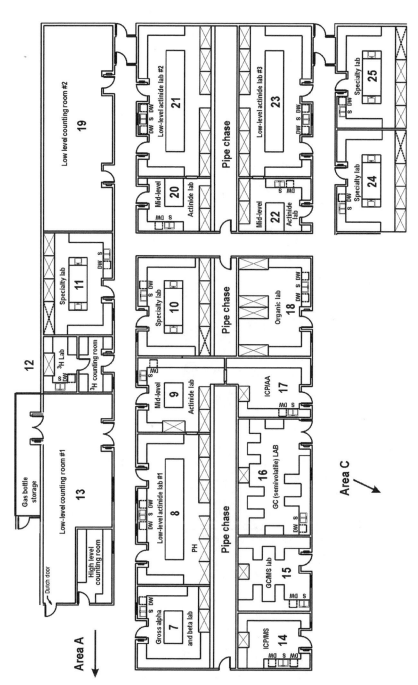

FIGURE 13.2. Laboratory and counting room space (Area B).

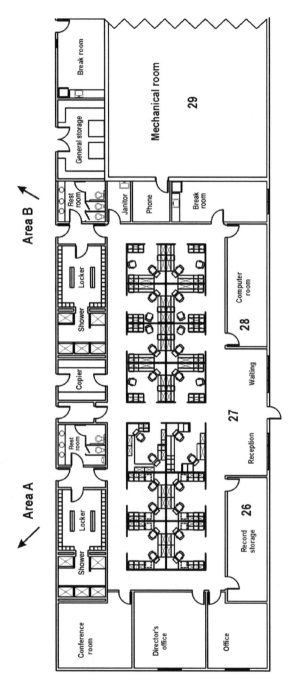

FIGURE 13.3. Administrative and support areas (Area C).

13.3. Sample Receiving and Initial Processing

The loading dock, (1) in Fig. 13.1, should adjoin the sample receiving area. The dock should accommodate shipments in large trucks and small cargo vans by being equipped with a height-adjustable bridge to link the dock area to a semi-trailer, and a ramp for moving dolly loads of packages onto the dock from a small vehicle. Ideally, the loading dock should extend across the back of the facility, but at a minimum it must be capable of handling two cargo vans at a time.

Sample containers off-loaded onto the dock should be moved immediately into the sample receiving area, (3). The sample receiving and storage areas must be adequately sized to facilitate receipt of daily incoming shipments. The Chain of Custody form (see Figure 11.1) should only be signed after all the laboratory sample acceptance criteria have been met. Signing the bill of lading to signify receipt of the shipment does not constitute acceptance of Chain of Custody. A rule of thumb is to design the sample holding area to accommodate five times the design throughput of the laboratory. Most samples are shipped in large (48-L capacity) coolers. The design should include stainless steel or high impact rolling shelving, walk-in coolers and freezers, seamless floors, and epoxy-based paint on the lower walls. The design incorporates surfaces that can be readily decontaminated on the assumption that some delivered containers may leak or have surface contamination on arrival.

Walk-in coolers and freezers are required to preserve samples prior to sample preparation and during the storage and analysis periods. Cold rooms and freezers can be installed on any wall that provides direct access to a pipe chase or outside wall so that condensate from the chillers can be drained into the waste liquid holding tanks. The chillers can be mounted in the mechanical room penthouse or on an outside wall.

Sample chain of custody is a critical path control tool for preparing legally defensible data for the client. As mentioned in Section 11.3.3, some of the problems associated with the Love Canal contamination issues resulted from an insufficiently documented chain of custody process. Whenever possible, the chain of custody process begins in the field with the assignment of a bar code designation, and is subsequently carried seamlessly into and through the laboratory.

Otherwise, the sample receiving room is where the internal chain of custody begins (see Section 11.2.5). The first step in accepting a sample is visual inspection of the container for signs of leakage or damage. Next, the level of radioactivity of the sample must be determined. Each container is scanned for external radiation to confirm that shipping label and papers are accurate, and its surface must be checked to assure absence of surface contamination. Accordingly, the receiving area must have radiation detection instruments (located in the pre-screening counting room (3) of Figure 13.1) to handle radiological scanning and smear counting for all incoming containers. Portable alpha particle and beta particle/gamma-ray detectors are used to scan the incoming packages and a low-background proportional counter is used to count smears for alpha and beta particles.

After placing the package in the fume hood, it is opened and the interior should be checked for signs of internal damage, leakage, or removable surface contamination. If a package does not meet acceptance criteria of the laboratory or Department of Transportation, the package must be moved to an area designated for damaged or leaking containers and processed to avoid further leakage. The laboratory director must decide whether the package is rejected, repackaged and returned to sender, or transferred to the waste storage building for disposal. Samples are scanned with a high purity germanium (HpGe) detector plus spectrometer to check whether the sample contents that emit gamma rays agree with the sample description.

This area also serves as the preliminary sample processing facility. It may contain muffle furnaces, drying ovens, grinding mills, evaporation bays, and fume hoods with dedicated exhaust lines. All exhaust from this area must be treated by pre-filters, HEPA filters, and wet scrubbing prior to discharge to the environment. Each muffle furnace and drying oven must be covered with a canopy exhaust system to remove the heat and fumes generated from the heated samples.

The facility should have a separate room for storing radioactive standard and stock solutions. This room usually is located near the sample receiving and processing area. Radioactive standards and solutions must be kept separate from other laboratory operations to prevent cross-contamination. The room should have cold storage capabilities and lockable cabinets. It should be designed to the same specifications as other sample preparation rooms, with a fume hood and computer access to permit dilution and other processing of radioactive standard reference materials and stock solutions.

Finally, samples that have been pre-screened and are prepared for processing can be moved through the air-lock doors into the analytical laboratory. The air-lock doors separate the analytical laboratories and counting rooms from the other parts of the facility. This minimizes contamination inadvertently transported to or released in the sample preparation areas.

13.4. The Radioanalytical Chemistry Laboratory

After initial sample processing, radioanalytical chemistry is performed in the individual laboratories, numbered 7–11 and 19–25 in Figure 13.2. The model laboratory unit is designed to accommodate two radioanalytical chemists. The sidewalls consist of ample processing surfaces and cabinet space, and the center island contains a double-sided worktable, covered with a seamless laminate that can be replaced when necessary.

Electrical outlets should be located along the center of the worktable to accommodate the use of mobile instruments such as pH meters. Two computer workstations are indicated for each laboratory; at least one of the personal computers (PC) should be connected to the main server to report information for samples that are being processed and add information to their internal chain of custody dossiers.

Each laboratory has a full complement of glassware, reagents, and storage cabinets. Glassware, utensils, and reagents should remain in the assigned room and not be interchanged with other laboratories. The dishwasher internals and sinks should be stainless steel to avoid deterioration from harsh chemicals left on the glassware.

The row of laboratories should be backed up to a physical/mechanical support chase that supplies water, electricity, vacuum, gas, air, and drain lines. The support chase should be large enough to accommodate maintenance personnel and equipment, but access should be restricted for other facility personnel. Some designs incorporate an overhead mechanical room or eliminate the limited-access support chase. All air supplies and ductwork should be overhead and directly accessible from a maintenance penthouse that holds filter banks and scrubbers.

Individual laboratories should be designed to be shut down individually and isolated from every other room in the event of an emergency. Valves for gas, water, and vacuum lines should be accessible through the support chase or the mechanical room penthouse. Electrical cut-off switches should be built-in at the door. These design features prevent a spill or contamination incident from affecting operations in other rooms or laboratories. Ideally, the contaminated laboratory is taken off-line, decontaminated, serviced, and brought back on line with only minor impact on other facility operations.

The laboratory design described in the following sub-sections considers effluent and waste control. Its aims are minimizing cross-contamination and release of radionuclides to the environment.

13.4.1. Floors and Walls

Floor and wall coverings should be selected for ease of cleaning and decontamination. Epoxy paint is recommended for walls and floors because it is relatively impervious to most common chemicals. Its surface is smoother and tougher than regular latex paints and will withstand the wear and tear of normal operations. Epoxy paint is more expensive and difficult to apply but will save time, labor and aggravation in the long run. If paint is not acceptable for floors, a seamless vinyl floor covering may be installed. The vinyl should have a smooth surface and should be turned up the walls for 10 cm (4″) to serve as baseboard for the room walls. Turned-up sides minimize the potential for spills seeping under the covering or walls.

Laboratories must not have floor drains to prevent discharge of liquids to a waste stream that flows directly outside the facility. Spills should be picked up with a HEPA-filtered wet-vacuum cleaner or should be absorbed on solids. The cleanup materials (liquid, solid, and filters) should be handled as radiological/hazardous liquid or solid waste.

13.4.2. Airflow

The primary airflow design feature for the laboratory is single pass air. The air is treated after it circulates through the laboratory and prior to its discharge to the environment. Air is not re-circulated through the laboratory after treatment.

Each laboratory should have an average air turnover rate of 6–8 room volumes per hour.

All air must be treated by filtration and/or scrubbing before it reaches the discharge plenum. This treatment may occur in several stages. First, inexpensive roughing filters (general collection efficiency of 50 percent or greater) are used to remove large particles. Medium filters (collection efficiency of 80 percent or greater) then collect smaller particle sizes, at a higher cost than the roughing filter. High efficiency particulate air (HEPA) filters (collection efficiency of 99.999 percent or greater) significantly reduce the amount of particulate materials above 1-micron particle size, but they cost considerably more. Without the roughing and intermediate pre-filters, frequent HEPA filter changes would incur a prohibitive cost.

Activated charcoal beds are used to capture volatile materials and delay the discharge of radioactive noble gases. Wet-scrubbers are used to capture acids and entrained liquids. In-line filters and wet-scrubbers should be located in the penthouse of the facility. Scrubbers should have an average flow of 20–30 L/min and should run continuously in any hood that evaporates mineral acids. The individual hood washdown system should have a separate switch to wash down the entire ductwork at the end of the day or at the end of a process.

Specific treatment combinations can be designed to match airborne contaminants from the physical or chemical operations, the type of sample (solid, liquid, gas), and the quantity that is processed. For example, sample preparation rooms where soil and vegetation are dried, ashed, ground and sieved require particle filter combinations but not charcoal beds and scrubbers. For treating laboratory air, the multiple stage filter system should be based on the expected maximum radionuclide concentration and airborne fraction of the processed samples. Typical combinations include pre-filters, HEPA filters and charcoal beds.

The individual laboratory is maintained under negative pressure relative to the hallways and adjacent accessible areas. Negative pressure must be maintained in each room to ensure that air is not drawn out of the fume hoods through the room and into the corridor when a door is opened. The appropriate negative room pressure is supported by sealing all wall joints and penetrations and by supplying make-up air for hoods.

Each laboratory should have the capacity for four radiochemical fume hoods, with one hood rated for perchloric acid. In Figs. 13.1 and 13.2, hoods appear as a large X. Each fume hood must be provided with external make-up air to assist in balancing the facility airflow. The make-up air should be at the same temperature as the room air and provide from 50 to 75 percent of the airflow across the face of the hood to maintain the hood face velocity at the required flow. This flow normally is 85 to 100 linear feet per minute (26–31 meters per minute). Newer designs may require different flow rates that should be verified by the facility industrial hygienist or safety officer. Proper ductwork and blower motor sizes will keep the noise level at the hood opening in the acceptable range of 85 to 92 decibels. The air discharged from each fume hood should be wet-scrubbed and filtered, and then vented to the

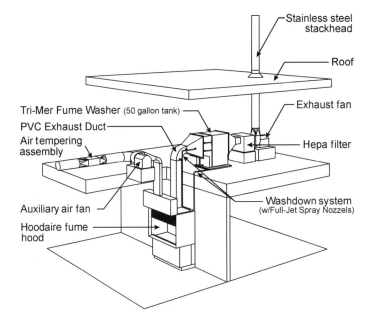

FIGURE 13.4. Total dissolution fume hood.

environment through a single stack (see Fig. 13.4). Normal stack heights should be roughly six times the stack diameter. In rooms without fume hoods, the room air should be ducted into one of the existing plenums where filtration is in-line.

New designs or major upgrades should incorporate energy conservation techniques. Design and construction of "green buildings" is gaining the support of government agencies and can reduce utility costs. The planning architect should include these techniques in the laboratory design. Examples of energy-conserving design elements are:

• Heat exchangers on exhaust air streams
• Low-flow fume hoods (under development)
• Night set-back controls on hoods with closed sashes
• Improved control systems to regulate air volumes and temperature on HVAC units
• A single air discharge plenum and fan for multiple hoods
• Water reuse and rainwater collection
• Task lighting to replace general lighting.

13.4.3. Liquids

The laboratory should be designed to discharge minimal amounts of radionuclides to the sewer. Releases must meet the limits specified in the Code of Federal Regulations Title 10, Part 20, Table 3 (see Section 14.8 for an explanation of the CFR titles) or DOE Order 5400.5. All liquids from the facility (including laboratories, change and shower facilities, rest rooms, lunch rooms, and offices) are directly piped into one of three holding tanks. Each tank should hold at least a 5-day volume of liquid waste, or 10,000 gallons (3.8×10^4 L) for the designed laboratory.

All fume hoods should have a water wash-down system. Each hood should be washed at the end of the day to prevent acid build-up and the resulting degradation of air ducts. The wash-down water is collected and re-circulated several times prior to discharge to the holding tanks. The wash-down water usually is acidic (pH <2) and needs to be neutralized daily. Eventually, solids will accumulate in the wash-down water container and should be flushed to the holding tanks.

The holding tanks are located in a separate bermed area that can retain 110% of the tank volume. The discharge lines from each laboratory (i.e., sinks, hoods, dish washers) and the holding tanks are constructed of acid- and solvent-resistant materials. The Teflon joints are heat-sealed to prevent leaks or separation. The immediate vicinity of the tanks is surveyed at regular intervals to check for leaks and associated radioactivity.

The holding tanks are monitored and agitated 24 hours a day, seven days a week. One tank is actively receiving liquid waste; the second tank holds the liquid waste off line for agitation, sampling, treatment, and discharge; and, the third tank is held in reserve, in the event problems occur with the other two tanks. Tanks should be discharged daily to prevent excessive accumulation; in no case should tanks be held for more than 7 days prior to treatment and discharge.

The liquid waste in the second tank is treated by adjusting the pH to >5.2 and then sampled for radionuclide analysis. If the concentration is acceptable, then the tank may be discharged to the sanitary sewage system. If the concentration exceeds release limits, several options may be considered. The three simplest are:

- Increase the dilution in the tank to reduce the concentration to an acceptable level for discharge (NOTE: Dilution as a disposal option is not permitted everywhere).
- Filter the tank contents to remove particulate matter for disposal as solid waste, then re-analyze and, if the concentration is acceptable, discharge the liquid.
- Treat the liquid by flocculation to remove dissolved materials and then treat the liquid as described in the second bullet.

Regardless of the chosen method, all employees must adhere to government regulations and site-specific license requirements. Because of the acids and solvents used in many laboratories, some liquid waste streams should be collected separately to avoid exothermic reactions that may lead to explosions.

13.4.4. Solid Radioactive Waste

Solid radioactive wastes are collected in each laboratory for transfer to the waste disposal building. Sludge from the holding tanks is dried and also transferred to the waste disposal building. The chemical and radiological contents of the wastes are characterized to provide information for the disposal and shipment records.

13.4.5. Residual Samples and Waste Disposal

Each sample that comes through the door for analysis brings with it a requirement for final disposal of the sample, the analyte, and the process waste residue. Rarely is a sample completely consumed in the analytical process. Several disposal options are available:

- Contract with a licensed commercial disposal service to remove all waste.
- Return the sample residue to the client and engage a licensed commercial disposal service for the waste residue.
- Return all samples and associated waste to the client.

Waste disposal is an integral part of the analytical process and a contributor to analytical costs.

Sample disposal options depend on the sample type and size. Liquid samples are mostly consumed during the analytical process and require minimal handling as waste. Sample residue should be archived in case re-analysis is required. When the sample results are validated and the client authorizes disposal, the sample residue is discharged to the holding tank.

After data validation and sample release for disposal by the client, solid samples can be returned to the client or logged into the sample disposal database and packaged for disposal at an approved waste disposal facility. Samples must be disposed as either radiological or mixed waste, and not commingled with sanitary sewer or general landfill discharges.

A mixed stream sample, which contains radionuclides and a hazardous chemical or a biological toxin, should be identified during sample receipt. Always review the shipping papers and container markings of submitted samples to identify hazardous chemical or biological contaminants. Mixed stream samples accepted for radionuclide analysis may require long holding times and create significant waste disposal problems.

13.5. The Counting Room

After processing, samples are moved to the counting rooms, numbered 13 and 19 in Figure 13.2. Only a small higher-level counting room area (inset to room 13) is provided because the majority of samples processed at most laboratories are low, environmental radiation, level samples.

Table 13.2 suggests the types and number of radiation detectors needed for the laboratory. The first column lists the equipment to process 1000 samples per month. For radioanalytical chemistry laboratories that processes level C and D samples, count times per sample are in the 50–1,000 min range. If higher-level A and B samples are processed, the brief count times require fewer detection instruments. The recommended equipment number provides for counting the radiation background, calibration standards and quality control samples as well as the actual samples.

TABLE 13.2. Counting room instrumentation

	High Level Counting Room		Low Level Counting Room	
	Large Scale Production Laboratory (1000 samples/month)	Minimum Design Basis Laboratory (125 samples/month)	Large Scale Production Laboratory (1000 samples/month)	Minimum Design Basis Laboratory (125 samples/month)
High Purity Germanium Detectors (HpGe)	2	1	16	2
Alpha Particle Spectroscopy Detectors	16	8	164	16
Liquid Scintillation Counter	1	1	4	1
Low Background Gas Proportional α-β Counter	1	1	4	1

The second column describes a smaller laboratory that can process up to 125 samples per month. Although some laboratories start by offering limited services and then try to grow into a full-service facility, a minimum number of samples must be processed monthly to produce a revenue stream that will sustain operation. This minimum must be calculated for each type and quantity of sample, number and types of analytes requested, and facility overhead.

The counting room houses the radiation detectors and counting systems indicated in Table 13.2. The counting room depicted in Figure 13.5 is a blueprint for an actual counting room. It is approximately 6 m × 9 m in the main body, with an extra leg of about 1.5 m × 3 m for additional detectors. A "dutch" door (a door divided horizontally, so that the upper portion can be opened without opening the lower portion) is used to pass samples from the sample preparation room (numbered 6) into the counting room (numbered 13). This door, located to the left in Figure 13.5, allows NO foot traffic. Authorized personnel may enter through the other doors in Fig. 13.5.

This counting room has sufficient detector capacity to analyze 1000 samples per month. The counting room should be located near the liquid nitrogen delivery system, which is generally installed in the gas storage room. A separate loading dock for the gas bottle storage room can be located directly behind the counting room, as shown in Figures 13.2 and 13.5. The counting room requires liquid nitrogen and P-10 counting gas, and possibly other gases that are used for chemical analyses.

Germanium detectors require liquid nitrogen for cooling, and a minimum footprint of 1 m by 1 m. Each Dewar should be directly connected to the liquid nitrogen feed line through a ball-cock on/off valve and special low-temperature piping. The Dewar is mounted on a scale to monitor liquid nitrogen content. Detectors are housed in a steel-clad lead shield to minimize the radiation background, which

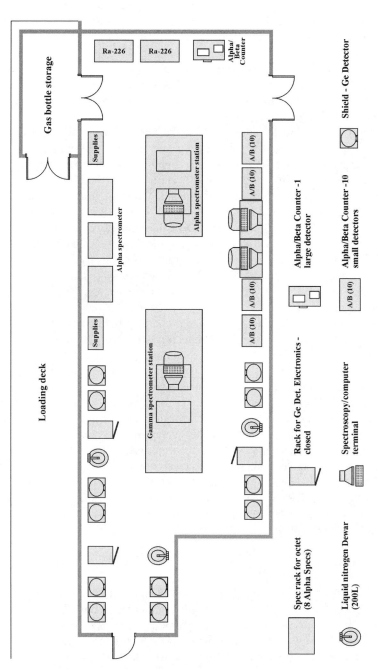

FIGURE 13.5. Counting room layout.

includes ambient and nearby sources of radiation. The high-voltage and signal cables can run overhead to each multi-channel analyzer. The low-level counting room should have additional shielding in the walls that separate it from the second room that is devoted to counting higher-level samples in the same area.

Because of the heavy weight of lead shields and steel stands, the counting room should always be on the ground floor. Even with this stipulation, the floor should be built of high density concrete with imbedded I-beams for increased support. Some laboratories use a thick steel plate under the equipment footprint to distribute the weight over a larger area.

Radiation detector shields must be of low-background materials (e.g., lead or steel). In regions where radon gas intrusion and accumulation present a high-background problem, the incoming air to the counting room may have to be passed through charcoal beds to reduce the radon concentration, and room's walls and floor interfaces should be sealed to avoid radon emanation from the ground. The intended location of the counting room should be checked to avoid elevated external radiation levels of natural or man-made origin. Once the counting room is built, nearby placement of facilities that will elevate external radiation levels must be prevented.

The counting room also houses alpha-particle spectrometry systems, which require a vacuum line or pumps. These detectors can be placed anywhere within the counting room and operated without interference from any elevated external radiation background.

Low-background alpha/beta particle gas proportional counters can be placed anywhere if a dedicated counting-gas cylinder is provided for each unit. If a central supply of counting gas is used, the units should be placed in direct access to the gas supply manifold. Liquid scintillation counters can be placed anywhere in the counting room because they can be operated as stand-alone instruments.

All counting systems are linked directly to the laboratory computer system to permit centralized control and for direct data transfer at the completion of counting. The counting room requires one computer workstation to control each type of counter, and additional terminals to support multiple users. Counting systems must be connected to a stable uninterruptible power supply (UPS) that will provide at least six hours of power in the event of power failure to the laboratory (see Section 13.7.6). The UPS system should be mounted in the penthouse to minimize equipment located on the counting room floor.

13.6. Storage

A busy laboratory utilizes large quantities of chemicals for sample preparation and processing. The chemical storage facility should stand alone from the main facility and be used to store and segregate hazardous materials such as flammables, acids and bases (see Fig. 13.6). Chemicals in these different categories should never be stored in the same room because inadvertent mixing could result in an explosion or fire.

Radioactive waste also should be stored separately from the main facility to avoid cross-contaminating other waste categories or supplies. Waste minimization

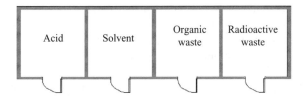

FIGURE 13.6. Segregation of chemicals in the hazardous materials storage building.

policy and procedures should be implemented and monitored to reduce expense.

The floor in a storage facility should be recessed at least 0.3 to 0.5 m or bermed to this depth in divided sections. The bermed area should have sufficient volume to collect twice the storage capacity of each section. Each section should be covered by an acid- and solvent-resistant grate (a material that will withstand aqua regia) and strong enough to support the weight of stored materials, moving dollies, and personnel. All storage shelves should be made of the same corrosion-resistant material. The building needs to be located near the loading dock to facilitate transfer of materials, yet detached and sufficiently distant from the main facility to minimize damage from fires or explosions.

By laboratory policy, an individual should notify the receptionist or safety staff before entering the building. The storage building should be under continuous surveillance by remote TV monitors, and should have direct telephone links (picking up the phone connects to the switchboard without dialing) on the outside of the building near each door. Individuals entering the building should have direct communication to safety support staff. Containers moved between buildings or laboratories should be carried in rubber-wheeled carts or special carriers to minimize spills because of glass breakage. All hallways and entry/egress ports should be designed for the use of these carts.

Ventilation in the chemical storage facility should provide a minimum of five room air exchanges per hour to minimize build-up of vapors and gases. The ventilation system should vent directly to the environment, well away from air intakes for other buildings.

Separate rooms in the storage facility should be heated and cooled based on the acceptable temperature ranges of the stored chemicals. All electrical circuits, including lights, switches, and thermostats should be made of non-corrosive explosion-proof materials.

Individual laboratories should be equipped with cabinets for storing relatively small amounts of chemicals that are frequently used. Separate cabinets should be available to segregate chemicals by hazard type, i.e., flammable, acid, or base. The cabinets should be labeled by type of contents and provided with locks.

13.7. Elements of Facility Management

A laboratory director faces a number of interrelated tasks in managing the accurate and timely analysis of the many samples submitted by clients. One is to place and direct competent scientists, engineers, and technicians in analytical and technical positions, and support personnel in management, maintenance, security, emergency response, health, and safety. A second task is to establish the framework—facilities, equipment and supplies, budgets, and schedules—of laboratory operation. A third task is to direct laboratory services to assure reliable results that meet all client requirements and are consistent with government regulations.

13.7.1. Analytical and Technical Staff

The organizational structure of the laboratory includes the key positions of director, quality assurance (QA) manager, radiation safety officer (RSO), radioanalytical chemist, counting room operator, information technology (IT) manager, facility manager, financial manager, and human resources manager. The specific staffing needs of the laboratory are primarily based on the number and types of samples that are processed. In the smallest laboratory, some of the listed functions are combined. A large laboratory has multiple chemists and separate radiation safety (or health physics—HP) and industrial safety officers. Figure 13.7 suggests the staffing requirements for a laboratory that processes about 1,000 samples per

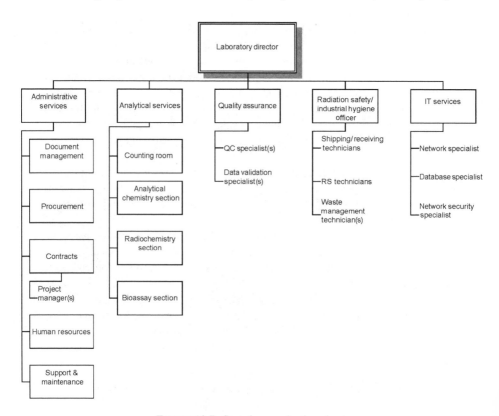

FIGURE 13.7. Sample organization chart.

TABLE 13.3. Recommended laboratory staff positions

Title	Education/ Experience	Responsibility
Laboratory Director	MS or PhD + 10–15 years experience	General Manager
Radiation Safety Officer	MS or PhD + ABHP Certification	Overall responsibility for safety and licensing
Senior Radioanalytical/Analytical Chemist	MS or PhD + 10–15 years experience	Overall responsibility for sample preparation
Counting Room Supervisor	MS or PhD + 5 years experience	General operations of counting room
IT Manager	BS or MS + 5 years experience	Maintains computer systems/servers and data security/integrity
Radioanalytical/ Analytical Chemist	BS or MS + 2 years experience	Day to day operations in individual laboratory
Technician	High School or Associate degree	Sample receiving, preparation, counting
QA Manager	MS or PhD + 5 years experience	Overall responsibility for quality assurance and quality control
Project Manager	Bachelors + 2 years experience	Client/laboratory point of contact

month. In comparison, the smallest viable laboratory processes about 125 samples per month.

Assurance of competent staff begins by preparing position descriptions that specify the educational requirements, operational experience, and duties of the laboratory personnel. The qualifications of the supervisors, analysts, and instrument operators who are hired must match these descriptions. Each staff position should have a set of core training specifications (e.g., radiation safety, chemical safety, waste minimization and disposal, quality assurance) plus specific training in its area of responsibility. Table 13.3 provides suggested position titles, educational requirements, and general responsibilities for the optimal laboratory staff.

The radioanalytical chemistry laboratory has some specific needs, notably that the chemists must have specific training and experience in radioanalytical chemistry. The skills of the radiation safety officer must match the duties described in Section 14.3.2. The project managers should be fully versed in the operations and capabilities of the laboratory to serve as client's advocate with the analytical staff and the laboratory's liaison with the client.

13.7.2. Support Staff

A routine facility maintenance program should keep the facility clean and in operating condition without inadvertent interruptions or discharges to the environment. The maintenance program includes (1) general housekeeping, (2) facility support features and (3) equipment. The HP staff oversees maintenance activities to minimize radiation exposure to personnel and spread of contamination. The HP staff must first review the proposed activity, then survey the tasks while in progress, perform a final survey before resuming normal operations, and finally record radiation exposure and radionuclide contamination levels.

Laboratories may employ either in-house or contractor janitorial staff. The housekeeping staff needs initial training and routine re-training for areas that are off-limits for routine cleaning or trash disposal. An individual laboratory may have several trash receptacles to segregate normal trash, radioactive waste, and hazardous chemical waste. Housekeeping staff should be trained to recognize the various hazard symbols and placards and the appropriate response to each.

13.7.3. Equipment Maintenance

Mechanical support services—heat, air, electricity, water, sanitary drains, holding tanks, scrubbing systems, and filter banks—occupy much space and require large expenditures for support and maintenance contracts. Maintenance support and laboratory staff must cooperate to minimize disruption of laboratory activities during routine and non-routine maintenance. Conflicts and downtime can be reduced by careful scheduling of routine maintenance for equipment such as wet scrubbers, filter banks, and motors.

Contracts with reliable and responsive service companies for maintaining facility support equipment (i.e., HVAC systems, electrical systems) are vital for efficient laboratory operations. Many modern systems are computer driven and so complex that only a specialist can repair them. When laboratory integrity is threatened, e.g., by flooding from a burst pipe or power failure, only fast response by a maintenance service company can protect against expensive losses and prolonged down time.

Maintenance contracts should include scheduled replacement of system components. In the air handling systems, blower motors need regular inspection, oiling, and belt replacement. The dust loading of the filters should drive filter replacement frequency. Replacement of roughing filters (quarterly recommended) will prolong the life of the HEPA filters and activated charcoal beds. The wet scrubbers need to be drained and re-filled on a routine schedule because the pH continues to drop when large quantities of acids are processed in the fume hoods. Scrubber water can be collected, neutralized with a strong base (sodium carbonate or sodium hydroxide), and recycled back into the scrubber. A contract maintenance program to service chillers, motor dampers, computers, and airflow measuring instruments can assist the in-house engineering staff.

The holding tanks should be on a weekly maintenance schedule to check for wet regions in the bermed area. These may indicate leaks, but could also be from rain or condensate formed on exterior tank surfaces. Pumps and agitator motors should be checked and serviced on the same schedule as sampling and discharge.

Radiation detection instruments require routine maintenance and occasional repair. Counting room staff is expected to perform initial trouble shooting, but specialists often are needed to diagnose problems and perform repairs. Most instruments can be obtained with maintenance and repair contracts at the time of purchase. Operation of a small laboratory may at times be at the mercy of a repair service for rapid turn-around in detector repair. A large laboratory may consider comparing the cost and turn-around time of an internal repair staff with contract repair. Rapid obsolescence of certain equipment may suggest a lease to be more cost-effective than a purchase.

13.7.4. Laboratory Security

Security has always been a consideration, but awareness has increased since September 11th 2001. The laboratory must be protected against illegal entry, damage, and theft. The samples must be protected as client property. Radioactive material requires heightened protection, especially if controlled material is included.

In smaller or less secure laboratories, checkpoints are established by locking all access doors—i.e., only those with a key may enter. In a more security-conscious laboratory complex, each area or room in the facility is individually controlled with a computer-activated system. The computer system can identify the person who enters, log entry and exit times and give management the ability to restrict unnecessary or unscheduled access (e.g., entry late at night). Each room has a list of approved access personnel, and only those personnel will be able to enter by use of their magnetic or bar-coded personal ID card.

Access to higher-security areas is restricted at designated boundaries. The area of lowest security in the building—the only open and directly accessible part of the building—should be the reception area, numbered 27 in Fig. 13.3. All visitors must sign in at the reception area, and should be given temporary ID cards. Visitors should be escorted throughout the remainder of the building.

Employees will use their ID card to pass through the reception area into more secure areas. This zone contains offices of the laboratory director and support staff, technical staff workstations and meeting space. Bathroom, shower and change facilities appropriate to the size and amenities of the facility should be located in this area. Kitchen and dining space should be provided here because eating and drinking are not allowed inside any laboratory work area.

The entire laboratory work area is the higher-security zone. Entry into the laboratory should be limited on a "need to be there" basis. Direct access to and from the laboratory should be minimized. The shower and change facility constitutes a buffer area for transit from the main laboratory to the administrative area, as shown in Fig. 13.3.

13.7.5. Data Security

Clients consider their submitted samples to be their property, and information gleaned from these samples to be proprietary. If the client is a government agency, some sample results may be considered classified information that must be protected. The laboratory and specified personnel must be certified as capable of handling such classified information, and personnel must have the appropriate security classification. Facilities regulated by the Departments of Energy and Defense have orders and guidance documents that specify the requirements for data security, integrity and authenticity. The client may specify individual security requirements.

Regardless of formal security requirements, all analytical data should be treated as a valuable and irreplaceable resource. Data security begins with physical security of the facility discussed above, and is maintained through protection of computer hardware, software and Internet accessibility.

The laboratory computer system is the backbone of operations. It is used for sample management and tracking, data storage, reduction, and presentation, and sample archival or disposal. The system also may process personnel records, payroll, chemical inventories, accounts receivable and payable, and waste management. A suitable computer system, at the time of writing, can consist of two high-end servers with a hot-swappable redundant array of independent disk (RAID) drives. Such drives vastly increase the throughput of the system. The computers must be equipped to perform automatic information backup with magnetic tape or by CD read/write. Ethernet high-speed interconnectivity and ultra-high-speed Internet connections should be available. Individual PC-style workstations with barcode readers should be installed where samples are received, stored, processed, counted or discarded.

The computer system should be installed in a secure limited-access room; in Fig. 13.3 the computer system is in the room numbered 28, conveniently located near the record storage room (26). The computer room should be equipped with a UPS that has a six-hour rating. The system requires low humidity ($<60\%$ relative humidity) and constant temperature (18–$21°C$).

Personnel access to the system should be controlled with a password authorization system, but password protection is only as good as the type of system implemented and maintained by IT personnel. Passwords should be changed quarterly, require a minimum of 8 characters and numbers, and IT personnel should routinely test the system with password-breaking software to identify and replace easily inferred passwords.

Commercial software (currently, Microsoft, Corel, AutoDesk, etc.) upgrades that are issued every year or two should be evaluated for backward compatibility, bugs, and security flaws. Lack of compatibility may prevent opening some files and corrupt others. Replacement of specific scientific software should be evaluated for consistency in aspects such as look-up tables, algorithms and calculation routines.

Computer security depends on limiting system breaches from the Internet. Commercial products are available to detect intrusion, i.e., unauthorized attempts to penetrate the system. These products alert IT personnel when an attack is detected. Commercial firewalls can minimize system penetration. Anti-virus programs can continually scan incoming material for viruses, worms, and trojans. None of these products can replace continual vigilance by IT personnel because a determined hacker can penetrate any commercial system and disrupt laboratory information handling activities, notably sample processing and data delivery.

13.7.6. Laboratory Emergency Support

The laboratory safety measures discussed in Chapter 14 go far toward preventing emergency situations, but some accidents and incidents are unavoidable. Wise management practices include planning for and preparing the facility and staff for such eventualities.

Maintaining electrical supply is vital during any emergency. First responders are aided in their efforts by the availability of light, communications, operating equipment, and active security controls. Electrical supply to the laboratory should be provided at three levels:

- Regular service is the primary feed to the entire facility. Preferably, the supply is conditioned to reduce the impact of voltage fluctuations.
- A standby generator supplies emergency power. The emergency supply operates emergency lights, cold rooms, freezers, refrigerators, and fume hoods. Two-speed blowers in fume hoods should automatically drop to the lower speed when the emergency generator is activated.
- UPS batteries and power converters to replace interrupted power for the computer systems and counting room instruments. The batteries are continually charged by the regular service system. During power loss, they support orderly shutdown of computers and instruments with preservation of accumulated data and instructions for calculations. An individual system is recommended for each piece of equipment instead of a large hard-wired system to avoid catastrophic data loss if the hard-wired system fails.

The safety and HP staff responsible for day-to-day laboratory support should be cross-trained as emergency responders (e.g., first aid, CPR, spill control, fire suppression). The safety and HP office should be conveniently located near the sample receiving and laboratory areas, as it is in the top left corner of Figure 13.1. The office should store emergency equipment and support materials. In addition, each hallway should have an emergency response cart equipped with a first aid kit, spill control materials, respirators (if needed), and portable survey instruments. A fire extinguisher should be in every laboratory and hallway. Emergency response supplies and carts should be inspected and restocked at designated intervals that are recorded on labels or tags at each location.

In addition, each hallway and each room should have an emergency ("panic") button that will set off an alarm and by-pass the computer-controlled access system on each door. This system allows anyone to enter a laboratory when an emergency occurs that requires immediate assistance, or to leave the laboratory. Because pressing the button triggers a series of events and emergency responses and subverts security, it should only be used for a true emergency.

One of the operating license criteria for commercial facilities is the preparation and testing of an Emergency Management Plan (EMP). Arrangements must be coordinated with the local hospital, fire department, police department, HAZMAT team, and the city or county governments. The EMP must itemize the potential hazards at the facility and specific instructions for emergency response personnel. For example, the Fire Department may be advised not to enter a building without respiratory protection or to perform fire suppression from a distance. The Police Department may be advised to enter a building only when accompanied by designated site personnel. The hospital may be advised that an incoming radionuclide-contaminated patient should be treated in a designated radiation area of the hospital. Communications must be concise, clear and rapid during an emergency, and should be limited to selected personnel. Sections 14.5 and 14.6 provide more information on emergency planning and accident response.

13.8. Controlling Regulations and Licensing Requirements

Some of the design for, and much of the work at, radioanalytical and mixed-waste laboratories is controlled by regulations and professional standards that must be incorporated into the decision making process. The architect, builder and design team members are expected to be familiar with such regulations. Laboratory managers are advised to familiarize themselves with the regulations and criteria pertaining to their laboratory.

13.8.1. Laboratory Design and Modification

New laboratories must be constructed in accordance with Federal, State, and local building codes. At the date of writing, more than 100 regulations are directly applicable to the building process. The many groups cited in Appendix B are considered important as regulators and criteria resources, but there will be others, depending on the nature of the facility and its location. The architectural firm that designs the facilities should submit all pertinent regulations as part of the bid specification package. At least one member of the laboratory planning team should review the bid specifications to ensure that they contain the current and applicable regulations and guides.

Design modifications for laboratories that are renovated or upgraded by replacing individual components must meet current regulations because they are rarely "grandfathered" under the old regulations. The Federal Facility Compliance Act of 1992 includes rigorous specific requirements for the final design of the facility. To meet these requirements, renovation of older structures may cost more than new construction. Once an existing structure is modified, any problems with the entire structure become the responsibility of the modifying party. One prominent example from newspaper headlines is the widespread requirement for asbestos remediation. In some cases, the NRC may require the facility to institute design modifications, called "back fitting," regulated under 10 CFR 50.109.

13.8.2. Operating Licenses

The Nuclear Regulatory Commission (NRC) regulates management and control of radioactive materials. The Atomic Energy Act of 1954, section 274, allowed responsibility to be transferred to individual states for controlling sources of radiation. One requirement for approval is that state controls be just as restrictive, if not more so, than those of the NRC. If approved by the NRC, controlling authority is transferred to the state, thereafter called an Agreement State.

Agreement State programs were developed to bring a modicum of control to the state level, but they do not cover all handling of nuclear materials. For instance, all import and export of nuclear materials is controlled by the NRC, as is the construction of any nuclear material production or utilization facility. Many waste disposal issues also fall under the aegis of the NRC. The NRC has the ultimate authority to control all nuclear materials deemed pertinent to the defense and security of the United States. More information on NRC and Agreement State regulations is online at http://www.hsrd.ornl.gov/nrc/index.html (Dec. 2005).

In the U.S, either the NRC or one of the currently 33 Agreement State radiological control programs issues the radioactive material license. This license is the controlling force for the design, staffing, and operation of a radioanalytical chemistry laboratory. License categories include Byproduct Materials, Special Nuclear Materials, and Generally Licensed Materials, with some sub-categories. The usual laboratory needs a Broad Scope Byproduct Materials license and a separate Special Nuclear Material license by the governing authority. A license application must be submitted to and approved by the licensing agency before operation can begin. License approval requires that the laboratory is fully organized, with a qualified Radiation Safety Officer; laboratory staff with demonstrably appropriate education, training, and experience; an acceptable waste management plan; spill countermeasures; and suitable radiological instrumentation. These items and many more must be fully described in a Radiation Protection Plan (see Section 14.2) that can be distributed as a Radiation Safety Manual.

The laboratory also may be required to report to the EPA its radionuclide quantities or to monitor airborne emission from operating stacks and by other discharges under the National Emission Standards for Hazardous Air Pollutants (NESHAPS). Subpart H protects the public and the environment from radioactive materials (other than radon) emission at DOE facilities and subpart I applies to other federal facilities, including NRC licensees. The basic criteria are that the annual effective dose equivalent to any individual

- must not exceed 10 mrem, but
- must not exceed 3 mrem from radioiodine.

The EPA provides the computer code COMPLY for calculating the radiation dose from release of radionuclides to air. The operator must use measurements of stack velocity/volume flow rate and radionuclide amounts to calculate radiation doses by COMPLY. A laboratory may qualify for an exemption from monitoring requirements if specific radionuclide amounts do not exceed limiting values.

Clients need a copy of the laboratory license to verify permission to receive samples that contain radionuclides. In turn, the laboratory should request a copy of the client's license. Samples not covered by the client's license become the responsibility of the laboratory that accepts them.

13.8.3. Laboratory Decommissioning

The Agreement States and the NRC require that, as a condition of granting the operating license, a decommissioning plan and a budget are available in the event that a laboratory is dismantled. An alternative to a sequestered budget is a surety bond. The decommissioning plan formulates the final release criteria, the survey methodology, survey instrumentation, and the final status survey report. This rule is intended to protect the public from the environmental hazards of an abandoned and decaying laboratory site, and to protect taxpayers from bearing the cost of the resulting cleanup.

Decommissioning of buildings, facilities and structures for government laboratories must be in accord with Federal regulations in 10 CFR 835, *Occupational Radiation Protection* and DOE Order 5400.5, *Environmental Radiation Protection* (online at http://www.directives.doe.gov as of Dec. 2005). All equipment valued over $5,000 is inventoried and tracked, and the final disposition is directed by regulations,

Commercial facilities are closely regulated in terms of release of facilities and equipment for unrestricted use. The radioactive materials license has specific requirements for disposal of radioactive sources and trace materials, even when the useful life of these radioactive materials has been exceeded.

13.9. Laboratory Service

Any laboratory can survive for a brief period, but long-range sustainability depends on the management skills of the director and the reputation of the laboratory. While commercial laboratories have come and gone, the successful ones have the following characteristics in common:

- A well-developed business plan with sufficient capitalization.
- Skilled managers, professional analysts, and support staff.
- State-of-the-science facilities, methods and instruments.
- Client-service orientation in analysis, scheduling, and reporting.
- A reputation for of institutional integrity.

Facility design was discussed earlier in this chapter while the other topics are addressed below.

13.9.1. Business Planning and Capitalization

Laboratory planning must begin by identifying the clients that have a high probability of sample submission, and estimating the expected sample analysis rate. The planner must match the sample type and analytical load to sample processing methods, qualified staff, number and type of radiation detection instruments, laboratory design, and ancillary staff and facilities. These factors must be considered in

TABLE 13.4. Estimated new laboratory
construction and operating costs

General Category	Cost ($1 M USD)
Design and Engineering	1.5
Land and Site Preparation	2.0
Construction	11.5
Equipment and Supplies	1.5
Personnel and Training	3.8
TOTAL	$20.3

terms of location, availability, controlling regulations, operating cost, and project capitalization.

Table 13.4 outlines project capitalization at the time of writing for a new laboratory designed to analyze approximately 1,000 samples per month for both radiological and hazardous chemical analyses. Initial capitalization is about twenty million dollars to build the laboratory and begin processing samples. A definite lag time should be expected until sufficient contracts are in place to fund operation at or near capacity. An estimated 65 percent capacity—the approximate break-even point in terms of net profit—may be reached after two years. Unless near-capacity sample loads are obtained sooner, an additional amount of about nine million dollars to pay two years of salaries, staff benefits, supplies, and maintenance should be added to the initial capitalization, pushing start-up costs to over $29 million dollars.

Not surprisingly, few new companies have the resources to construct a new facility and keep it operational until it reaches profitability. A tempting alternative is the "bootstrap" approach of starting a much smaller laboratory—one that processes a minimum feasible 125 samples per month—and expanding it as business warrants. Few such attempts have been successful. The more common approach is construction of the new, full-scale, laboratory as part of an ongoing program.

The decision to move forward on constructing a new laboratory must be fully developed in an operational business plan, with sufficient detail to apply for and receive adequate project funding. A large part of this effort is determining the analytical cost structure and refining the operating plan of the laboratory.

13.9.2. Determining Analytical Costs

A set of default parameters must be estimated at the outset of operation to provide the client with cost of analytical procedures. A sample cost list for radioanalytical chemistry services is illustrated in Table 13.5. The earlier cost data are from a 1990 business plan developed by the authors and the 2004 cost data were extracted from cost proposals by a commercial laboratory. The magnitude of the cost for other analyses can be inferred from their similarity in processing to the cited methods.

Costs did not change greatly during these 14 years. The modest changes may reflect increased costs of instruments, salaries, and supplies, balanced by cost savings

TABLE 13.5. Radioanalytical chemistry costs

Analysis	Comments	Estimated Cost (USD) 1990	2004
Gamma-ray Spectral Analysis	Any geometry, up to 2 h count time	$73	$105
Alpha-particle Spectral	Uranium or Thorium	$170	$156
Analysis—Soil	Plutonium	$170	$166
Alpha-particle Spectral	Uranium or Thorium	$170	$156
Analysis—Water	Plutonium	$170	$166
	Total Uranium		$40
Liquid Scintillation Analysis	Water, swipes or air samples	$30	$20
	Soil (tritium distillate)	$67	$83
	Soil (direct count)	$30	$20
Screening Smears	Gross alpha/beta		$13
	Beta (LSC, up to 2 isotopes)		$20
Screening—Gross Alpha and	Soil	$67	$39
Gross Beta	Water	$67	$39
	Air Filter	$51	$51

with improved analytical methods. The laboratory cost list should be updated periodically to reflect such changes.

The basic cost of a radioanalytical chemical method is estimated from the person-hours devoted to analysis from sample receipt to result submission. The associated cost of supplies, supporting efforts, instrument use, and overhead is apportioned among the various analyses that are performed during the same period. Personnel cost is affected by the extent of processing needed to preserve and dissolve the sample and to remove interfering radionuclides. In essence, the simpler the process, the lower the cost. The degree of precision and magnitude of detection level are other cost determinants to the extent that they affect analytical skill, instrumentation, and time requirements.

The quality assurance program, whether the normal effort or with special items added by the client, is a major ancillary cost. The usual overhead costs of items such as management, structures, maintenance, employee benefits, and taxes always must be considered. Special costs may be invoked by studies to test or improve methods of purification and measurement, explain unanticipated results, or answer questions that arise during a project.

The above-cited factors suggest the potential benefits in achieving the most appropriate cost by interacting with the client in preparing the Quality Assurance Project Plan (see Section 11.3.2). Once the information needs are understood, analytical efforts may be expanded or reduced with associated cost changes. Results of initial screening analyses may suggest cost-reducing specification changes. Scheduling the sample flow to the laboratory and the turnaround time for data reporting may reduce cost by optimizing assignment of laboratory staff and instruments.

13.9.3. Minimum Detectable Activity or Concentration

A defined minimum detectable activity (MDA) or concentration (MDC), as discussed in Section 10.4, should be estimated for the result of each procedure, based on the planned sample matrix, sample volume or mass, analytical method, counting method, and counting time. If the client requests an MDA or MDC below the default value, the laboratory needs to consider adjusting the method, with an associated increase in cost. For example, a larger sample volume can reduce the MDA/MDC, but the larger sample volume may increase the analytical cost and time. A longer counting time will reduce the MDA/MDC but can impact the turnaround time, and require more counting equipment. Judicious planning and communication with the client can help meet individual client needs, but the laboratory director may be limited in the options offered to one client by commitments made to other clients.

13.9.4. Turnaround Time

Every client wants sample results delivered as soon as possible. The laboratory operation plan should include a reasonable estimate of analysis completion time for every method to present the client with the turnaround time. The specified period begins when the sample enters the laboratory door and ends when the analytical result is transmitted to the client. This period must include sufficient time for ancillary activities related to the sample and for unscheduled events that occasionally incur delay. The time requirement for the individual method then must be placed into the context of other analyses for this and other clients.

As an example, consider a client who requires total uranium analysis of 600 samples to be delivered at the beginning of each quarter. If the analysis schedule indicates a processing rate of 60 samples per day, then the sample load requires a minimum of 10 days for analysis. This period must include the following activities:

- Receiving and logging 600 samples into the system
- Preparing the samples for analysis
- Preparing reagents and maintaining instruments
- Incorporating quality control samples and measurements into each batch
- Processing samples and performing measurements
- Reviewing the data
- Checking the QC samples for acceptability
- Performing additional analyses or data review as needed
- Approving results report for delivery to the client

Turn around time can be increased by situations such as workload distribution (i.e., delay while other samples are processed), analysis difficulties, weather interruptions, employee absence, and external audits. These issues may increase the minimum time by a factor of two or three. Such eventualities must be considered in providing your client with a reasonable turnaround time.

A fine line separates operating the laboratory at capacity and being over-extended; unanticipated events often cause a laboratory to cross the line between full capacity processing and missed sample deadlines. Each project manager must coordinate with the laboratory director and technical staff to ensure that no one project or event detrimentally affects another client or overall laboratory operation. If these disruptions occur consistently, one or more of the following changes can be made:

- Increase staff levels.
- Add off-shift work.
- Install more counting instruments.
- Reduce the project load.
- Reorganize the sample-processing schedule with selected priorities.

A longer-term option is to increase the level of automation in the laboratory. Introduction of automated processes can benefit the processing schedule and engage staff in more efficient and creative activities. Automation is discussed in Chapter 15.

One final caveat: samples that contain hazardous materials such as volatile organic compounds (VOC's) or short-lived radionuclides must be analyzed within a limited time period. VOC's, regulated under 40 CFR because of their high volatility, must be analyzed within 7 days of collection. If the deadline is exceeded, the sample is invalid. Short-lived radionuclides must be analyzed before they decay below the detection limit.

13.9.5. Integrity

A reputation for integrity is vital to the successful operation of the radioanalytical chemistry laboratory, as for any analytical facility. The client depends on the validity of the reported analytical results in operating a program or performing a task. In turn, the client provides information based on the analytical results to various stakeholders, including workers, the public, and regulators, to demonstrate safe operation or absence of excessive risk. Even a minor instance of unreliable data can call into question entire reports of results and thus, the operation of the client's program.

Laboratory management must institute elaborate controls to guarantee data reliability (See Sections 10.5 and 10.6). Supervisors' must overview the work of analysts and instrument operators. A quality assurance program that typically includes comparisons with other laboratories, calibrations, quality control analyses and measurements, and numerical comparisons of results should be instituted to ensure accurate results. Periodic audits, questioning staff, and searching records should confirm that all personnel are following written instructions. These reviews have greatly expanded over the past several years in response to falsified results; the cause of falsification ranged from minor errors to major incompetence and even deceit.

Institutional controls can reassure the client and other users of data that any problems will be found and corrected, but the main quality assurance effort must be devoted to instilling integrity in laboratory staff. Managers and supervisors must preach integrity at every opportunity; more importantly they must demonstrate integrity in their practices and orders. While emphasizing production and timeliness, they must place greater value on following specified procedures and achieving correct analytical results. Chemists, instrument operators, and support staff must be trained to follow procedures, report any deviations from protocol or suspicions of unreliable instrumentation, and consider a questionable result to be unacceptable. Integrity can be defined as always doing the right thing, especially when no one appears to be watching. The laboratory director's responsibility is equating a correct result to "the right thing."

A final aspect of integrity involves the interaction of the facility with the community in which it operates. The laboratory needs to maintain good will within the community by being open, honest, and responsive to concerns as they are raised. Most commonly, the concerns will center on laboratory waste and effluent, but may also involve questions about the security of hazardous material and the soundness of the facility's emergency planning. Unless the facility is viewed as a good neighbor and partner in the community, it risks becoming a target of animosity that will inevitably damage its operations and morale.

14
Laboratory Safety*

ARTHUR WICKMAN[1,5], PAUL SCHLUMPER[2], GLENN MURPHY,[3]
and LIZ THOMPSON[4]

14.1. Introduction

Laboratory instruments and the chemicals used in preparing samples can create conditions that range from relatively benign to highly hazardous. If not identified and controlled or eliminated, these hazards can expose the laboratory worker to injuries and illnesses and cause damage to property and the environment. When even an experienced radioanalytical chemist begins to work in a laboratory, it behooves the supervisor to present clearly and with emphasis on safety the documented procedures of the laboratory and the behavioral practices that are expected of the employee. The chemist should respond with attention and concern. Each individual in the laboratory must be given and accept the primary responsibility for safety. The safety culture must flow from management to laboratory worker and must be embraced by each individual.

The reader was introduced to concern for a safe working environment and responsible behavior in the laboratory at his/her first experience with laboratory activities. This chapter is intended to build on that initial training, by providing the reader with a comprehensive discussion of safe laboratory practice from prevention measures to emergency protocols. The discussions in this chapter focus on the radioanalytical chemistry laboratory, whether it is a government, contractor, or academic research laboratory. Safety planning is discussed first to lay the groundwork for more specific discussions to follow. Safety officers are identified and their duties are described, together with the safety plans they administer. Workplace hazards,

---

* This text owes the genesis of its content to Dr. Isabel Fisenne, Department of Homeland Security. The authors would like to thank Dr. Alena Paulenova of Oregon State University for her careful review of the text.

[1] Health Sciences Branch, Georgia Tech Research Institute, Georgia Institute of Technology, Atlanta, GA 30332

[2] Safety Engineering Branch, Georgia Tech Research Institute, Georgia Institute of Technology, Atlanta, GA 30332

[3] The Matrix Group, Inc., 188 Hidden Lake Dr., Hull, GA 20646

[4] Environmental Radiation Branch, Georgia Tech Research Institute, Georgia Institute of Technology, Atlanta, GA 30332

[5] Arthur Wickman, Georgia Institute of Technology, Atlanta, GA 30332-0837; email: art.wickman@gtri.gatech.edu

safety precautions, and environmental safety in these laboratories are addressed, noting when necessary the special considerations that accompany a specific laboratory environment. The regulatory environment is discussed to indicate to the reader the laws that enforce safe work practices in the U.S.

14.2. Safety and Health Management Systems

A successful safety structure must be well-planned and well-documented. By developing and implementing an integrated safety and health management system, supervisors are supported in ensuring a safe and healthy working laboratory. A safety and health management system has the following main elements:

- management commitment and employee involvement;
- hazard identification and control;
- laboratory worker training; and
- periodic program review and revision.

Commitment of management and involvement of workers (in this case, laboratory analysts), are essential. Management commitment to worker safety requires:

- training workers to work safely;
- implementing radiological protection policy and practices based on the ALARA (as low as reasonably achievable) principle for radiation workers;
- identifying pertinent Occupational Safety and Health Administration (OSHA) standards, as per the Title 29, 1910, series of the Code of Federal Regulations (CFR 2004);
- emphasizing individual accountability for compliance with safety standards;
- involving workers in identifying existing and potential workplace hazards and controlling them;
- implementing a process to manage hazards through prevention, mitigation and control;
- empowering workers to exercise their Stop Work authority in response to hazards;
- ensuring compliance with applicable requirements and regulations; and
- providing adequate resources for safety management.

Involvement by workers requires their consideration of safety and health as an integral component of the work in the laboratory, for their own sake as well as that of other workers in the laboratory. A safety culture is necessary to ensure that proper procedures are followed. In addition to conducting laboratory operations in a safe and responsible manner, the radioanalytical chemist can involve himself in the safety culture by:

- participating in the planning process prior to selection and assignment of tasks;
- aiding in the preparation and review of job hazard analyses;

- making the immediate supervisor aware of the hazards and their possible controls for proposed methods; and
- communicating constructively with safety staff by identifying concerns and responding to guidance.

Hazard identification and control are important aspects of safety in a laboratory. Most hazards in a laboratory environment involve either unsafe conditions or behavior. Conditions can be controlled through proper analysis and inspection of the work environment, and implementation of controls to reduce or eliminate the exposure to these hazards. A formal job hazard analysis, where individual tasks are observed, broken down into their individual components, and analyzed for existing and potential hazards is necessary for hazard identification and corrective action. This activity must be followed by periodic formal inspections and hazard assessments.

Unsafe behavior may be difficult to predict and control. Laboratory worker training is thus an important component of a safety and health management system. Before they even set foot in the laboratory, workers must be aware of the potential hazards to which they might be exposed, the protective measures they should take regarding those hazards, and the management procedures and policies regarding safety and health. Training is usually conducted for new workers as an orientation and then periodically in refresher courses. Training also is required whenever a change occurs in the laboratory environment related to procedures, materials, or equipment. Finally, training is an important aspect of corrective action, should observation indicate that a laboratory worker is performing a task unsafely. Individuals who demonstrate *unsafe* work practices require additional training to ensure they have proper knowledge of *safe* work practices. If this remedy is ineffective, disciplinary action must be instituted.

To close the loop in the safety and health management system, periodic assessment and feedback are necessary. Indicators should be chosen that can assess the overall performance of the laboratory with respect to safety and health. Whenever possible, "leading" indicators such as behavioral observations should be measured and reviewed, as well as "trailing" indicators such as the type and number of injuries and illnesses and loss of working time. The purpose of this assessment is to determine the overall effectiveness of the safety and health management system and to correct any areas of deficiency.

This safety and health management system constitutes a safety framework with elements that should be implemented in all laboratories. More specific tools to delegate the responsibility of safety management in the radioanalytical chemistry laboratory are described in the following section.

14.3. Chemical Hygiene and Radiation Safety Plans and Staffing

Day-to-day oversight for maintaining and operating a safe work environment is delegated to two groups: the Industrial or Chemical Hygiene Office and the

Radiation Safety or Radiological Control Office. The industrial hygienist (IH) or chemical hygiene officer (CHO) is concerned with the overall safety and comfort of the laboratory workforce. The radiation safety officer (RSO) is concerned more specifically with radioactive materials. The IH/CHO and RSO staffs function under different laws and regulatory agencies, but have parallel responsibilities that organizationally may be either combined or separate.

Management must support the efforts of these offices with necessary funding as well as respect for their expert opinion on workplace concerns. As a corollary, management must support rational efforts to ameliorate identified issues or areas of concern with regard to worker/workplace safety and health. As a basic principle, the laboratory supervisor must not countermand a decision to halt a work plan because of health, safety, or radiation concerns by the IH/CHO or RSO, and the work plan must be revised to meet these concerns. Descriptions of the duties and responsibilities of the CHO and RSO, and the plans they administer, are given below.

14.3.1. Chemical Safety Plans and Staffing

All laboratories must have a written plan—the Chemical Hygiene Plan (CHP)—which describes the provisions that have been made for safety by the laboratory managers. This requirement is regulated by OSHA under standard 29 CFR 1910.1450, "Occupational Exposure to Hazardous Chemicals in Laboratories." The CHP sets out the specific procedures, work practices, safety equipment and personal protective equipment that have been selected to provide employee protection for the hazards found in each laboratory. An individual university or college laboratory will follow the CHP of the institution, which applies to all laboratories on campus; the laboratory may also have a CHP that is specific to its individual conditions. College students should follow the provisions of the applicable CHP to ensure their safety. The CHP must include:

- standard operating procedures for safety and health;
- engineering controls needed to limit exposures;
- operational and maintenance criteria for engineering controls
- provisions for safety training;
- identification of hazardous chemicals; and
- any permits or special procedures required for the laboratory.

Additionally, the CHP may contain the emergency plans for the laboratory (see Section 14.5).

The CHP must be administered by a CHO, an individual qualified by training and experience to handle safely the chemicals and processes in the laboratory who also has the authority to take corrective actions when needed. The CHO must be familiar with all safety and environmental regulations that apply to the laboratory. The CHO must understand the health and safety hazards of the chemicals in use. The CHO must be aware of any biological monitoring required for employees or students who use regulated chemicals. To be effective, the CHO must have the support of managers and administrators, and must have a clear mandate to enforce safety requirements. Hazardous conditions identified by the CHO must be corrected, and the CHP must be revised to prevent any recurrence of the hazard. The CHO must review and accept or modify any changes to chemicals or procedures in the laboratory.

14.3.2. Radiation Safety Plans and Staffing

A radioanalytical chemistry laboratory requires a Radiation Safety Manual (RSM) or Radiation Protection Plan (RPP) in addition to a Quality Assurance Plan (QAP). The format is controlled by the licensing agency for the facility; the Nuclear Regulatory Commission (NRC) requires the RSM, while the Department of Energy (DOE) requires the RPP. These plans are stand-alone documents that deal specifically with radiation safety issues and practices to set safe operating parameters in the laboratory.

Both the NRC and DOE provide guidance documents to aid the user in the development and implementation of the RSM or RPP. Topics concerning radionuclides and radiation that are routinely controlled under the program include:

- ALARA principles;
- individual training requirements;
- individual dosimetry requirements;
- air concentration levels;
- residual contamination levels;
- exposure rate levels;
- instrumentation;
- routine operating procedures;
- emergency procedures;
- waste disposal options; and
- record-keeping and reports.

Each worker in the laboratory must undergo initial radiation safety training that prepares him/her to work with radioactive materials. Refresher training must brief the employee about changes to operating procedures and regulatory requirements.

Radiation safety staffing levels may vary widely among facilities on the basis of facility size, radioactivity levels, and number of samples processed. Each facility is required to have an RSO that meets the education, training and experience requirements by the NRC or DOE. The DOE protocol is for government laboratories while NRC and Agreement State licenses control other radionuclide-using entities, including commercial and academic laboratories.

The NRC and Agreement States have specific license requirements for RSO education, training and experience. To practice this specialty, the RSO must meet the requirements of 10 CFR 35.900 series Subpart J—Training and Experience Requirements. Three different sets of experience can qualify an individual to become an RSO:

- certification as a Certified Health Physicist by the American Board of Health Physics in comprehensive health physics;
- demonstration of successful completion of 200 hours of specified classroom and laboratory training, plus one year of full-time experience at a medical institution under the supervision of a qualified RSO; or
- work experience for an authorized user under an NRC license.

The DOE does not use the license concept, so there are no formal requirements to become a Radiological Control Manager (RCM)—the equivalent of an RSO. Regardless of title and regulatory mechanism, the RSO or RCM is responsible for implementing a safety program for the use and control of radioactive materials in the laboratory. The RSO/RCM is responsible for the following activities:

- maintaining the facility license (for the NRC or the Agreement State) or the Radiological Protection Program Plan (for the DOE);
- procuring, receiving, and delivering the radioactive material to the individual user;
- training the user in routine handling and emergency response procedures;
- implementing an internal and external dosimetry program;
- performing routine surveys in radioactive materials use areas;
- disposing of radioactive waste; and
- controlling any other item pertinent to radiation protection (for example, a position description may specify "5% of time spent on other duties as assigned").

The RSO is responsible for working with the radiation laboratory worker in planning any program that may result in radiation exposure to persons and radionuclide contamination of the work place or the environment. The radioanalytical chemist should be able to depend on the RSO for guidance in minimizing the potential for such exposure and contamination. In brief, radioactive material must be stored, handled, and discarded separately from and differently than the usual laboratory reagents. The RSO must inform the regulator, management, and the worker of personnel radiation exposure and any unusual radiation exposure conditions.

14.3.3. Responsibility of the Radioanalytical Chemist

The newly employed analyst must read carefully the laboratory CHP and RPP to assimilate information vital to the safe operation of the laboratory and the safety of its workers. The analyst should make certain that the CHP and RPP reflect any changes in procedure, and are reviewed and revised on a scheduled basis, usually biannually, to address changes in legislation and institutional policies. The following discussions consider aspects of worker safety that are included explicitly in the CHP and RPP of a given laboratory.

14.4. Hazards in the Workplace

Laboratory work must be conducted with the realization that safe operation is only achieved by constant focus on safety. The point is not to be *afraid*, but to be *aware* of potential hazards and *responsible* for one's behavior. Hazard identification is crucial for accident prevention. Adverse effects can be avoided when both the manager and the worker make safety a priority. This section identifies the primary hazards in the radioanalytical chemistry laboratory, and ways to minimize the possibility of accidents.

14.4.1. General Hazards and Precautions

Hazards common to any environment include tripping over an obstacle or slipping in a spill. Floors must be kept clear of any equipment or stored chemicals. Any spills on floors must be immediately cleaned. Any power cords or flexible hoses that cross floors must be secured so that they do not present a tripping hazard.

The chemistry laboratory has additional hazards. Open flames, ovens and furnaces can cause fires and burns. Cold storage rooms and their contents, notably dry ice and liquid nitrogen, may cause freeze burns. Corrosive acids and other chemicals can cause chemical skin burns and internal damage from inhalation, ingestion and absorption through the skin. Many chemicals are poisonous: chemicals should not be tested by taste or smell. Compressed gases constitute inhalation, explosion and fire hazards if handled improperly.

The radioanalytical chemistry laboratory is subject to all of these hazards with the addition of radioactive solids, liquids and gases. The potential risks to laboratory workers from radioactive materials include exposure to ionizing radiation, both from external sources and from internal sources that were inhaled, ingested, or absorbed through the skin. Contamination of the immediate work area is a safety concern that becomes amplified if the contamination is not removed promptly and is subsequently spread over a much wider area. Each of these hazardous situations may result from an unintended release or from improper handling of radioactive materials. Often, the person who is affected is not the one who caused the problem in the first place.

Radiation hazards should be placed in perspective by distinguishing between lower and higher levels of radionuclides in samples submitted for analysis, although these are qualitative terms not defined by regulation (see Section 13.2). Many man-made radionuclides in the laboratory are at levels so low that they are comparable to naturally-occurring radionuclides in soils, rocks, biota, and urine, such as ^{40}K in potassium salts and uranium and thorium isotopes and their progeny in uranium and thorium salts. At these lower levels, the radiation hazard in the laboratory can be viewed as minor compared to chemical hazards. Samples with significantly higher radionuclide levels should have been described appropriately in the shipping and/or chain of custody documents and must be processed to minimize radiation exposure to the analyst on the basis of guidance by the RSO or RCM staff.

Carelessness, absentmindedness and roughhousing increase the likelihood that a potential hazard will occur. Avoid pranks, horseplay, and other inappropriate behavior in the laboratory. Be respectful of the presence of others in the laboratory, and do not crowd, push, or gesture carelessly.

14.4.2. Design and Maintenance

The work environment should be designed and maintained for safety. The laboratory described in Chapter 13 represents the ideal: spacious, well organized and easy to clean. In many laboratories that are older or have outgrown their space, workers must be particularly vigilant to avoid spills and breakage in cramped operating and storage spaces, keep aisles clear, and ensure unimpeded access to the accident response equipment described in Section 14.6. Laboratory exits and electrical power disconnect panels must never be blocked. In all laboratories, but especially in older ones, management should regularly inspect the condition of plumbing, electrical hardware, lighting fixtures and ventilation systems, and repair what is defective.

Good housekeeping augments good design in reducing the potential for accidents. Analysts should keep work spaces clean and organized, have only a limited supply of chemicals on hand and store the remainder safely, label chemicals clearly, and dispose of waste promptly and correctly. Only the equipment needed for an active procedure should be set up, and any unused equipment should be stored away from the laboratory bench. At the conclusion of an assignment, all equipment should be dismantled, cleaned, and stored.

14.4.3. Glassware and Chemical Handling

Glassware should be inspected prior to use to remove chipped, cracked, or broken items from service and dispose them in a designated container. Soiled glassware should be taken to the laboratory sink designated for cleaning. Clean glassware promptly with detergent and water. Vacuum glass apparatus should be shielded to contain contents and glass fragments in the event of an implosion. Except for glass tubing and stirring rods, all glassware in the laboratory should be composed of borosilicate (e.g. Pyrex).

Laboratory refrigerators must be explosion-proof and specifically designed for chemical storage. Standard household appliances do not meet these criteria. All containers placed in refrigerators should be labeled with the identity of the contents, the name of the person placing the container in the refrigerator, and the date of storage. Containers in refrigerators should be placed on storage trays with sides of sufficient height to contain any spilled or leaking containers. No food or beverages shall be stored in the laboratory refrigerator.

Once chemicals have been logged into the receiving area, they must be moved immediately to a specified storage area, as described in Section 13.3. Large containers that could break during transport must be transported in a protected shipping container or a secondary protective container large enough to hold the contents in the event of a spill. Chemicals must be segregated by hazard classification and compatibility, and stored in a secure area equipped with local exhaust ventilation. The storage area should be continuously ventilated and under slightly negative air pressure, well lighted, supervised, and capable of being locked to prevent unauthorized access. Chemicals should be stored at or below eye level. Large bottles should be no more than two feet from ground level, and not stored on the aisles or floor. Mineral acids should be stored on acid-resistant trays, separately from flammable and combustible materials. Acid-sensitive materials, such as cyanides and sulfides, should be protected from contact with acids. At least annually, the inventory of stored chemicals should be examined to identify evidence of container deterioration or damage and to identify any chemicals to be replaced based on length of service. Chemicals that are no longer needed should be disposed of properly.

Chemicals taken from the storage area for use at the laboratory bench should be transported in durable secondary containers of sufficient size and composition to retain any spill of the chemical. Preparation or repackaging of chemicals should take place in a designated area of the laboratory, outside of the main storage area. Chemicals stored at the laboratory bench should be limited to the amounts necessary for current application. Bench chemicals should be stored in an orderly manner, away from sunlight and external sources of heat. Containers should be protected from falling by placing them well away from the edges of counter tops. Containers may not be stored on the floor. Chemicals may not be removed from the laboratory without permission. Unnecessary items should not be stored at the laboratory bench.

Material Safety Data Sheets (MSDS) are compiled and available for all chemicals. These documents explain the inhalation and contact health risks, as well as the flammability and toxicity of the chemical they describe. They also outline special handling and storage considerations. When handling a new chemical, it is prudent to examine its accompanying MSDS.

With or without an MSDS, however, caution in chemical handling should be observed. As a general guideline, worker and student exposures to laboratory chemicals should be kept to a minimum. Because so many laboratory chemicals are hazardous to humans in some way, conservative risk assessment should be employed. Persons in the laboratory should assume that personal protection is required whenever they are working with chemicals. Chemicals of unknown toxicity initially should be treated as toxic with respect to exposure during work performed in the laboratory. For work with chemicals of known toxicity, appropriate precautions should be taken. When working with mixtures of chemicals, the risk of the mixture as a whole should be assumed at least to equal the risk for the most toxic component of the mixture.

Skin contact with chemicals should be avoided. Personnel must wash all areas of exposed skin prior to leaving the laboratory. Mouth suction for pipettes or for starting siphons is not allowed. Prohibitions on eating, drinking, smoking, gum chewing, or the application of cosmetics in the laboratory should be posted, and the rules concerning these activities should be enforced by the supervisor. Hands and face should be thoroughly washed prior to eating or drinking. Protective aprons, jackets, or coats should be removed prior to entry into areas where food may be consumed.

14.4.4. Personal Protective Equipment

All laboratory workers must wear appropriate personal protective equipment (PPE). Safety goggles must be worn in laboratory areas. Face shields, laboratory coats or aprons, latex, rubber or heat-resistant gloves, dust masks and safety shoes may be required for specific work. Understanding the hazards of the materials at hand is important for wearing the appropriate PPE, but also for shunning excessive use of PPE. Overprotection may cause awkwardness and increase the possibility for injury. To strike a balance between reasonable and inordinate safety measures, the safety officer and analyst should communicate on how best to perform a given operation. In some laboratories, there may be a HAZMAT-certified PPE technician that can help in making these choices.

Proper dress for the laboratory includes shoes with leather or synthetic covering over the feet. Open toed shoes, sandals, high heels, or shoes with woven fabric coverings are not allowed. Minimize skin exposure by avoiding the use of shorts or cutoffs; pants are preferable. Minimize the use of jewelry, remove any hanging jewelry, and secure loose hair and clothing.

Street clothing must be protected by the use of chemically resistant laboratory coats, jackets, or aprons. Laboratory jackets should be equipped with snaps, not buttons, so that they can be removed quickly in case of an emergency. Protective clothing must be fire-resistant and impervious to chemical splashes.

Gloves must be selected based on the chemical or hazard of concern, and on the extent of expected exposure. Factors to consider in selecting protective gloves include proper fit, durability, cut resistance, chemical breakthrough (permeation), chemical resistance, comfort, and cost. Manufacturers of protective gloves are required to test glove materials and to make data available concerning permeation, diffusion, and chemical degradation of glove products. Some gloves cover only the hands. In other applications, gloves should cover the forearms or the entire arms. Certain individuals may have allergic reactions to the proteins found in latex gloves, and should use suitable substitutes made of non-latex materials. Gloves that allow permeation or diffusion of specific chemicals should not be reused. Remove gloves prior to handling objects which are not part of the laboratory procedure, and always remove them prior to leaving the laboratory. Wash hands thoroughly whenever gloves are removed.

Chemical splash goggles must be worn at all times by all personnel in the laboratory. Visitors to the laboratory must don goggles prior to entry. Safety glasses, face shields, or prescription glasses may not be substituted for goggles. Contact lenses may be worn under goggles; they should never be adjusted or removed during laboratory procedures except in emergency conditions. When a vigorous reaction, implosion, or splashing is possible, a face shield should be worn in addition to goggles.

14.4.5. Safety Equipment

Safety equipment includes trays, hoods, glove boxes and radiation shields. Shallow metal or glass trays are used for nonvolatile substances to contain potential spills and leaks of materials such as radionuclides. They isolate specific substances that can react in a hazardous manner with others processed on the same laboratory bench. Trays also serve to define work stations for specific purposes, and to distinguish them from clean areas for notebooks and reagents.

Laboratory hoods must be used for all procedures that might release hazardous chemical vapors or dusts. Wet chemistry for volatile substances—notably certain organic solvents and radionuclides—should be performed in a hood to reduce inhalation risks. Chemicals that pose a substantial risk of explosion must be handled in a hood designed to contain explosions.

Hood fans must be operated whenever hoods are in use, and the fans should continue to run following the end of work for a sufficient time to clear residual air contaminants from the ductwork. Hoods should be equipped with a meter to record air flow or air pressure, and users should confirm that hoods have proper face velocity. The laboratory must be supplied with sufficient make-up air to satisfy all air exhausted through the hoods. When work at the hood is concluded, the sash should be lowered.

Hoods should not be designed to recycle exhaust air into the laboratory. "Hoods" that do not vent to the outside, but rather filter in-laboratory air, are called laminar flow boxes. These are important for working with dusty radioactive materials, like soils, but are not appropriate for most other laboratory operations. Laminar flow boxes should be clearly labeled so that laboratory policy can limit the analytical operations that can be performed in them.

Hoods should not be used to store chemicals. Because they often are a common workspace for several analysts, hoods tend to accumulate glassware, instruments such as heater/stirrers, reagents, waste and unfinished samples. Extended storage of such materials in the hood constitutes bad housekeeping that increases the likelihood of accidental spills. Stored materials can disrupt airflow which causes the hood to operate less efficiently. If chemicals are left temporarily in the hood, the fan should continue to operate with the sash slightly opened.

Laboratory hoods should be inspected at least every three months. The CHO, industrial hygienist, or designated technician should measure the face velocity of hoods periodically to ensure that 75 to 125 feet per minute (0.4–0.6 ms^{-1}) air velocity is maintained. The sash opening height should be properly marked to indicate the face velocity. A transport velocity in the exhaust duct of the hood of 3500 feet per minute (18 ms^{-1}) is recommended. A record of the inspection should be dated and posted at the hood. Prior to a change in chemicals or procedures, the adequacy of the ventilation system and the suitability of the hood for the new process must be evaluated.

While many operations with chemicals must take place in a hood, rarely would glove boxes or shields be required in the radioanalytical chemistry laboratory dedicated to processing environmental or bioassay samples. A glove box or shield may be needed when an analyst is working with radioactivity standards before dilution (which may be as much as a million times as radioactive as low-level samples), or for an unusually radioactive sample. An environmental laboratory generally would not accept a highly radioactive sample because of concern about contamination, but an emergency situation may necessitate its analysis. Should this unusual situation arise, the responsibility of the laboratory analyst and RSO/RCM staff is to insist on issuance of suitable protective equipment. Failure by the supervisor to provide equipment for safe operations should automatically trigger a "Stop Work" notification by the analyst. The "Stop Work" notification applies to all phases of laboratory work, not just radionuclide analysis. The RSO/RCM or IH/CHO staff should document such an incident in writing, and promptly submit the report to management.

14.4.6. Flammable and Combustible Liquids

A wide variety of chemicals can be used in any laboratory environment, including those chemicals that have the properties of being flammable or combustible. Before discussing some of the general precautions to take in handling and storing flammable and combustible liquids, the following definitions are necessary:

- *Flashpoint*: The minimum temperature at which a liquid gives off vapor within a test vessel in sufficient concentration to form an ignitable mixture with air near the surface of the liquid.
- *Combustible liquid*: Any liquid with a flashpoint at or above 100 degrees Fahrenheit. The two classes of combustible liquids are Class II liquids with a flashpoint between 100 and 140 degrees Fahrenheit and Class III liquids with a flashpoint above 140 degrees Fahrenheit.
- *Flammable liquid*: Any liquid with a flashpoint below 100 degrees Fahrenheit. Flammable liquids are also known as Class I liquids. There are three subcategories of Class I liquids. Class IA liquids have a flashpoint below 73 degrees F and a boiling point below 100 degrees F. Class IB liquids have a flashpoint below 73 degrees F and a boiling point at or above 100 degrees F. Class IC liquids have a flashpoint at or above 73 degrees F but below 100 degrees F.

Basic precautions taken for flammable or combustible liquids consist of separating the liquids from sources of ignition or from oxygen. This approach can prevent or extinguish combustion by separating from each other the fuel, source of ignition, and oxygen. The following are a few general precautions to be considered when storing or using flammable or combustible liquids:

- *Minimizing stored quantities*: According to the OSHA requirements in 29 CFR 1910.106, Class IA flammable liquids must be limited to a maximum of 25 gallons (95 L) of liquids in containers in storage in any one fire area of a facility. Other flammable or combustible liquids must be limited to a maximum of 120 gallons (450 L) of liquids in containers.

- *Eliminating or isolating sources of ignition:* Among the most common sources are smoking, electrical arcing or static, flames such as Bunsen burners, and heat-producing equipment. These potential sources of ignition should be eliminated in areas where flammable or combustible liquids are stored or used. Smoking must be prohibited in all laboratory areas. Electrical equipment in areas where flammable vapor-air mixtures could occur must be suitable for these locations. Sources of ignition such as flames or heat-producing equipment must be separated from areas where flammable or combustible liquids are stored or used. A good rule of thumb is to have a distance of at least 20 feet (6 m) between any source of ignition and flammable or combustible liquids, but more stringent requirements may apply in specific situations.
- *Utilizing flammable storage cabinets or rooms*: Larger quantities of flammable or combustible liquids should be stored in approved flammable liquid storage cabinets or storage rooms. Construction specifications must be met for different sizes of storage container and class of chemical stored therein; 29 CFR contains the specifications for construction of flammable liquid storage buildings (29 CFR 1910.106(d)(5)(vi)) and for construction of flammable liquid storage cabinets (29 CFR 1910.106(d)(3)).
- *Separating incompatible chemicals:* Chemicals that may react with each other violently should be stored separately. For example, acids should be separated from bases, and oxidizers should be separated from flammable and combustible liquids.

14.4.7. Compressed Gas Cylinders

Compressed gas cylinders utilized in a laboratory environment can be hazardous. The hazard can be posed by both the type of gas in the cylinder and the pressurized nature of the gas. Sudden release of this pressure can result in a cylinder being thrust into the laboratory as if it were a rocket. This can occur if the valve is broken off when the cylinder is knocked over.

The main precaution with respect to compressed gas cylinders is to train laboratory workers in their proper handling, storage, and use. Workers should respect the potential for hazard of the specific gases that might escape and of sudden pressure releases. Cylinders in storage should be securely attached to a permanent object or wall, or kept strapped on carts that are designed for holding cylinders. Valve protection caps, where cylinders are designed to accept them, must be kept on cylinders when they are in storage. Flammable gas and oxygen cylinders should be stored away from each other by at least 20 feet (6 m) or be separated by a $1/2$-hour-rated fire wall which is at least 5 feet (1.5 m) high.

Cylinders should be stored in the upright position, especially flammable—gas cylinders that are designed to be stored in this manner. Because OSHA generally considers cylinders that are not used within a 24-h period to be "in storage," oxygen and acetylene cylinders that are not in daily use should be taken off the cart and stored properly. As an alternative, the above-cited firewall to separate the cylinders can be made an integral part of the cart. When moved, cylinders should always be strapped securely to a cart designed for carrying them.

14.4.8. Electrical Hazards

The following is a list of the more likely hazards related to electricity that may be encountered in a laboratory environment:

- *Lack of a permanent and continuous ground*: This hazard can result from a broken or removed ground plug or a wiring problem at an electrical receptacle. With an open ground, a fault in equipment can result in the path of electricity through a laboratory worker instead of the desired ground path. Laboratory inspections must ensure that ground pins are correctly in place for equipment designed to have them and that receptacles are tested periodically for correct wiring.
- *Unused openings or missing covers in electrical equipment or boxes*: Electrical equipment can become damaged or neglected over time, with exposure to energized electrical parts through unused openings such as knock-out closures or missing covers. Exposure to energized electrical parts, especially parts operating at over 50 Volts, must be prevented. Electrical parts should be serviced or maintained only after de-energizing the circuit and following the established lockout/tagout procedure. When de-energizing is not feasible, only qualified electricians operating after training and with personal protective equipment and insulating tools should attempt to work in the vicinity of energized electrical parts.
- *Damaged insulation*: Damaged insulation must be repaired immediately by a qualified electrician. Simply covering insulation with electrical tape will most likely not be sufficient to protect individuals from contact. The repaired section must have the same mechanical strength and insulating properties as the original wiring.
- *Wet or damp environments*: Water can be a good conductor of electricity, hence working with electrical equipment in a wet environment can be a hazardous activity. Electrical equipment should be isolated from moisture. Electrical receptacles in wet or damp environments must be designed for this type of environment, with ground-fault-circuit-interrupter (GFCI) protection. This type of receptacle should be tested periodically in accordance with the manufacturer's recommendations.
- *Circuit overload*: Electrical equipment should not be utilized in a manner not intended by its manufacturer. Overloaded wiring and receptacles can damage the equipment and result in electrical shock, electrocution, and fire. Equipment installers and users should always understand the intended use of the equipment and follow manufacturer's recommendations.

14.4.9. Unattended Operations

Unauthorized procedures and working solo in the laboratory are not permitted. Laboratory equipment that operates unattended for continuous or overnight processing may pose a risk of fire, explosion, or other unintended consequences. Any persons planning to use unattended operations in the course or an experimental series must receive written approval from the laboratory supervisor. The approval must provide sufficient detail to show that the planned procedure will operate safely to completion. The permit should include the anticipated completion time and the name and telephone number (or other means of communication) of the person who is responsible for operating the equipment and dismantling it upon completion. Entrances to the laboratory must be posted to provide warning of the laboratory procedure that is in progress, as well as the name and telephone number of the responsible person. Lights in the laboratory should be left illuminated. Precautions, such as fail-safe operational mode, must be integral to the equipment being used, so that an interruption of utility services such as water, gas, compressed air, or electricity will still allow the process to shut down safely.

14.4.10. Monitoring

In the radioanalytical chemistry laboratory, higher- and lower-level radiation areas must be clearly designated. When specifying the radiation level of an area, signs akin to those in Figure 14.1 would be posted. There are many variations of this type of sign, as they are placed anywhere radiation is used or found—including hospitals, laboratories, chemical plants and waste clean-up sites. Each of these locations has different radiation concerns, and a different audience viewing the signage. What all signs will have in common, however, is the tri-foil. Depicted in

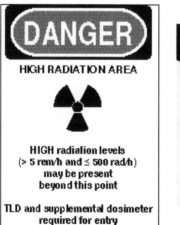
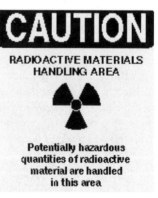

FIGURE 14.1. High-level radiation alert signage. Online at:
http://www. epa.gov/radiation/students/symbols.html (Dec. 2005).

FIGURE 14.2. The tri-foil. Online at:
http://www.epa. gov/radiation/students/symbols.html (Dec. 2005).

Figure 14.2, the tri-foil is the international symbol for radiation. The symbol can be magenta or black, on a yellow (or sometimes white) background.

Access between the two appropriately labeled radiation areas should be restricted, in order to avoid contaminating the low-level area. In laboratories where both "hot" and "cold" operations are being performed in a small amount of space, it is important to set up well-defined and clearly labeled radioactive material work stations. In this way, radionuclide analysis is confined so that its impact on the surrounding laboratory is minimal. The presence of radiation throughout the laboratory should be monitored as discussed below.

Because the five senses are useless for detecting radiation, each facility must have readily available portable radiation detection instruments. These instruments should be selected to detect and quantify the three basic types of radiation: alpha particles, beta particles, and gamma rays, as discussed in Chapter 2. In some cases, neutron detectors may be required. The RSO/RCM generally is responsible for calibrating the instruments at selected intervals, typically six months. The individual user is responsible for daily operational and source checks prior to each use.

Each radiation safety program should have an established survey program to monitor the radiation and contamination levels in the laboratory so that the individual user can maintain a safe working environment. The measurements can identify and measure individual radionuclides or measure radiation dose. The dosimetry program should monitor exposures to individuals from external radiation or internal—ingested or inhaled—radionuclides, and also in areas of the laboratory.

External exposure levels generally are monitored with a thermoluminescent dosimeters (earlier, with film badges) or electronic dosimeters. Either method provides individual exposure information for a worker or a location in terms of exposure over a pre-set monitoring period, after which the dosimeters are exchanged for new ones. The used dosimeter is read to determine the radiation exposure during the interval. In laboratory areas that process samples with higher radionuclide concentrations, wrist and ring badges or direct-reading radiation dosimeters also may be worn, but they usually are unnecessary when processing environmental or bioassay samples. An exception may be made for persons who receive mixed samples at the dock or handle radioactivity standards before dilution.

The internal dosimetry program generally relies on *in vitro* (outside) or *in vivo* (inside) monitoring capabilities. *In vitro* programs use an external radiation

detection system to monitor radiation being emitted from inside the body and detected on the outside of the body. *In vivo* programs use a sample from the body
(usually urine, but sometimes feces, blood, hair, or breath) to monitor contamination within the body.

Urine samples for bioassay are collected from workers who may have inhaled
or ingested radionuclides or hazardous chemicals, either while routinely handling
samples with higher radionuclide or chemical concentrations, or after accidental
exposure. The samples are analyzed for the radionuclides or chemicals that may
have been taken in by the worker and the measured concentrations are used to
calculate radiation exposures or to compare to radionuclide or chemical concentration limits. The frequency of sampling for bioassay depends on the occurrences
of exposure and the rate of turnover of the radionuclides or chemicals in the body.

A large facility usually has an infirmary with medical staff for treating injuries
and performing physical examinations at the beginning and end of employment,
and at regular intervals. The type and frequency of routine physicals may depend on
the probability of adverse effects associated with the work environment. Employee
medical records should be maintained to track the occurrence or absence of health
effects due to the work environment.

Management must promptly report to the worker and the regulatory agency
any unusually elevated radiation exposures—both external and internal—or concentrations of hazardous chemicals, and institute changes in the work environment to prevent continuing exposures that reach exposure limits. The IH/CHO or
RSO/RCM staff must be instrumental in achieving such changes. The exposure
must be recorded with a description of the causes and instituted remedies.

14.5. Emergency Response

Among the various types of emergencies that can occur in a laboratory environment, occurrence of a fire is a major probability. The best way to minimize the
effect of emergencies is to prevent them. For this reason, fire prevention plans
should be instituted. If a fire or other emergency does occur, emergency plans
must be in place to protect the laboratory and its workers. These plans may be
a part of the laboratory CHP, or they may stand alone. Various issues associated
with life safety must be considerd to maximize the occupant's ability to escape the
facility during an emergency.

14.5.1. Fire Prevention

Each laboratory facility must have a fire prevention plan, preferably in writing.
According to OSHA 29 CFR 1910.39, the fire prevention plan should include at
least the following elements:

- A list of all major fire hazards, proper handling and storage procedures for
  hazardous materials, potential ignition sources and their control, and the type of
  fire protection equipment necessary to control each fire hazard;

- Controls to prevent accumulation of flammable and combustible waste materials;
- Procedures for regular maintenance of safeguards installed on heat-producing equipment to prevent accidental ignition of combustible materials;
- The name or title of those individuals responsible for maintaining equipment to prevent or control sources of ignition or fires; and
- The name or title of those individuals responsible for the control of fuel source hazards.

Laboratory workers must be aware of the fire prevention plan and must be trained in the elements of the plan and its location within the laboratory.

14.5.2. Emergency Planning

Emergency action plans are especially necessary in laboratories. Plans should be prepared for response to situations such as fire, explosions, flooding, severe weather (tornadoes, hurricanes), earthquakes, medical emergencies, violent acts or threats of violence (e.g., bombs), and release of hazardous materials. Accidents at nearby locations such as adjacent laboratories and transportation facilities may affect the laboratory worker. According to OSHA 29 CFR 1910.38, the minimum elements of an emergency action plan include:

- Procedures for reporting a fire or other emergency;
- Procedures for emergency evacuation, including type of evacuation and exit route assignments;
- Procedures to be followed by laboratory workers who remain for necessary laboratory operations before they evacuate;
- Procedures to account for all laboratory workers after evacuation;
- Procedures to be followed by laboratory workers performing rescue or medical duties; and
- The name or title and address (e.g., telephone or e-mail) of every individual to be contacted by laboratory workers who need more information about the plan or an explanation of their duties under the plan.

14.5.3. Life Safety

The National Fire Protection Association (NFPA) produces an entire text to address the complex issues of life safety in the Life Safety Code (NFPA 101) that should be read for detailed information. The following are general items to ensure that laboratory workers can safely evacuate the laboratory in an emergency:

- At least two exit routes should be available to permit prompt evacuation during an emergency. The two routes should be located as far as practical from each other in case one is blocked. More than two exit routes may be necessary, depending on the number of occupants, size of the building, building occupancy, or arrangement of the workplace, while a single route may be allowed in some instances.
- Occupants must be able to open an exit route door from the inside at all times without keys, tools, or special knowledge of door operation. Exit route doors

must be free of any device or alarm that could restrict emergency use of the exit route if the device or alarm fails.
- Exit routes must support the maximum permitted occupant load of each floor and the capacity (width) of an exit must not decrease in the direction of travel.
- The minimum exit width is 28 inches (0.7 m), but should be wider in most circumstances.
- Exit routes must be arranged so that occupants will not have to travel toward a high-hazard area (such as a flammable liquid storage room), unless the path of travel is effectively shielded from the high-hazard area by physical barriers.
- Exit routes must be free and unobstructed.
- Exit routes must be adequately lighted so that an occupant with normal vision can see along the exit route. The lighting must be impervious to power outage, so backup lighting must be on an independent power supply (generator or batteries). Automatic illumination by these lights should be tested periodically.
- Each exit must be clearly visible and marked by a sign reading "Exit".
- Laboratories must install and maintain an operable alarm system that has a distinctive signal to warn occupants of fire or other emergencies.

14.6. Accident Response and Equipment

Most safety guides emphasize hazard identification and accident prevention. When prevention has failed, personnel must be familiar with the laboratory CHP and RPP accident protocols and the emergency plans described in Section 14.5.2. These plans instruct the worker about what to do and whom to notify. This information includes the facility evacuation routes. Qualified personnel must be prepared to use the accident response equipment placed in the laboratory to ameliorate an accident situation or to aid in treating the injured. Examples of this equipment include fire extinguishers, eyewash stations and safety showers.

14.6.1. Fire Extinguishers

Fire extinguishers should be accessible in every laboratory, mechanical support area and electrical room. Several extinguishers may be warranted in a large space; the OSHA rule is that one should not have to travel more than 75 feet (23 m) to reach an extinguisher (29 CFR 1910.157(d){1}). Fire extinguishers also must be located in the hallways outside the laboratory. The employer is responsible for proper selection and distribution of fire extinguishers on the advice of the safety officer.

Fire extinguishers must match the type and class of fire hazards associated with the particular laboratory. Four fire hazard classes are defined by the U.S. Department of Labor: Classes A, B, C and D.

- Class A fires result from the burning of ordinary combustible materials, such as paper or cloth. Class A extinguishers may be water, loaded stream (a water-based extinguisher that contains one of several added chemical components that increase permeation of the stream into the fire), foam or multipurpose dry chemical. The numerical rating refers to the amount of water or other fire extinguishing agent that the extinguisher holds.

- Class B fires are fueled by flammable liquids and gases, or grease. Class B extinguishers may be Halon 1301, Halon 1211, carbon dioxide, dry chemicals, or foam. Here, the numerical rating indicates the approximate number of square feet of Class B fire the average user can be expected to extinguish using that extinguisher.

- Class C fires are electrical fires, and may be extinguished with Halon 1301, Halon 1211, carbon dioxide, or dry chemical fire extinguishers. Class C extinguishers have no additional numerical rating.

- Class D fires involve flammable metals, such as magnesium, sodium, potassium, titanium and zirconium. Extinguishers that contain water, gas, or certain dry chemicals cannot extinguish or control this type of fire and may actually fuel it. Class D fire extinguishers utilize agents including Foundry flux, Lith-X powder, TMB liquid, pyromet powder, TEC powder, dry talc, dry graphite powder, dry sand, dry sodium chloride, dry soda ash, lithium chloride, zirconium silicate, and dry dolomite. There is usually no numerical rating for these extinguishers.

An excellent resource for fire safety and fire extinguishers can be found at http://www.hanford.gov/fire/safety/extingrs.htm. This website is maintained by the Hanford Fire Department in Richland, WA, which oversees fire safety and operations for the DOE's Hanford facility.

The old symbols for fire extinguishers geared for each type of fire are shown in Figure 14.3. Figure 14.4 contains the new symbols. Note that the new pictograms have no symbol for Class D extinguishers, which are often specific to the type of metal burning.

Among Classes A–C, crossover in the active firefighting agent results in some extinguishers that are Class A/B (for example, foam) or Class A/B/C (some dry chemical extinguishers). All fire extinguishers are labeled with their class hazard rating to insure that the extinguishers are properly used. Before attempting to put out a fire, check the label to confirm that the extinguisher is appropriate. In the old format, the symbol for each Class of fire that the extinguisher is designed to address will be on the extinguisher bottle. In the new format, all three pictograms are on the extinguisher bottle, with a red strike through any Class that the extinguisher

Ordinary Combustibles   **Flammable Liquids**   **Electrical Equipment**   **Combustible Metals**

FIGURE 14.3. The old symbols, representing Class A, B, C, and D fire extinguishers. Online at: http://www.hanford.gov/fire/safety/extingrs.htm. (Dec. 2005).

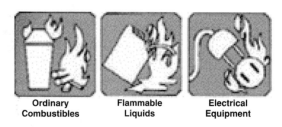

| Ordinary Combustibles | Flammable Liquids | Electrical Equipment |

FIGURE 14.4. The new pictograms, representing Class A, B, and C fire extinguisher. There is no pictogram for Class D extinguishers. Online at: http://www.hanford.gov/fire/safety/extingrs.htm. (Dec. 2005).

is NOT designed to address. *One should remember that water is only useful in putting out Class A fires.* A mixed-use extinguisher will never contain water, and the user should never improperly use a water-based extinguisher (Class A) on any other type of fire.

Fire extinguishers must be inspected routinely to insure that they are in operating condition. Employees must be trained in the use of fire extinguishers unless immediate evacuation is the facility policy. New employees should check the CHP to determine their responsibility (some facilities may designate only some employees as fire officers), and inquire about the laboratory fire safety training program.

In addition to fire extinguishers, a sprinkler system in the ceiling is almost always available. These are, of course, useful in extinguishing Class A fires. However, these sprinklers activate automatically, in response to the presence of smoke and/or debris in the laboratory air. The automatic addition of water to a class B, C or D fire would be counter-productive and should be avoided. Therefore, the fire-producing potential of material in each room should be evaluated, prior to installation of the sprinkler system.

14.6.2. Eyewash Stations and Washdown Showers

Chemical laboratories must be equipped with a combination emergency eyewash and shower. The unit should be located at a distance of no more than 10 seconds of level, unobstructed travel time from anticipated exposure points. As an estimate, 100 feet (30 m) can be traveled in 10 seconds if the pathway has no obstacles. The location of the emergency wash must be clearly marked, well lighted, and easily accessible. No obstacles or doorways should be located on the travel path, which should have few turns.

Initiation of the emergency wash must be accomplished by one action using one hand. Once initiated, the flow must continue, leaving both hands free. Contaminated clothing should be removed as quickly as possible while the emergency wash is operating. For contaminants in eyes, the unit should be operated for 15 continuous minutes, with both hands used to hold the eyelids open. Emergency washes should be flushed weekly for a minimum of 3 minutes. Bump tests should be conducted on eye washes daily and on showers weekly. A full flow test should be conducted monthly. Emergency washes should provide tepid, potable water. Eyewashes should supply 3 gallons (12 L) of water per minute for at least 15 minutes. Showers should provide 30 gallons (120 L) per minute for 15 minutes. Fountain heads for eyewashes should be covered to protect them from collecting dust, debris, or chemical residues.

Eyewash stations and washdown showers should be checked periodically for proper flow because they are infrequently used. The saline solution at stand-alone eyewash stations should be replaced at its expiration date, or sooner if it becomes dried out.

14.6.3. Accident Protocols

All workers must know where to go and what to do when an accident happens. The laboratory CHP and RPP contain the emergency plan with which all laboratory workers must be familiar. In a serious accident, workers should respond in the following order unless a different protocol has been specified for the laboratory:

1. Stop a fire, accidental spill, or gas leak at its source, if possible. This means extinguishing or removing the flammable source of the fire, stopping the chemical leak, or turning off the gas. *If these activities are too dangerous, do not attempt them; isolate the room and proceed to the next step.*
2. Raise the alarm for those in the area to evacuate.
3. Assess injuries and assist the injured, removing them from the area for medical attention, if possible.
4. Notify the designated response group promptly and provide as much information as can be obtained in a brief period. The designated responder will contact the fire department, police, and HazMat, based on the information provided. Be as calm and accurate as possible when relaying the details of the accident.

Note: The OSHA regulations do not prescribe a specific order of operations in an emergency, in recognition of the differences among laboratories; considerations of location, accessibility and institutional protocol must be addressed in determining a plan. The above recommendations, generally accepted among university CHPs, are published online, but are not a substitute for familiarity with the accident protocol in one's own laboratory.

HazMat (see Section 14.9) publishes a guide to aid first responders in (1) quickly identifying the specific or generic classification of the material(s) involved in the incident, and (2) protecting themselves and the general public during this initial response phase of the incident. This publication is called the Emergency Response Handbook, and can be downloaded online. The handbook is updated every three to four years to accommodate new products and technology. The next version is scheduled for 2008.

14.7. Environmental Safety

Protecting the worker is the primary concern of the Radiation Safety and Industrial Hygiene staffs, but their roles also extend to protecting the environment and members of the public. It would seem obvious that a program that is designed to protect the individual working in close proximity to the radioactive materials will also protect the general public and the environment. However, more restrictive protection levels are designated for members of the public. To ensure that these standards are met, laboratories must control, and in some instances monitor, liquid and airborne effluents as well as radiation levels that may affect the public.

An environmental safety program begins with the construction of the laboratory and development of the chemical and radiation safety plans. Good pre-planning will go a long way toward minimizing waste generation, releases to the environment, and resulting pollution levels. A chemical and radioactive material tracking system should be used to determine where and how reagents are used and how much waste is generated. The results should be reviewed periodically to identify items or quantities of materials that can be reduced. Since waste disposal costs are a significant part of the operating budget, waste minimization can help the budget.

As a part of the operating license, both the NRC and DOE specify airborne and liquid release quantities to the air or the sanitary sewer system, respectively. Programs must be established that track discharges relative to the limiting concentrations. Effluent and waste control regulations must be understood by a staff expert and communicated to the users of radioactive materials and chemicals. Compliance with these regulations must be established on an institution-wide basis. Unusual or accidental discharges must be reported promptly to the supervisor and RSO/RCM staff. Management must inform the appropriate regulator of any discharges that reach reporting requirements. If the facility fails to comply, legal actions can result, including fines or loss of license for on-site activities.

14.8. The Regulatory Environment

Laboratory management, the IH/CHO and RSO/RCM staffs, and those responsible for receiving and shipping materials must be familiar with the regulations and underlying laws that control the operation of the laboratory, handling of hazardous and radioactive material, and release of these materials, as well as workers' rights and responsibilities. The radioanalytical chemist should have a reasonable understanding of regulations that are pertinent to the work, and thus should be familiar with the Code of Federal Regulations (CFR) titles already mentioned in this chapter.

The CFR is the comprehensive regulatory framework of the US. It is kept online, current and in its entirety, by the National Archives and Records Administration (NARA) at http://www.gpo.gov/nara/cfr/cfr-table-search.html#page1 (Dec. 2005). When Congress passes a law that dictates a change of policy, the agency under whose aegis the policy falls is prompted to produce a corresponding regulation. These agencies (e.g., EPA, NRC, DOE, OSHA, DOT) write the regulations and submit them for expert and public comment; the result is modification or approval. Each of the 50 CFR titles contains the regulations pertaining to a particular subject and indicates the government agency (or agencies) responsible for administering those regulations.

For instance, Title 29 (29 CFR) concerns labor regulations and defines the purpose and responsibilities of the Department of Labor and OSHA. Title 10 defines energy regulations and the role of the DOE and NRC. Title 40 describes the regulations for protection of the environment and the EPA. The NARA website above is a useful resource in researching all 50 titles; the radiochemist should be conversant with pertinent aspects of titles 29, 10 and 40, in particular. The following paragraphs are a brief overview of these titles.

All private sector research laboratories, whether academic or commercial, are subject to U.S. government regulation under OSHA in 29 CFR 1910.1450, "Occupational Exposures to Hazardous Chemicals in Laboratories." For public sector research laboratories, OSHA defers regulatory authority to the applicable state government and its designee, such as the University Board of Regents. The descending levels of regulation are state, municipal and institutional or local requirements. In general, state regulations must at least meet, but may be more stringent than, the OSHA standards. Any state program must be approved and monitored by OSHA.

Title 10 CFR Part 20—"Standards for Protection Against Radiation"—establishes the legal framework for protecting non-government workers from radiation. The NRC uses Parts 1 through 199 to govern the use and control of radioactive materials. The DOE uses Parts 200 through 1099 to govern the use and control of radioactive materials. Title 10 CFR Part 835—"Occupational Radiation Protection" establishes the legal framework for protecting government workers from radiation.

Nuclear materials are specifically regulated under 10 CFR 1.42, administered by the U. S. NRC, Office of Nuclear Material Safety and Security (NMSS). The duties and responsibilities of the NMSS include protecting the public health and safety, national defense and security by licensing, inspection, and environmental impact assessment for nuclear facilities and activities involving nuclear materials, and for the import and export of special nuclear materials.

Emissions from specific sources of air pollution linked to potentially serious health problems are limited by the National Emissions Standards for Hazardous Air Pollutants (NESHAPS) in 40 CFR 61. The NESHAPS regulations are a subset of the EPA Clean Air Act. Hazardous air pollutants (HAPS) are identified as chemical or radioactive hazards. The radiological NESHAPS regulations pertain to pollutants that emit radiation from sources such as academic institutions, hospitals, industry, vehicles, and nature (such as radon).

Solid wastes are managed under the requirements of the EPA Resources Conservation and Recovery Act (RCRA). The provisions of RCRA pertain to solid chemicals as well as waste mixtures that are hazardous on the basis of ignitability, corrosivity, reactivity, or toxicity. Some states regulate RCRA-related activities under approval by EPA.

The US Department of Transportation (DOT) controls transportation of hazardous and radioactive materials under the Hazardous Materials Transportation Act by regulations in 49 CFR 171–179. Shipments of samples to the laboratory and of waste from the laboratory must meet these regulations.

14.9. Available Guidance

Individuals responsible for monitoring safety and health in a radioanalytical chemistry laboratory have a variety of sources from which to obtain information. Some of the organizations that are relevant to this topic are included in the following list. Since the most recent information available can be found by searching online, a web address has been included for each entity. All were viewable as of December 2005.

- The American Chemical Society (ACS)—A membership organization that consists of individuals working in all fields relating to chemistry. The ACS has developed a series of booklets to convey the essentials of the OSHA regulations for chemical laboratories in a concise and readable form. To assist in understanding the regulation, defining the responsibilities for implementation of 29 CFR 1910.1450 and guidance in audit and inspection, three ACS booklets are available (ACS 1990, 1998 and 2000). The ACS also publishes a guide specifically designed for academic laboratories (ACS 2002). Online at http://www.chemistry.org/portal/a/c/s/1/home.html.
- American Industrial Hygiene Association (AIHA)—A professional membership organization comprised of industrial hygienists or occupational health specialists. The AIHA publishes a protocol guide entitled "Laboratory Chemical Hygiene" (AIHA 1995) to establish the framework to implement the OSHA Standard. The guide presents an overall strategy for the laboratory chemical hygiene plan (CHP) to ensure that all individuals potentially at risk for exposure performing laboratory tasks are informed about the hazards involved and the means to minimize their exposures. Online at http://www.aiha.org/.
- American National Standards Institute (ANSI)—A non-governmental industrial consensus institute responsible for developing and distributing a variety of standards, including those related to safety and health standards (see Appendix A-3 and Section 6.5). Online at http://www.ansi.org/.
- American Society of Safety Engineers (ASSE)—A professional membership organization comprised of safety engineers. The ASSE publishes findings that advance the knowledge base for accident reduction and illness prevention. It also contributes to the development of standards, often working with ANSI. Online at http://www.asse.org/.
- Environmental Protection Agency (EPA)—A branch of the United States government responsible for developing and enforcing regulations that ensure the protection of the environment. Online at http://www.epa.gov/.

- Health Physics Society (HPS)—A professional scientific organization whose mission is to promote the practice of radiation safety. Online at http://www.hps.org/.
- National Fire Protection Association (NFPA)—A non-governmental professional association whose members specialize in safety and health standards related to fire protection, including the electrical standards in the National Electrical Code. Online at http://www.nfpa.org/index.asp.
- Occupational Safety and Health Administration (OSHA)—A branch of the United States Department of Labor (or a delegated State Agency in some states), this organization is responsible for overall regulation of safety and health as it relates to occupational exposure. Online at http://www.osha.gov/.
- The Office of Hazardous Materials safety (HAZMAT) is a branch of the United States Department of Transportation. Their mission is to minimize the risks to life and property inherent in the commercial transportation of hazardous materials. To that end, HAZMAT sponsors a variety of training courses across the U.S. These include 1–2 day seminars and longer courses designed to certify technicians in a wide variety of areas. Online at http://hazmat.dot.gov/training/training.htm.

In addition to the information available from the organizations listed above, an extensive website containing Health Physics resources is available from Oak Ridge Associated Universities (ORAU) (http://www.orau.org/ptp/infores.htm) (Jan. 2006). The ORAU web site contains links to a number of radiation safety training programs. Louisiana State University offers an online resource guide to hazardous materials and safety (http://www.lib.lsu.edu/sci/chem/guides/srs103.html) (Jan. 2006).

Finally, the Health Physics and Radiological Health Handbook(Shleien 1998) and the AIHA White Book (DiNardi 2003) are commonly used as safety reference guides.

15
Automated Laboratory and Field Radionuclide Analysis Systems

HARRY S. MILEY and CRAIG EDWARD AALSETH

15.1. Introduction

Many radiochemical analyses consist of a series of identical steps with little or no variation from sample to sample. Operator fatigue in the execution of repetitive steps leads to increased process variability and execution errors. Such tasks are attractive candidates for automation because they can be made more efficient, consistent, and cost-effective. This trend comes to radiochemistry at a time when schoolchildren use computers regularly, typing or dictating homework assignments to a word processor, with other tools such as "spell check" and "word count" to complete the task. This digital age has also ushered in complex computers predicting molecular, biological, and other highly intricate processes. While an automated device can perform in a moment the manual computations of a scientist's lifetime, however, the same computer would continue computing in error the equivalent of a thousand lifetimes, while a scientist would perceive the error and stop! Thus, despite these tremendous information age advances, the automation of physical and chemical processes must be done thoughtfully to avoid repetition of errors in quality, lapses in safety, and other pitfalls that a scientist would instantly perceive and halt.

Automated systems have several potential advantages over classical technical labor. Analytical functions are reproduced on a programmed schedule, reducing the turnaround time of the samples and enabling a more reliable estimate of analytical costs. In addition to the accuracy of the equipment, costs are readily estimated, the process is easier to scale up or down, and process parameters can be more easily varied in an understandable way. Automated systems can produce data from worldwide locations and may be readily standardized and compared. Manual operation remains preferable for procedures that require considerable judgment in the selection of alternatives.

Aspects of chemistry for which automation has been a popular and effective improvement over manual processes include continuous-flow separation processes, instrumental analysis, and continuous monitoring. Radiometric measurements are a good example, as numerous samples are counted, one after the other, each for a

Pacific Northwest National Laboratory, Richland, WA 99352

318

specified time period. Automation can replace manual placement of the samples for measurement, with a large number of sequential measurements possible without operator intervention. In fact, the sample collection process can be automated in some cases and then mated to an automated measurement system.

Most radioanalytical chemistry processes may be automated to some degree, depending on the type of physical manipulation the sample requires and the complexity of the chemical process. In some cases, the essential mechanical manipulation can be achieved with commercially available components originally designed to automate manufacturing operations. For instance, the pick-and-place automation developed for the semiconductor and other industries can be readily adapted to sample measurement. In other cases, automation relies on specially designed commercial hardware and software tools. Examples of special computer-controlled automated systems developed for atom-at-a-time detection and analysis of actinide and transactinide element isotopes with half-lives as short as a few seconds are shown in Figs. 16.4–16.6. Such techniques have been used to perform the very first studies of the chemical properties of $_{104}$Rf through $_{108}$Hs.

Comprehensive knowledge of the radioanalytical chemistry procedure is needed to devise successful mechanical substitutes for the manual steps of the process and to ensure that the controlling software responds appropriately and predictably. As might be expected, handling special cases and identifying problematic samples are key elements. Examples of automated sampling, analytical processing, and measurement are presented in this chapter to illustrate what may be achieved by combining measurement and engineering expertise. These are illustrative of the principles of automation and are arranged in order of increasing complexity. While the engineering aspects of automation are beyond the scope of a radioanalytical chemistry text, it is useful to explore some of the ideas needed for a successful experience.

15.1.1. Automated System Control and Data Handling

One advantage of automated radionuclide analysis systems is the potential for leveraging the benefits of rapidly developing information technology. Despite the revolution in electronic data processing and associated technology, it is not uncommon today to find world-class measurement systems controlled manually with handwritten logbooks for data storage, data transfer via human-originated email, or physical transfer of storage media. The authors are familiar with a world-leading system managed by use of paper note cards and "carbon-based automation," known as laboratory technicians.

By comparison, significant advantages can be obtained from rather ordinary measurement systems properly linked and employing modern information technology. An example is the International Monitoring System (IMS), in particular the network aspect of the automated monitoring systems of 80 stations that comprise the radionuclide measurement component. In the aerosol monitoring part of the IMS network, numerous stations with good automated reporting and communications provide redundancy. This allows remarkably effective operation even in the case of tampering, forced outages, or national withdrawal of some stations.

Automation adds value to the results recorded. The IMS network provides a vast database of signals recorded under essentially identical conditions and reflecting regional variations in backgrounds. This allows temporal filtering to eliminate false alerts based on the frequency of similar signatures in the past and detailed characterization of system state-of-health, thus enabling higher confidence in radionuclide spectral data. The remainder of Section 15.1 examines aspects of control and data handling for automated systems in more detail.

15.1.2. Spectral and Other Data Types

Several of the monitoring systems discussed in this chapter were designed to use automated data handling. One feature of this design is the use of specific formats for various data types. Many good standard data formats exist; a standard format for an automation application greatly eases integration of disparate software components. For example, the IMS formats support email transmission of several spectral data types: normal spectra, calibration spectra, quality assurance measurement spectra, and system blank spectra. These formats include metadata or extra data-capturing details of the measurement such as time, calibration values, and sample number.

Other types of data useful in automated systems are state-of-health, log entry data (automatically generated and human-originated entries), and alert messages. Alert messages are typically limited to reporting out-of-bounds state-of-health values for loss of power, overheated motors, detector failure, and similar urgent conditions. Planning adequate state-of-health monitoring, such as the desired frequency of status messages and the automatic actions initiated by state-of-health data, are important design steps.

The recorded operating parameters captured by state-of-health data have great utility; when a problem exists, the operator can review the most recently collected state-of-health data to diagnose the problem and determine corrective action. In addition, state-of-health data can be scanned automatically to attempt failure prediction. Reanalysis of state-of-health data from the systems described in this chapter has identified subtle hardware failures not detected by system operators until months later.

15.1.3. Command and Control

Transmission of state-of-health information, particularly maintenance-related data, suggests the next aspects of effective automated systems: command and control. This is most important for remote field systems, but can make operation of laboratory equipment easier as well.

Several scenarios exist for creating command and control systems for automated devices or networks. Ubiquitous and reliable email systems provide a way to issue, receive, and archive commands. Commands arrive as formatted emails; a local software interpreter parses the email and initiates the desired function.

Interactive processes, such as setting the energy calibration of a gamma-ray spectrometer, do not lend themselves to email or text commands, which provide no graphical feedback. The most favorable human–computer interface for highly interactive operations is a graphical user interface. These may require larger communications bandwidth but provide much greater speed of user interaction. Existing schemes for supporting such interfaces over low-bandwidth connections include methods for compressing X-Windows protocol traffic. Another solution is to operate a graphical tool on the operator's end of the connection, with only sparse command-response transmissions with the instrument. Well-designed graphical user interfaces require more software development, balanced by greatly reduced user training.

The broad categories of remote control activities are

1. system inspection commands, or data requests,
2. parameter changes and commands, including Stop and Reboot,
3. interactive sessions, such as energy calibration, and
4. software uploads.

These categories are listed in order of increasing intrusiveness of the remote control. Reckless system modification can easily result in a disabled instrument configuration with, worse yet, no remote way to recover normal operation. Even in the early days of automation, it is surprising how often this circumstance resulted in a technician making travel arrangements. The first and second categories can be handled via email commands or a low-bandwidth graphical user interface command system. The third category requires a higher bandwidth interface and effectively must either transmit graphics or use interactive graphical tools located on the operator's console. The last category can be implemented most easily with a standard and secure file transfer protocol.

15.1.4. Data Surety and Authentication

The potential for computer attack against an automated system or conflicts between poorly coordinated operations arises from allowing connection to remote machines. An attack can be resisted by operating the system on an independent network with no possible connection from unauthorized parties. Use of public infrastructure, such as public voice or data networks is more cost-effective, but open to mischievous or malicious attack.

The threat of inappropriate remote control can be addressed by requiring a validated electronic key to be transmitted with the command (Harris et al., 1999). In the case of the first and second of the command types discussed in Section 15.1.3, this can be done by first calculating a hash or checksum values for the command message. A checksum or hash is a value calculated from the transmitted data, which is sensitive to any modification of the original data. Many such algorithms exist in the public domain (Schneier, 2000), e.g. CRC, RMD160.

Digital signatures and encryption made possible with a public-/private-key infrastructure (PKI) (Schneier, 2000) also address command authentication and offer protection and authentication for potentially sensitive data, e.g., results that could be used within a legal framework or stringent quality assurance regime. Such techniques can allow immediate authentication that data were not changed after original encryption or signing. This effectively addresses the problem of data or command tampering.

The third and fourth command types discussed in Section 15.1.3 can be authenticated and secured via a standard protocol such as secure shell (ssh) (Schneier, 2000). This uses PKI to secure network traffic and provide secure interactive communications and file transfer.

While these concepts address the threat of data or command tampering, the threat cannot be eliminated altogether. It can be said that the risks from unscrupulous, mischievous, or malicious actions in automated systems can be made less than in classical, hand generated, and authenticated laboratory data.

15.2. Manual and Automated Systems Compared: Real-Time Aerosol Radionuclide Monitoring

To motivate the drive toward automation, consider the example of environmental radiological monitoring for aerosol fission-product debris in the atmosphere. When performed manually, as described in Section 15.2.1, the monitoring procedure requires considerable labor. A robust long-term monitoring program can be maintained only through procedural controls, training, and low staff turnover over many years. In Sections 15.2.2 and 15.2.3, automation of the manual system is described. These automated methods reduced cost and free staff for other tasks.

15.2.1. Traditional Air-Filter Analysis

Manual analysis of aerosol fission debris began with the advent of nuclear weapons testing and involved the collection of atmospheric aerosols on filter paper from whole-air samples of at least 1000-m^3 volume. The filters were then folded into a good geometry for radiometric measurement, and gamma-ray spectral analysis was performed with NaI(Tl) or Ge detectors. When greater sensitivity was desired, larger volumes of air were sampled. Eventually, 10,000 m^3 of air per day was filtered for each sample. At the end of a 7-day collection period, the filter was removed from the filter holder, carefully bagged, and stored for a period of a few hours to a few weeks to allow short-lived gamma-ray-emitting daughters of ^{222}Rn and the longer-lived progeny of ^{220}Rn to decay. The filter was then hydraulically compressed into a small right-circular cylinder 0.5-cm tall and 5 cm in diameter to provide good measurement geometry. The prepared sample was placed for measurement between two large NaI(Tl) detectors operating in coincidence or, more recently, against a single Ge detector. After a measurement period selected to provide sufficient precision and sensitivity, the resulting gamma-ray spectra were analyzed for fission and activation products.

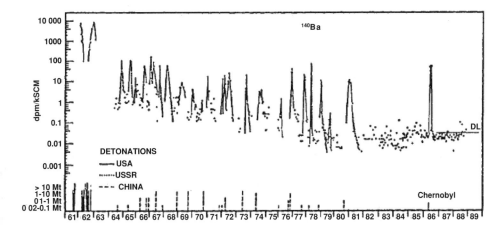

FIGURE 15.1. Temporal distribution of ^{140}Ba measured in the atmosphere of Richland, WA. Figure from Perkins et al. (1989). (By permission of Pacific Northwest National Laboratory)

The concentrations of some 30 airborne radionuclides were measured between 1961 and 1989. Figure 15.1 (Perkins et al., 1989) shows the airborne concentration of ^{140}Ba over this period with a detection limit (DL) of about 0.02 disintegrations per minute (dpm) per m^3.

15.2.2. Automating Spectral Analysis

An automated method of analysis for germanium spectra was eventually adopted with the advent of sufficient computing power in an ubiquitous platform. This method was improved until it eventually eliminated all but periodic human QA checks. This automated method (Gunnink and Niday, 1972) divides analysis into five tasks:

- energy calibration;
- smooth background computation and subtraction;
- peak location and integration;
- matrix-inversion computation of radionuclide contributions;
- activity calculation/reporting.

Many versions of this basic approach exist, the most significant variation being whether matrix inversion is used to connect a library of radionuclides to observed peak intensities or whether a list of energies of interest is used to make key calculations. Thus, the two main types today are matrix inversion and list directed. Automated spectral analysis software is available from commercial and academic sources with a mix of national and international quality certifications, specialized capabilities, and user control. Programs of this type can handle thousands of automated analyses per day and run on most types of computers.

15.2.3. Real-Time Sampler/Analyzer System

Although advances in computer software allowed for easier methods of spectral data analysis, the actual collection and preparation of the sample remains a time-consuming and repetitive task. The integration of collection and analysis steps was accomplished in the design of the real-time aerosol radionuclide analyzer collector (R-TARAC). The R-TARAC was created to conduct continuous radiation monitoring as part of the program to search for airborne species associated with nuclear proliferation (Smart, 1998). The R-TARAC incorporates a large (140% relative efficiency) HPGe detector placed in the air stream directly behind an automatically changed air filter.

This system has been used inside a research aircraft where outside air is ported into the aircraft and through the system, as well as in an under-wing pod with direct frontal airflow, as seen in Fig. 15.2. The system continuously measures and records the concentration of airborne photon-emitting radionuclides. The R-TARAC stores all collected data and can be automatically or manually controlled from a remote console to provide instant data reporting to the flight crew and an operations center.

FIGURE 15.2. The R-TARAC, configured for under-wing pod use (top) and for fuselage- or automobile-based use (bottom). U.S. Patent 6,184,531. (By permission of Pacific Northwest National Laboratory)

The crew can use such instant data availability to maneuver the aircraft for locating and monitoring radioactive plumes.

As with any radiation detection system, background is a concern. Ambient airborne radionuclides (mostly radon daughters) accumulate on the filter and lessen the detection sensitivity for the radionuclides of interest as compared to a decayed sample. This accumulation is addressed with a filter carousel, which rotates a fresh filter into place either at regular intervals or when the background activity level on the filter becomes excessive.

Another potential background source is contamination of internal and external surfaces of and near the instrument from exposure to ambient airborne radionuclides. These can be deposited from the air stream onto surfaces near the filter and detector. This background can be expected to increase while the R-TARAC is exposed to an air stream until the surface contamination reaches mechanical or decay equilibrium. Tracer tests showed that because of the geometry of surfaces susceptible to contamination relative to the detector, these deposits contribute generally less than 1% of the measured radiation. A larger contribution should be expected after the aircraft has moved through a high-radiation plume into a region of low-radionuclide concentration.

15.3. Automated Laboratory System Discussion and Examples

To give a flavor for the kind of features and challenges encountered in automating laboratory systems, this section discusses the automation of some common laboratory processes and collects a number of examples of automated systems that have been developed.

15.3.1. General Automation Techniques

System automation is made easier by the availability of many subsystems that are easily controlled by computer. Process instrumentation of all types and radiometric measurement equipment is available with standard computer interface options. Computer hardware and software is available both for simple and complex control systems. Mechanical equipment for automating the handling of multiple samples includes pumps, valves, heaters, shakers, vibrating plates, and stirring systems for mixing samples.

15.3.2. Automated Sample Changers for Counting Radionuclides

Automated sample-changing equipment has been available commercially for many years. In liquid scintillation counting (LSC) systems, several hundred vials may be placed in a train (see Section 8.5.2) for dark adaptation to allow decay of delayed

FIGURE 15.3. Automated gamma-ray spectral analysis system. (By permission of Georgia Institute of Technology)

fluorescence, and then moved, one at a time, to an elevator that brings the sample between two photomultiplier tubes for counting. After counting, the sample is returned to the train.

Similarly, planchets for automated gas-flow proportional-counter systems are stacked in plastic holders. The bottom holder is moved beneath the detector for counting and then removed to a second stack for storage or recycling.

For gamma-ray spectral analysis, a set of bulk samples can be placed on a rotating tray that moves each sample in turn next to the massive shield that encases the detector (see Fig. 15.3). The door in the shield opens and a mechanical arm places the sample on top of the detector. After counting, the sample is lifted and returned to the tray. Alpha-particle spectral analysis generally uses no automation because the samples are counted for a long time.

15.3.3. Automated Laboratory Radionuclide Column Separation System

Use of ion-exchange columns is a common chemical separation strategy. Elution of radionuclides from an ion-exchange column occurs as a function of elutriant volume. For a fixed continuous flow, radionuclides are eluted from the column as a function of time. The output concentration of a particular radionuclide can be described by a peak defined by the number of column volumes passed between the beginning of its elution and the end. Each radionuclide may follow at its own column-volume parameters. These parameters are a reproducible function of process characteristics such as column dimensions, flow rate, type of ion-exchange resin, and elutriant reagent. To automate the collection of column-separated radioisotopes, an automatic changer for collection containers can be used. The changer can be linked to an elutriant flow controller for volume-based separation. For constant-flow columns, the changer can be controlled by timing. The result is the radionuclide or radionuclides of interest in separate containers.

Flow into columns can be controlled by computer-activated valves from reservoirs. One reservoir contains the dissolved sample with the radionuclides to be separated, another holds a wash solution, and yet another holds the elutriant. First the sample reservoir is drained into the column, next a timed or measured volume of wash solution is admitted to the column, and finally a timed or measured volume of elutriant is added to elute the radionuclide from the column. Additional reservoirs may hold other reagents for subsequent elution of other radionuclides and column regeneration.

15.3.4. Automated ^{99}Tc Separation and Measurement by Flow-Injection Analysis

Technetium-99 is a fission-product radionuclide that is analyzed widely in the environment and in radioactive waste. Because of its long (213,000-y) half-life and high fission yield (6.1%), it can remain at detectable levels after shorter-lived radionuclides have decayed. The emitted low-energy, beta particles (0.294 MeV, maximum) are usually measured by LS counting after purification to remove other radionuclides and interfering salts. ^{99}Tc also is measured with proportional counters (see Section 6.4.1) or by ICP-MS (see Section 17.8).

Radiochemical analysis of ^{99}Tc is challenging because, unlike radionuclides such as ^{137}Cs, it cannot be measured directly and nondestructively by gamma-ray spectral analysis. The usual processes of dissolution, purification, and preparation for counting can be complex for ^{99}Tc because it has multiple possible oxidation states, notably cationic Tc^{4+} and anionic pertechnetate TcO_4^- (see Section 6.4.1).

Because of these characteristics, ^{99}Tc provides a good example of automated laboratory analysis. Integration of modern selective chemical separation procedures, radiation detectors, and fluid handling instrumentation in a single functional unit allows the development of an automated radionuclide analyzer and radiation sensor, as shown by Egorov et al. (2003) and by Grate and Egorov (2003), among others. The fully automated radioanalytical chemistry system developed for rapid analysis of ^{99}Tc in aged nuclear waste is shown in Fig. 15.4. The instrument executes fluid handling steps that acidify the caustic sample, microwave-assisted sample oxidation to pertechnetate Tc(VII) with peroxidisulfate, separation of ^{99}Tc(VII) from radioactive interferences on an anion-exchange column, and delivery of the purified pertechnetate to a flow-through scintillation detector.

The automated radiochemical process is performed in a single functional unit. The instrument design incorporates advanced digital fluid handling techniques with multiple zero dead volume digital syringe pumps and multiple valves for sample and reagent delivery. Comprehensive multithreaded control software was

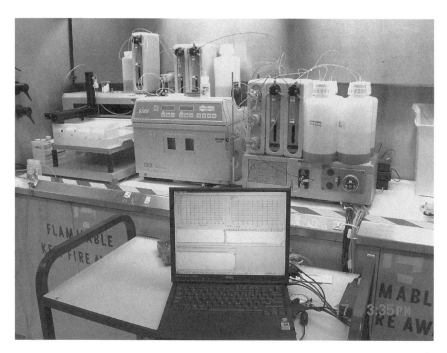

FIGURE 15.4. Automated total ^{99}Tc analyzer based on classical radiochemical measurement principles. (By permission of Pacific Northwest National Laboratory)

developed to enable fully automated asynchronous operation of the instrument components as well as data processing, storage, and display.

The automated sample treatment protocol begins with sample acidification for the first digestion step. This initial treatment ensures removal of nitrites, which are abundant in basic waste solution matrixes and interfere with subsequent oxidation. In addition, initial heating promotes rapid dissolution of the $Al(OH)_3$ precipitate, which forms during acidification of the caustic matrix. Nitric acid concentration and volume are selected to ensure complete dissolution of $Al(OH)_3$ upon heating and later maintaining high pertechnetate uptake on the anion-exchange material during sample loading. A second digestion treatment with sodium peroxidisulfate as the oxidizing reagent converts any reduced technetium species to pertechnetate.

Pertechnetate is retained on a macroreticular, strongly basic, anion-exchange resin (AGMP-1, Biorad). This anion-exchange material was selected because of its long column life under elevated back-pressure conditions. Pertechnetate is eluted rapidly from the anion-exchange column with strong nitric acid solution by reversing the direction of flow through the column. The pertechnetate separation by anion exchange in nitric acid medium offers adequate separation selectivity from ^{90}Sr/^{90}Y and ^{137}Cs for determining ^{99}Tc. To remove interfering anionic radionuclides such as ^{106}Ru, ^{125}Sb, and ^{126}Sn, a comprehensive column wash sequence with 0.2 mol/l nitric acid, 1 mol/l sodium hydroxide, 0.2 mol/l nitric acid, 0.5 mol/l oxalic acid, and 2 mol/l nitric acid is required prior to pertechnetate elution.

A flow-through scintillation detector equipped with a lithium glass solid scintillator flow cell is used to detect the eluted ^{99}Tc. The glass scintillator enables an absolute detection efficiency of $\sim$55% and is stable in the 8 mol/l nitric acid medium for pertechnetate elution.

An automated standard addition technique is part of the analytical protocol. An aliquot of a ^{99}Tc standard solution is added to a duplicate of the sample during acidification. The ^{99}Tc standard is in a nitric acid solution of the same concentration used for sample acidification. To perform the standard addition measurement, the sample acidification monitor instrument automatically substitutes a given volume of the ^{99}Tc standard solution for an equal volume of the nitric acid. The volume of the standard solution is selected by the software to yield an estimated threefold higher signal relative to the analysis of an unspiked sample.

The total effective analytical efficiency (product of the recovery efficiency and the detection efficiency) is calculated based on the difference in analytical response obtained by the analysis of the spiked and unspiked samples. This approach provides a reliable method for remote, matrix matched, instrument calibration. Automated standard addition can be used for each sample or batch of samples. The automated radiochemical analysis procedure is rapid, with a total analysis time of 12.5 min per sample. The total analysis time for the standard addition measurement is 22 min (including analysis of both unspiked and spiked samples). For low-level samples, a much longer counting time can be expected.

The analyzer instrument was successfully tested with various waste solution samples from the US DOE Hanford site, including those with high organic content. Quantification was verified by independent sample analysis with ICP-MS.

15.4. Automated Field System Examples

15.4.1. Aerosol Monitors: Radionuclide Aerosol Sample Analyzer

The Caribbean Radionuclide Early Warning System and the verification technology proposed as part of the Comprehensive Nuclear Test Ban Treaty (CNTBT) (see http://www.ctbto.org/) (Jan 2006) presented a challenge to automation capabilities. The CNTBT features as its verification arm the International Monitoring System, a network for daily monitoring of the atmosphere for fission debris, such that the sensitivity to ^{140}Ba is <30 μBq (<0.0018 dpm) per m^3 of air (Schulze et al., 2000). This sensitivity requirement is the primary driver for the design of either manual or automatic technology for the 80-station network.

For practical source–detector geometry, a compressed filter sample can improve the detection efficiency by about a factor of five over an uncompressed filter. To eliminate the need for sample compression, several solutions could be pursued.

Increasing the size and hence the bulk efficiency of the detector is one possible but expensive avenue. Increasing the volume of air sample is another avenue. But since the air contains both the radionuclides of interest and obscuring background radionuclides, the improvement factor due to volume increase is only approximately proportional to the square root of the sample volume increase. The approach applied for substantial improvement in detector–source geometry and thus counting efficiency was automatic folding or layering of the filter material.

Several sensitivity-enhancing techniques were applied simultaneously in the Radionuclide Aerosol Sampler Analyzer (RASA) (Miley et al., 1998). This system employs a Ge detector with about twice the efficiency of contemporary manual systems (90 vs. 40% relative efficiency) and twice the airflow of manual systems (24,000 vs. 12,000 m^3 per day). A simple layering mechanism provides a large filter (0.25 m^2) during sampling and a moderately small filter volume (~400 cm^2 and 0.5-cm thick) for measurement.

Filter volume is minimized for efficient radionuclide measurement with a segmented sampling head and six independent, continuous filter rolls, as shown in Fig.15.5. These rolls store more than a 1-year supply of daily filter changes for drawing to and through the sample head. The six simultaneously exposed filters (each 10 cm × 40 cm) are brought together after exposure and sealed between two layers of sticky polyester tape that is fed from two rolls. This single filter bundle then rests in a decay position for 24 h to eliminate gamma rays from the ^{222}Rn daughters, ^{214}Pb and ^{214}Bi, and reduce those from the ^{220}Rn daughters (mainly ^{212}Pb, ^{212}Bi, and ^{208}Tl). The filter package is then pulled into a loop around a rotating drum that circles the germanium detector. The filters remain stationary for about 24 h for gamma-ray spectrometric analysis and are then advanced by drive rollers forward 50 cm into the next position. Thus, in normal operation the following processes occur: (1) today's 0.25 m^2 filter collects aerosol, (2) the previous day's filter is held for decay, and (3) the 2-day-old filter is being measured by the detector within a small lead cave.

The system is automated with software modules that (1) collect gamma-ray spectra from the sensors, (2) control relays to activate motors for advancing the

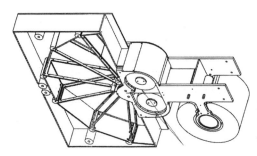
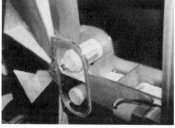

FIGURE 15.5. Segmented sample head, showing filter baffles, rolls of encapsulating polyester strips, and the wraparound path between the detector and lead shield. U.S. Patent 5,614,724. (By permission of Pacific Northwest National Laboratory)

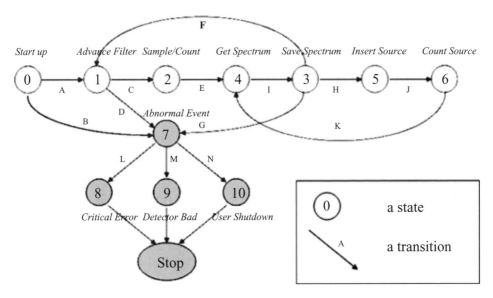

FIGURE 15.6. State Machine showing states and transitions. (By permission of Pacific Northwest National Laboratory)

filters through the aerosol collection, storage, and analysis process, and (3) control the associated processes. In all, a dozen such software devices can completely control, monitor, and manipulate all the features of the RASA. Because the RASA performs a sequential set of steps, from startup to shutdown, a State Machine—a software construct that is the equivalent of a process flow diagram—was chosen as control, as shown in Fig. 15.6.

Each decision point in the diagram corresponds to a transition from one state to another. The State Machine most favorably is written such that a noncomputer programmer can adjust the operation of the device by editing an English language file that describes the transition from each state to the others. For example, the RASA always recognizes its current state via electronic signals like pressure, temperature, voltage, and timers. Upon a state change such as a loss of electrical power or the time of sample measurement, the state is shifted and new functions, such as filter advance and calibration, or data retrieval, are automatically performed.

An important feature of any automated system is its behavior at the application and loss of power. The initial and final conditions need to be known to prevent, say, loss of a sample or sample analysis data. This is easily accomplished by saving the final condition at loss of power, but battery backup is needed to monitor and record this state. A catastrophic condition could result if the automated system fails to shut down or restore properly, such as system damage or bystander harm from unexpected mechanical actions.

The spectral data collected daily by the system can confirm successful operational functioning of the RASA, including important features such as start and stop time, detector resolution, and gain. To assure that the results are correct and can be used as legal evidence, a robust quality assurance (QA) program also is required. This begins with system certification at the factory or production laboratory and includes site-specific documentation of local operating procedures. The system performs daily wide-range energy calibration, which also serves as a check on the stability of efficiency values.

Regional laboratories that conform to national standards practices perform external QA measures of the network of RASA systems. An international testing procedure assures that the laboratories remain proficient. Randomly selected filters from the network of monitoring stations, automatic and manual, are sent to regional laboratories to determine if the station results are in control. This can be partially accomplished by measuring the level of ^{7}Be ($t_{1/2} = 53.28$d), a cosmic-ray spallation product of atmospheric nitrogen and oxygen that is always in the atmosphere and is easily measurable on air filters.

One of the advantages of this automatic system is that the state-of-health data recorded for the numerous subsystems, including blowers, component states, temperatures, and other critical information, allow remote failure detection, diagnosis, and possibly prevention. As an example, variation in detector temperature may show the onset of failure of a mechanical cooler. Remote diagnostics are used to schedule repair trips and minimize down time.

15.4.2. Gas Monitors: Real-Time ^{133}Xe and ^{135}Xe Sampler/Analyzer

The Automatic Radioxenon Sampler/Analyzer (ARSA) (McIntyre, 2001) was designed to measure radioxenon produced in nuclear explosions. Observation of radioxenon has other uses, including the ability to indicate releases from a reactor or a medical facility. Gaseous radioxenon is an important indicator of leakage from underground nuclear tests because gases are more likely to escape than aerosols. One potential problem is measuring the relatively low gamma-ray energies of the two radionuclides, which can be obscured by Compton continua for higher-energy gamma rays.

Stable xenon comprises only 80 ppb of the atmosphere or 0.08 cc/m^3 of air at STP. A series of chemical and physical processes are thus required to remove contaminants, including the major gases N_2, O_2, and CO_2, the other rare gases (especially radon), as well as atmospheric moisture. Separation of radioactive gases in a series of traps held at the condensation temperature for each gas had earlier been applied manually to analyze fission-produced xenon and krypton radionuclides in air (Momyer, 1960). The first trap is cooled to a temperature at which one elemental gas is sorbed or condensed while the others pass through, and then heated to release the trapped gas for further purification and radiation detection. The gases that passed through the first trap are pumped to a second trap, held at a temperature suitable for trapping a second gas but not the other gases, and the second gas is then released by heating the trap and processed for counting. The ARSA design is an improvement on these early techniques because it uses selective

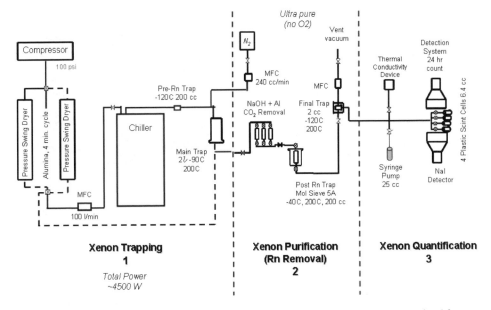

FIGURE 15.7. ARSA process diagram. Some duplicate process lines have been omitted for clarity. (By permission of Pacific Northwest National Laboratory)

condensation with dryers, molecular sieve traps, and charcoal beds, as shown in Fig. 15.7.

In Area 1 of Fig. 15.7, both trapping and regeneration occur; regeneration steps are shown by the dashed line. Air is first forced though a pair of pressure-swing dryers that consist of powdered alumina. While one is drying the incoming sample air, the second dryer is being regenerated with dry waste air from elsewhere in the process. These dryers switch from drying to regeneration every few minutes on a timer, or more frequently, on detection of break-through moisture. The dried air is then chilled.

The total flow rate is controlled by a commercial mass flow controller (MFC), which contains an internal servo mechanism that links a mechanical valve to a resistance thermal device (RTD). The RTD measures mass flow (rather than gas velocity) by the change of electrical resistance in a sensing wire heated by an adjacent hot wire. Because this measurement is affected by the specific heat of the gas, the MFC must be calibrated for each individual gas. The desired MFC flow is set by applying a voltage to the MFC that corresponds to the voltage generated by the RTD at that flow. A comparator in the MFC opens or closes the internal valve to balance the RTD and the applied voltage.

The cold sample gas flows at 100 l/min into a 0.2-l charcoal trap that is cooled to −90°C to adsorb radon. Xenon flows through the trap and is then collected on the 2-l main charcoal trap at −120°C. Nitrogen, oxygen, and the lighter rare gases helium, neon, argon, and krypton pass through this trap.

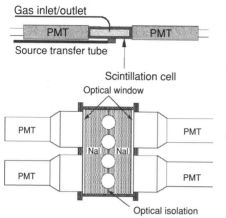

FIGURE 15.8. Xenon detector and internal gas cell. (By permission of Pacific Northwest National Laboratory)

Xenon is next eluted in Area 2. At the end of the sampling period, the main trap is valved off from the system and heated to 200°C to release the xenon. The outflow from the main trap is carried by ultrapure nitrogen (no CO_2) through a MFC at 0.24 l/min. Desorption is timed to release most xenon; flow is diverted when the remaining radon is expected to desorb. This slow flow takes the desorbed xenon gas with N_2 carrier gas though one of a series of disposable chemical traps (NaOH + Al) to remove most CO_2. The now purified gas is collected on a 0.2-15 Å molecular sieve trap at –40°C, while the carrier N_2 escapes. The trap is then heated to 200°C to desorb xenon and transferred to a tiny cold trap that is cooled to –120°C to sorb xenon in preparation for loading the counting cell. This step is necessary to transfer the gas into a scintillation cell, as shown in Area 3 of Fig. 15.7 and also in Fig. 15.8.

The tiny final trap, loaded with the equivalent of 10 cc (at STP) Xe/CO_2 product, is heated to 200°C and the product gas is allowed to expand into one of four counting cells. Transfer efficiency is boosted by use of a syringe pump, as shown in Area 3 of Fig. 15.7. The exact ratio of the Xe and CO_2 mix is determined with a thermal conductivity device (TDC). This is an important measurement because the xenon separation efficiency depends on various factors, e.g., the conditioning of the traps and the ambient temperature. The ARSA typically produces a product gas that is about 50% xenon.

The ARSA system is designed to operate continuously in remote areas with minimal maintenance and consumables limited to replacement bottles of nitrogen carrier gas and CO_2 removal traps. Dryers and gas traps were chosen that could be automatically regenerated in the field. The system has many components and a complex path for the analyte gas that traverses the system. The numerous valves ($\sim$100) for controlling the gas flow and trap regeneration processes require a control system about one order of magnitude more complex than that for the RASA system described in the preceding section, if the number of parts is taken as a measure of

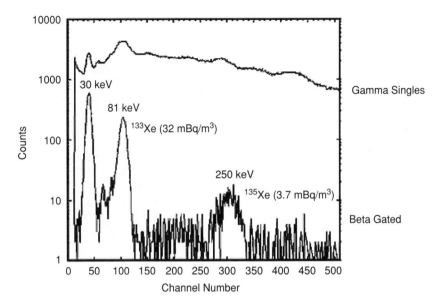

FIGURE 15.9. Measurement of Xe-135 in the atmosphere, made possible by automated sampling and analysis. (By permission of Pacific Northwest National Laboratory)

complexity. In addition, because of the use of elevated temperatures, particularly in regenerating charcoal traps, the ARSA has a module specifically designed to monitor safety parameters.

The sample should be processed quickly because of the short half-life of ^{135}Xe (9.10 h) compared to ^{133}Xe (5.234 d). The ARSA performs three sample analyses per day. The fair match of this frequency to the resolving time (6 h) of meteorological measurements made worldwide facilitates coordination with atmospheric transport models. The rapid analysis capability allowed the measurement of ^{135}Xe in the Earth's atmosphere, see Fig. 15.9 (Bowyer et al., 1999).

This system suppresses radon by several orders of magnitude, but the fact that every atom of radon is radioactive means that a specialized radiation coincidence detection apparatus is still required to resolve radioxenons from radon. Moreover, the radiations emitted by the two xenon radioisotopes must be resolved by spectral analysis. This approach yields an impressive sensitivity of about 100 μBq (0.006 dpm or 3,000 atoms of ^{135}Xe) per m^3 of air based on sampling of about 40 m^3 of air sample and a 24 h radiometric measurement.

The ARSA system has a significant advantage over manual methods because large numbers of technicians would be required for the time-intensive gas separation steps. Prior manual radioxenon monitoring efforts attempted to circumvent this problem by sampling for longer collection periods. This came at the cost of much poorer time resolution and a loss of ability to measure the shorter-lived ^{135}Xe. Measurement of ^{135}Xe is useful for determining the ratio of ^{135}Xe to ^{133}Xe to distinguish between incidental formation in nuclear weapons and continuous formation in nuclear reactors.

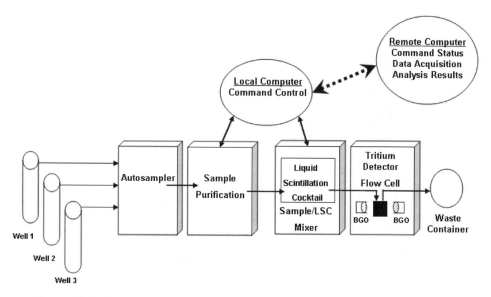

FIGURE 15.10. Conceptual diagram of FDTAS, showing process components from sampling to analysis. (By permission of Pacific Northwest National Laboratory)

15.4.3. Tritium in Water

Tritium is a major radionuclide in effluent at many nuclear facilities. In the form of HTO, it moves essentially as water does in groundwater, surface waters, and airborne moisture. The radionuclide emits only low-energy beta particles. Tritium is typically measured by mixing purified water with liquid scintillation cocktail and performing LS counting (see Section 6.4.1). Water that contains tritium at concentrations not far below regulatory limits and can reach drinking water supplies should be analyzed frequently. Automated sampling, analysis, and reporting address this concern.

The field deployable tritium analysis system (FDTAS) (Hofstetter et al., 1999) samples water, purifies the sample, and performs Compton-suppressed LS counting to achieve laboratory-quality analyses. The system described by Fig. 15.10 has remote reporting and control capability.

The sampling and analysis aspects of the system follow earlier continuous monitors of tritium in water (Brown et al., 1993). Typically, two photomultiplier tubes in a lead shield face the sample cell between them. On a cyclical basis, a sample is collected from the body of water and injected into the sample cell. An equal volume of LS cocktail is mixed with the sample in the cell. The sample and cocktail are then counted and discharged. The cell is flushed with water. The next sample, which had been collected during the counting period, then is injected into the cell. The counting data are stored, processed to calculate the count rate, and reported.

Periodically, pure water is mixed with the cocktail and counted to provide a background count rate. Factors related to energy distribution, counts in non-tritium channels, and pulse shape can be recorded to correct for the presence of other radionuclides and the effects of quenching or luminescence (see Section 8.3.2). The counting period is selected to achieve the required detection limit. Measurements may be at the rate of 1 per 100 min or less frequent if longer counting times are required.

The FDTAS design adds water purification and improved detection sensitivity. Laboratory tests of the various methods of purifying environmental samples applicable to unattended field deployment led to selection of a single use, commercially available, mixed bed resin column for water purification. By employing active background suppression and pulse shape discrimination with bismuth germanate (BGO) guard scintillators, lead shielding, and low-background cell components, the FDTAS achieves tritium detection sensitivity that almost rivals laboratory LS instruments. The combination of a shield, an active BGO guard detector, and low-background materials attains a routine background of ∼1.5 count/min in the tritium energy channel. A special low background quartz counting cell, containing 11 ml of a 50:50 mixture of the sample and the liquid scintillation counting fluid, yields a tritium detection efficiency of ∼25%.

The combined low background, high detection efficiency, and 5.5-ml sample volume lead to a detection limit of 10 Bq (600 dpm) per liter for a 100-min count (95% confidence limit). These results are achieved routinely in field tests of the FDTAS at monitoring wells, surface streams, ground water remediation facilities, sewage treatment plants, and the Savannah River. Results have been confirmed by parallel sampling and laboratory analyses.

The automation of the FDTAS is of a classic and cost effective type that mostly uses commercial hardware and support from manufacturers to create sophisticated custom radiation detection apparatus. QC capabilities are built-in to validate the results. Nearly real-time reporting allows operational authorities a cushion of time to end a major tritium release and mitigate its effects.

15.5. Summary

A variety of radiochemical measurements have been successfully automated, both in the laboratory and in the field. It might appear to the student that the effort involved in automating a process is more trouble than the manual measurement of the activity. To an experienced practitioner, it is exactly this economic and quality decision point that makes the selection of automation or manual approaches interesting. When the expected variations in sample type and process parameters allow, automation should be evaluated for ongoing capabilities and field measurements. But it should also be said that the interactions of scientists, technicians, and engineers on an automation project can be truly rewarding in many ways, as the fusion of disciplines tends to accelerate the productivity of all.

16
Chemistry Beyond the Actinides

DARLEANE C. HOFFMAN

16.1. Introduction

Humans have been fascinated ever since ancient times with trying to understand the composition of the terrestrial world around them and even the stars beyond their reach. In the 4th century B.C., Aristotle proposed that all matter could be described in terms of varying proportions of four "elements"—air, earth, fire, and water. Elements, including gold, silver, and tin, that are found relatively pure in nature were isolated and used over the period of the next several hundred years. The alchemists of medieval times isolated and discovered additional elements and used secret formulas and rituals in the attempt to find the "philosopher's stone" and attain their goal of transmuting lead into gold. The development of experimental science and the scientific method in the 18th century accelerated the pace of the discovery of new elements, but uranium, discovered in 1789 by Klaproth in pitchblende from Saxony, Germany, remained the heaviest known chemical element for 150 years.

After the discovery by Becquerel in 1896 that uranium salts blackened photographic plates due to radiations from the natural decay chains of uranium, Marie Curie and Pierre Curie (Curie and Curie, 1898) began extensive studies of the new phenomenon that they dubbed "radioactivity." They were successful in isolating and identifying the first radioactive elements Po (84) and Ra (88) in pitchblende. They shared the Nobel Prize in Physics with Becquerel in 1903 for their studies of radioactivity. Marie Curie continued her investigations of radioactive species and isolated macro quantities of Ra to be used for medical purposes. She received the Nobel Prize in Chemistry in 1911 for the isolation of pure radium and the determination of its molecular weight. She is recognized as the founder of the subdiscipline of "radiochemistry," which can be defined as the study of radioactivity and radioactive elements; she could rightfully be acknowledged as the founder of nuclear medicine as well.

In the mid-1930s, the new breed of "nuclear" scientists, including both chemists and physicists, became intrigued with the possibility of actually synthesizing new "artificial" elements not found in nature. The discovery of "artificial" radioactivity by Joliot–Curies in 1934, the invention of the cyclotron by E. O. Lawrence in

Nuclear Science Division, Lawrence Berkeley National Laboratory, Department of Chemistry, University of California, Berkeley, CA 94720

1929, and the subsequent operation of larger cyclotrons at Berkeley facilitated production of artificial elements. The new developments rather quickly culminated in the realization of the ancient alchemists' dream of transmutation when the first "artificial" element technetium ($Z = 43$) was identified in 1937 in the products of deuteron bombardment of molybdenum in the cyclotron (Perrier and Segrè, 1937). Then astatine (85) was identified in 1940 in cyclotron bombardments of Bi with He ions (Corson et al., 1940).

Shortly thereafter, the first transuranium element Np (93) was identified (McMillan and Abelson, 1940) in experiments conducted to investigate the newly reported phenomenon of nuclear fission in neutron irradiations of U. The production and chemical identification of element Pu (94) in the form of ^{238}Pu produced in deuteron bombardments of uranium in the cyclotron at Berkeley followed in 1941, but it was not published until 1946 (Seaborg et al., 1946). The discoveries were voluntarily kept secret because it was recognized that the highly fissionable ^{239}Pu produced at about the same time (Kennedy et al., 1946) might have potential military uses. The artificial element Pm (61) was separated and positively identified (Marinsky et al., 1947) among the fission products of uranium in 1945 during the World War II Plutonium Project at the Oak Ridge, Tennessee Laboratory. Although Tc, Pm, At, and Pu are often called "artificial" elements, all of them are found in nature in exceedingly small amounts, either as uranium fission or capture products or from natural decay chains. The very long-lived ^{244}Pu was also reported in 1971 to exist in nature in minute quantities. Thus the distinction between "artificial" and "natural" elements is in itself somewhat artificial.

During what might be called the "golden age of discovery" between 1940 and 1955, all the transuranium elements from Np through Md (101) were discovered. They were produced by either neutron or alpha bombardments of suitable targets and were identified with little controversy, probably because their half-lives were long enough to permit chemical separation and identification. Beginning with No (102), identification of new elements became ever more difficult and controversial as the half-lives became shorter and the production rates even smaller. The first positive identifications of the elements beyond Md were based on physical techniques rather than on chemical techniques, and many controversies ensued. However, with the discovery of longer lived isotopes of No, Lr (103), and Rf (104), it was possible to confirm Seaborg's hypothesis (Seaborg, 1945) that the actinide series ends with Lr and that the chemistry of Rf will be quite different from that of the actinides. More complete discussions of these discoveries, the experimental methods involved, and some of the ensuing controversies can be found elsewhere (Seaborg, 1967; Hoffman and Lee, 1999; Hoffman et al., 2000).

The transactinide elements are defined simply as those elements with atomic number greater than 103 (Lr), which completes the actinide series with the complete filling of the inner 5f shell. The schematic periodic table of 2004 given in Fig. 16.1 shows the lanthanide (4f) series, the actinide (5f) series, and the transactinides as a 6d transition series that begins with element 104 (rutherfordium, Rf). According to results of atomic relativistic calculations (Pershina and Fricke,

1																	18
1 H	2											13	14	15	16	17	2 He
3 Li	4 Be											5 B	6 C	7 N	8 O	9 F	10 Ne
11 Na	12 Mg	3	4	5	6	7	8	9	10	11	12	13 Al	14 Si	15 P	16 S	17 Cl	18 Ar
19 K	20 Ca	21 Sc	22 Ti	23 V	24 Cr	25 Mn	26 Fe	27 Co	28 Ni	29 Cu	30 Zn	31 Ga	32 Ge	33 As	34 Se	35 Br	36 Kr
37 Rb	38 Sr	39 Y	40 Zr	41 Nb	42 Mo	43 Tc	44 Ru	45 Rh	46 Pd	47 Ag	48 Cd	49 In	50 Sn	51 Sb	52 Te	53 I	54 Xe
55 Cs	56 Ba	57 La	72 Hf	73 Ta	74 W	75 Re	76 Os	77 Ir	78 Pt	79 Au	80 Hg	81 Tl	82 Pb	83 Bi	84 Po	85 At	86 Rn
87 Fr	88 Ra	89 Ac	104 Rf	105 Db (Ha)	106 Sg	107 Bh	108 Hs	109 Mt	110 Ds	111 Rg	112	113	114	115	116	(117)	(118)
(119)	(120)	(121)	(154)														

LANTHANIDES	58 Ce	59 Pr	60 Nd	61 Pm	62 Sm	63 Eu	64 Gd	65 Tb	66 Dy	67 Ho	68 Er	69 Tm	70 Yb	71 Lu
ACTINIDES	90 Th	91 Pa	92 U	93 Np	94 Pu	95 Am	96 Cm	97 Bk	98 Cf	99 Es	100 Fm	101 Md	102 No	103 Lr
SUPERACTINIDES	(122)	(123)	(124)	(125)	(126)									(153)

FIGURE 16.1. 2004 periodic table showing placement of transactinides through element 154.

1999), this transition series ends at element 112 with the filling of the 6d shell. The p shells are filled in elements 113 through 118, which are expected to be the heaviest members of the noble gas group. The 8s shell is expected to be filled in elements 119 and 120, making them homologs of groups 1 and 2. Based on relativistic calculations, element 121 should have an 8p electron in its ground-state configuration in contrast to the 7d electron expected by simple extrapolation from the elements in group 3 of the periodic table. A 7d electron would then be added in element 122, giving it the configuration $[118]8s^27d8p$, different from that expected from its homolog thorium, which has the configuration $[Rn]7s^26d^2$. The situation becomes even more complicated beyond element 122 because the energy spacings of the 7d, 6f, and 5g levels as well as those of the 8p, 9s, and 9p levels are so close that the chemical properties of these elements will be nearly impossible to characterize on the basis of currently known properties.

16.1.1. Transactinide Elements and Controversy

Throughout the period from 1960 to 1977, controversies were associated with claims to priority of discovery and proposals for the naming of elements 104 and 105, the first of the transactinides, by research groups working at Lawrence Berkeley Laboratory (LBL) in the United States and the Joint Institutes for Nuclear

Research (JINR) in Dubna, the USSR. In 1974, the International Union of Pure and Applied Chemistry (IUPAC) and the International Union of Pure and Applied Physics (IUPAP) appointed an *ad hoc* committee of neutral experts to consider the competing claims and to facilitate cooperation between the groups in reaching agreement on these issues. However, the committee never met as a whole, and it was finally disbanded in 1984, although a few smaller meetings with some of the Berkeley and Dubna researchers were sponsored, observers were exchanged, and some reports were issued (Hyde et al., 1987).

Pending settlement of the competing claims to discovery and approval of names, the Commission on Nomenclature of Inorganic Chemistry (CNIC) of the IUPAC mandated (Chatt, 1979) the use of three-letter "systematic names" based on 0=nil, 1=un, 2=bi, 3=tri, 4=quad, 5=pent, 6=hex, 7=sept, 8=oct, and 9=enn for elements with $Z > 100$. Thus, elements 104 and 105 officially became unnilquadium (unq) and unnilpentium (unp), although these names never found common usage among heavy element researchers. In the meantime, the names kurchatovium (Ku, 104) and nielsbohrium (Ns, 105) and rutherfordium (Rf, 104) and hahnium (Ha, 105) continued to be used by the Dubna and Berkeley groups, respectively.

Among the reasons for the controversies over discovery of elements 104 and 105 are their short half-lives and small production rates that made it necessary to study and identify them "online" at the accelerators where they were produced on the basis of their decay properties rather than on "conventional" radioanalytical methods. New radioanalytical techniques had to be developed for unequivocally establishing that a new element had, indeed, been produced.

The Berkeley group developed the α–α correlation technique in which the unknown element's α-decay is correlated with that of the α-decay of known daughter and/or granddaughters. In some cases, the daughters can be identified by radiochemical separations. The measurement of the characteristic X-rays of a new element is another definitive method. It was used at Oak Ridge National Laboratory by Bemis et al. (1973, 1977) to confirm the Berkeley group's discoveries of elements 104 and 105. However, this method requires detection of relatively "large" numbers of events in order to obtain statistically significant measurements of the X-ray energies. In contrast to these methods, the Dubna group relied primarily on the detection of spontaneous fission (SF) decay. This is an extremely sensitive technique, but detection of only the fragments from the fission process makes it extremely difficult to deduce the identity of the fissioning species, especially based on detection of only a few events.

Indirect methods such as half-life systematics, excitation functions for the production reactions, and cross bombardments have been used to reinforce this information. In order to positively identify the atomic number of a spontaneously fissioning nuclide from detection of the fragments, the atomic numbers of both *primary* fragments from the same SF event must be determined in coincidence and added together to determine the Z of the new, unknown fissioning nuclide. Detection of only SF decay has resulted in much controversy concerning discovery and identification of the transactinide elements.

16.1.2. Naming of Elements 104 Through 109

After the reported discoveries of elements 107 through 109 between 1981 and 1984 (Münzenberg et al., 1981, 1982, 1984) by researchers using the Separator for Heavy Ion Reaction Products at the Gessellschaft für Schwerionenforschung mbH (GSI) in Darmstadt, Germany, the IUPAP in 1986 decided to appoint a Transfermium Working Group (TWG) to examine the priority of discovery of elements 101 through 109 so that official names could be proposed and approved for these elements. The IUPAC asked that they also be represented on the Committee because naming of the chemical elements was clearly within their historical jurisdiction. The study consisted of two phases: (1) establishment of criteria for discovery of new chemical elements and (2) application of these criteria in practice to the transfermium elements.

Phase 1 of the committee's report was accepted by both the IUPAP and the IUPAC and published by Barber et al. (1991) in the IUPAC journal in late 1991. The judgmental phase was quickly approved by the IUPAC Bureau in August 1991 and the IUPAP Council in September 1991, and both Phase 1 and Phase 2 conclusions were published in the IUPAP journal in 1992 (Barber et al., 1992).

Because the U.S. groups were not given an opportunity to view and comment on the accuracy of any draft reports from the TWG after their visit to Berkeley in 1998 and their subsequent visit to Dubna in 1999, they were totally taken aback by this rapid publication of the TWG conclusions without the "iterative" process that they had understood would take place with the involved groups at LBL in the United States, GSI in Darmstadt, Germany, and JINR in Dubna, USSR. In an attempt to counteract this criticism, responses from the Berkeley, Dubna, and GSI groups were invited and published in the IUPAC journal immediately following the TWG "Discovery" article of 1993 (Barber et al., 1993). The IUPAC then stated that this TWG report could now be considered by the CNIC, which had the responsibility for recommending names. However, it was pointed out that the TWG report had not been subjected to the external and internal review required by the IUPAC prior to publication.

Another long period of controversy ensued in which the CNIC at one point unanimously declared that *an element should not be named after a living person.* The name "seaborgium" for element 106 had been proposed by its undisputed discoverers at the March 1994 National Meeting of the American Chemical Society (ACS) and had been quickly accepted by the ACS Committee on Nomenclature and approved by the Board of Directors in November 1994. But at that time Glenn T. Seaborg was obviously alive and the CNIC of the IUPAC declared that the name "seaborgium" could not be used. Furthermore, they decided that the names for elements 104–109 would be chosen from among those suggested by the three groups involved, but not necessarily for the elements for which they had been originally suggested! They proposed the following names: 104, dubnium; 105, joliotium; 106, rutherfordium; 107, bohrium; 108, hahnium; 109, meitnerium. This resulted in complete confusion and the decoupling of the names from those

TABLE 16.1. CNIC/IUPAC compromise recommendation for names of transfermium elements. Approved by IUPAC, August 30, 1997, Geneva, Switzerland

Element	Name	Symbol	Year of discovery
101	Mendelevium	Md	1955
102	Nobelium	No	1958
103	Lawrencium	Lr	1961
	Transactinides		
104	Rutherfordium	Rf	1969
105	Dubnium (Hahnium)[a]	Db (Ha)[a]	1970
106	Seaborgium	Sg	1974
107	Bohrium	Bh	1981
108	Hassium	Hs	1984
109	Meitnerium	Mt	1982

[a] Many publications of chemical studies prior to 1997 use hahnium (Ha) for element 105.

suggested by the discoverers for the elements they claimed to have discovered and which had been in common use.

A worldwide storm of protest and criticism resulted and the IUPAC convened a series of meetings to try to agree upon a compromise set of names. The claims to discovery and the subsequent naming controversies surrounding elements 104–109 are discussed in detail in Chapters 9, 10, and 13 of Hoffman et al. (2000). Finally, after several unacceptable naming proposals, the IUPAC backed down on their edict disqualifying living persons, and the so-called compromise list of names and symbols shown in Table 16.1, including "seaborgium, Sg" for element 106, was approved in August 1997 (CNIC, 1997). This was only a year before Glenn Seaborg suffered the stroke and fall at the 1998 Boston ACS meeting, which ultimately resulted in his death on February 25, 1999.

16.1.3. Elements Beyond Mt

Discoveries of element 110 have been reported by groups at LBNL (Ghiorso et al., 1995a,b), GSI (Hofmann et al., 1995a), and a combined Dubna/Lawrence Livermore National Laboratory (LLNL) group working at Dubna. Discovery of elements 111 and 112 was reported by the GSI group in 1995 and 1996 (Hofmann et al., 1995b, 1996) and in 2002 (Hofmann et al., 2002) reported confirmatory experiments. A procedure for naming of new elements was then proposed by the IUPAC to try to avoid future naming controversies and the use of different names by competing research groups (Koppenol, 2002). A joint working party of the IUPAC/IUPAP (Karol et al., 2001) examined the various claims to discovery and gave credit for discovery of element 110 to the GSI group and invited them to propose a name. They proposed the name Darmstadtium with symbol Ds for element 110 after the place in Germany, where the discovery experiments were conducted. The IUPAC Commission on the Nomenclature of Inorganic Chemistry (CNIC) considered the proposal and recommended to the IUPAC Bureau that it be accepted (Corish and Rosenblatt, 2003). It was officially approved by the IUPAC Council in Ottawa, Canada, in August 2003.

In mid-2003, the joint working party assigned credit for discovery of element 111 to the GSI group and asked them to propose a name (Karol et al., 2003), but the evidence for discovery of element 112 was still deemed insufficient. The name roentgenium, symbol Rg, in honor of Wilhelm Conrad Roentgen, who discovered X-rays in 1895, was proposed for element 111 by the discovery group. In May 2004 a provisional recommendation for approval of the proposal (Corish and Rosenblatt, 2004) was sent to the IUPAC Bureau and Council, and it was approved in November 2004.

Evidence for elements 112 through 116 has been published between 1999 and early 2004 by a Dubna/LLNL group and a Dubna/international group (Oganessian et al., 1999a,b,c, 2000a,b, 2002, 2004a,b,c; Lougheed et al., 2000; Oganessian, 2001, 2002), who conducted experiments at Dubna. These reports have not yet been confirmed by other groups, and the elements are shown in *italics* in the periodic table given in Fig. 16.1. Positive mass and atomic number assignments are extremely difficult because the reported decay chains for these elements do not connect to previously known nuclides.

An LBNL group reported discovery of element 118 in 1999 (Ninov et al., 1999), but the claim was retracted when the original discovery could not be confirmed (Ninov et al., 2002) and Gregorich et al. (2002) have set limits on the production cross section of the originally reported 118 decay chain of less than a picobarn. The isotopes of Rf through Mt and of elements 110 through 116 reported as of early 2003 are shown in Figs. 16.2 and 16.3, respectively.

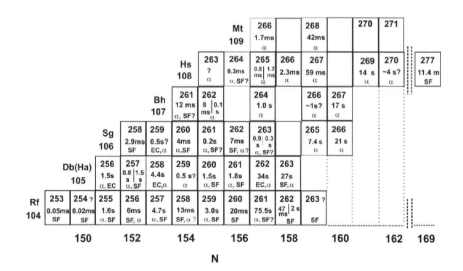

FIGURE 16.2. Isotopes of Rf through Mt (2003).

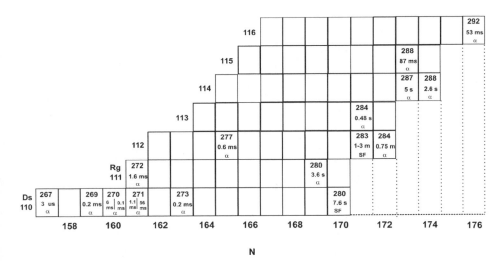

FIGURE 16.3. Isotopes of elements 110 through 116 (2003).

16.1.4. Challenges Involved in Studies of Chemical Properties of the Transactinides

As mentioned earlier, the ever shorter half-lives and decreasing production rates of the transactinides, the presence of a plethora of unwanted reaction products, the necessity for production at accelerators with high-intensity beams, and the need for special radiochemistry laboratories and detection facilities have posed formidable challenges to studies of their chemical properties. These demands have stimulated the development of new radioanalytical methods suitable for use in "atom-at-a-time" studies in which the production and detection rates range from a few atoms per minute for Rf to only an atom per week for Hs. The isotope to be used must have unique decay characteristics if it is to be positively identified as belonging to the element whose chemistry is being studied on an atom-at-a-time basis. The kinetics of the chemical procedures to be used must be fast enough to be accomplished in times comparable to the half-lives of the isotopes involved and must give the same results for one atom as for a large number of atoms. For example, precipitation reactions that depend on exceeding a given solubility product are obviously not suitable. Chromatographic methods in which a single atom undergoes many identical chemical interactions, such as in two-phase extraction systems with rapid kinetics that reach equilibrium quickly, have been shown to be valid.

Theoretical studies (Adloff and Guillaumont, 1993; Guillaumont et al., 1989, 1991) have shown that it is valid to combine the results from many identical experiments in which only a single atom was present to obtain information about chemical behavior although it may be necessary to repeat the identical experiment hundreds of times to obtain statistically significant results. Thus, computer-controlled automated systems become especially attractive. Although they are not necessarily

TABLE 16.2. Isotopes used in first chemical studies of elements Rf through Hs

Isotope (half-life)	Production reaction	Cross section (nb)	Year
^{261}Rf (75 s)	^{248}Cm + ^{18}O → ^{261}Rf + 5 n	~5	1970[a]
^{262}Ha (35 s)	^{249}Bk + ^{18}O → ^{262}Ha + 5 n	~6	1988[b]
^{265,266}Sg (21 s,7 s)	^{248}Cm + ^{22}O → ^{265,6}Sg + 5 n, 4 n	~0.3	1997[c]
^{267}Bh (17 s)	^{249}Bk + ^{22}Ne → ^{267}Bh + 4 n	~0.06	2000[d]
^{269,270}Hs (2–7 s, 14 s)	^{248}Cm + ^{26}Mg → ^{269,270}Hs + 5 n, 4 n	~0.005	2002[e]

[a] Silva (1970); [b] Gregorich (1988); [c] Schädel (1997a); [d] Eichler (2000); [e] Düllmann (2002a,b); Kirbach (2002).

faster than manual operations, they usually give more reproducible results and are nearly essential for around-the-clock experiments lasting weeks at a time.

To avoid dissolution of the highly radioactive and rare targets listed in Table 16.2 that must often be used to produce the isotopes of interest, foils sometimes are placed directly behind the target to catch the activities recoiling from the target after irradiation and then they are processed. Now it is more common to attach the recoiling activities to aerosols in a flowing stream of inert carrier gas such as He in which they can be rapidly transported outside the irradiation site within the accelerator shielding to the site of the chemical system to be used.

16.2. Radioanalytical Techniques

16.2.1. Gas-Phase Chemistry

Both solution and gas-phase studies of the chemical properties of the transactinides have been conducted. Gas-phase studies are especially useful for short-lived isotopes because they avoid the lengthy process of evaporating the liquid samples produced in most aqueous chemistry procedures prior to detection of SF and α-decay with solid-state detectors. Both *gradient* gas chromatographic (GGC) and *isothermal* gas chromatographic (IGC) methods have been used. In early GGC studies pioneered by Zvara et al. (1972, 1974, 1976), a negative temperature gradient was established along a chromatographic column (usually quartz) through which a gas stream containing volatile species of the isotopes of interest was conducted. These species deposited on the surface of the column according to their volatilities and later the deposition zones and temperatures were determined from fission tracks registered in detectors positioned along the column. The deposition temperatures were then correlated with the adsorption enthalpy. The advantage is that the process can be applied to half-lives as short as a few seconds.

The disadvantages of these early GGC techniques are that the deposition positions are determined only after the experiment is finished and only SF is detected. Real-time observation of the nuclear decay is not possible. Therefore, the half-lives cannot be determined and since it is necessary to correct for the half-lives of the different isotopes involved, interpretation of the results is extremely difficult.

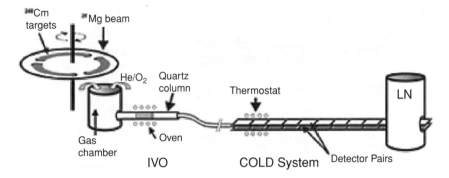

FIGURE 16.4. Schematic of IVO-COLD system for study of gas-phase properties of HsO_4 and lighter homologs. Adapted from Düllmann et al. (2002b).

There also is the usual problem of positively determining the atomic number of the fissioning nuclide whose chemistry was being studied.

Recently, these problems were solved in a very imaginative manner by forming the chromatographic column from opposing (Si) photodiode detectors upon which the volatile radioactive species were directly deposited. In this way, both the radiations (α and SF), and their deposition positions as a function of the temperature gradient along the Si column, were determined simultaneously and continuously recorded and stored via a computer system. A gradient cryogenic version of this system, the Cryo On-Line Detector (COLD), was used to perform the first successful chemical studies of Hs (108) with the α-emitting isotopes ^{269,270}Hs, which were identified by their decay to known α-emitting daughters. The COLD system was used together with the In situ Volatilization and On-line (IVO) detection system to study the volatile oxides of Hs and Os. A schematic diagram of the IVO-COLD system (Düllmann et al., 2002a,b) is shown in Fig. 16.4. The study indicated that Hs formed a volatile oxide similar to that of the tetroxide of Os, its lighter group 8 homolog.

Online *isothermal* chromatographic systems, such as the On-Line Gas Analyzer (OLGA) developed by Gäggeler and coworkers (Gäggeler, 1994) and the Heavy Element Volatility Instrument (HEVI) developed by Kadkhodayan et al. (1996), have been used to study volatile halides and oxyhalides of Rf, Db (Ha), Sg, and Bh. The α-emitting isotopes shown in Table 16.2 were used in these studies. In these systems, the entire length of the chromatographic column (usually quartz) is kept at a constant temperature and the volatile species pass through the column in a carrier gas such as He and undergo numerous sorption/desorption steps. The yield through the column is determined by measuring the α-radiations from the exiting species that are either reattached to aerosols and transported to a detection device or condensed directly on a detector system.

Experiments are conducted at a series of isothermal temperatures to study the chemical yields of the volatile species as a function of temperature. Retention time is indicative of the volatility at the given isothermal temperature. It is equal

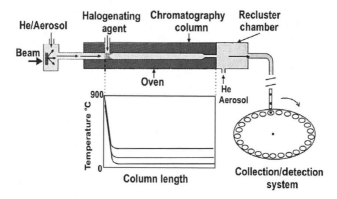

FIGURE 16.5. Schematic diagram of the HEVI system.

to the half-life of the isotope at the temperature at which 50% of the plateau or steady-state yield is obtained. This $T_{50\%}$ retention time can then be used as a relative measure of volatility. A Monte Carlo program taking account of all the experimental parameters, such as gas flow rate, length of column, and the half-lives of the isotopes in the volatile species, is used to deduce the adsorption enthalpies for the volatile species. A schematic diagram of the HEVI system, used for isothermal gas-phase studies, is shown in Fig. 16.5. The recoiling products from the reaction are attached to either KCl, KBr, or MoO_3 aerosols, and transported in He gas to the entrance to HEVI where they are deposited on a quartz wool plug and halogenated at 900°C. The volatile products are transported through the quartz column in flowing He gas and are again attached to aerosols in the recluster chamber and transported via another gas-jet to the Merry Go-around (MG) system. They are then deposited on thin polypropylene films placed around the periphery of its horizontal wheel, which is rotated to position the foils successively between pairs of surface barrier detectors for α and SF spectroscopy.

Photos of HEVI, OLGA, and the MG rotating wheel and detection system are shown in Fig. 16.6(a)–(c); parts (d) and (e) of Fig. 16.6 will be discussed in the following section.

16.2.2. Solution Chemistry

Examples of automated computer-controlled systems developed for studies of solution chemistry of the actinides and transactinides are the Automated Rapid Chemistry Apparatus (ARCA) and the microcentrifuge system SISAK (Special Isotopes Studied by the AKUFE technique). They are quite different in that although ARCA can be used to perform rapid, repeated, high-pressure liquid chromatography column experiments on the seconds-time scale to determine distribution coefficients, the collected liquid samples must be dried prior to the measurement of α-particle or SF decay with high-resolution solid-state detectors that limits detection to nuclides with half-lives of about 30 s or longer. However, with suitably designed separation

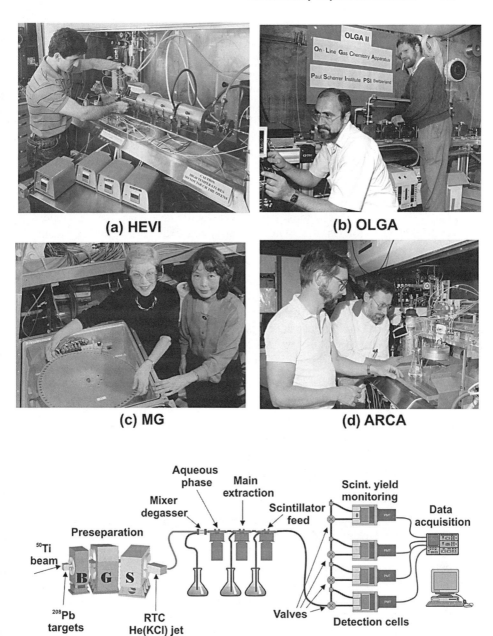

(a) HEVI

(b) OLGA

(c) MG

(d) ARCA

(e) SISAK

FIGURE 16.6. (a) B. Kadkhodayan with HEVI (1992); (b) J. Kovacs and H. W. Gäggeler with OLGA (1988); (c) D. C. Hoffman and D. M. Lee with MG rotating wheel system; (d) J. V. Kratz and M. Schädel with Automated Rapid Chemistry Apparatus (1988); (e) Schematic diagram of a typical SISAK liquid–liquid extraction configuration with Berkeley Gas-filled Separator as a preseparator.

procedures, longer lived daughter products can be detected and used to deduce the properties of the parent element. The ARCA shown in 16.6(d) has been used successfully in separations of Rf through Sg. Several reviews of the results have been published (Hoffman, 1994; Schädel, 1995; Hoffman and Lee, 1999; Kratz, 1999a; Kratz, 1999b).

The SISAK system, which had previously been used to perform liquid–liquid extraction studies of γ-emitting nuclides with half-lives as short as a second, was adapted for use with α-emitters by coupling it to a flowing liquid scintillation system. This provides continuous separation and measurement of α–α correlations and detection of SF decay of nuclides with half-lives of only a few seconds. This permits chemical studies of short-lived nuclides but the energy resolution is not as good as with solid-state detectors. Nevertheless, a recent experiment by Omtvedt et al. (2002) showed that rapid preseparation in the Berkeley Gas-filled Separator (BGS) achieved sufficient decontamination from the extremely high background of unwanted activities. Detailed information was obtained about the chemical properties of Rf for 4.7-s ^{257}Rf produced in the ^{208}Pb(^{50}Ti,n) reaction. Similarly, the chemistry of Db (Ha) can be studied with 4.4-s ^{258}Db and still heavier elements can be investigated although the production rates are steadily decreasing. Another limitation is the requirement to choose extraction systems that have kinetics that are rapid enough that equilibrium can be attained in the short extraction contact times associated with the SISAK system. A schematic diagram of the SISAK system as configured for these experiments is shown in Fig. 16.6(e).

16.2.3. Relevance and Applications to Other Fields

These exotic, frontier studies of the heaviest elements attract many undergraduate and graduate students to nuclear and radiochemistry. Research in this field provides excellent education and training for future contributions and careers in radioanalytical chemistry and a wide variety of other applied areas and research fields. Graduates find employment at U.S. national, federal, and state laboratories, regulatory agencies, private industry, and universities.

The rapid, high-yield, computer-controlled automated techniques that have been developed for separation, detection, data storage, and analysis of hundreds of identical experiments of short-lived species, using both gas-phase and solution chemistry, are described in this chapter. These systems can certainly be adapted for use in radioanalytical chemistry and a wide variety of other applied fields, as well as other research areas, when large numbers of samples must be analyzed with reproducible results for long-lived as well as short-lived species (see also Chapter 15). Applications in the following areas might be envisioned: (1) environmental studies including long-term monitoring, modeling, and prediction of the behavior of actinides and other radionuclides in the environment; (2) studies of nuclear waste isolation and remediation of both radioactive and other toxic sites; (3) treatment, processing, and minimization of wastes; (4) ultrasensitive analyses and instrumentation; (5) computer-controlled automated remote processing for hazardous materials; (6) studies of radionuclide chemistry in nuclear reactors and in the nuclear fuel cycle; (7) surveillance of clandestine nuclear activities and evaluation of terrorist threats using gas-analysis techniques; (8) radioisotope production and separation; (9) radiopharmaceutical synthesis, and detection systems for both diagnostic and therapeutic nuclear medicine procedures; (10) biochemical and agricultural research; and (11) geochemical dating studies.

16.3. Results of Experimental Studies of Transactinides

Chemical studies of the transactinides from Rf (104) through Hs (108) have now been performed although Bh (107) and Hs have been studied only in the gas phase. Studies of Mt (109) await discovery of longer lived isotopes than 40-ms ^{268}Mt, currently the longest known isotope of Mt. Predictions of the half-lives of ^{270,271}Mt range from tenths of a second to a few seconds based on Smolańczuk's calculations (Smolańczuk, 1997). Initial attempts to study element 112 have been unsuccessful so far. Placement of the transactinides in the periodic table has been based on comparison of the behavior of the individual transactinide with that of its lighter homologs in a given group in the periodic table. Several papers (Kratz, 1999a,b; Pershina and Hoffman, 2003) that give detailed reviews of the experimental and theoretical studies of the transactinides are available.

16.3.1. Chemistry of Elements 104(Rf) and 105(Db/Ha)

The first chemical experiments on the transactinides were designed to determine their placement in the periodic table and whether the 5f actinide series actually ended with Lr as predicted by Glenn Seaborg (Seaborg, 1945). Then elements 104 and 105 should have different chemistry from the actinides, and as members of the 6d transition series should be placed as the heaviest members of groups 4 and 5 and have similar properties.

Soon after the discovery of element 104, Silva et al. (1970) studied its solution chemistry for the known isotope 75-s ^{261}Rf. Positive identification was made by measuring its known half-life and decay characteristics. Several hundred elutions with α-hydroxyisobutyrate solutions from cation exchange resin columns were performed to compare the behavior of ^{261}Rf with those of No^{2+}, trivalent actinides, Hf^{4+}, and Zr^{4+}. The behavior of Rf was similar to those of Zr and Hf, which did not sorb on the column and entirely different from No^{2+} and the trivalent actinides that did sorb on the column at pH 4.0. These studies showed unequivocally that Rf should be placed under Zr and Hf in the periodic table.

The first computer-controlled automated system for performing very rapid solution chemistry experiments on an atom-at-a-time basis was used in later pioneering experiments (Hulet et al., 1980). Results showed that the anionic–chloride complexes of Rf were clearly similar to Hf and much stronger than those of the trivalent actinides, again confirming the position of Rf in group 4 of the periodic table.

In 1972, Zvara et al. (1972), in the earliest gas-phase chemical studies of element 104, reported that its tetrachloride volatility was similar to $HfCl_4$ and greater than

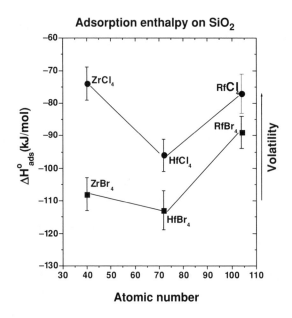

FIGURE 16.7. Adsorption enthalpy values and relative volatilities on SiO₂ for Zr, Hf, and Rf chlorides and bromides.

actinide and Sc chlorides. They concluded that element 104 should be placed in group 4 of the periodic table. The results were not considered conclusive (Kratz, 1999b) because only SF events were measured, making it impossible to identify positively the element being studied.

After these initial studies, a rather long time elapsed before additional experiments were conducted in the late 1980s. These were motivated by theoretical calculations, suggesting that relativistic effects might alter the chemical properties of Rf and Ha from those expected by simple extrapolation of properties within a given group in the periodic table. Subsequent studies of the gas-phase chemistry of Rf, Zr, and Hf chlorides (Kadkhodayan et al., 1996; Türler et al., 1996) and bromides (Sylwester et al., 2000) were conducted with OLGA and HEVI at the 88-Inch Cyclotron at LBNL. Contrary to expectations based on simple extrapolation of group 4 volatilities, these studies showed (Türler et al., 1996) that the Rf halides were more volatile than those of Hf. In agreement with relativistic calculations (Pershina and Fricke, 1999), the bromides were all found to be less volatile than their respective halides as shown in Fig. 16.7

Many additional studies to compare the solution chemistry of Rf with Zr, Hf, Th(IV), and Pu(IV) have been conducted and include both manual column and extraction studies, and automated studies using ARCA. Extractions from aqueous solutions into triisooctylamine, tributylphosphate, and thenoyltrifluoroacetone have been studied in considerable detail. They have shown that, although Rf behaves generally like a group 4 element, its behavior varies with different complexing agents.

Zvara et al. (1974, 1976) reported early gradient thermochromatographic studies comparing the volatilities of the chlorides and bromides of element 105, using 1.8-s ^{261}105 with Nb and Hf. They concluded that the volatility of the 105 bromide was less than that of the bromide of Nb and about the same as Hf. Again, only SF events were detected so it is not certain that the properties of element 105 were actually measured.

Studies of the chemistry of element 105 languished until the very first studies of its behavior in solution were reported by Gregorich et al. (1988). These were undertaken simply to determine its most stable oxidation state in aqueous solution. It had been postulated (Keller, 1984) that because of relativistic effects Ha might have a $7s^2 6d7p^2$ valence configuration that could result in a 3+ oxidation state if the $7s^2$ electrons were sufficiently stabilized by relativistic effects. But the $6d^3 7s^2$ was expected by analogy to Ta, which has a $5d^3 6s^2$ configuration and a most stable state in aqueous solution of 5+. The reaction shown in Table 16.2 was used to make 35-s ^{262}Ha. It was identified by measurements of the energy and time distribution of the α-decay and of time-correlated pairs of α-decays from ^{262}105 and its 4-s ^{258}Lr daughter. The recoiling products from the production reaction were attached to aerosols and transferred via He jet to a collection site in a hood located outside the radiation area where manual "glass" chemistry was performed with only a few microliters of solutions.

The sorption of Ha on glass cover slips was compared with that of tracers of the group 5 elements Nb and Ta produced online under similar conditions. These group 5 elements are known to sorb on glass surfaces after fuming with nitric acid while the group 4 elements and actinides do not. About 800 such manual extractions taking 50 s each were conducted and a total of 26 α-decays and 26 SFs were detected. Element 105 was found to sorb on the glass as did Nb(V) and Ta(V) while Zr, Hf, and the trivalent actinides did not, thus demonstrating that element 105 should be placed in group 5 of the periodic table. However, in extractions into methylisobutyl ketone from mixed HNO_3/HF solutions, Ta extracted but Ha remained behind with Nb, again indicating that details of complexing behavior cannot be predicted on the basis of simple extrapolation from group trends.

These experiments provided the impetus for future explorations of the complex behavior of element 105, using ARCA II, an improved version of ARCA, to carry out the many thousands of experiments required to get statistically significant results. A collaboration of groups from GSI/Mainz, PSI/Bern, and LBL/UCBerkeley subsequently carried out many experiments at LBL and GSI to study the chemistry of element 105 in more detail. Guidance from theoretical chemists (Pershina, 1998a,b; Pershina and Bastug, 1999) has been invaluable in designing appropriate experimental conditions for elucidating the influence of relativistic effects.

An international group (Gäggeler et al., 1992; Türler, 1992) used OLGA II and 35-s ^{262}Db to perform the first online isothermal gas chromatography of the bromides of element 105. The volatile species were deposited on a moving tape and transported in front of six large area passivated implanted planar Si detectors for measurement of α-particles and SFs. The adsorption enthalpies were nearly the same for Nb and Ta, but Db appeared to be significantly less volatile and it was postulated that an oxybromide might have formed from traces of oxygen present. It is predicted to be less volatile than the pentabromide (Pershina et al., 1992). Later experiments in which the partial pressure of oxygen was varied showed that both volatile and less volatile species were formed and additional measurements are needed to try to produce the pure pentabromide.

16.3.2. Chemistry of Element 106 (Sg)

The first discovered isotope (Ghiorso et al., 1974), 0.9-s ^{263}Sg, remained the longest lived known isotope of Sg for 20 years. In 1994, a Dubna-LLNL collaboration (Lazarev et al., 1994; Lougheed et al., 1994) reported production of the longer lived isotopes ^{266}Sg and ^{265}Sg (Fig. 16.2) with maximum cross sections of approximately 0.08 and 0.26 nb, respectively. Sg is expected to be the heaviest member of group 6 in the periodic table. Considerations based on both relativistic molecular–orbital calculations (Pershina and Fricke, 1996) and empirical relationships (Eichler et al., 1999) predict that SgO_2Cl_2 should be the most stable of the Sg oxychlorides and that the order of volatility of the dioxydichlorides of the group 6 elements should be Mo > W > Sg.

Attempts were made at Dubna in 1996 (Timokhin et al., 1996; Yakushev et al., 1996) to use the 0.9-s ^{263}Sg produced in the ^{249}Cf (^{18}O, 4n) reaction to study the volatility of the oxychloride in experiments with the online quartz gradient thermochromatographic column (TC) technique discussed earlier. They attributed 29 fission tracks found in a temperature zone of 150–250°C (close to the ^{166}W deposition temperature) to decay of the very small SF branch of ^{263}Sg, and claimed that this was the first chemical identification of Sg. However, *no positive identification* based on the much more abundant, known α-decay chain of ^{263}Sg, was made and the detected fissions could have originated from other elements such as 104 and 105. Kratz (1999b) reviewed these reports in detail and concluded that the data did not support these claims to the first chemical identification of Sg.

The behavior of the oxychlorides of short-lived isotopes of Mo and W produced with a reactive gas mixture of O_2, Cl_2, and $SOCl_2$ was investigated using OLGA III (Gärtner et al., 1997) in preparation for studies of Sg. The adsorption enthalpies of –90 kJ mol^{-1} and –100 kJ mol^{-1} deduced for Mo and W from the isothermal yield curves were in agreement with the predictions. Subsequently, the first chemical experiments on Sg were performed (Schädel et al., 1997a) with OLGA III by an international collaboration in 1995 and 1996 at the UNILAC at GSI using ^{266,265}Sg produced as shown in Table 16.2. The volatile species exiting OLGA III were again attached to aerosols and transported and deposited sequentially on thin polypropylene foils placed in 64-collection positions around the periphery of a rotating wheel system. It was stepped every 10 s to place the collected activity between pairs of passivated ion-implanted planar silicon (PIPS) detectors for measurement of α- and SF activities. Times, energies, and positions of all events were registered in list mode with a computer system. A mother–daughter stepping mode was used to avoid interference from the ubiquitous ^{212}Po^m 8.8–MeV α-activity. Three events attributable to ^{265}Sg (7 s) were identified: two in daughter mode and one triple correlation. One α–α correlation between ^{266}Sg (21 s) and its ^{262}Rf daughter was detected. This gas-phase study of SgO$_2$Cl$_2$ constituted the first chemical study in which the detected Sg activities were positively identified as belonging to Sg.

In a later experiment (Türler et al., 1999), the isothermal temperature was varied to obtain the adsorption enthalpy for Sg. The yield of short-lived W isotopes from reactions on Gd incorporated in the target was also measured at the same time. An adsorption enthalphy of –96 $\pm$ 1 kJ mol^{-1} was obtained for the dioxydichloride of W in agreement with the previous measurement. Based on 11 events attributable to Sg, an adsorption enthalpy of –100 $\pm$ 4 kJ mol^{-1} or 96 $\pm$ 1 kJ mol^{-1} was obtained depending on the Sg half-life used in the Monte Carlo calculations. Thus, the experimental results confirmed the theoretically predicted volatility sequence of Mo > W > Sg. From an analysis of the decay properties of the correlated decay chains detected in these experiments, Türler et al. (1998) were able to recommend better half-lives and cross sections of 7.4 + 3.3/ − 2.7 s and approximately 0.24 nb for ^{265}Sg and of 21 + 20/ − 12 s and approximately 0.025 nb for ^{266}Sg for ^{22}Ne beam energies between 120 and 124 MeV.

Schädel et al. (1997a,b) also reported the first successful studies of the aqueous chemistry of Sg in the same series of collaborative experiments conducted at the UNILAC at GSI. In order to favor production of the longer lived ^{266}Sg formed by the 4n out reaction, 121-MeV ^{22}Ne projectiles were used. A helium gas-transfer system was used to transport the activities from the irradiation site to the ARCA II system, where 3,900 collection and elution cycles were performed. A solution of 0.1 M HNO$_3$/5 $\times$ 10^{-4} M HF was used to elute the activity sorbed initially on columns (1.8 mm i.d. $\times$ 8 mm long) filled with the cation exchange resin Aminex A6. This eluant was chosen because previous online tracer experiments had shown that the formation of neutral and anionic complexes with F$^-$ ions is characteristic of the group 4, 5, and 6 elements, but that there are distinct differences in behavior between the groups. The procedure was designed to provide rapid decontamination from the interfering high-energy Bi and Po α-activities and trivalent actinides also formed in the irradiation, and efficient separation of the Rf and No daughter activities from their Sg precursors. The mean time for separation of Rf and No from Sg was only approximately 5 s, but α-particle and SF measurements were not begun until approximately 38 s after the end of collection because of the time required to evaporate the eluted samples and manually transport them to the PIPS detectors. The energies, times, and detector positions were recorded in list mode on magnetic disk and tape for later data analysis. Collection times of 45 s were used for most of the cycles.

Three correlated α–α-events were identified as belonging to the ^{261}Rf (78 s) $\rightarrow$ ^{257}No (26 s) $\rightarrow$ decay sequence, thus indicating that the 7-s ^{265}Sg parent had been present in the chemically separated Sg (W) fractions and decayed to Rf and No during the time interval before measurements began. No evidence for the longer lived ^{266}Sg was found, presumably because of its much smaller production cross section. Later investigations were performed (Strub et al., 2000) to verify that Rf, Th, Hf, and Zr would not elute from the cation columns under these conditions, and it was concluded that ^{261}Rf could have been only in the Sg fraction as a result of the decay of ^{265}Sg and that Sg, like its lighter group 6 homologs Mo and W, formed neutral species of the type MO_2F_2.

Another similar series of more than 4,500 cycles was conducted (Schädel et al., 1998) under the same conditions using 0.1 M HNO_3 but without any *fluoride ions* to determine whether Sg would then behave in a manner different from that of W as suggested by theoretical examination of the hydrolysis of group-6 elements and U(VI) (Pershina and Kratz, 2001). Indeed, the Sg activity remained on the cation-exchange resin columns while W was eluted with 0.1 M HNO_3. This behavior was tentatively attributed to the weaker tendency of Sg to hydrolyze in dilute HNO_3 so that it remained as a 2+ or 1+ complex while hydrolysis of W (and Mo) resulted in the neutral species $MO_2(OH)_2$. In the earlier experiments containing fluoride ion, Sg may have been eluted from the cation-exchange columns as neutral or anionic fluoride complexes of the type SgO_2F_2 or $[SgO_2F_3]^-$ rather than as $[SgO_4]^{2-}$. Theoretical calculations of complex formation in HF solutions (Pershina and Hoffman, 2003) indicate that complex formation competes with hydrolysis in aqueous solutions and the dependence on pH and HF concentration is extremely complicated and reversals in trends may occur. The theoretical predictions provide experimenters with valuable guidance for planning future experiments. These studies illustrate the fruitful synergism that can result from close interactions and iterations between theory and experiment.

16.3.3. Chemistry of Element 107 (Bh)

The discovery of 17-s ^{267}Bh (Wilk et al., 2000) produced via the ^{249}Bk(^{22}Ne, 4n) reaction provided an isotope long enough for the first chemical studies of Bh. Online IGC experiments were conducted with OLGA at the PSI cyclotron to compare the volatility of the oxychloride of Bh with those of Re and Tc, its expected homologs in group 7 of the periodic table. The recoiling reaction products were transported to OLGA on carbon particles suspended in flowing He gas. The volatile chlorides were produced by treatment with HCl and oxygen, and the species passing through the quartz chromatographic column were reattached to aerosols and transported to a rotating wheel system for measurement of α-particles and SFs. Over a period of about a month of irradiation time, six decay chains were attributed to the decay of ^{267}Bh from observation of its decay to known daughter activities.

A Monte Carlo program based on a microscopic model of the adsorption process was used to deduce an adsorption enthalpy of −75 kJ mol^{-1} with a 68% confidence interval of −66 to −81 kJ mol^{-1} for BhO_3Cl, predicted to be the most probable oxychloride under these conditions. The values for Tc and Re were −51 and −61 kJ mol^{-1}, respectively. These results were in agreement with the calculations (Pershina and Bastug, 2000) that predicted a volatility sequence for the trioxychlorides of Tc > Re > Bh. Thus Bh can properly be placed in group 7 of the periodic table.

16.3.4. Chemistry of Element 108 (Hs)

Chemical separation procedures for Hs were developed on the basis of its expected similarity to the group 8 elements Ru and Os. Fully relativistic density functional calculations have been performed for the tetroxides of these elements (Pershina et al., 2001). The calculated adsorption enthalpies for the tetroxides were -36.7 ± 1.5, -38 ± 1.5, and $-40.4 \pm 1.5 \, \text{kJ mol}^{-1}$ for Hs, Os, and Ru, respectively, giving the volatility sequence Hs $\geq$ Os > Ru.

Prior to experiments with Hs, a cryo-thermochromatographic separator (CTS) was developed with a channel formed by two facing rows of silicon PIN (Positive Implanted N-type silicon) diode α-particle detectors at Berkeley by Kirbach et al. (2002). They performed online studies of the tetroxides of α-emitting Os activities produced at the LBNL 88-in. cyclotron. It was operated with a negative temperature gradient ranging from 247 K at the entrance to 176 K at the exit of the channel. The volatile species condense on the surface of the detectors at a characteristic position along the channel. Temperatures are determined from thermocouples placed at intervals along the channel and from measured resistances of the PIN diodes. Data are stored continuously on disk or tape via a computer system for future offline data analysis. The radioactive nuclides in the volatile species are identified from their measured α-decay energies and half-lives.

The BGS was used as a preseparator to remove unwanted transfer reaction products and scattered beam particles. The separated Os ions were stopped in a gas mixture of 90% He and 10% O_2 and then transferred in a flowing He/O_2 mixture to a quartz tube heated to 1,200 K where OsO_4 was produced and then transported to the CTS. An adsorption enthalphy of $-40.2 \pm 1.5 \, \text{kJ mol}^{-1}$ was obtained for OsO_4 on the quartz (SiO_2) surface from Monte Carlo fits to the measured adsorption distributions.

The IVO-COLD system shown earlier in Fig. 16.4 was used in the Hs experiments that were conducted at the GSI/UNILAC. Three ^{248}Cm targets on a rotating wheel were used to optimize the production rate. The α-emitting isotopes ^{269,270}Hs were produced as shown in Table 16.2 and identified by measurement of their decay to known α-emitting daughters. The detection efficiency was about 77% for a single α-particle. The yield of short-lived Os isotopes was checked by online measurements before and after the end of the Hs experiments. The products recoiling from the nuclear reaction were thermalized and stopped in the recoil chamber of IVO containing He gas, while the high-energy beam particles passed on through to the beam stop and thus removed from the thermalized recoil products. The recoiling products were transported to a quartz wool plug where Hs and Os were oxidized to the tetroxides by treatment with oxygen at 600°C. They were then transported by the carrier gas to the COLD TC, which is formed by 12 pairs of opposing Si PIN photodiodes positioned so the gas flow is confined to the active detector surfaces coated with an outermost layer of Si_3N_4 (Düllmann et al., 2002b).

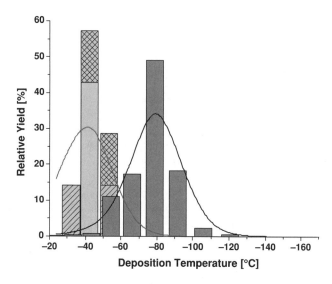

FIGURE 16.8. Merged thermochromatogram of HsO_4 and OsO_4. Adapted from Düllmann et al. (2002b).

A negative temperature gradient from −20°C to −170°C was established along the TC and measured at five positions.

Three α–α correlated decay chains detected in the 64-h experiment were assigned to the decay of ^{269}Hs. Two others possibly due to a new nuclide ^{270}Hs or an isomer of ^{269}Hs were also detected. The merged thermochromatograms for OsO_4 and HsO_4 are shown in Fig. 16.8. Indicated are the relative yields of HsO_4 and OsO_4 as a function of the deposition temperature. Measured values are represented by bars—HsO_4: light grey; OsO_4: dark grey. The distributions of the seven decay chains attributed to Hs isotopes are indicated by the pattern—^{269}Hs: blank; ^{270}Hs: hatched; Hs (isotope unknown): cross-hatched. For Os, the distribution of $1 \cdot 10^5$ events of $^{172}OsO_4$ is given. The maxima of the deposition distributions were evaluated as (-44 ± 6)°C for HsO_4 and (-82 ± 7)°C for OsO_4 where the uncertainties indicate the temperature range covered by the detector that registered the maximum of the deposition distribution. Solid lines represent results of a Monte Carlo simulation of the migration process of the species along the column with standard adsorption enthalpies of −46.0 kJ mol^{-1} for $^{269}HsO_4$ and −39.0 kJ mol^{-1} for $^{172}OsO_4$.

As noted above, the best Monte Carlo fits to the Os data gave an adsorption enthalpy of -39 ± 1 kJ mol^{-1} for OsO_4 on the Si_3N_4 surface in good agreement with previous measurements on SiO_2. Using only the three events assigned to ^{269}Hs and a half-life value of 11 s in the Monte Carlo analysis resulted in an adsorption enthalpy of -46 ± 2 kJ mol^{-1} for HsO_4. This is considerably lower than the calculated value of -36.7 ± 1.5 kJ mol^{-1} and suggests that it may be less volatile than OsO_4. Additional experiments to measure more ^{269,270}Hs decay chains and to clarify the nuclear properties are clearly needed. However, the study does show that Hs forms a volatile oxide similar to that of the tetroxide of Os, and should be placed in group 8 of the periodic table.

No studies of Bh or Hs in solution have yet been attempted, but they should be quite interesting and could furnish valuable information about most stable oxidation states in aqueous solution and redox and complexing reactions in different media. They might be expected to show oxidation states ranging from 3 to 7 for Bh and 1 to 8 for Hs, as do their homologs in groups 7 and 8. A summary of the chemical properties predicted for Bh and Hs was given in a review by Seaborg and Keller (1986).

16.4. Future

16.4.1. Prospects for Chemical Studies of Mt (109) Through Element 112

Very little attention has been paid to theoretical predictions of the chemical properties of Mt through element 120 since the 1986 summary of Seaborg and Keller. Based on the positions of Mt and element 112 in the periodic table, they should be noble metals like Pt and Au, and volatile hexafluorides and octafluorides might be produced and used in chemical separation procedures. Early relativistic molecular calculations (Waber and Averill, 1974; Rosen, 1998) suggested that $110F_6$ should be similar to PtF_6.

Because the maximum of relativistic effects in the ns shell in group 11 is at element 111, there has been considerable interest in the electronic structures of its compounds. Theoretical studies (Seth et al., 1996; Liu and van Wüllen, 1999) of the simplest molecule 111H show that the bonding is considerably increased due to relativistic contraction of the 7s orbital. Theoretical studies (Seth et al., 1998a,b) of the stability of higher oxidation states support earlier predictions that the 3+ and 5+ oxidation states will be more common for element 111 than they are for Au and that the 1+ oxidation state may be difficult to prepare.

Element 112 is the most interesting of the heaviest elements from the chemical point of view because the maxima of both the relativistic effects on the 7s shell in the entire 7th row and within group 12 occur at element 112. The strong relativistic contraction and stabilization of the 7s orbitals and its closed shell configuration should make it rather inert, and the predicted rather small interatomic interaction in the metallic state may even lead to high volatility as seen in the noble gases (Pitzer, 1975). A simple extrapolation of the sublimation enthalpies of its lighter group 12 homologs, Zn, Cd, and Hg, leads to a similar conclusion.

Recent relativistic molecular orbital calculations (Pershina et al., 2002) confirmed that element 112 can form rather strong metal–metal bonds 112M and deposit on some transition metals as 112M in which M = Cu, Pd, Ag, Pt or Au. The stability of the 2+ state is predicted to decrease from Hg to element 112, and $112F_2$ may tend to decompose but $112F_4$ may be stable. The species $112F_3^-$

and/or $112F_5^-$ may form in solution in an appropriate polar solvent, but they will probably quickly hydrolyze and bromides or iodides may be more stable and better for experimental studies of solution chemistry. A recent review of the results of theoretical studies of the properties of Mt through element 112 has been published (Pershina and Hoffman, 2003) and should be consulted for more detailed and up-to-date information and complete references.

Some preliminary chemical experiments on element 112 have been reported (Yakushev et al., 2001) using the spontaneously fissioning nuclide $^{283}112$ ($\sim$3 min) reportedly formed in the $^{238}U(^{48}Ca, 3n)$ reaction with a cross section of about 5 pb. The experiments were designed to test whether element 112 behaved more like Hg, which had been shown to deposit on Au- or Pd-coated silicon surface-barrier detectors or was a noble gas like Rn and stayed in the gas phase. Eight SF events were detected in the gas phase, which would indicate Rn-like behavior. However, the results cannot be considered definitive since it was not shown that the SFs belonged to element 112 nor has the original report of the approximately 3-min SF activity attributed to $^{283}112$ been confirmed. Additional experiments are planned, but use of an α-decaying isotope of 112, such as $^{284}112$ (10 to 70 s) reported previously (Oganessian, 2001) to be the daughter of $^{288}114$, may be required to ascertain whether or not element 112 was actually being observed.

Chemical studies of Mt depend on the discovery of longer lived isotopes than 42-ms ^{268}Mt, currently the longest known isotope of Mt, shown in Fig. 16.2. Longer lived isotopes might be expected around the deformed nuclear shell at 162 neutrons. The half-life of the 162-neutron nuclide ^{271}Mt can be estimated from interpolation of Smolańczuk's calculations (Smolańczuk, 1997) for even-proton, even-neutron nuclides to be a few seconds and should decay primarily by α-emission. However, ^{270}Mt with only 161 neutrons may be only tens of microseconds, too short for chemistry. Possible production reactions for ^{271,270}Mt include $^{238}U(^{37}Cl, 4n,5n)$ and $^{249}Bk(^{26}Mg, 4n,5n)$. The cross sections for these reactions may be a few pb and only tenths of pb, respectively. The BGS at LBNL is capable of separating and positively identifying such a new nuclide based on measurement of the known α-decay chain of its daughters even if the ^{271}Mt half-life is as short as tenths of seconds. Rotating, multiple targets of ^{238}U could be used to increase the production yields for the first reaction but use of multiple targets of the highly radioactive ^{249}Bk would be extremely difficult.

The half-lives of 7.5 s and 1 min reported for $^{280,281}110$ produced as granddaughters of the $^{288,289}114$ decay chains are long enough for chemical studies. However, the reports need to be confirmed and the yields via this decay route must be high enough to permit statistically significant atom-at-a-time studies. Currently, there appear to be no suitable reactions for direct production of these very neutron-rich isotopes of element 110.

Some isotopes of element 111 should have half-lives of seconds or more. Promising production reactions need to be investigated first using an online separator and detection system to measure half-lives and production cross sections before chemical studies can be contemplated.

16.4.2. SuperHeavy Elements

The experimental discovery that the elements Bh through element 112 decay primarily by α-emission, rather than SF as predicted earlier, sparked interest in renewing the quest for the long-sought island of stability originally predicted to be around atomic number 114 and neutron number 184. Although recent theoretical calculations (Smolańczuk, 2001a,b) predict that isotopes with half-lives of microseconds or longer will exist all along the route to the region of SuperHeavy Elements (SHEs), the half-lives predicted for this spherical SHE region have decreased dramatically since the 1970s (Hoffman and Lee, 2003). Smolańczuk has recently predicted that the doubly magic spherical SHE nuclide $^{298}114$ will decay predominantly by α-emission with a half-life of only about 12 min, but that $^{292}110$ will be more stable and α-decay with a half-life of about 50 years. Others have predicted that the strongest spherical shell effects might be at atomic numbers of 124 or 126 and neutron number of 184.

It is unclear how many more elements can exist, but it now appears probable that many more nuclides with half-lives long enough for chemical studies can exist. The question is not only what reactions to use to produce and positively identify them, but how to enhance their production rates so chemical studies will be feasible. New target arrangements to take advantage of higher beam intensities must be developed. Dedicated beam lines for use by international groups of scientists at suitable accelerators with preseparators or other special instrumentation prior to continuous automated, online systems for chemical studies might be envisioned.

If longer lived new elements are discovered, then methods for "stockpiling" them for offline chemical studies by producing them as by-products of other experiments might be devised. It is tantalizing to consider the possibility of extending chemical studies beyond Hs in order to fully explore the influence of relativistic effects and to further define the architecture of the periodic table—or perhaps even be forced to abandon the current form of the periodic table. Much ingenuity, dedication, and perseverance will be required to explore this exciting new frontier.

17
Mass Spectrometric Radionuclide Analyses

JOHN F. WACKER[1,2], GREGORY C. EIDEN[1], and SCOTT A. LEHN[1]

17.1. Introduction

Like most analytical instrumentation, the mass spectrometer (MS) initially found use as a research system. These complex laboratory-built research instruments were subsequently refined for commercial sale and widespread use, and are now available for routine sample assay.

In recent years, the MS has become a useful alternative to the radiation detector for measuring longer-lived radionuclides. It provides a second and completely independent method of measurement to confirm results; it also gives the analyst a choice, because some measurements are easier or more reliable by one method than the other. The MS has been used to measure radioactive atoms with half-lives greater than 10 years because the number of these atoms relative to their decay rate is proportional to the half life; for half lives greater than 10^9 y, even the conventional measurements of analytical chemistry are applicable.

Development of new MS systems focuses on the ability to measure smaller samples accurately, either for greater sensitivity or shorter-lived radionuclides. Advances in both construction and design have been applied, with the effect that the modern MS operator can measure ever-smaller numbers of atoms. A related development is the ability to individuate those samples more thoroughly by increasing MS resolution to decrease interference from isobars—atoms and molecules with the same atomic mass number but with minute differences in mass.

The distinction in sensitivity between measurements of radiation and atoms (as ions at a given mass-to-charge ratio by MS) can be demonstrated by comparing intensity in terms of the radioactive decay Eqs. (2.4) and (2.7) for a radionuclide with an atomic mass of 100 amu and a half-life of 10 y (3.16×10^8 s). If the sample under assay emits beta particles at the relatively low rate of 1×10^{-2} d/s, this corresponds to 4.6×10^6 atoms or, for a 100 amu radionuclide, 7.6×10^{-14} g. Alpha particles can be measured at an approximately 1,000-fold lower decay rate, for which the number of atoms and the mass correspondingly are 1,000-fold lower.

---

[1] Pacific Northwest National Laboratory, Richland, WA 99352
[2] To whom correspondence should be addressed. John F. Wacker, MS P8-01, 902 Battelle Blvd., Richland, WA 99354; email: john.wacker@pnl.gov

The radiation of this test isotope can be measured at these levels but the number of atoms (4.6×10^3) falls below the detection limits for quantification of most mass spectrometers.

The efforts in applying mass spectrometry to the measurement of radionuclides are devoted to achieving detection of small numbers of atoms. The crucial question is whether an isotope that has, say, 10^6 atoms in a gram of solid or liquid can be detected and quantified by MS in the presence of possibly 10^{21} atoms and molecules of other isotopes, including many with nearly the same mass. For some elements (e.g., Pu), detection limits below 10^6 atoms have been demonstrated for a variety of sample matrices (soil, water) using several mass spectrometric techniques (Beasley et al., 1998b; Oughton et al., 2004; Sahoo et al., 2002). Instrumental detection limits, in terms of counts (i.e., detected ions) per atom, have exceeded 5 counts per hundred atoms under optimum conditions.

This chapter provides an overview of mass spectrometer function and operation. It describes specific instrument types with demonstrated or potential application for measuring radionuclides and surveys the application of these instruments to radionuclide detection. Finally, it discusses the circumstances under which use of mass spectrometers is advantageous, the type of mass spectrometer used for each purpose, and the conditions of sample preparation, introduction and analysis. Its perspective is from a national laboratory active in environmental and non-proliferation monitoring. It emphasizes isotope ratio measurements, but mass spectrometric measurements also provide isotope mass information. Several recent books describe elemental and isotope ratio mass spectrometry in far greater detail than is presented here (Barshick et al., 2000; De Laeter, 2001; Montaser, 1998; Nelms, 2005; Platzner, 1997; Tuniz et al., 1998). High-resolution mass spectrometry forms the basis of the mass scale used for elemental and isotopic masses (Coplen, 2001), but this application of MS falls outside the scope of this chapter.

17.2. Instrumentation and Analytical Procedures

Mass spectrometers analyze gaseous ions by separating ions with differing mass-to-charge ratios (m/z; where z is the ion charge) and detecting these separated ions. The data produced are in the form of ion intensity at specified ion mass.

The first step in this analysis is the preparation of the sample. The point is to put the analyte into a form that is both compatible with the instrument and free of impurities. Next, a system is needed to introduce the sample into the instrument.

The mass spectrometer itself is composed of the following equipment, through which the sample is moved in order to achieve the analysis:

- an ion source to volatilize and ionize the analyte of interest
- a mass analyzer to perform the mass separation of the ions
- an ion detector to detect the separated ions
- a vacuum chamber with associated vacuum pumps and hardware to house instrumental components

Each of the instrumental components has electronic control for operation and data acquisition.

17.2.1. Sample Preparation and Insertion

Almost all types of samples require preparation to convert the analyte of interest into a form that is readily inserted into a mass spectrometer for isotopic assay. Although gaseous samples are injected directly into the source for the instrument, purification and separation with reactive metals (also called getters), gas chromatography, and cryogenic separation means may be required. A few relatively pure liquids and volatile solids can be vaporized directly, but most liquids and solids are processed by chemical means to prepare solutions for assay. For solids, steps must be taken to dissolve the sample; ashing may also be required to remove organic material. Finally, reducing the sample mass may be desirable to avoid instrumental problems, such as perturbing the plasma in an inductively-coupled plasma mass spectrometer (ICP/MS).

In addition to the preparation of the physical form of the sample, it must often be chemically processed to remove impurities. In traditional radiochemistry that uses radiation detection methods, the term "impurity" refers to the presence of another radionuclide that emits radiation of a type and energy that overlap emission from the radionuclide of interest. When using MS as the detection method, an impurity is defined as a material that has a mass close to that of the analyte of interest. These isobaric interferences obfuscate the identification of characteristic mass peaks and the measurement of the number of ions from the analyte of interest at these peaks. Interferences can be caused by isotopes of other elements (e.g., ^{238}U interferes with ^{238}Pu) or molecular species (e.g., both ^{208}Pb^{27}Al and ^{200}Hg^{35}Cl can interfere with ^{235}U). Isobaric interferences are particularly serious for assaying radionuclides because of their (usually) low abundance relative to the interfering species. Thus, the process of chemically preparing the sample stresses purification and contamination minimization; this means controlling both the analyte in question as well as other, often non-radioactive, species.

Chemical separations may be specific for the analyte of interest (see Chapter 3), such as liquid or gas chromatography, or scavenging (such as by precipitation) to remove the major interfering substances. Addition of carrier, as practiced in radioanalytical chemistry to assist in purifying radionuclides, usually is not appropriate for mass spectrometric analysis. Such addition undermines the isotopic ratio measurements that are often at the heart of this procedure, and also overloads the system for ion generation and peak resolution (but carrier addition is used for accelerator mass spectrometry). Addition of tracers, known as isotope dilution, is often employed for yield determination (see Section 17.2.9). Interferences are distinctly different in radiometric and MS analyses of radionuclides, and may be the deciding factor in selecting one method *versus* the other.

The need for extensive sample purification depends in part on the instrument. Typically, the inductively-coupled plasma mass spectrometer (ICP/MS) does not require the degree of separation needed in thermal ionization mass spectrometry (TIMS). TIMS offers greater detection sensitivity at the cost of more elaborate sample preparation and instrument control. An instrument such as an ICP/MS, which is operated at higher mass resolution, may require less sample preparation. Operation at high mass resolution, however, generally comes at the expense of overall sensitivity. For most radioanalytical analyses, sensitivity is crucial, hence high resolution analysis is rarely used. The collision/reaction cell interference reduction method recently introduced in ICP/MS does not suffer from the trade-off between interference reduction and sensitivity (see Section 17.4.5).

Cleanliness in the laboratory where samples are prepared is critical to maintaining a low background for high sensitivity analysis. Radioanalytical chemists are used to assessing chemistry blanks for radiation detection, but in MS analysis, both radioactive and non-radioactive isotopes can add to the background against which the sample atoms are measured.

The prepared sample must then be introduced into the MS. Liquid samples, usually weakly acidified solutions, can be converted into an aerosol by aspiration into the ICP/MS carrier gas in a mixing tube and inserted as a vapor. Alternatively, the prepared solutions can be applied to a TIMS filament and dried (see Section 17.5.1). The filament is subsequently inserted into the mass spectrometer. Direct volatilization from a furnace is an alternative for some elements. Ablation of a solid surface by a laser followed by direct injection of the aerosol is discussed in Section 17.7.2.

17.2.2. Ion Source

Because mass spectrometers operate by separating ionized atoms and molecules, they require an efficient means of producing ions. The ion source ionizes the sample atoms or molecules and accelerates the ions into the mass analyzer. Table 17.1 lists the commonly used ionization methods and Table 17.2 summarizes the benefits and problems associated with each method. Generally, the ion source produces a well-collimated, mono-energetic ion beam that is injected into the mass analyzer. Poor collimation degrades the ability of the mass analyzer to separate ions of different mass. A significant spread in the kinetic energy of the ions also degrades mass analyzer efficiency and resolution. Both positive and negative ions can be produced, depending on the nature of the sample and the need to separate the analyte species from potentially interfering species in the sample.

TABLE 17.1. Ionization methods used for mass spectrometry

Source	Ionization mechanism	Sample type
Electron impact (EI)	e^- collisions	Gas or vapor
Thermal ionization (or emission)	Evaporation from a hot surface	Solid on metal filament
Inductively Coupled Plasma (ICP)	RF plasma at atmospheric pressure	Solution, aerosol (includes laser ablated solid), or gas
Glow discharge (GD)	DC plasma at ~1 Torr	Solid
Ion bombardment	Ion sputtering	Solid
Laser ablation	Plasma	Solid
Resonance ionization	Photons	Gas, solid, or vapor

TABLE 17.2. Advantages and disadvantages of the various ionization methods

Source	Advantages	Disadvantages
Electron Impact (EI)	Universal ionization; simple & robust	Large energy spread, multiple charged ions formed
Thermal ionization (or emission)	High sensitivity and precision possible	Requires extensive sample chemistry, applicable to limited number of elements
Inductively Coupled Plasma (ICP)	Universal ionization; solutions easily analyzed	Ion-molecule reactions create interferences; ions have significant energy spread
Glow discharge (GD)	Universal ionization, simple sample preparation	Poor rejection of isobaric interferences
Ion bombardment	Universal ionization; imaging possible, high spatial resolution	Complex, ion-molecule reactions create interferences; poor rejection of isobaric interferences; calibration is difficult
Laser ablation	Universal ionization; high spatial resolution	Low precision, poor reproducibility, isotopic fractionation
Resonance Ionization	Element– and isotope–specific ionization	Expensive and complex

The following is a brief description of the various ionization sources used in mass spectrometry. Those sources most commonly used in the analysis of radionuclides will be described in greater detail later in the chapter.

Electron impact (EI) ionization is useful for elements that are either volatile or form volatile compounds. Typical ionization efficiencies are in the range of 0.1 to 1 percent. Suitable elements include noble gases and light elements such as C, N, O. Other elements include those that form volatile compounds (e.g., uranium in the form of UF_6). Except for specialized applications (e.g., noble gas analysis), EI is no longer widely used for elemental and isotopic analysis.

Thermal ionization is useful for elements with ionization potentials less than about 7 electron volts (eV). The thermal ionization process forms positive ions when an analyte is evaporated from a hot metal filament. The filament can be a single filament, where ionization and evaporation occur together, or a double or triple filament, where the sample is evaporated from one filament and the vapor is ionized on a separate filament. Common filament materials include rhenium, tungsten, and tantalum. Whatever the arrangement of the filament, it is always located above the case plate, shown at the top of Fig. 17.1.

Ions are generated at the filament and accelerated downward towards the exit plate. In a typical mass spectrometer, the filament is at high voltage (typical voltages are +5,000 V for positive ions and –5,000 V for negative ions; voltages from 1,000 to 15,000 can be used) and the exit slit plate is at ground. Typical ionization efficiencies range from 0.1 to 10 percent. Suitable elements include alkali, alkaline earth, many transition elements, the rare earths, and actinides, as well as lead and boron. Negative thermal ionization has become important in the past 15 years and it has increased the range of elements available for analysis by thermal ionization.

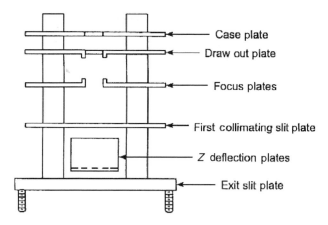

FIGURE 17.1. Thermal ionization source. Figure from Smith (2000), pg. 9.

As with the positive ion case, typical ionization efficiencies are 0.1 to 10 percent. Suitable elements include the halogens and oxides of transition elements such as Mo, Re, and Os.

Inductively–coupled plasma (ICP): An argon ICP will ionize all but a few elements (e.g., fluorine) and, therefore, is a nearly universal ion source. Figure 17.2 shows the ICP source; its ionizing coil and torch are on the right, while the vacuum interface is shown on the left.

Samples are introduced into the plasma as an aerosol formed by nebulization of a solution, by vaporization of solids in a furnace, or using laser ablation. Analyte vapor or aerosol is carried by flowing argon in the central tube in the torch. Argon also flows in the annular region of the torch. The two argon flows are balanced to give a stable plasma, which is driven by radiofrequency induction from the load coil. Analyte ions are formed by collisions with argon ions and other species in the plasma; these are subsequently drawn into the mass spectrometer through a differentially pumped gas inlet. Typical ionization efficiencies range upwards to 100%, but this figure is misleading as the transport of the ions from atmospheric pressure to the vacuum of the mass spectrometer is inefficient. Overall efficiencies are typically 0.1% or less (ions detected compared to atoms introduced into the plasma). Nominal voltages are near ground for a quadrupole mass analyzer, whereas for a magnetic sector analyzer the entire ICP source (torch and vacuum interface) is held at high voltage (thousands of volts).

Secondary ionization mass spectrometry (SIMS) is used mainly for the analysis of solid compounds and materials. A focused ion beam (called the primary ion beam) is directed against a surface containing the analyte of interest and ions are formed by the sputtering action of the impacting ion beam. Figure 17.3 shows a SIMS source, the actual dimensions of which are exaggerated for clarity. Note that some of the ions from the incoming beam remain embedded in the sample. Secondary ions are extracted by ion optics and accelerated into a mass analyzer for measurement. The sample is normally at high voltage for efficient extraction

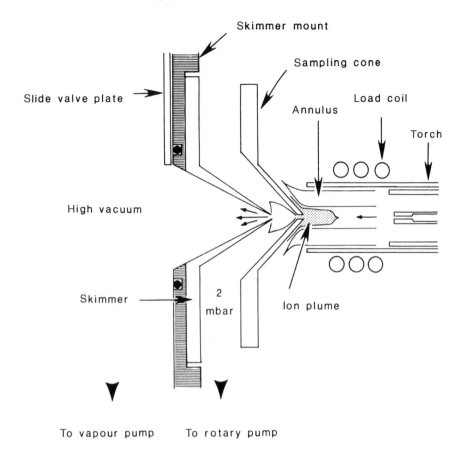

FIGURE 17.2. ICP source. Figure from Gray and Date (1983), Fig. 1.2.

and focusing of the secondary ions. Typical ionization efficiencies range from <0.1 to about 1 percent. Virtually the entire periodic table can be analyzed with this method. Both positive and negative ions can be produced, depending on the composition and polarity of the primary ion beam.

Laser ablation can be used to volatilize samples. An intense laser pulse is focused onto a solid at sufficient pulse power and energy (e.g., milliJoule energy in a pulse of ~10 nanosecond, or less, duration) to remove material from the surface. Typical conditions result in crater formation after one or a few laser pulses. If performed in flowing argon at atmospheric pressure, the ablated material can be fed into an ICP for ionization. Ions may also be formed by an ablation pulse in a vacuum and directly focused and extracted into a mass analyzer. Laser ablation/ionization is used primarily for analysis of solid compounds and materials. Virtually the entire periodic table can be analyzed with this method.

Resonance ionization mass spectrometry (RIMS) uses absorption of narrow bandwidth laser light to ionize an element of interest. The wavelength is chosen

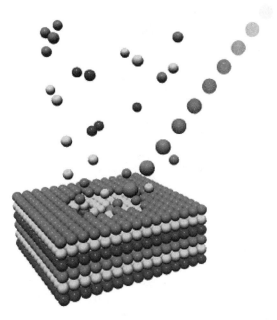

FIGURE 17.3. SIMS source. Figure courtesy of G. Gillen, National Institute of Standards and Technology, http://www.simsworkshop.org/graphics.htm (Jan. 2006).

to coincide with an allowed electronic transition of the atom (or isotope) of interest. Single photon transitions are exploited as well as "multi-photon" transitions. Isotopic selectively can be achieved with sufficiently narrow wavelength tuning. This method offers very high selectivity compared to other ionization methods, but usually with relatively low ionization efficiencies (<1%). Under specialized conditions, such as specific ion source geometry, narrowband excitation by continuous-wave lasers, rapid heating, pulsed laser operation, and well-matched laser diameter to sample size, much higher ionization efficiencies are possible. RIMS systems usually are found only in laser research laboratories because the lasers needed for resonance ionization are complex and expensive.

A notable exception regarding cost and complexity is the method of resonant laser ablation (RLA). In RLA, a low-fluence pulsed laser beam is focused onto the solid sample (Eiden et al., 1994). The laser pulse first ablates or desorbs a small amount of sample. On the timescale of the laser pulse (typically a few nanoseconds), the atoms liberated from the surface absorb laser photons and are ionized. By using a laser wavelength resonant with an atomic transition in the atoms of interest, ionization is highly selective. This method has been extensively demonstrated with relatively low-cost YAG pumped dye lasers.

The ions produced by any of the preceding sources are then detected by the mass analyzers described below.

TABLE 17.3. Common mass analyzers

Analyzer	Mass range (amu)	Mass resolution (M/ΔM)	Duty cycle	Electromagnetic fields[1]
Sector field	1–10,000	300–10000	High to continuous	DC M or E
Quadrupole mass filter (QMF)	1–1000	100–1000	Low	DC & RF E
Ion trap (IT)	1–10,000	300–3000	Medium	DC & RF E
Time-of-flight (TOF)	1–1,000,000[2]	100–5000	Low	Pulsed DC E

[1] M: magnetic, E: electric, DC: direct current (continuous), RF: radiofrequency
[2] Greater limitation is usually poor response of ion detectors to heavy, slow moving ions.

17.2.3. Mass Analyzer

An MS is constructed with one or more of the mass analyzers listed in Table 17.3. All of these analyzers separate ions based on mass, momentum, or velocity. Many, such as the sector and TOF analyzers, require monoenergetic ions. Mass analyzers of different types can be combined to create instruments with specific capabilities, such as high mass resolution or abundance sensitivity (i.e., the ability to detect a small-abundance mass peak near a large-abundance one). Combinations include instruments with multiple magnetic and electrostatic sectors, multiple quadrupole mass filters (QMF), quadrupole-time–of-flight (TOF) combinations, ion trap (IT) - TOF combinations, and quadrupole-IT combinations. Mass spectra are generated by varying an operating parameter (e.g., magnetic field for a sector field analyzer) and observing the ion intensity as a function of this parameter.

Most mass analyzers for ICP/MS instruments are operated in a scanning mode. These mass analyzers operate as bandpass filters, passing a single mass-to-charge ratio at a time for detection by the ion collection system. The mass analyzers are scanned either by "peak-hopping" or by continuous scan.

Most modern instruments employ the "peak-hopping" (also called peak switching) method; this allows the analyst to "hop" among the mass peaks of interest during the assay. The ion intensity is integrated for a period of time at each peak. The time spent "on peak" is maximized relative to the time between peaks. The peak switching mode requires more sophisticated and accurate electronics but offers the advantage that a maximum amount of time is spent measuring ions of interest and minimal time is spent analyzing the unwanted parts of the mass spectrum between the peaks. The time spent at each peak translates into more counts; recording the maximum possible number of counts is desirable for detecting ultratrace analytes.

At lesser sensitivity requirements, or if a measurement of the peak shape is desired, a continuous scan of mass to record full peaks may be preferred. In modern instruments, control is really a digital process, so that the continuous scan mode is essentially peak hopping with a small mass interval—0.1 amu or less per step. Very high dynamic range instruments, in which the peak "tails" are to be measured, might use 50 or more steps per amu.

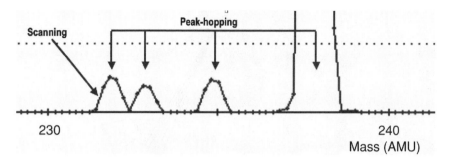

FIGURE 17.4. ICP/MS mass spectrum of U and Th isotopes.

Figure 17.4 is an ICP/MS mass spectrum (with a QMF mass analyzer) of U and Th isotopes; both scanning methods are illustrated. The small dots along the spectrum show a continuous scan (actually consisting of many sub-amu shifts). The large arrows show peak switching where the scan is halted and the QMF is tuned to the centers of the mass peaks. In this figure, masses 232, 233, 235, and 238 have significant ion mass peaks.

To measure an isotopic ratio, the analyzer is set to the first mass of interest, the signal is acquired for a selected duration or "dwell" time, and the analyzer is then switched to the next mass of interest for signal acquisition. The optimum dwell time is a function of many factors, but the main considerations are instrument noise and ion counting statistics. The dwell time should be long enough to average out the high frequency components of the noise, but not so long as to suffer from long-term drift or low frequency noise. Longer dwell times are needed for the less abundant isotopes to register sufficient counts for the desired level of precision. Quadrupole systems can be scanned much faster than magnetic sectors; dwell times for the former range from fractions of a ms to perhaps 100 ms whereas magnetic sector dwell times range from 0.01 to 1 s. TOF and ion trap mass analyzers are scanned in ways that are not easily characterized as "peak hopping" or continuous; these scan modes are described in the sections on TOF and ion traps.

The magnetic sector field represents the classical mass analyzer. In operation, a magnetic sector disperses ion masses in a manner analogous to an optical prism creating a light spectrum. Sector angles of 60° and 90° are most common. As shown in Figure 17.5, the ion source (labeled O) is at the top of the figure and the ion collector (labeled C) is at the bottom. The arrangement depicted is the most common one; the magnetic sector is designed to be symmetric, with the source–to–magnet and collector–to–magnet distances being equal. Other design configurations are possible.

Sectors require precisely collimated mono-energetic ion beams, which places additional requirements on the ion source and ion optics. Given a mono-energetic beam, the equation for a mass spectrometer with a sector of radius R and with ions accelerated to a kinetic energy V is:

$$m/z = eR^2B^2/2V \tag{17.1}$$

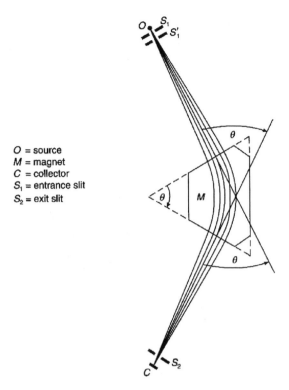

O = source
M = magnet
C = collector
S_1 = entrance slit
S_2 = exit slit

FIGURE 17.5. Magnetic sector analyzer. Figure from Platzner (1997), pg. 5.

where:

R = radius
m = ion mass
e = electron charge
z = ion charge (in units of electron charge e)
V = ion accelerating voltage
B = sector magnetic field

A single sector can achieve abundance sensitivity (i.e., the ability to detect a small abundance mass peak adjacent to a large abundance one) in the range of 1 ppm (i.e., an isotope ratio of 1×10^{-6}). Abundance sensitivity can be critical for rare isotopes; an example is ^{236}U, which occurs only at extremely low abundance (ratio relative to ^{238}U < 10^{-10}) in natural U, but is produced by irradiation of uranium in a reactor (^{236}U abundances vary from 10^{-10} to 10^{-2} compared to the abundances of all U isotopes present). Common sector geometries include 60° and 90° angles, and common radii include 15 and 30 cm.

A larger radius leads to higher dispersion of the masses (e.g., more separation between adjacent masses). High dispersion leads to better mass resolution and abundance sensitivity and enables the use of larger detectors and ion sources, which simplifies the design and construction of some component. High dispersion results in a larger overall instrument because the size of the instrument scales directly with the sector radius. This is especially true of the magnet, which for large instruments can weigh hundreds to thousands of kilograms. Large electromagnets are expensive and require large power supplies and control electronics for stable operation. Modern commercial sector instruments are designed with so-called extended geometries, which use specialized magnetic focusing to achieve the equivalent dispersion of a large radius sector with a smaller magnet. Extended-geometry instruments offer high dispersion at smaller size.

In its simplest form, a mass spectrum is created by scanning the magnetic field and measuring the ion intensity. Alternatively, the magnetic field is held constant and the ion source voltage (which controls the ion energy) is scanned. In modern instruments, peak switching is performed by setting either the magnetic field or the ion source voltage to values that correspond to the mass peaks of interest. Sectors have the advantage that multiple ions can be detected simultaneously with multiple detectors for each mass of interest or with an imaging detector. Early mass spectrometers used a photoplate for ion detection to measure the entire mass range at once, but photoplate ion images were difficult to measure quantitatively for relative ion intensities. In modern instruments, electronic detectors (see Section 17.1.4) are used to measure ions directly.

Sector mass analyzers offer many advantages, including high precision isotopic analysis, simultaneous mass detection (with multiple detectors), and stable operation. Disadvantages include large size, magnets that are heavy and require sophisticated electronics to operate in a stable manner, and high voltages of more than 10,000 V.

The quadrupole mass filter (QMF), shown in Fig. 17.6, represents an entirely different approach to selecting ions by mass. In its typical configuration, the QMF consists of four parallel cylindrical rods that are set on a square. The opposite corners are connected electrically to both DC and RF voltages. Ions traverse along the axis of the rods in a complicated motion and only ions of a selected ion mass-to-charge ratio can successfully transit the quadrupole. The (a) portion of the figure shows the conceptual layout, with electrodes with perfect hyperbolic cross-sections. The (b) portion shows a schematic of the QMF in practice with electrodes of circular cross-sections. Also shown are the ion source lens, collector, and a generalized version of the drive electrical circuitry. The QMF uses a combination of DC and RF electrical potentials to filter ions within a narrow range of m/z. In effect, the QMF is operated as a bandpass filter.

The equations governing the operation of the QMF are derived from solutions to Laplace's equation, which for the geometry of the QMF are described by the Mathieu equation (Duckworth et al., 1986; March and Hughes, 1989). The following discussion of the Mathieu equation is adapted from March and Hughes (1989).

$$\frac{d^2u}{d\xi^2} + (a_u - 2q_u \cos 2\xi)u = 0$$

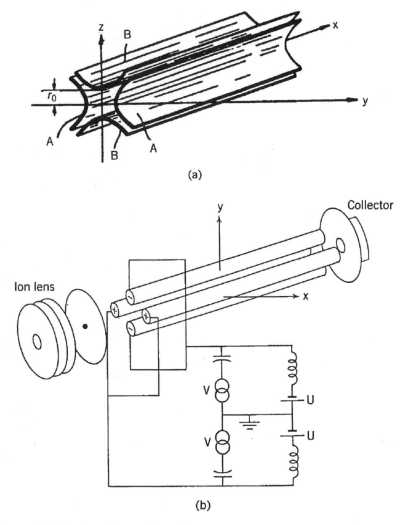

FIGURE 17.6. Quadrupole mass filter (QMF). Figure from March and Hughes (1989), pg. 3.

where

$$\xi = \frac{\omega t}{2}$$

$$a_u = a_x = -a_y = \frac{8eU}{m\omega^2 r^2}$$

$$q_u = q_x = -q_y = \frac{4eV}{m\omega^2 r^2}$$

(17.2)

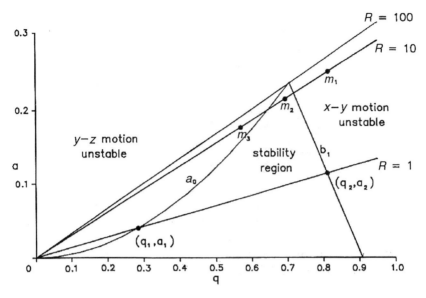

FIGURE 17.7. Mathieu stability diagram. Figure from March and Hughes (1989), pg. 50.

where:

r = inner radius between the quadrupole rods
m = ion mass
e = electron charge
V = RF voltage
U = DC voltage
ω = RF frequency
t = time
u = x or y position coordinate

The solutions to the Mathieu equation translate into a series of regions of stable motion for various values of a and q, which depend on the parameters r, m, V, U, and ω. A common approach to visualizing the operational parameters of the QMF comes from the use of a Mathieu stability diagram (March and Hughes, 1989), which plots a versus q. Figure 17.7 shows a Mathieu stability diagram, with the operating lines for various resolutions, R (which is $M/\Delta M$), overlayed on the diagram. For the line marked "$R = 10$," only mass M_2 is stable and will be transmitted through the QMF; the other masses will be ejected.

The apex of the stability curve for the primary stable region occurs at $a_y = 0.23699$ and $q_y = 0.70600$. For typical QMF analyzers, a given mass m is analyzed by use of set values for the parameters r and ω (r and ω typically 1 cm and 1–3 MHz, respectively) and then varying the parameters U and V (the DC and RF voltages, respectively). Mass selection is achieved for the desired mass resolution, R, given

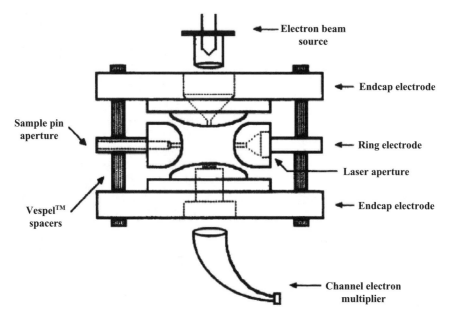

FIGURE 17.8. Schematic diagram of an ion trap mass spectrometer analyzer. From Gill et al. (1991), Fig. 1a.

in the usual form as $m/\Delta m$, where Δm is the desired mass range for transmission through the QMF. A mass spectrum is created by scanning the DC and RF voltages U and V, respectively, while maintaining a constant ratio of U/V. Resolution is controlled by how close to the apex one scans the U/V ratio.

QMF analyzers offer advantages of light-weight, small size, and rapid response (an entire elemental mass spectrum can be scanned in less than a second). They find application in low resolution ($\sim$1 amu), low abundance sensitivity ($>$1 ppm), and low measurement precision ($\sim$1%) applications (exception: precision of 0.1% can be achieved with state-of-the-art ICP/MS instruments with QMF analyzers). QMF analyzers have distinct advantages over magnetic sectors as the QMF does not require a large magnet or high voltages, and the QMF is more compact. A disadvantage of the QMF is that it operates as a bandpass filter and can essentially measure one ion mass at a time, whereas a magnetic sector is capable of simultaneous multi-mass detection.

The RF quadrupole ion trap mass spectrometer (ITMS) is a close relative of the QMF and ideally can be thought of as a three-dimensional quadrupole (see Fig. 17.8). The close relationship of these two devices is evident by the fact that ion motion in the two devices is governed by essentially the same mathematical equations. As with the QMF, the ITMS uses DC and RF electric fields and the operation of the IT is described by solutions to the Mathieu equation. Unlike the QMF, ITMS analyzers trap ions within the mass analyzer. Ions are trapped, ejected to select the mass of interest, and then ejected in a controlled manner for detection.

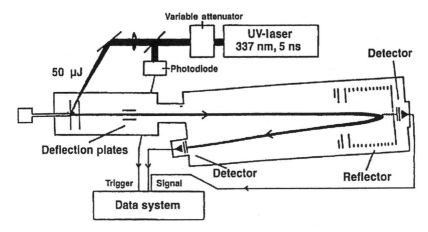

FIGURE 17.9. Time of flight MS analyzer. Figure from Larsen and McEwen (1998), pg. 20.

Modern ITMS analyzers utilize milli-Torr pressures of helium to cool ions, which enables higher resolution operation and long (millisecond to second) storage times. The ITMS is applied in organic and biochemical analysis, and in a limited fashion for isotopic and elemental analysis. The ITMS in Fig. 17.8 is configured to use a laser pulse to vaporize and ionize metal atoms inside the mass analyzer. Unique features of ion traps include the ability to store selectively specific m/z ratios (ion species).

Many other kinds of ion traps are known, but apart from the ITMS, only the ion cyclotron resonance (ICR) trap has been used for elemental or isotopic analysis. In an ICR trap, the ions are trapped by a combination of RF and DC fields applied to the walls of a box. The box is held inside a very strong magnetic field and the combined RF and magnetic fields result in ions undergoing cyclotron motion whose frequency depends on the ion's mass-to-charge ratio. The mass spectrum is determined by measurement of these characteristic frequencies. First, the time domain signal associated with ion motion is measured. The mass spectrum is generated by inverting the time domain waveform to a frequency domain representation with a Fourier Transform. The method is thus known as Fourier transform, ion cyclotron resonance mass spectrometry or FT-ICR-MS. A key feature of FT-ICR-MS is extremely high resolution.

The time-of-flight (TOF) mass spectrometer utilizes a pulsed ion beam for mass analysis. In practice, a packet of ions is accelerated to the same energy and injected into a field-free drift tube (Fig. 17.9). At constant energy, light ions move faster than heavy ones, and mass separation is achieved by measuring the arrival times of ions at the ion detector. Given a constant drift distance L (the length of the drift tube from ion injection to ion detector), the time, t that an ion of m/z requires to travel the distance L is given by:

$$t = L\sqrt{(m/z)/2eV} \qquad (17.3)$$

where:

m = ion mass
e = electron charge
z = ion charge (in units of electron charge e)
V = ion accelerating voltage
L = drift tube length
v = ion velocity

Figure 17.9 shows a schematic diagram of a reflectron-type TOF used for laser ablation mass spectrometry. Ions are generated in the source region, which consists of a sample mount between two parallel plates (the plate on the right has a small hole for transmitting ions into the drift tube on the right). The laser generates a pulse of ions that are quickly accelerated into the larger drift region on the right. Lighter ions move faster than heavier ions. The reflector (also known as a reflectron, see discussion below) diverts the ions towards the ion detector at the bottom. A mass spectrum is produced initially as a plot of ion arrival time versus ion intensity. The laser ablation system is well-matched to the TOF since DC voltages can be used for ion extraction and acceleration. In TOF systems that have a continuous source of ions, pulsed electronics must be used to generate the rapid pulse of ions for the TOF.

The TOF analyzer is capable of measuring a wide mass range of ions, but only on a pulsed basis. In simple designs and operating modes, the spread in ion energies must be small ($<1\%$) and the ions must initially be tightly packed (within 1 mm or less), otherwise the resolution of the TOF will be severely degraded. Methods are applied for compensating for both initial ion energy spread and spatially extended ion sources. The duty cycle (the fraction of time that ions are being generated and analyzed) is usually short, hence the overall throughput of a TOF is low. For longer duty cycles, specialized high speed pulse generators and high throughput data acquisition systems are used (such systems were used for a brief time in a commercial ICP-TOFMS instrument). Because data are acquired with high-speed digitizers (typically 1×10^9 samples/sec) with deep memories ($>1 \times 10^6$ samples), digitizing resolution is usually limited to 8 bits. If signal averaging can be used, the effective resolution can be improved, but the usual 8-bit data severely limits abundance sensitivity of the overall instrument. The abundance sensitivity is thus much less than other mass analyzers, typically in the 100–1000 ppm range (i.e., isotope ratios less than 10^{-4} are unresolvable).

Mass resolution is dependent upon drift tube length, ion energy, extractor design, and timing resolution. High resolution is easily achieved, but TOF analyzers are not often used for applications that require high mass resolution. A variant of the TOF analyzer, called the reflectron, uses a gradient electric field to reflect the ions back to an ion detector located near the ion source. The reflectron has energy-focusing properties that correct for the energy spread (i.e., velocity distribution) of the ions to result in higher mass resolution.

In a TOF analyzer, a mass spectrum is created by measuring the ion intensity as a function of time of arrival after the ions are injected into the TOF analyzer.

Ions may be pulsed either by pulsing the ion source itself (e.g., a pulsed laser) while the TOF-MS ion extraction electrodes are held at constant DC potential, or by pulsing the ion extraction electrodes. Pulsing the electrodes is difficult because the pulse rise time must be short (typically <10 nanoseconds) and the potentials are high (typically 1–5 KV). Depending on whether the ions to be analyzed are formed in a continuous ion source (hot filament, continuous plasma) or in a pulsed ion source (laser, pulsed glow discharge), the TOF analyzer may detect all of the ions formed or just a portion of the ions formed. If all the ions formed are contained in the extraction region of the TOF-MS, as in a typical laser source, then they can be accelerated and delivered to the detector with high efficiency. The TOF-MS analyzer is capable, in principle, of simultaneous detection of different isotopes. In practice, isotope ratio precision is limited by the coupling between the ion extraction process and the continuous ion source and by the poor digitizing resolution and analog detection schemes (i.e., not ion counting).

TOF analyzers offer advantages of compactness (they can be made very small—cm-length analyzers are possible), simple construction (the TOF is basically a tube), and their wide mass range (they obtain the full mass spectrum, usually 1–300 amu for elemental and isotopic analysis) for every measurement. Drawbacks include the difficulty of efficiently coupling to various ion sources, limited mass resolution, relatively low precision (1–10 % are typical) and the need to use expensive high precision and high speed electronics for pulse generation, ion detection, and data acquisition.

17.2.4. Ion Detector

Mass-analyzed ions are detected by either Faraday cup (FC) or electron multiplier (EM) detectors. Each is described below.

The FC collects the ion charge for direct current measurement of the ion beam. The FC is essentially a metal bucket that collects the ion beam. Figure 17.10 shows a schematic diagram of a Faraday Cup with associated repeller plates and slits. The ion beam enters from the left and is captured in the Faraday cup on the right. The biased plate to the left of the cup has a negative voltage ($\sim$100 V or less) to repel secondary electrons back into the cup.

Modern FC detectors are deep and narrow so that ions and secondary electrons are captured and contained within the cup. The current can be measured directly by a picoammeter; for high sensitivity measurements, the current is measured as a voltage across a high resistance (10^9 ohms or greater). High sensitivity measurements often utilize an integrating circuit in which the current (or voltage) is integrated for a period of time (e.g., 10 s) and the integrated value is used to determine the total ion charge collected during the integration interval. FC detectors are limited to ion beam currents of 10^{-14} Ampere or above ($>10^5$ ions/s). The FC detector is stable and is used for high precision isotope measurements when large amounts of analyte are available. Some quadrupole instruments and secondary ion mass spectrometers employ FC detectors, especially for high current measurements. FC detectors are also used for multiple mass detection and high

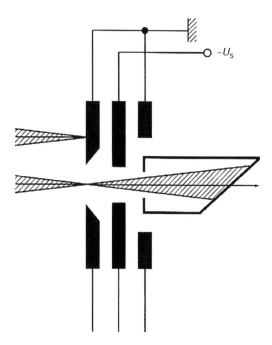

FIGURE 17.10. Faraday Cup ion detector. Figure A from Platzner (1997), pg. 34.

precision isotopic analyses on magnetic sector analyzers. The main disadvantage is low sensitivity and slow response.

The EM converts ions to electrons and amplifies the resultant electron signal by factors of 10^6 or more. In operation, ions strike a conversion dynode or surface at the front of the EM to generate a few electrons. The secondary electrons are accelerated to the next dynode and each in turn generates more electrons. The process is repeated at each dynode to produce a large number of electrons after multiple stages. Typical EM detectors have 15 to 20 dynodes or their equivalent, depending on the design. Figure 17.11 shows a drawing of a commercial detector. The cutout oval at the top shows the single incoming ion beam, which is converted to an electron signal that is multiplied as it strikes each dynode down the right side of the instrument.

Variants of the EM include the discrete dynode, channeltron (or continuous dynode), the microchannel plate, and the Daly detector. The EM can detect ion currents below 1 ion/s. EM detectors are often operated in the pulse counting mode, in which the pulse produced by a single ion is amplified and detected as an individual events. Pulse counting offers excellent signal-to-noise ratio, high sensitivity, and insensitivity to change in EM gain.

The EM also can be operated in the current or analog mode (these terms are synonymous) in which the ion current is amplified and measured. EM detectors are available that operate simultaneously in pulse counting and analog modes. Part way along the amplifying dynode chain, a portion of the signal is diverted to an

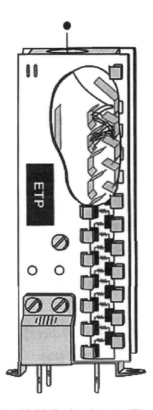

FIGURE 17.11. Electron Multiplier ion detector. Figure from (ETP 2005).

operational amplifier (op amp) that ultimately produces an "analog" signal. The remainder of the electron beam in the dynode chain continues to the end of the chain to produce a pulse counting signal. When the ion beam becomes too intense to operate the detector in pulse counting mode (the anode current must be kept below a critical value to avoid damaging the detector), the pulse counting part of the detector is switched off (by switching the bias to one of the dynodes) and the analog signal is used. The response of these two sections of the detector can be cross-calibrated to yield an overall dynamic range in excess of 10^8.

A variant of the EM is the Daly detector (Daly, 1960) in which ions are accelerated by $\sim$30 KV (this is called post-acceleration because it occurs after mass analysis) into a conversion dynode, which generates a significant number ($\sim$10) of secondary electrons. These electrons are accelerated into a scintillator and converted into light, which is detected with a photomultiplier. The Daly detector offers high gain, low noise, and excellent stability. Other variants of the post-acceleration detector exist; a simple configuration uses a metal plate to convert the ions into electrons for detection with an EM. Another variant of the EM is the microchannel plate (Coplan et al., 1984; Odom et al., 1990; Wiza, 1979). Microchannel plate EM detectors have excellent sensitivity but poor gain stability. When operated in an analog mode (measuring amplified ion current), the gain of some detectors can drift by several percent during an analysis. Ion counting eliminates most gain-drift problems. EM detectors have limited lifetimes due to the accumulated damage at the ion conversion dynode from the impact of detected ions.

Besides the ability to count ions, the EM detector responds to changes in the ion beam intensity instantaneously (on a nanosecond timescale), whereas the high feedback resistance associated with FC detectors results in a long time constant for the measurement (typically 1 to 10 seconds). Fast scanning mass analyzers such as QMF, ITMS, and TOF analyzers almost always use EM type detectors because the FC response is too slow.

17.2.5. Vacuum

Mass spectrometers require various degrees of vacuum in different parts of the instrument. A vacuum is produced by a range of vacuum pumps, including rotary vane, diaphragm, scroll, turbo-molecular, diffusion, ion, and getter pumps. Most ion sources operate at low pressure, some at high vacuum, while the ICP operates at atmospheric pressure. The mass analyzer and ion detectors of a mass spectrometer almost always operate in high or ultrahigh vacuum. Because gases flow and behave differently in these various vacuum/pressure regimes, understanding vacuum system design and operation requires some familiarity with the properties of gases in such systems. An excellent reference for vacuum systems is Dushman and Lafferty (1962).

The various mass spectrometer performance requirements drive vacuum system design. These include correct operation of the ion source (arcing can occur in the high voltage components if the operating pressure is too high), and good transmission of ions through the ion optics and mass analyzer (scattering by background gases reduces transmission). Rough vacuum is required on the foreline of high-vacuum pumps such as turbo-molecular pumps and diffusion pumps and in the interface between atmospheric pressure and the vacuum of an ICP/MS. Most modern instruments use oil-sealed rotary vane pumps for these applications, although oil-free or so-called "dry" pumps (also known as scroll and diaphragm pumps) are used because of their inherent cleanliness. Dry pumps do not insert oil vapor into the vacuum system, a high-cost component of a high-performance mass spectrometer system.

17.2.6. Data Acquisition and Instrument Control

The "user friendliness" of major MS methods for radionuclide analysis varies widely. The commercially available ICP/MS instrument is highly automated. Minimally trained staff can operate it for routine work when supported by trained maintenance staff. The TIMS instrument also can be obtained as a highly automated system, but greater operator skill generally is required than for ICP/MS to achieve a comparable level of data quality. Likewise, the SIMS instrument is complex, both from an operational aspect, as well as for calibration and interpretation of results. The accelerator mass spectrometer (AMS) instrumentation is still highly specialized and requires expert operators.

Instrument control typically is limited in commercial instruments to the extent that non-routine uses of the instrument for development purposes may be difficult. The data acquisition software is not always optimized for isotope ratio measurements. These issues should be considered when purchasing an instrument for work that includes both production and development.

17.3. Measurement Methodologies

17.3.1. Mass Bias

All mass spectrometers experience some level of mass bias in which different isotopes of the same element are measured with different sensitivity. The main causes are fractionation in the ion source and mass discrimination in the ion optics and mass analyzer. Although the effect is usually small, corrections for mass bias (sometimes called mass fractionation or mass discrimination) are needed to obtain correct values of isotopic compositions from mass spectrometric data (Barshick et al., 2000; De Laeter, 2001; Montaser, 1998; Nelms, 2005; Platzner, 1997; Tuniz et al., 1998). In most cases, mass bias is corrected by running suitable standards for which the isotopic composition is well known. Because of the prevalence of ICP/MS, a detailed discussion is given below.

Mass bias in ICP/MS is 10 times more severe than in other types of mass spectrometry, on the order of 1%/amu. The observed bias is sensitive to both sample composition and instrument ion optical tuning conditions. The source of this bias is dominated by space charge effects in the ion beam extracted from the ICP (Douglas and Tanner, 1998).

When the plasma from the ICP passes into the first vacuum chamber, a free jet expansion results. In such jets, there is a mass dependence relative to the centerline flux, with intensity at a given distance downstream varying as $m^{-1/2}$ (Miller, 1988). This mass dependence applies to uninterrupted, skimmed free jet expansions. A weaker mass dependence is expected in ICP/MS because the free jet effect only dominates in the field free region near the skimmer tip. The first strong electric field of the mass analyzer ion optics is experienced by the skimmed plasma flow as it passes through the next stage of differential pumping. This electric field focuses ions and repels electrons. The resulting net positively charged beam typically has a charge density many orders of magnitude greater than what can be focused by conventional optics (the ion number density greatly exceeds the limit imposed by the Child-Langmuir law). The result is a rapid and severe "space charge explosion" due to the mutual electrostatic repulsion of positive ions in the beam.

This effect has a much stronger mass dependence than the gas dynamic effect and dominates the overall bias. Anything that affects this space charge field, affects the observed bias. Mass bias in ICP/MS thus depends on sample preparation (as more sample matrix is put into the ICP, the composition and intensity of the ion beam changes) and instrument tuning (changing the potential of an optical element affects the total ion flux and its spatial distribution). These effects have been recently discussed by Praphairaksit & Houk (2000).

The ICP ion source has been utilized in combination with most types of mass analyzer (RF quadrupole ion filters and traps, time-of-flight, and magnetic sector field), although commercial instruments utilize either quadrupoles or magnetic sectors as mass analyzers. The observed mass bias is comparable in these different instruments (Encinar et al., 2001). The precision attainable does not depend (strongly) on the mass analyzer employed per se, but rather on whether the instrument is a multi-collector/detector instrument or not. The attainable precision with single detector instruments is much more limited than what can be achieved with multi-collectors (MC). Subtle corrections are, therefore, of much greater interest in the latter instrument type. Mass bias corrections are derived by measuring the bias with an isotopic abundance standard material and fitting measured bias to a function. In most quadrupole instruments, a linear fit suffices. In higher-accuracy work (MC-ICP/MS), the functional forms are non-linear (power law or exponential). The isotopes used for mass bias correction can be measured in the same solution as the sample (for internal bias correction) or in a separate solution (for external bias correction) when correction isotopes and the sample isotopes of interest may overlap.

Extrapolation of the bias function from the masses of the reference isotopes to the masses of the analytes of interest is typically kept as small as is practical. Ideally, the analytes of interest are bracketed by the mass bias calibration isotopes, i.e., the bias correction is interpolated. For example, a Tl isotopic standard is used to correct unknown Pb ratios, but Tl would not be a good choice for correction of Pu isotopes. If appropriate isotopic abundance reference materials are available and appropriate care is used, MC-ICP/MS is among the most versatile (applies to most of the Periodic Table) and accurate methods for determining unknown isotopic composition.

17.3.2. Precision and Accuracy

Precision is a measure of how close repeat measurements of the same sample are to one another. There are two main types of precision, internal and external. When performing a mass spectrometric analysis, several measurements of the elements of interest are made. How close these determinations are to each other defines internal precision. Internal precision generally is a measure of the stability of sample introduction to the mass spectrometer and the stability of the ion source and mass analyzer/detector. Internal precision in mass spectrometry is often counting-statistics limited (see Section 10.3). When the mass spectrometric results are Poisson distributed, the best internal precision that is achievable is $N^{-1/2}$ where N is the total number of counts. Thus, if 100 counts are registered for some isotope of interest, then this result will only be reproducible, on average, to 10% relative standard deviation. To achieve 0.1% precision, at least 1,000,000 counts are required; this level or better is the precision/counts regime of most modern mass spectrometry. External precision is determined by running duplicate samples. External precision is a measure of internal precision combined with how reproducibly sampling and sample preparation are performed.

Accuracy of a method of analysis generally is determined by analyzing standards with matrices the same or similar to the samples of interest. These standards have known isotopic abundances and/or concentrations. The limit to this approach often is the accuracy of the reference technique that was used to characterize the standard.

17.3.3. Isotope Dilution or Traced Analysis

Quantitation by mass spectrometry or radiometric counting requires reference to known standard material. This reference can be "internal" to the sample, wherein the reference material is added to the sample at an appropriate stage of processing, or it can be "external" when the response of the analyte in the sample is compared to the response measured for the reference material. A commonly employed method, isotope dilution mass spectrometry, is to add a known amount of an isotopically altered tracer (sometimes called a spike) to the sample.

Isotope ratios, as measured by mass spectrometry, along with the known amount of the tracer, are used to determine the amount of the analyte in the sample. Isotopically altered elements (sometimes consisting of nearly pure stable isotopes) and radioactive isotopes are commercially available for use as tracers. Nearly every element can be purchased as a solution whose atom concentration is known and traceable to NIST. Standard materials are also available with isotopic composition determined to a high degree of accuracy. Radioactive tracers, such as ^{233}U and ^{244}Pu, are available for tracing actinide elements that do not have stable isotopes. Because many of the samples analyzed in radionuclide determinations already have altered isotopic ratios, many times it is possible to use a natural isotope as the tracer.

Isotope dilution mass spectrometry (cf. (Heumann, 1992; Yu et al., 2002)) has two main requirements. The first is that the element being analyzed must have more than one isotope. The second is to have a well-characterized and pure tracer solution that has a significantly different isotopic composition from the element under analysis. In practice, a known amount of the tracer is added to the sample, which is then treated by any necessary chemical separations before being inserted into the mass spectrometer. The tracer must be isotopically equilibrated with the sample by forcing them into a common valence state, as discussed in Section 4.7. For elements with multiple valence states (such as uranium or plutonium) this is a crucial requirement. Failure to achieve isotopic equilibration will lead to erroneous results. Sample quantitation by isotope dilution can be determined by use of the following general equation:

$$N_s = N_t \frac{(R_m B_t - A_t)}{(A_s - R_m B_s)}$$

(17.4)

where:

N_s = the number of atoms of analyte in the sample
N_t = the number of atoms of the tracer added to the sample
R_m = the measured ratio of isotopes A to B for the traced sample
A_t, B_t = the atom fractions of isotopes A and B in the tracer
A_s, B_s = the atom fractions of isotopes A and B in the untraced sample

All relevant isotopes of the analyte element are monitored and the amounts of the unknown isotopes are referenced to the known amount of isotope tracer. If the tracer isotope is an isotope of interest in the sample, both traced and untraced sample analyses are performed, the former analysis providing elemental concentrations and the latter providing isotope ratios.

Equation (17.4) is a general form that is applicable to any element and tracer combination. For quantitation of radionuclides in which isotopically pure tracers are available that are not present in the sample at significant abundance, a simplified version of Eq. (17.5) will suffice:

$$N_s = N_t \times R_m \qquad\qquad (17.5)$$

where:

N_s = the number of atoms of analyte isotope in the sample
N_t = the number of atoms of the tracer isotope added to the sample
R_m = the measured ratio of analyte to tracer isotope A for the traced sample

For the case in which the tracer also contains the radionuclide of interest,

$$N_s = N_t \times (R_m - R_t) \qquad\qquad (17.6)$$

where:

N_s = the number of atoms of analyte isotope in the sample
N_t = the number of atoms of the tracer isotope added to the sample
R_m = the measured ratio of analyte to tracer isotope for the traced sample
R_t = the ratio of analyte to tracer isotope in the tracer

Equations (17.5) and (17.6) are commonly used to quantify radionuclides by isotope dilution mass spectrometry.

17.4. Inductively Coupled Plasma Mass Spectrometer

The ICP/MS is an elemental and isotopic analysis method that was first developed in the early 1980s. The ICP had been used only as a source for emission spectroscopy until it was adapted for producing ions for a mass analyzer (Douglas and French, 1981; Houk et al., 1980; Houk et al., 1981; Houk and Thompson, 1982). Since 1983, several manufacturers have sold ICP/MS instruments that incorporate various mass analyzer systems, such as quadrupole mass filter, magnetic sector field, time-of-flight, Paul ion trap, and ion detection systems such as the electron multiplier (EM), Faraday cup (FC), and Daly-type (post-acceleration and scintillation). These systems are used alone or in combinations of "multi-collector" instruments. In addition, researchers have coupled these and other mass analyzers, such as Fourier transform-ion cyclotron resonance, and multiple analyzers as in quadrupole-quadrupole configurations, to the ICP.

The power of this MS technique has driven the development of methods to interface ICP/MS instruments with various sample introduction systems. Specialized sample introduction systems include ion chromatography (Seubert, 2001), gas chromatography (Vonderheide et al., 2002), and capillary electrophoresis (Costa-Fernandez et al., 2000). Other techniques are hydride generation (used to volatilize selected species and obtain some matrix/elemental separation) (Reyes et al., 2003) (Bings et al., 2002) laser ablation (Gonzalez et al., 2002; Heinrich et al., 2003; Russo et al., 2002), and electrothermal vaporization (Richardson, 2001; Vanhaecke and Moens, 1999).

17.4.1. Instrument Types

Commercially available instruments almost exclusively use quadrupole or magnetic sector analysis; an ICP-ion trap instrument is sold in Japan and attempts have been made to market an ICP-TOF-MS. Quadrupole-based instruments generally cost less than magnetic-sector-field-based instruments and provide adequate resolution, stability, speed, precision, and accuracy. High-resolution magnetic sector instruments can resolve many of the spectral interferences encountered in ICP/MS, but are more often purchased for their high sensitivity in low-resolution modes, as well as their ability to perform multi-isotope collection. To address the problem of isobaric interferences, an alternative to high mass-resolving power is the collision/reaction cell (CRC), which is inserted between the ion source and mass analyzer to chemically remove interfering species, as discussed in Section 17.4.5. The CRC-ICP/MS method opens new application areas for ICP/MS, such as biological studies involving phosphorous, sulfur, and other elements that were difficult to resolve from interferences.

17.4.2. The ICP and the ICP-to-Vacuum Interface

Since the ICP was introduced by Fassel in the late 1960s, a great deal of research has been performed to optimize the inductively coupled plasma as an emission source (Fassel, 1971; Fassel and Kniseley, 1974; Montaser, 1992; Montaser, 1998). Variations in torch design, gas flows, gas composition, and applied power were just some of the variables that were studied. The end result by the early 1980s was a source that remains essentially unchanged today. The ICP has flow velocities through the torch such that a small amount of injected sample aerosol moves through the plasma in ~10 milliseconds.

At this plasma intensity and timescale, sample aerosol droplets below a certain threshold (~10 microns) are desolvated and the remaining solid particle is vaporized and atomized. The resulting atoms are ionized with an efficiency that depends

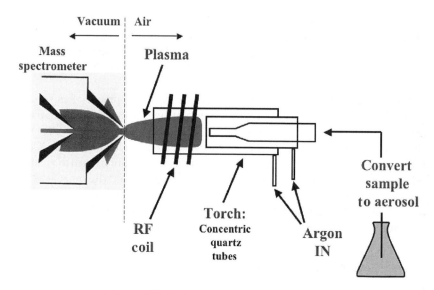

FIGURE 17.12. Plasma transfer in the ICP/MS.

on factors such as the ionization potential of the atom and the temperature of the plasma. This source has an ionization efficiency of around 95% with high excitation temperatures for inducing optical emission for the majority of the elements in the periodic table. Because of the high efficiency of ionization, researchers such as Houk and Fassel (1980) developed the ICP as an ion source for mass spectrometry. Early attempts were unsuccessful due to formation of a boundary layer that obstructed the ion-sampling aperture. Progress in overcoming this complication, among other key developments, led to the powerful technique in use today.

The key to the development of ICP/MS was a technique to transfer ions from the plasma into the high vacuum of the MS. This was accomplished with a differentially pumped atmospheric-pressure-to-vacuum interface shown in Fig. 17.12. An aerosol is injected into the central argon flow, where aerosol particles are desolvated and vaporized. The ions created in the plasma are drawn into the MS at the vacuum interface. The plasma density and temperature permit isotropic flow through the first orifice (typically ~1 mm diameter aperture), and the plasma undergoes free jet expansion (Scoles, 1988) in the first vacuum chamber.

This free jet structure is "skimmed" as it passes through a second aperture into the first high vacuum stage. The interface contains a skimmer cone (made of temperature-resistant alloy) with a small orifice through which the plasma (consisting of positive ions, electrons, and neutral gas) is drawn into the vacuum interface. High capacity vacuum pumps remove the gas, while ions are electrically injected into the mass spectrometer using ion optical elements and accelerating voltages.

The skimmed beam is accelerated and focused by ion optical elements (typically, cylindrically symmetric lenses). The strong electric field in these lenses (~ 1 keV/cm) leads to rapid charge separation in the skimmed beam. That is, the electrons are rapidly repelled by this field and the positive ions are accelerated and focused. The intensity of the beam is such that once the electrons are removed, the self-repulsion of the positive ions due to space charge effects (the so-called space charge explosion) leads to significant loss of ion beam intensity, an important source of analyte loss. Loss of ion current in this atmospheric-pressure-to-vacuum interface is the main contributor to low efficiency in ICP/MS.

17.4.3. Analytical Performance

The virtues of ICP/MS include efficient production of positive atomic ions; robustness of the plasma (tolerance of sample matrix material in the ion source); sample introduction at atmospheric pressure; ability to handle solid, liquid or gaseous sample forms; and speed of analysis. Sample preparation requires much less chemical purification than for TIMS, AMS or radiation detection by beta- or alpha-spectrometry.

Most elements are efficiently ionized at the high temperature of the plasma. For most metals (ionization potential <8 eV), ionization is typically greater than 90%. Many elements have an ionization efficiency >98%. The few elements not analyzed by ICP/MS are H, He, C, N, O, F, Pm, Ne, Ar, Kr, and short-lived radionuclides. These exceptions have high ionization potentials, severe spectral interference, or are better measured by radiometric counting or other mass spectrometric methods.

While a number of gases have been used for the plasma support gas, argon is the gas of choice in nearly all instruments in use today. Argon has properties that offer the best compromise in cost, robustness, thermal conductivity, ionization potential, interferences, reactivity, solvent load handling, and ease of plasma operation. Argon generally provides greater ionization for a wider range of elements due to the higher ionization potential leading to more energetic electrons.

Molecular gases such as nitrogen, oxygen, and carbon dioxide offer advantages over argon in higher thermal conductivity and better solvent load handling, but they also tend to have higher levels of interferences, more reactivity, and require higher flow and applied power to sustain a stable plasma. Helium has a higher ionization potential than argon with more energetic electrons in the discharge, but a helium plasma cannot tolerate high solvent loads.

Sensitivity is assessed by comparing signal level and stability of the background with analyte response. In early ICP/MS instruments, response was on the order of 10^6 ion counts/s per part-per-million (ug/mL) of analyte in a solution introduced at a flow rate of 1 mL/min. This is an analyte introduction rate of 1 ug/min, or 10^{14} atoms/s for the analyte ^{100}Mo. The overall efficiency is thus 10^{-8} given 10^6 counts/s *vs.* 10^{14} atoms/s from the sample. Research and development has led to efficiencies that have steadily increased over the years. State-of-the-art instruments and high efficiency sample introduction systems can now achieve 10^{-3} (0.1 percent) overall efficiency in terms of counts detected per atom in the sample. Champion efficiencies as large as 0.5 percent have been reported (Rehkamper et al., 2001). For comparison, TIMS champion efficiencies are ~1 and ~5 percent, for uranium and plutonium, respectively.

TABLE 17.4. ICP/MS precision for various instruments and sample amounts

Isotope ratio	Method*	Precision	Sample size	Efficiency	Reference
**$^{235}U/^{238}U \sim 1$	ICP/QMS	0.028%	255 pg	Not Avail	Platzner
**$^{235}U/^{238}U \sim 1$	DF-ICP/MS	0.026%	255 pg	Not Avail	Becker
$^{238}U/^{235}U = 137.9$	ICP/QMS	0.07%	663 pg	0.13%	PNNL
$^{238}U/^{235}U = 31.8$	ICP/QMS	1.4%	307 fg	0.15%	PNNL
$^{235}U/^{234}U = 160.5$	ICP/QMS	3.8%	9.42 fg	0.15%	PNNL
$^{236}U/^{234}U = 1.05$	ICP/QMS	7.4%	324 ag	0.16%	PNNL

* QMS: QMF mass analyzer, DF-ICP/MS: double-focusing sector-field mass analyzer
** From (Becker and Dietze, 2000).

The biggest performance difference in instruments is associated with the number of detectors, not the type of ion source. Single-detector instruments yield precision 0.01% to 0.5% (relative standard deviation or RSD) for large samples. For small samples, say where fewer than 10^6 counts are obtained for the minor isotope(s) of interest, precision is limited by counting statistics. With larger samples and multiple detector instruments, an RSD somewhat below 100 ppm can be achieved. This "zeroth order" picture is true for both TIMS and ICP/MS, but TIMS is the more precise and accurate method when detailed comparisons are made. Bias is significantly different for TIMS versus ICP/MS and remains a challenge for multi-collector ICP/MS (Becker, 2002). Obtaining accuracy that is commensurate with precision at the sub-100 ppm range is difficult regardless of the method used, but ICP/MS has more fundamental limits at present. Internal precision is limited for low-level samples by counting statistics, as shown in Table 17.4.

Detection limits are most often reported as concentration detection limits because most ICP/MS applications are not limited by sample amount. Absolute mass detection limits are more relevant when the amount of analyte available is highly limited. The lowest mass detection limit for ICP/MS is $\sim 10^5$ atoms, or ~ 0.1 femtogram for actinide elements.

17.4.4. Isobaric Interference

Isobaric interference is the overlap of ion peaks with nominally the same m/z. Interferences occur between isotopes of different elements (e.g., $^{54}Cr/^{54}Fe$, $^{90}Sr/^{90}Zr$, $^{187}Re/^{187}Os$), and between isotopes and molecular ions ($^{56}Fe/ArO^+$, $^{80}Se/Ar_2^+$, $^{239}Pu/^{238}UH^+$). The resulting signal at the m/z of interest represents a generally unknown combination of analyte signal and interferent signal. Molecular ions usually arise in one of two ways: they are formed in the ion source or by reaction of the ion beam with background gases in the instrument's vacuum chamber.

The nature of an isobaric interference is illustrated in Fig. 17.13, where element B is the analyte of interest for which the isotope ratio indicated by the two spectral lines is to be measured. A lighter element, A, has isotopes whose diatomic oxides AO^+ have three masses indicated as "molecular ions" in the figure, two of which occur at the element B masses of interest. The spectrum observed at the element B masses of interest is thus a composite of signals from the element B atomic

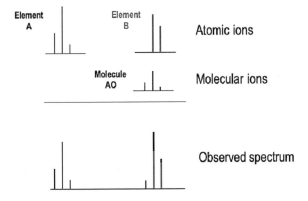

FIGURE 17.13. ICP/MS spectral analysis with isobaric interference.

ions and the element A molecular ions. In this case, interference by A in B can be corrected because an AO^+ ion exists that does not overlap any of the B isotope masses. From the intensity of the lightest isotope and the abundance of each of the three isotopes of A, one can calculate the contribution of the other AO^+ isotopes to the signals observed at the B isotope masses.

Each of the major types of MS instruments for radionuclide measurements has unique mechanisms by which isobaric interference are formed within them. In an ICP/MS, the most common interference is from atomic isobars and the diatomic molecular oxide, hydride, nitride, and argide ions. In the typical argon support gas for the ICP, argide-based interferences abound, from ^{40}Ar interfering with ^{40}Ca to $^{195}Pt^{40}Ar$ interfering with ^{235}U. The usual aqueous aerosol introduced into the ICP provides oxygen and hydrogen, which form oxides and hydrides. Examples of these interfering ions are listed in Table 17.5 with the analyte(s) of interest and the mass resolving power needed to separate them.

Considerable effort has been expended to reduce interference without losing analyte signal. These efforts include chemical removal of interference-causing atoms prior to mass analysis, with sample introduction by hydride formation at elevated temperature, and with selective removal of water vapor from the sample aerosol (Montaser, 1998). Reduction of molecular ion interferences involving high ionization potential (IP) atoms, such as argides, has been achieved by reducing the plasma temperature (via lower RF power). Reduction of plasma temperature reduces the intensity of high-IP species much more than that of lower-IP species. This "cold" plasma has been shown to reduce Ar^+ and ArO^+ faster than the reduction in intensity of the species with which they interfere (Ca and Fe, respectively).

17.4.5. Collision/Reaction Cells

Introduction of a collision/reaction cell (CRC) into the ICP/MS is a recent technique for interference reduction (Koppenaal et al., 2004). Interfering ions are eliminated by their gas phase reaction in a gas-filled cell inserted between the

TABLE 17.5. Isobaric Interference in ICP/MS and mass resolution
required to separate them

Interferent	Isobaric analyte	M/ΔM
^{40}Ar	^{40}Ca	190,000
^{40}Ar^{14}N	^{54}Fe, ^{54}Cr	2,090; 2,030
^{40}Ar^{16}O	^{56}Fe	2,500
^{35}Cl^{16}O (in Cl matrix)	^{51}V	2,600
^{40}Ar^{35}Cl	^{75}As	8,000
Br	^{79}Se ($\tau_{1/2} = 650,000$ a)	487,000
Ar$_2$	^{80}Se	9,700
Kr (impurity in Ar)	^{84}Sr	43,600
Zr	^{90}Sr ($\tau_{1/2} = 28.78$ a)	29,600
^{90}ZrO	^{106}Pd	27,400
^{59}Co^{40}Ar	^{99}Tc ($\tau_{1/2} = 213,000$ a)	9,270
Xe (impurity in Ar)	^{129}I ($\tau_{1/2} = 1.57 \times 10^7$ a)	625,000
Ba	^{135}Cs ($\tau_{1/2} = 2.3 \times 10^6$ a)	613,000
^{195}Pt^{40}Ar	^{235}U ($\tau_{1/2} = 7.04 \times 10^8$ a)	2,010
^{238}UH	^{239}Pu ($\tau_{1/2} = 24,100$ a)	37,500

ion source and the mass analyzer. The reacting ions undergo a change in mass-to-charge ratio in the cell. If the interfering ions react at a much greater rate than the ions of interest, the analyte of interest is detected with reduced background. Typically, interferences can be reduced by three to nine orders of magnitude; there is some accompanying loss of analyte, from a few percent to 10-fold loss.

Collision/reaction cells used in ICP/MS consist of an enclosure to contain the gas, apertures at each end to allow ions in and out of the cell, and a set of electrodes (e.g., octopole, hexapole, or quadrupole rod sets) to guide ions through the cell. The main considerations in the use of the method are the reagent gas (whether to use a "buffer" gas), gas number density, ion kinetic energy, cell length, multipole design, use of auxiliary fields (in addition to the ion guiding field), and reaction time. Other considerations include reagent gas purity, side reactions, and chemical resistance of the sample ions (including desired analyte, interferences, and matrix ions) to the reagent gases. The operating principles and design of RF multipole ion guides and gas collision cells have been described in detail (Gerlich, 1992).

The dramatically different reactivities of gas phase ions provide an opportunity to reduce or eliminate many isobaric interferences. Identification of suitable gaseous reagents is aided by the advances in the understanding of gas phase ion-molecule chemistry achieved in recent years (Armentrout, 2004). Hydrogen gas is the most selective reagent gas currently in use. Other reagent gases, including NH_3, H_2O, CH_4, and O_2, have been found to be much less selective, and usually require that selective ion storage/reaction methods be employed to avoid generation of new interferences.

Collision/reaction cells are typically operated at the lowest possible kinetic energy for several reasons. First, the benefits of collisional focusing are realized at low kinetic energy. Second, ions at lower energy remain longer in the cell and have more opportunity to react. Last, lower energy enables control over endothermic reactions.

Most manufacturers of ICP/MS instruments now offer a model that incorporates a collision cell. Interference reduction by sample purification, use of a collision cell (also called chemical resolution), and mass separation—are said to be orthogonal, i.e., gains in each method are independent and multiplicative with gains in the others. The ultimate instrument in this regard—a collision cell equipped, high resolution, magnetic sector MS—has not yet been built, but would enable the nearest approach yet to detection without interference of any isotope at ultra-trace levels.

17.5. Thermal Ionization Mass Spectrometry

Thermal ionization mass spectrometry, or TIMS, is sometimes called thermal emission mass spectrometry or surface ionization mass spectrometry. It is one of the mainstays of isotopic analysis. Historically, the emission of ions from a heated surface was observed nearly 100 years ago, and was used as a mass spectrometer ionization method in 1918 (Smith, 2000). Modern TIMS grew out of research in the 1920s on thermionic emission from surfaces (Langmuir and Kingdon, 1925) and has continued to be the method of choice for high precision isotopic analysis for actinides and lanthanides, as well as transition elements. TIMS is popular for isotopic analysis because the ionization method produces a relatively low energy spread ($\sim$0.5 eV), thus high performance analyses can be performed with single stage magnetic sector instruments. By comparison, other ionization techniques (e.g., electron impact) produce ions with higher energy spreads, as well as multiply-charged ions, which often requires additional focusing to obtain well-separated and well-behaved ion beams.

Langmuir and Kingdon (1925) developed the theoretical basis for thermal ionization and showed that positive ion emission was well represented by:

$$\frac{N^+}{N_0} = A e^{\frac{W-I}{kT}} \tag{17.7}$$

where:

N^+ = emission rate of positive ions
N_0 = emission rate of neutral species
W = work function of the filament metal (eV)
I = ionization potential of element or molecule (eV)
k = Boltzmann's constant (eV/K)
T = absolute temperature (K)
A = constant (see text)

This equation is now called the Saha-Langmuir equation. The constant, A, depends on the quantum energy levels of the element or molecule and of the ionizing filament material. Efficient production of positive ions occurs from filament material made of high work function metals and for analyte species with low ionization potentials. Most metals have work functions in the range of 4 to 5 eV, which effectively limits TIMS to elements (or molecular species) with ionization potentials less than 8 eV. In practice, TIMS is generally used for elements with ionization potentials of 6 eV or less. Rhenium is the filament material of choice because of its high work function (4.98 eV) and high melting point (3180 C). Commercial rhenium is available in ultra-high purity (99.999% or better), which minimizes unwanted ion emission. Other metals commonly used for filaments include tantalum and tungsten. Although platinum has the highest work function (5.13 eV), its relatively low melting point (1772 C) limits its utility for TIMS use.

Although primarily associated with positive ions, negative TIMS is also possible. The governing equation (17.8) is similar:

$$\frac{N^-}{N_0} = A e^{\frac{EA-W}{kT}} \tag{17.8}$$

where:

N^- = emission rate of negative ions
N_0 = emission rate of neutral species
W = work function of the filament metal (eV)
EA = electron affinity of element or molecule (eV)
k = Boltzmann's constant (eV/K)
T = absolute temperature (K)
A = constant

Unlike the positive ion case, negative TIMS is most efficient with low work function surfaces. Such surfaces are created by coating the ionizing filament with a low work-function material such as LaB_6 or $Ba(NO_3)_2$. Elements with high electron affinities (hence easily ionized) include the halogens, selenium, and tellurium. Oxides of some metals (e.g., Mo, Os, Re) also have high electron affinities and also have high volatilities, especially when compared to their reduced, metallic forms. Since the mid 1980s, negative TIMS has become the method of choice for high precision and high sensitivity isotopic analysis of refractory Group VIb, VIIb and VIII elements. For instance, Heumann (1995) developed and demonstrated a viable method for analyzing the geochronometer Re-Os using negative TIMS. The Re-Os system has also recently been analyzed by MC-ICPMS (Malinovsky et al., 2002; Yin et al., 2001).

Whether the ionization is positive or negative, TIMS requires careful sample preparation, often involving considerable chemical processing to separate and purify the element of interest. TIMS finds applications in geoscience, environmental analysis, cosmochemistry, biosciences, medicine, material science, and physics. Samples generally include soil, minerals, meteorites, and biological tissue. More information on the specifics of the TIMS technique and its applications is in the monograph by De Laeter (2001).

17.5.1. Instrumentation

At the heart of the TIMS ion source are one or more hot filaments that serve to vaporize and ionize atoms or molecules of interest. Once generated, the ions are accelerated, focused, and directed into the mass analyzer for measurement. The classic TIMS instrument consists of an ion source, a single magnetic sector mass separator, and an ion detector. Such an instrument is capable of measuring isotope ratios as small as 1×10^{-6}, sufficient for the isotopic analysis of most elements. For radionuclide analysis, smaller isotope ratios are often encountered. Specialized mass spectrometers include multiple magnetic and electric sectors and sector instruments with retarding quadrupole lenses (Smith, 2000) to measure down to the 10^{-9} range.

TIMS instruments tend to be table-top-sized, with total path-lengths from ion source to detector in the 1 to 2 m range. Popular sector geometries include 60° and 90° deflection, which is the nominal deflection produced by the magnetic field. The deflection radius is normally in the 10 to 30 cm range. Typical operating parameters for actinide analysis are 10 kV acceleration voltage and ~0.5 to 1 Tesla magnetic field to produce ion beams that are separated by mm-scale distances (1 amu mass difference at 238 amu ion beam mass). Ions are detected by either a Faraday cup or an electron multiplier. Because of the small mass separation, detector size can be a critical issue, especially for simultaneous detection of multiple ion masses.

Historically, most TIMS instruments were equipped with a single ion detector. In operation, the ion mass region of interest is either scanned or peak-stepped, by changing either the ion accelerating voltage or the magnetic field. Peak-stepping requires very stable electronics and magnet control (if an electromagnet is used) to select accurately and with good repeatability a suite of ion masses. The ion beam must also be of high quality and reproducibility.

A common requirement for TIMS instruments is to have flat-topped ion peaks. Flat-top peaks are produced when the ion beam is focused to a fine line and the analyzer slit (usually the slit in front of the detector) is wider than the ion beam; therefore, the entire ion beam is captured by the ion detector. Flat-top peaks are desired for high precision isotopic analysis because the measured ion beam signal is insensitive to slight instabilities in the instrument magnetic field, which causes the beam to shift position. Flat-top peaks are achieved at the cost of reduced mass resolution, but such reductions are usually insignificant compared to the gain in isotopic measurement precision. Figure 17.14 shows the TIMS peak shape, acquired by high-resolution mass scans for several neodymium isotopes. Note that several lines are slightly shifted from each other. The data were acquired with an Isoprobe-T instrument and multi-collector FC detection system.

Modern TIMS instruments are equipped with multiple ion detectors (see Fig. 17.15). In the 1980s, Faraday cup arrays became commercially available, and these provided significant improvements in isotopic precision and sample utilization. In the 1990s, arrays of pulse counting ion detectors with very compact EM

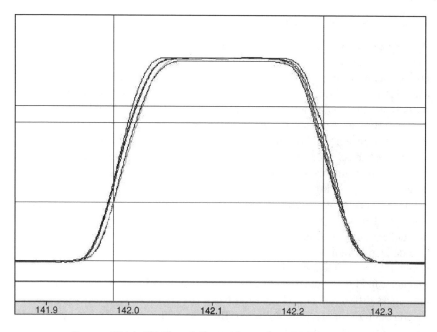

FIGURE 17.14. TIMS peak shape. Figure from GV Instruments.

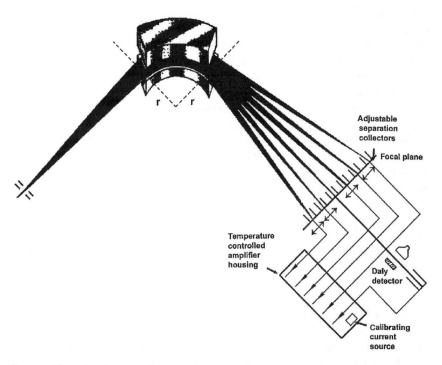

FIGURE 17.15. A schematic diagram of a multi-collector mass spectrometer. Figure from Smith (2000), pg. 14.

or microchannel plate technology (De Laeter, 1996; De Laeter, 2001) were developed for high sensitivity and high precision isotopic analysis. These are becoming commercially available.

17.5.2. Sample Preparation

Isotopic analysis of an element by TIMS is preceded by extensive chemical preparation to separate and purify the element to be assayed. Most TIMS analyses require picogram to microgram quantities of the separated and purified element, although modern instruments and sample preparation techniques have lowered detection limits to the sub-femtogram-range (Dai et al., 2001; Stoffels et al., 1994). Detailed description of sample processing techniques is beyond the scope of this discussion, but most sample preparation utilizes dissolution, gross separation, detailed purification, and preparation of the analyte into a form suitable for loading onto a mass spectrometer filament. The separation and purification steps are usually performed by procedures similar or even identical to those used in conventional radiochemical techniques (ion exchange, redox extraction, precipitation, etc.). The final purification and preparation is often an art because the chemical form on the filament is critical to producing efficient ionization and low background (Beasley et al., 1998b; Delmore et al., 1995; Kelley and Robertson, 1985; Perrin et al., 1985; Rokop et al., 1982; Smith and Carter, 1981; Walker et al., 1981).

17.5.3. Performance and Application

Many applications focus on the isotopic analysis of elements in environmental samples (water, rocks, soil, minerals), meteorites, biological tissues, and nuclear materials (such as nuclear fuels). A relatively new application has been the use of TIMS in nuclear safeguards and arms control treaty verification (IAEA, 2003). Of most interest to this discussion is the high quality isotopic analysis of actinide elements, especially U and Pu. Typical performance parameters are given in Table 17.6 for isotopic analysis of actinide elements and additional examples are given in Section 17.8.

TABLE 17.6. Performance Parameters for uranium and plutonium analysis by TIMS

Element	Measured isotope ratios	Detection limit	Comment
U	$^{234}U/^{238}U$ $^{235}U/^{238}U$ $^{236}U/^{238}U$	$\sim$1pg	Detection limit is usually limited by background U
Pu	$^{238}Pu/^{239}Pu$ $^{240}Pu/^{239}Pu$ $^{241}Pu/^{239}Pu$ $^{242}Pu/^{239}Pu$	<1 fg	Detection usually limited by sample quantity

17.6. Accelerator Mass Spectrometry

Accelerator mass spectrometry (AMS) was developed for analyzing ^{14}C in environmental and archeological specimens. It is now used to measure isotopes spanning the periodic table from tritium to plutonium. AMS represents the marriage of high-energy physics with radiochemistry and finds applications in geoscience, environmental analysis, archeology, cosmochemistry, biosciences, medicine, material science, and physics (Fifield, 1999; Hellborg et al., 2003; Herzog, 1994; Hotchkis et al., 2000; Kutschera, 1994; Liu et al., 1994; Rucklidge, 1995; Skipperud and Oughton, 2004; Tuniz et al., 1998).

AMS consists of a high-energy particle accelerator and an ion detector. Typical atom ratios for AMS isotope measurements are in the range of 10^{-10} to 10^{-15}. These ratios fall well below the 1 to 10^{-9} range measured with other mass spectrometers, except for laser-based systems with resonance ionization techniques, which can match the range of the AMS.

The accelerator began as a source of energetic ions to bombard targets. Researchers (Tuniz et al., 1998) recognized that the techniques used to create, select, and direct ion beams in the MeV energy range could be applied to measuring extremely small isotope ratios with high sensitivity and low background. The technique provides extremely high selectivity and efficient rejection of interferences, with quantitative elimination of interference by molecular species. By comparison, all other mass spectrometric systems described in this chapter suffer from varying degrees of isobaric interference by ionic molecules with nearly identical mass. High mass resolution can be applied with other MS systems to resolve molecular and isotopic ion species with small mass differences, but at the cost of lower sensitivity and increased instrument complexity.

17.6.1. Instrumentation

The AMS consists of an ion source, an input mass or momentum selector (usually a magnetic sector), a high–energy accelerator, a "stripper" (a low-pressure gas or thin foil placed in the middle of the high voltage region), post accelerator ion optics and mass selectors, and one or more detector systems. Figure 17.16 shows a generalized diagram. The high-energy accelerator is usually a tandem accelerator, which consists of two back-to-back high-energy accelerators with a stripper between the accelerators. The voltage increases from zero at one end to the middle and decreases to zero at the other end. Low-energy negative ions injected into one end of the accelerator emerge as high-energy positive ions at the other end. This arrangement allows the ion detector region to be operated at or near ground potential to simplify construction and electrical operation of the instrument.

Ions typically are produced by a secondary ion or sputter source (the usual choice for the AMS), often with a beam of Cs ions to sputter the sample to produce negative ions. These ions are extracted, focused, and accelerated to energies of 10–100 keV.

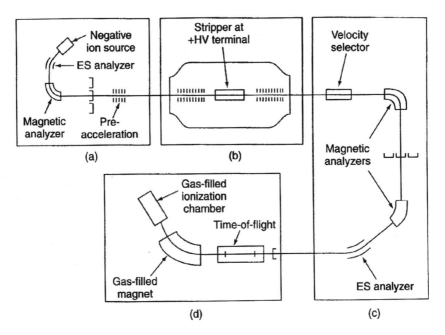

FIGURE 17.16. Schematic diagram of an AMS instrument. Figure from De Laeter (2001), pg. 62.

The input mass selector directs a specific ion mass into the tandem accelerator, which then accelerates the negative ions to MeV energies. A stripper in the central region of the accelerator dissociates molecular ions and removes multiple electrons from the ions to convert ingoing negative ions (elemental or molecular) into outgoing multi–charged positive ions. These positive ions are further accelerated *by the same potential* and emerge from the back-end of the accelerator with energies that can reach above 30 MeV, depending on the accelerator base voltage (or potential) and the degree of ionization. Charge states of +10 or more can be produced, especially for heavy ions such as the actinides. After acceleration, additional ion optics and switching magnets are used to select the ion mass and charge state and to direct these ions into detectors for measurement.

By its nature, an AMS instrument is large. Path-lengths from ion source to detector can be tens of meters. The high ion energy results in large differences in ion path angles and trajectories. After selection by mass, the ions are separated at cm-scale distances, in contrast to mm-scale separations produced in sector-based mass spectrometers. A Faraday cup detector is a common choice for the major isotope mass, whereas nuclear particle detectors (which include electron multipliers) count individual ions of the rare isotope. The high ion energies allow the use of detectors that can measure ion time-of-flight, ion energy, ion energy loss rate, and ion charge. Measurement of the ion charge/mass ratio provides additional selectivity for identifying the rare isotope.

In operation, the major and rare isotopes are measured sequentially by peak switching from one isotope to the other, as with other mass spectrometers, to determine atom ratios. The major isotope may produce a current in the microampere range, whereas the rare isotope produces a signal as low as a few counts per minute. The raw isotope ratio consists of counted pulses divided by integrated current. Calibration by measurement standards with known isotope ratios provides the factor to convert the raw ratio into an atom ratio.

Unique features of the AMS include ion energy in the MeV range compared to keV range or lower for other mass spectrometers, conversion of negative ions into positive ions, and generation of multi–charged ions. The rare and major isotopes may be measured in different charge states, depending on the element under analysis, the need to eliminate isobaric interference, and the configuration of the instrument. On the other hand, the high energy poses safety and health considerations. The ions in an AMS have sufficient energy to produce radionuclides and radiation by nuclear reactions. Ions that strike various system parts (e.g., slits, deflection plates, the inner walls of the vacuum system) can produce neutrons and x-rays. High voltages are associated with the ion source, deflection plates, and other parts of the instrument. These features require strict attention to radiological and electrical safety protocols and protection measures.

17.6.2. Sample Preparation

The AMS is used to measure rare isotopes in samples that include air, water, ice, soil, minerals, meteorites, biological tissue, and archeological artifacts. Sample sizes generally are in the mg range to produce adequate rare isotope count rates for analysis.

Samples usually are processed before analysis to extract and purify the target element and its corresponding rare isotope. Some elements require the addition of a carrier for processing and yield determination, such as iodine for ^{129}I analysis in seawater, but care must be taken to measure the rare isotope background in the carrier. By comparison, carrier addition is avoided in the analysis of ^{14}C because samples have sufficient carbon for analysis and the goal is to measure the $^{14}C/^{12}C$ ratio in the sample. The purified element is converted into a solid form suitable for generating negative ions by cesium sputtering. The element of the rare isotope under analysis can be analyzed in molecular form or in reduced elemental form, depending on the ionization characteristics of the element. For instance, carbon is analyzed as graphite whereas iodine is analyzed as silver iodide. The resultant solid is pressed into a metal holder, commonly aluminum or copper (Tuniz et al., 1998) and placed in the AMS ion source for analysis.

17.6.3. Performance and Application

The rare isotopes ^{14}C, ^{10}Be, ^{26}Al, ^{36}Cl, ^{41}Ca and ^{129}I provide the bulk of the work performed by AMS laboratories. Recently, AMS has been applied to heavy element analysis, especially for U and Pu. Other isotopes analyzed by AMS include

TABLE 17.7. Experimental Parameters for AMS, after Tuniz et al. (1998)

Isotope	Sample form	Measured isotope ratio	Detection limit	Carrier	Comments
^{14}C	Graphite	^{14}C/^{12}C	5×10^{-16}	None	—
^{10}Be	BeO	^{10}Be/^{9}Be	3×10^{-15}	Sometimes	—
^{26}Al	Al$_2$O$_3$	^{26}Al/^{27}Al	3×10^{-15}	None	—
^{36}Cl	AgCl	^{36}Cl/^{35}Cl, ^{36}Cl/^{37}Cl	1×10^{-15}	Sometimes, Chloride	Both ratios measured
^{41}Ca	CaH$_2$ or CaF$_2$	^{41}Ca/^{40}Ca	6×10^{-16}	None	—
^{129}I	AgI	^{129}I/^{127}I	3×10^{-14}	KI	—
U	U + Nb metal	^{236}U/^{238}U	1×10^{-12}	Nb metal	(Berkovits, 2000)
Pu	Pu + Fe–oxide	^{240}Pu/^{239}Pu, ^{241}Pu/^{239}Pu, ^{242}Pu/^{239}Pu	<1 fg	Fe–oxide	Detection limit in atoms; multiple ratios analyzed

tritium, ^{59}Ni, ^{90}Sr, and ^{205}Pb. Detection limits for AMS are typically given as limiting isotope ratios, which is the measured quantity in AMS. The overall ion transmission of a typical AMS is such that the absolute atom detection efficiencies are lower than either TIMS or ICPMS—on the order of 1 count per 100,000 atoms injected into the instrument. Nevertheless, the extremely low background of the AMS technique permits the detection and quantification at the sub-femtogram-level, which is comparable to ICP/MS and TIMS analyses. Some experimental parameters are given in Table 17.7 for the most commonly analyzed isotopes.

For some radionuclides, only the direct measurement is needed, based on calibration with a radionuclide standard and reference to the measured sample mass. For other radionuclides, the isotope ratio to its stable element is needed. An example of a more complex situation is measurement of the ^{14}C/^{12}C isotope ratio of an environmental or archeological sample in comparison to the modern atmospheric CO$_2$ value. This value must be adjusted for anthropogenic ^{14}C produced by atmospheric testing of nuclear weapons and by other nuclear operations, and also for changes in atmospheric CO$_2$ with cosmic-ray flux fluctuations over time. For an element such as Pu, which has no stable isotope, the total quantity is measured by isotope dilution mass spectrometry in which the sample is traced (or spiked) with ^{242}Pu or ^{244}Pu (see Section 17.3.3).

17.7. Other Mass Spectrometers

17.7.1. Secondary Ionization Mass Spectrometer

Secondary ionization mass spectrometry (SIMS) is yet another method to analyze solid samples directly. SIMS uses a focused beam of ions (called the primary ion beam) to sputter atoms from the surface of a sample. A small fraction of the sputtered atoms are ionized by the sputtering action, hence the term secondary ionization. The secondary ions are extracted, accelerated, and analyzed by a mass analyzer. Two main configurations exist: conventional SIMS, which uses electric

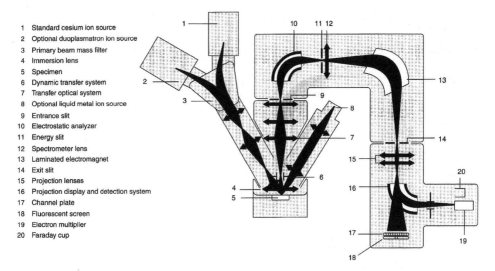

1 Standard cesium ion source
2 Optional duoplasmatron ion source
3 Primary beam mass filter
4 Immersion lens
5 Specimen
6 Dynamic transfer system
7 Transfer optical system
8 Optional liquid metal ion source
9 Entrance slit
10 Electrostatic analyzer
11 Energy slit
12 Spectrometer lens
13 Laminated electromagnet
14 Exit slit
15 Projection lenses
16 Projection display and detection system
17 Channel plate
18 Fluorescent screen
19 Electron multiplier
20 Faraday cup

FIGURE 17.17. The Cameca IMS-6F secondary ion mass spectrometer. Figure from De Laeter (2001), pg. 49.

and magnetic sectors to perform the mass separation, and TOF-SIMS, which uses a time-of-flight mass analyzer. Both variants are available as commercial instruments, as in the conventional SIMS instrument shown in Fig. 17.17.

The primary ion beam is formed on the left and accelerated and focused onto a sample held at high voltage. Secondary ions are extracted and accelerated into the analyzing mass spectrometer, which consists of an electrostatic sector (for energy focusing) and a magnetic sector (for mass separation). Among the detectors used to analyze the ions is an imaging detector. The ion optics on the Cameca instrument are specifically designed to permit imaging of the sample with ions.

Production of secondary ions depends on many factors, but the composition of the primary ion beam is vital to the efficient production of ions. Cesium or oxygen ions (positive or negative, depending upon whether negative or positive secondary ions are desired) are used (Cristy, 2000; De Laeter, 2001). Other primary ions include noble gases, especially Ar, and more exotic species such as rhenium oxide (Groenewold et al., 1997). The primary ion beam usually is mass selected with a magnetic sector or quadrupole mass analyzer, and focused to a micrometer-sized spot on the sample under analysis. Normally, the beam can be raster-scanned across the sample with magnetic and electrostatic deflection in a manner similar to the electron optics used in an electron microscope.

An advantage of SIMS is that the method provides micro-scale analysis of the sample. The primary ion beam can be set to analyze a small area or raster-scan across the sample. Some instruments are designed to produce ion images analogous to images obtained in an electron microscope. The imaging capability allows for high-resolution isotopic (and elemental) analysis of the sample, with spatial resolution of 1 μm or better that depends on the instrument. The sputtering action of the primary ion beam can be controlled to permit calculation of the depth of the material removed during the analysis. Depth resolution to a few nanometers is possible. Because of these micro-scale and imaging capabilities, a SIMS instrument is sometimes referred to as an ion microprobe or ion microscope.

Production of secondary ions is non-selective in that all elements can be ionized by this method. Thus, the entire periodic table is accessible using SIMS. The relative and absolute efficiencies for the production of ions for a given element are highly dependent upon the composition of the sample, as well as the configuration and operating parameters of the instrument. Generally, isotopic compositions are easily measured, whereas elemental compositions, which are derived from the analysis of isotopes of a given element, are more difficult to determine. Good quantitative analysis requires the use of calibration standards that closely resemble the sample under analysis.

Preparation of samples for analysis resembles that used to prepare samples for electron microscopy. In general, preparation consists of polishing small specimens and mounting them on a planchet. A common planchet consists of a cylinder of vitreous carbon with polished ends. The planchet must provide a conductive pathway to drain the charge imparted by the impact of the primary ion beam. Without some means to dissipate the charge, the sample will become charged, which can deflect the primary ion beam and can interfere with the extraction of secondary ions.

Operation of a SIMS instrument resembles both that of an isotope ratio mass spectrometer and an electron microscope. Most SIMS instruments include an optical microscope so that the sample can be directly viewed during analysis, which allows for accurate positioning of the area of interest on the sample. Data can be in the standard mode used for other types of mass spectrometers in which ions are produced and the mass spectrum is analyzed by scanning or peak-hopping. This mode is sometimes called the microprobe mode in SIMS. Another application for SIMS is the acquisition of ion-images. This mode is called the microscope mode because the SIMS is operated as an ion microscope.

17.7.2. Laser Ablation Mass Spectrometer

Laser ablation (LA) is a powerful tool for mass spectrometric sampling of radionuclides from solids. When sufficiently intense laser light strikes a solid, material is removed from the surface. This material can be in the form of atoms, molecules, ions, electrons, or small particles. Several approaches can be used to analyze the surface for these species. The light that usually also is emitted can be resolved by wavelength to identify atomic or molecular species associated with the sample surface. Depending upon the wavelength of the laser, the atoms and ions in the laser plume can be excited, and the resulting emission can be detected.

In laser ionization or laser ablation mass spectrometry, the ions formed by the interaction of the laser with solid are directly sampled into a mass spectrometer. Another option is that the material removed from the surface can be swept into an inductively coupled plasma (ICP) for atomic analysis either by light emission from the ICP or by analyzing the ions produced in the ICP. This latter technique is referred to as laser ablation ICP/MS. Numerous mass spectrometer analyzer types have been used for LA-ICPMS including time-of-flight (TOF), quadrupole, sector field, and ion trap. The analyzer type is determined by the application.

One of the problems that has been noted in laser ablation is fractionation between the sample surface and the material removed. For any given sample type, certain elements will be preferentially removed, giving these elements a higher relative abundance in the material analyzed by the mass spectrometer. If fractionated, this material is not representative of the sample, and biased data results. Research has shown that LA-ICPMS with deeper UV wavelengths (e.g., 157 nm, 193 or 213 nm) results in less fractionation. A pulsed laser with short pulse widths—on the order of picoseconds to femtoseconds—provides even less fractionation.

The most common LA-ICP/MS work to date is with flash-lamp pumped Nd:YAG lasers that produce a light pulse of 3–10 nanoseconds in width and are relatively inexpensive. Shorter pulse width lasers (picosecond to femtosecond) are widely available but more expensive. A wavelength other than IR or UV is used only for resonant laser ionization or ablation. In this case, the goal is to excite or ionize a specific element preferentially, and the laser wavelength is chosen to correspond to a particular transition of an element, as discussed in Section 17.7.4.

The laser typically used for laser ablation is the frequency quadrupled Nd:YAG (266 nm) with a pulse width of a few nanoseconds. With this laser, some fractionation will occur. Unless a good external standard is used, quantification of the elements in the sample is difficult if not impossible. This is not an impediment to use of laser ablation in the field of radiochemistry or radionuclides. One of the most important aspects of radionuclide analysis is determining isotope ratios of elements of interest. All isotopes of a particular element behave the same when removed from a solid sample by a laser pulse. Ionization of those isotopes can depend on the laser bandwidth and polarization so care must be used when using lasers for direct ionization of atomic species.

The purpose of LA-ICP/MS use is to remove material from the surface and transport it into the ICP for ionization to obtain isotope ratios quickly with little sample preparation. Inter-element isotope ratios are also important in certain radionuclide applications, but are compromised by fractionation issues. Laser ablation provides an overview of which analytes and isotopes are present and approximates the concentration of each.

Literature reviews and primary literature are available for a more thorough analysis (Durrant, 1999; Gunther et al., 2000; Russo et al., 2002; Winefordner et al., 2000). (Gastel et al., 1997) measured long-lived radionuclides in a concrete matrix using LA-ICPMS. The radionuclides investigated were ^{99}Tc, ^{232}Th, ^{233}U and ^{237}Np. With a quadrupole mass spectrometer, detection limits on the order of 10 ng/g were achieved while a double-focusing sector instrument was able to deliver sub ng/g detection limits. (Gibson, 1998) used resonant laser ablation mass spectrometry to investigate actinide oxides. Oxides of Th, U, Np, and Am were imbedded in a copper matrix and analyzed by resonant LA-MS. Since actinides are present as oxides in many common forms such as glass, ceramics, soils, and others, the study was designed to determined whether resonant laser ablation would provide any benefit over normal LA in the analysis of these elements. Due to the many closely spaced energy levels of these elements, a significant advantage of resonant laser ablation over normal laser ablation was not observed, but the study showed the ability to use laser ablation for these types of samples.

Becker and Dietze (2000) used laser ablation to measure long-lived radionu-clides in geological samples, high-purity graphite, and concrete. An effort was made to improve quantification by using geologic standards and synthetic stan-dards of the graphite and concrete. Solution nebulization was also used as a cali-bration method. The isotopes studied were ^{99}Tc, ^{232}Th, ^{233}U, ^{235}U, ^{237}Np and ^{238}U. The detection limits that resulted were in the low pg/g range. Boulyga et al. (2003) used LA-ICP/MS to determine plutonium and americium in mosses. To improve quantification, isotope dilution was used. Detection limits in the single fg/g range were demonstrated for both elements.

17.7.3. Glow Discharge Mass Spectrometer

Another method to analyze solid samples directly is glow discharge MS (GD-MS). A glow discharge is a reduced-pressure plasma ($\sim$1 torr) that is formed between two electrodes. The two main categories of glow discharge are direct current (DC) and radiofrequency (RF). The DC glow discharge requires the solid, electrically conducting sample as one of the electrodes. An RF glow discharge can analyze non-conducting samples. The GD has been used as a source for both atomic emission and mass spectrometry, but our focus is on the glow discharge as an ion source for mass spectrometry of radionuclides. The reader is directed to additional literature for more information concerning glow discharges (Marcus and Broekaert, 2002) (Baude et al., 2000).

Because the only commercially available GD-MS systems use a DC glow dis-charge, the range of samples that can be analyzed has been somewhat limited, especially in the field of radionuclide determination. Once an RF GD-MS is avail-able to purchase, it is possible that GD-MS could become more popular for these applications.

Betti (1996) and co-workers used GD-MS for sample screening in isotopic measurements of zirconium, silicon, lithium, boron, uranium, and plutonium in nuclear samples. The results obtained from the GD-MS were compared with re-sults from thermal ionization mass spectrometry (TIMS). For boron and lithium concentrations from μg/g to ng/g levels, isotopic ratios determined by GD-MS were comparable to TIMS in terms of accuracy and precision. Uranium isotopic ratios determined by GD-MS were also in good agreement with values measured by TIMS with regards to accuracy. Chartier et al. (1999) used GD-MS to analyze erbium and uranium in molybdenum-uranium fuel samples. The ratio of ^{166}Er to ^{238}U was then compared to numbers determined by thermal ionization mass spec-trometry. The ratio of erbium to uranium was accurate to within 3% of the number determined by TIMS.

Pajo et al. (2001a) used GD-MS to measure impurities in uranium dioxide fuel and showed that these impurities could be used to identify the original source of confiscated, vagabond nuclear materials. De las Heras et al. (2000) used GD-MS to determine neptunium in Irish Sea sediment samples. The sediment samples were compacted into a disk that was used with a tantalum secondary cathode in the glow discharge. Using a doped marine sediment standard for calibration, detection limits down to the mid pg/g level were determined.

17.7.4. *Resonance Ionization Mass Spectrometer*

A resonance ionization mass spectrometer (RIMS) uses a tunable, narrow band-width laser to excite an atom or molecule to a selected energy level that is then analyzed by MS. The selective ionization often is accomplished by absorption of more photons from the exciting laser, but can also be effected by a second laser or a broadband photon source. Multiple photon absorption can result in direct ioniza-tion or in production of excited species that can then be ionized with a low-energy photon source (IR laser) or by a strong electric field. Resonance ionization methods have been applied to nearly all elements in the periodic table and to many radionu-clides, including Cs (Pibida et al., 2001), Th (Fearey et al., 1992), U (Herrmann et al., 1991), Np (Riegel et al., 1993), Pu (Smith, 2000; Trautmann et al., 2004; Wendt et al., 2000), radioxenon and radiokrypton (Watanabe et al., 2001; Wendt et al., 2000), and [41]Ca (Wendt et al., 1999).

Resonance ionization MS is a sensitive and accurate method for determining the ionization potential of atoms. Erdmann et al. (1998) recently reported improved measurements of the ionization potentials for 9 actinide elements by RIMS.

17.8. Applications

Mass spectrometry finds applications in many fields. A partial list of fields includes:

- archeology
- bioassay
- biosciences
- cosmochemistry
- geochemistry
- health physics
- environmental monitoring
- nuclear non-proliferation treaty monitoring
- nuclear science
- radiochemistry

All of these fields have requirements for sensitive analysis of radionuclides. In general, mass spectrometry is used for measuring long-lived radionuclides at low abundances, where radiation-counting techniques provide insufficient sensitivity. Modern mass spectrometers can detect individual atoms, which translates into

TABLE 17.8. Suitable mass spectrometric techniques for analyzing radionuclides

Isotope	ICP/MS	TIMS	AMS	Other
Tritium			Yes	SIMS, ^{3}He-NGMS
^{10}Be			Yes	
^{14}C			Yes	
^{10}Be			Yes	
^{26}Al			Yes	
^{36}Cl			Yes	
^{41}Ca		Yes	Yes	SIMS
^{59}Ni			Yes	
^{90}Sr	Yes		Yes	RIMS
^{99}Tc	Yes	Yes	Yes	SIMS
^{129}I	Yes	Yes	Yes	SIMS
^{135}Cs	Yes	Yes		RIMS
U	Yes	Yes	Yes	SIMS
Pu	Yes	Yes	Yes	SIMS, RIMS
Fission products	Yes			RIMS
Activation products	Yes			RIMS

sub-femto-gram sensitivities in real-world samples. A useful rule-of-thumb is that mass spectrometry is generally more sensitive than radiation counting for isotopes with half-lives greater than 100 years. This rule depends on the specific radionuclide in question, its decay scheme, and the radiation counting method.

A survey by radionuclide group and application is given below. Table 17.8 lists most of the radionuclides measured by various mass spectrometric techniques. Specific examples are given below. The interested reader is directed for further information to the many excellent reviews on specific mass spectrometric techniques, including Platzner (1997), Montaser (1998), Tuniz et al. (1998), Barshick et al. (2000), de Laeter (2001).

17.8.1. Uranium

ICP/MS and TIMS are the methods of choice for analyzing uranium isotopes. TIMS can measure all of the long-lived isotopes of uranium, including ^{236}U, with high precision (0.01 percent or better) and high sensitivity (10^{-12} g or less) with multiple collectors (Adriaens et al., 1992; Becker and Dietze, 1998; De Laeter, 2001; Delanghe et al., 2002; Efurd et al., 1995; Pajo et al., 2001b; Platzner, 1997; Sahoo et al., 2002; Smith, 2000; Stoffels et al., 1994; Taylor et al., 1998; Yokoyama et al., 2001). The TIMS requires extensive chemical processing to isolate uranium from the sample. In contrast, ICP/MS, typically with a quadrupole mass analyzer—although multi-collector, sector-based ICP/MS instruments are becoming increasingly popular—is used for various samples, often without chemically separating the uranium. The ICP/MS can provide isotopic information at the 0.01 to 1 percent precision level for the major isotopes, ^{235}U and ^{238}U (Aldstadt et al., 1996; Becker et al., 2002; Becker et al., 2004a; Becker and Dietze, 2000; Bellis et al., 2001; Boulyga and Becker, 2001; Boulyga et al., 2000; Haldimann et al.,

2001; Kerl et al., 1997; Magara et al., 2002; Manninen, 1995; Platzner et al., 1999; Schaumloffel et al., 2005; Uchida et al., 2000; Wyse et al., 1998). Recently, AMS has been used for uranium analysis, especially for measuring rare isotopes such as ^{236}U (Brown et al., 2004; Danesi et al., 2003; Fifield, 2000; Tuniz, 2001; Zhao et al., 1997).

17.8.2. Other Actinides

As with uranium, TIMS is the system of choice for measuring actinides such as plutonium, americium, and neptunium (Beasley et al., 1998a; Beasley et al., 1998b; Beasley et al., 1998c; Dai et al., 2001; Kelley et al., 1999; Kersting et al., 1999; Lewis et al., 2001; Magara et al., 2000; Poupard and Jouniaux, 1990; Shen et al., 2003; Stoffels et al., 1994). Very high sensitivity ($<10^{-15}$ g) can be achieved for plutonium and other actinides due to the low backgrounds for these elements in typical environmental samples. Isotopic data obtained from measurement of several actinides (e.g., neptunium, plutonium, americium, curium) can provide diagnostic information on the conditions under which these elements were formed.

New mass spectrometric methods have been developed for high sensitivity actinide analysis. ICP/MS is routinely used for actinide analyses in various samples (Agarande et al., 2001; Alonso et al., 1993; Alonso et al., 1995; Becker, 2003; Becker and Dietze, 1998; Becker et al., 2004b; Boulyga and Becker, 2002; Boulyga et al., 2003; Chiappini et al., 1996; Henry et al., 2001; Jerome et al., 1995; Kenna, 2002; Kim et al., 1991; Kim et al., 2000; Magara et al., 2000; Moreno et al., 1997; Muramatsu et al., 2001; Muramatsu et al., 1999; Pappas et al., 2004; Rodushkin et al., 1999; Rondinella et al., 2000; Sturup et al., 1998; Taylor et al., 2001; Ting et al., 2003; Wyse and Fisher, 1994; Wyse et al., 1998; Zoriy et al., 2004). Since the late 1990s, AMS methods have been developed to measure plutonium and other actinides (Brown et al., 2004; Lee et al., 2001; McAninch et al., 2000; Oughton et al., 2000; Oughton et al., 2004; Priest et al., 1999; Skipperud and Oughton, 2004). RIMS has also been used for ultrasensitive actinide analyses (Gruning et al., 2004; Trautmann et al., 2004).

17.8.3. Fission Products

Fission-produced isotopes that are currently measured by MS are ^{129}I, ^{99}Tc, ^{90}Sr, and long-lived noble gases. Both ^{129}I and ^{99}Tc are measured with negative ionization TIMS, ICP/MS and AMS (Beals and Hayes, 1995; Becker and Dietze, 1997; Becker et al., 2000; Fifield et al., 2000; Hidaka et al., 1999; McAninch et al., 1998; Probst, 1996; Raisbeck and Yiou, 1999; Roberts and Caffee, 2000; Yamamoto et al., 1995; Yiou et al., 2004). Methods for analyzing long-lived ^{135}Cs have been developed using RIMS, TIMS, ICP/MS, and AMS (Karam et al., 2002; Lee et al., 1993; Litherland and Kilius, 1997; Meeks et al., 1998; Pibida et al., 2001; Song et al., 2001). Strontium-90 can now be measured by a RIMS technique that uses three lasers to resonant excitation of ^{90}Sr, followed by mass analysis by quadrupole MS (Wendt et al., 1997). Detection limits are in the 10^5 atom range at ^{90}Sr/^{88}Sr isotope ratios as small as 10^{-10}.

17.8.4. Activation Products

Radionuclides with the potential for detection by MS include tritium, ^{14}C, and ^{36}Cl. AMS is the method of choice for ^{14}C and ^{36}Cl (Fifield, 1999; Fifield, 2000; Herzog, 1994; Hotchkis et al., 2000; Rucklidge, 1995; Skipperud and Oughton, 2004; Tuniz et al., 1998). Tritium can be analyzed by measuring its ^{3}He decay product using noble gas mass spectrometry (NGMS) (De Laeter, 2001; Eaton et al., 2004; Faure, 1986; Love et al., 2002), but high sensitivity measurements may require several months to produce adequate levels of ^{3}He for detection. AMS has been successfully developed for tritium analyses and is routinely used (Chiarappa-Zucca et al., 2002; King et al., 1987; Love et al., 2002; Tuniz et al., 1998).

17.9. Instrumental Advancements

As mentioned in the Introduction to this chapter, development of new MS systems has focused on the ability to accurately measure smaller samples, either for greater sensitivity or shorter-lived radionuclides, and to more effectively individuate these samples. In addition to these advancements, refinements of the instrumentation are now allowing manufactures to scale down the size of their MS workstations.

The MS systems discussed in this chapter are large, laboratory-scale instruments. Recent developments in building small mass analyzers, in miniaturized electronics, and in compact vacuum systems, have stimulated efforts to build a portable, or fieldable, MS. Chances are good for producing suit-case sized instruments that can be used in a remote office, mounted on aircraft or trucks, or placed in a remote location for environmental monitoring. Prototypes of such instruments have been demonstrated and commercial versions are becoming increasingly available. Several companies are producing portable, battery-powered systems that are used for screening analyses on-site (i.e., not in the laboratory). One drawback is mass; these instruments are still quite heavy, at 20–25 kg for an instrument the size of a duffle bag.

An internet keyword search for "portable mass spec" reveals a great deal of advancement in these endeavors. More is certain, as lighter and stronger materials are found and engineered to produce more portable and rugged systems. Ongoing improvements in battery technology will also have an impact. These efforts parallel the remote automated systems discussed in Chapter 15, as the need for immediate and/or ongoing analysis of samples becomes ever greater in our society.

Appendix

Appendix A1: ASTM Radiological Standard Test Methods, Practices, and Guides (ASTM 2005)

Committee	Publication	Title
C26.01	C0859-92b	Standard Terminology Relating to Nuclear Materials
C26.05	C0696-99	Standard Test Methods for Chemical, Mass Spectrometric, and Spectrochemical Analysis of Nuclear-Grade Uranium Dioxide Powders and Pellets
C26.05	C0697-98	Standard Test Methods for Chemical, Mass Spectrometric, and Spectrochemical Analysis of Nuclear-Grade Plutonium Dioxide Powders and Pellets
C26.05	C0698-98	Standard Test Methods for Chemical, Mass Spectrometric, and Spectrochemical Analysis of Nuclear-Grade Mixed Oxides $((U, Pu)O_2)$
C26.05	C0758-98	Standard Test Methods for Chemical, Mass Spectrometric, Spectrochemical, Nuclear, and Radiochemical Analysis of Nuclear-Grade Plutonium Metal
C26.05	C0759-98	Standard Test Methods for Chemical, Mass Spectrometric, Spectrochemical, Nuclear, and Radiochemical Analysis of Nuclear-Grade Plutonium Nitrate Solutions
C26.05	C0761-96	Standard Test Methods for Chemical, Mass Spectrometric, Spectrochemical, Nuclear, and Radiochemical Analysis of Uranium Hexafluoride
C26.05	C0799-99	Standard Test Methods for Chemical, Mass Spectrometric, Spectrochemical, Nuclear, and Radiochemical Analysis of Nuclear-Grade Uranyl Nitrate Solutions
C26.05	C1342-96	Standard Practice for Flux Fusion Sample Dissolution

C26.05	C1344-97	Standard Test Method for Isotopic Analysis of Uranium Hexafluoride by Single-Standard Gas Source Mass Spectrometer Method
C26.05	C1347-96a	Standard Practice for Preparation and Dissolution of Uranium Materials for Analysis
C26.05	C1380-97	Standard Test Method for Determination of Uranium Content and Isotopic Composition by Isotope Dilution Mass Spectrometry
C26.05	C1413-99	Standard Test Method for Isotopic Analysis of UF_6 and UNH Solutions by Thermal Ionization Mass Spectrometry
C26.05	E0219-80(1995)	Standard Test Method for Atom Percent Fission in Uranium Fuel (Radiochemical Method)
C26.05	E0244-80(1995)	Standard Test Method for Atom Percent Fission in Uranium and Plutonium Fuel (Mass Spectrometric Method)
C26.05	E0267-90(1995)	Standard Test Method for Uranium and Plutonium Concentrations and Isotopic Abundances
C26.05	E0318-91	Standard Test Method for Uranium in Aqueous Solutions by Colorimetry
C26.05	E0321-96	Standard Test Method for Atom Percent Fission in Uranium and Plutonium Fuel (Neodymium-148 Method)
C26.05.01	C0998-90(1995)e1	Standard Practice for Sampling Surface Soil for Radionuclides
C26.05.01	C0999-90(1995)e1	Standard Practice for Soil Sample Preparation for the Determination of Radionuclides
C26.05.01	C1000-90(1995)e1	Standard Test Method for Radiochemical Determination of Uranium Isotopes in Soil by Alpha Spectrometry
C26.05.01	C1001-90(1995)e1	Standard Test Method for Radiochemical Determination of Plutonium in Soil by Alpha Spectroscopy
C26.05.01	C1163-98	Standard Test Method for Mounting Actinides for Alpha Spectrometry Using Neodymium Fluoride
C26.05.01	C1205-97	Standard Test Method for The Radiochemical Determination of Americium-241 in Soil by Alpha Spectrometry
C26.05.01	C1284-94	Standard Practice for Electrodeposition of the Actinides for Alpha Spectrometry
C26.05.01	C1387-98	Guide for the Determination of Technetium-99 in Soil

C26.05.01	C1402-98	High-Resolution Gamma-Ray Spectrometry of Soil Samples
C26.05.02	C1108-99	Standard Test Method for Plutonium by Controlled-Potential Coulometry
C26.05.02	C1165-90(1995)e1	Standard Test Method for Determining Plutonium by Controlled-Potential Coulometry in H_2SO_4 at a Platinium Working Electrode
C26.05.02	C1168-90(1995)e1	Standard Practice for Preparation and Dissolution of Plutonium Materials for Analysis
C26.05.02	C1206-91e1	Standard Test Method for Plutonium by Iron (II)/Chromium (VI) Amperometric Titration
C26.05.02	C1235-99	Standard Test Method for Plutonium by Titanium(III)/Cerium(IV) Titration
C26.05.02	C1267-94	Standard Test Method for Uranium by Iron (II) Reduction in Phosphoric Acid Followed by Chromium (VI) Titration in the Presence of Vanadium
C26.05.02	C1307-95	Standard Test Method for Plutonium Assay by Plutonium (III) Diode Array Spectrophotometry
C26.05.02	C1411-99	Standard Practice for the Ion Exchange Separation of Uranium and Plutonium Prior to Isotopic Analysis
C26.05.02	C1414-99	Standard Practice for the Separation of Americium from Plutonium by Ion Exchange
C26.05.02	C1415-99	Standard Test Method for Pu-238 Isotopic Abundance by Alpha Spectrometry
C26.05.03	C1109-98	Standard Test Method for Analysis of Aqueous Leachates from Nuclear Waste Materials Using Inductively Coupled Plasma-Atomic Emission Spectrometry
C26.05.03	C1310-95	Standard Test Method for Determining Radionuclides in Soils by Inductively Coupled Plasma-Mass Spectrometry Using Flow Injection Preconcentration
C26.05.03	C1317-95	Standard Practice for Dissolution of Silicate or Acid Resistant Matrix Samples
C26.05.03	C1345-96	Standard Test Method for the Analysis of Total and Isotopic Uranium and Total Thorium in Soils by Inductively Coupled Plasma-Mass Spectrometry
C26.05.03	C1379-97	Standard Test Method for Analysis of Urine for Uranium-235 and Uranium-238 by Inductively Coupled Plasma Mass Spectrometry

C26.05.03	C1412-99	Standard Practice for Microwave Oven Dissolution of Glass Containing Radioactive and Mixed Wastes
C26.05.04	C1204-91(1996)	Standard Test Method for Uranium in the Presence of Plutonium by Iron(II) Reduction in Phosphoric Acid Followed by Chromium(VI) Titration
C26.05.04	C1287-95	Standard Test Method for Determination of Impurities in Uranium Dioxide by Inductively Coupled Plasma Mass Spectrometry
C26.05.04	C1295-95	Standard Test Method for Gamma Energy Emission from Fission Products in Uranium Hexafluoride
C26.05.05	C1110-88(1997)e1	Standard Practice for Sample Preparation for X-Ray Emission Spectrometric Analysis of Uranium in Ores Using the Glass Fusion or Pressed Powder Method
C26.05.05	C1254-99	Standard Test Method for Determination of Uranium in Mineral Acids by X-Ray Fluorescence
C26.05.05	C1255-93(1999)	Standard Test Method for Analysis of Uranium and Thorium in Soils by Energy Dispersive X-Ray Fluorescence Spectroscopy
C26.05.05	C1343-96	Standard Test Method for Determination of Low Concentrations of Uranium in Oils and Organic Liquids by X-Ray Fluorescence
C26.05.05	C1416-99	Standard Test Method for Uranium Analysis in Natural and Waste Water by X-Ray Fluorescence
C26.06	C1215-92(1997)	Standard Guide for Preparing and Interpreting Precision and Bias Statements in Test Method Standards Used in the Nuclear Industry
C26.08	C1009-96e1	Standard Guide for Establishing a Quality Assurance Program for Analytical Chemistry Laboratories Within the Nuclear Industry
C26.08	C1068-96e1	Standard Guide for Qualification of Measurement Methods by a Laboratory Within the Nuclear Industry
C26.08	C1128-95e1	Standard Guide for Preparation of Working Reference Materials for Use in the Analysis of Nuclear Fuel Cycle Materials
C26.08	C1156-95e1	Standard Guide for Establishing Calibration for a Measurement Method Used to Analyze Nuclear Fuel Cycle Materials

C26.08	C1188-91(1997)e1	Standard Guide for Establishing a Quality Assurance Program for Uranium Conversion Facilities
C26.08	C1210-96E1	Standard Guide for Establishing a Measurement System Quality Control Program for Analytical Chemistry Laboratories Within the Nuclear Industry
C26.08	C1297-95e1	Standard Guide for Qualification of Laboratory Analysts for the Analysis of Nuclear Fuel Cycle Materials
C26.10	C1030-95	Standard Test Method for Determination of Plutonium Isotopic Composition by Gamma-Ray Spectrometry
C26.10	C1133-96	Standard Test Method for Nondestructive Assay of Special Nuclear Material in Low Density Scrap and Waste by Segmented Passive Gamma-Ray Scan
C26.10	C1207-97	Standard Test Method for Nondestructive Assay of Plutonium in Scrap and Waste by Passive Neutron Coincidence Counting
C26.10	C1221-92(1998)	Standard Test Method for Nondestructive Analysis of Special Nuclear Materials in Homogeneous Solutions by Gamma-Ray Spectrometry
C26.10	C1268-94	Standard Test Method for Quantitative Determination of Americium-241 in Plutonium by Gamma-Ray Spectrometry
C26.10	C1316-95	Standard Test Method for Nondestructive Assay of Nuclear Material in Scrap and Waste by Passive-Active Neutron Counting Using a Cf-252 Shuffler
D19.04	D1890-96	Standard Test Method for Beta Particle Radioactivity of Water
D19.04	D1943-96	Standard Test Method for Alpha Particle Radioactivity of Water
D19.04	D2460-97	Standard Test Method for Alpha-Particle-Emitting Isotopes of Radium in Water
D19.04	D2907-97	Standard Test Methods for Microquantities of Uranium in Water by Fluorometry
D19.04	D3084-96	Standard Practice for Alpha-Particle Spectrometry of Water
D19.04	D3454-97	Standard Test Method for Radium-226 in Water
D19.04	D3648-95	Standard Practices for the Measurement of Radioactivity
D19.04	D3649-91	Standard Test Method for High-Resolution Gamma-Ray Spectrometry of Water

D19.04	D3865-97	Standard Test Method for Plutonium in Water
D19.04	D3972-97	Standard Test Method for Isotopic Uranium in Water by Radiochemistry
D19.04	D4107-91	Standard Test Method for Tritium in Drinking Water
D19.04	D4785-93	Standard Test Method for Low-Level Iodine-131 in Water
D19.04	D4922-94e1	Standard Test Method for Determination of Radioactive Iron in Water
D19.04	D4962-95e1	Standard Practice for NaI(Tl) Gamma-Ray Spectrometry of Water
D19.04	D5072-92	Standard Test Method for Radon in Drinking Water
D19.04	D5174-97	Standard Test Method for Trace Uranium in Water by Pulsed-Laser Phosphorimetry
D19.04	D5411-93	Standard Practice for Calculation of Average Energy Per Disintegration (E) for a Mixture of Radionuclides in Reactor Coolant
D19.04	D5811-95	Standard Test Method for Strontium-90 in Water
D19.04	D6239-98a	Standard Test Method for Uranium in Drinking Water by High-Resolution Alpha-Liquid-Scintillation Spectrometry
D19.05	D5673-96	Standard Test Method for Elements in Water by Inductively Coupled Plasma-Mass Spectrometry
D22.05	D6327-98	Standard Test Method for Determination of Radon Decay Product Concentration and Working Level in Indoor Atmospheres by Active Sampling on a Filter
E01.05	E0402-95	Standard Test Method for Spectrographic Analysis of Uranium Oxide
E10.03	E1893-97	Standard Guide for Selection and Use of Portable Radiological Survey Instruments for Performing In Situ Radiological Assessments in Support of Decommissioning
E10.05	E0181-93e1	Standard Test Methods for Detector Calibration and Analysis of Radionuclides
E10.05	E0266-92	Standard Test Method for Measuring Fast-Neutron Reaction Rates by Radioactivation of Aluminum
E10.05	E0343-96	Standard Test Method for Measuring Reaction Rates by Analysis of Molybdenum-99 Radioactivity From Fission Dosimeters
E10.05	E0393-96	Standard Test Method for Measuring Reaction Rates by Analysis of Barium-140 From Fission Dosimeters
E10.05	E0481-97	Standard Test Method for Measuring Neutron Fluence Rate by Radioactivation of Cobalt and Silver
E10.05	E0523-92	Standard Test Method for Measuring Fast-Neutron Reaction Rates by Radioactivation of Copper

| E10.05 | E0692-98 | Standard Test Method for Determining the Content of Cesium-137 in Irradiated Nuclear Fuels by High-Resolution Gamma-Ray Spectral Analysis |
| E10.05 | E0704-96 | Standard Test Method for Measuring Reaction Rates by Radioactivation of Uranium-238 |

Appendix A2: Standard Methods, part 7000—Radioactivity (Standard Methods 2005)

Method no.	Contents
7010	Introduction
	A. General Discussion
	B. Sample Collection and Preservation
7020	Quality Assurance/Quality Control
	A. Basic Quality Control Program
	B. Quality Control for Wastewater Samples
	C. Statistics
	D. Calculation and Expression of Results
7030	Counting Instruments
	A. Introduction
	B. Description and Operation of Instruments
7040	Facilities
	A. Counting Room
	B. Radiochemistry Laboratory
	C. Laboratory Safety
	D. Pollution Prevention
	E. Waste Management
7110	Gross Alpha and Gross Beta Radioactivity (Total, Suspended, and Dissolved)
	A. Introduction
	B. Evaporation Method for Gross Alpha-Beta
	C. Coprecipitation Method for Gross Alpha Radioactivity in Drinking Water
7120	Gamma-Emitting Radionuclides
	A. Introduction
	B. Gamma Spectroscopic Method
7500 Cs	Radioactive Cesium
	A. Introduction
	B. Precipitation Method
7500 I	Radioactive Iodine
	A. Introduction
	B. Precipitation Method

7500 Ra   Radium
           A. Introduction
           B. Precipitation Method
           C. Emanation Method
           D. Sequential Precipitation Method
           E. Gamma Spectroscopy Method

7500 Rn   Radon
           A. Introduction
           B. Liquid Scintillation Method

7500 Sr   Total Radioactive Strontium and Strontium 90
           A. Introduction
           B. Precipitation Method

7500 ^{3}H   Tritium
           A. Introduction
           B. Liquid Scintillation Spectrometric Method

Appendix A3: Applicable ANSI Standard Procedures (ANSI 2005).

Procedure no.	Description
ANSI N42.31-2003	American National Standard for measurement procedures for resolution and efficiency of wide-bandgap semiconductor detectors of ionizing radiation
ANSI N42.25-1997	American National Standard calibration and usage of alpha/beta proportional counters
ANSI N42.23-1996	American National Standard measurement and associated instrument quality assurance for radioassay laboratories
ANSI N42.22-1995	American National Standard—traceability of radioactive sources to the National Institute of Standards and Technology
ANSI N42.16-1986	American National Standard specifications for sealed radioactive check sources used in liquid-scintillation counters
ANSI N42.15-1997	American National Standard check sources for and verification of liquid-scintillation counting systems
ANSI N42.14-1999	American National Standard for calibration and use of germanium spectrometers for the measurement of gamma-ray emission rates of radionuclides
ANSI N42.12-1994	American National Standard calibration and usage of thallium-activated sodium iodide detector systems for assay of radionuclides

Appendix A4: Applicable ISO Standard Procedures (ISO 2005).

Regulation ID	Description
ISO 9696:1992	Water quality: Measurement of gross alpha activity in non-saline water (Thick source method)
ISO 9697:1992	Water quality: Measurement of gross beta activity in nonsaline water
ISO 9698:1989	Water quality: Determination of tritium activity concentration (Liquid scintillation counting method)
ISO 10703:1997	Water quality: Determination of the activity concentration of radionuclides by high resolution gamma-ray spectrometry
ISO 15366:1999	Nuclear energy: Chemical separation and purification of uranium and plutonium in nitric acid solutions for isotopic and dilution analysis by solvent chromatography

Appendix B: Regulation- and Standard-Setting Agencies

AGENCY	ADDRESS	CONTACT	WEB SITE
American Institute Of Steel Construction (AISC)	One East Wacker Drive, Suite 3100 Chicago, IL 60601-2001	Ph: 312-670-2400	E-mail: pubs@aisc.org Internet: http://www.aisc.org
American National Standards Institute (ANSI)	1819 L Street, NW, 6th Floor Washington, DC 20036	Ph: 202-293-8020	E-mail: info@ansi.org Internet: http://www.ansi.org/
American Society Of Civil Engineers (ASCE)	1801 Alexander Bell Drive Reston, VA 20191-4400	Ph: 703-295-6300 800-548-2723	E-mail: marketing@asce.org Internet: http://www.asce.org
American Society Of Heating, Refrigerating And Air-Conditioning Engineers (ASHRAE)	1791 Tullie Circle, NE Atlanta, GA 30329	Ph: 800-527-4723 404-636-8400	E-mail: ashrae@ashrae.org Internet: http://www.ashrae.org
ASTM International (ASTM)	100 Barr Harbor Drive, P.O. Box C700 West Conshohocken, PA 19428-2959	Ph: 610-832-9500	E-mail: service@astm.org Internet: http://www.astm.org
Builders Hardware Manufacturers Association (BHMA)	355 Lexington Avenue 17th Floor New York, NY 10017	Ph: 212-297-2122	E-mail: assocmgmt@aol.com Internet:http://www. buildershardware.com
International Code Council (ICC)	5203 Leesburg Pike, Suite 600 Falls Church, VA 22041	Ph: 703-931-4533	E-mail: webmaster@iccsafe.org Internet: http://www.intlcode.org

AGENCY	ADDRESS	CONTACT	WEB SITE
National Electrical Manufacturers Association (NEMA)	1300 North 17th Street, Suite 1847 Roslyn, VA 22209	Ph: 703-841-3200	E-mail: webmaster@nema.org Internet: http:// www.nema.org/
National Fire Protection Association (NFPA)	1 Batterymarch Park P.O. Box 9101 Quincy, MA 02269-9101	Ph: 617-770-3000	E-mail: webmaster@nfpa.org Internet: http:// www.nfpa.org
National Institute For Occupational Safety And Health (NIOSH)	Mail Stop C-13 4676 Columbia Parkway Cincinnati, OH 45226-1998	Ph: 800-356-4674	E-mail: pubstaff@cdc.gov Internet: http://www. cdc.gov/ niosh/ homepage.html
National Roofing Contractors Association (NRCA)	10255 West Higgins Road, Suite 600 Rosemont, IL 60018		

AGENCY	ADDRESS	CONTACT	WEB SITE
Ph: 847-299-9070	Internet: http://www.nrca.net		
Sheet Metal & Air Conditioning Contractors' National Association (SMACNA)	4201 Lafayette Center Drive, Chantilly, VA 20151-1209	Ph: 703-803-2980	E-mail: info@smacna.org Internet: http://www.smacna.org
Underwriters Laboratories (UL)	333 Pfingsten Road Northbrook, IL 60062-2096	Ph: 847-272-8800	E-mail: northbrook@us.ul.com Internet: http://www.ul.com/
U.S. Army Corps Of Engineers (USACE)	3909 Halls Ferry Rd. Vicksburg, MS 39180-6199	Ph: 601-634-2664	E-mail: mtc-info@erdc. usace.army.mil Internet: http://www. wes.army.mil/SL/ MTC/handbook.htm
U.S. Environmental Protection Agency (EPA)	Ariel Rios Building 1200 Pennsylvania Avenue, N.W. Washington, DC 20460	Ph: 202-260-2090	Internet: http://www. epa.gov

Mr. Doug Ashley, Architect, with Bullock Tice Associates, Pensacola, FL, provided the information given in this appendix.

Appendix C

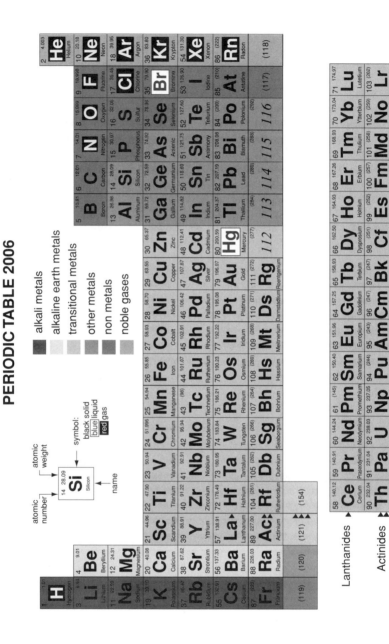

PERIODIC TABLE 2006

Glossary

ablation: In this context, volatilization of a solid surface by localized heating, as by a laser.

abundance: Amount of atoms or molecules present.

abundance sensitivity: A measure of the ability to detect a minor peak next to an abundant one.

accelerator: Machine used to produce ions at high energy for nuclear experiments. See *Cockcroft–Walton accelerator*.

actinides or actinide series: The fourteen elements with the atomic numbers 90 through 103, inclusive, in which filling of the 5f inner electron shell occurs. Name is derived from the immediately preceding element actinium ($Z = 89$). Analogous to the lanthanide series, atomic numbers 58–71, inclusive, in which electrons are added to the 4f electron shell.

activity: See *disintegration rate*.

Agreement States: States that have programs to assume partial NRC regulatory authority under the Atomic Energy Act of 1954. This authority includes licensing and regulating radioactive byproduct materials; source materials (uranium and thorium); and special nuclear materials.

ALARA: As low as reasonably achievable for controlling radiation exposure.

aliquot: A measured fraction of a solution.

alpha particle (α particle): A particle, consisting of two protons and two neutrons, that is emitted from the nucleus of an unstable radionuclide. Equivalent to a helium nucleus.

amu: Abbreviation for atomic mass unit or Dalton. The unified atomic mass unit (symbol u) finds increasingly common usage, and is equivalent to the amu. See *Dalton*.

annihilation (process): The process of interaction between matter and antimatter to produce electromagnetic radiation. For example, the interaction between an electron and a positron.

antineutrino: In the process of negatron particle emission, the antineutrino carries away the residual energy, momentum, and spin to conserve these quantities. This sub atomic particle has zero charge, near-zero mass, and 1/2 spin.

aqua regia: A mixture of nitric and hydrochloric acids. The term literally means "royal water," because of its ability to dissolve gold and platinum, which are noble metals.

artificial or "synthetic" elements: Elements not found in nature; sometimes even refers to elements first synthesized in the laboratory before their discovery in nature, e.g., technetium ($Z = 43$) or plutonium ($Z = 94$).

ashing: Burning a sample until only ash remains.

asymmetric fission: Fission process where fission products are distributed around a light mass and a heavy mass.

Atomic Energy Act of 1954: One of the fundamental laws governing civilian uses of nuclear materials and facilities.

atomic number (Z): The number of protons in the nucleus of an atom.

atomic orbital: A representation of the three-dimensional region in space in which the electron is likely to be found. The orbital functions, which define the space, are derived by quantum mechanics.

attribution: Ascribing responsibility. In the context of this textbook, attribution studies refer to research conducted in order to determine the origin of a particular set of detected radionuclides.

audit: The complete and methodical review of laboratory performance and results.

Auger electron: Pronounced *O-jhay* and named after French physicist Pierre Auger, Auger electron emission is one possible consequence of an internal conversion. When an inner core electron vacancy is filled by an electron from a higher energy level, energy must be released. This energy release sometimes occurs as the emission of an atomic electron, the Auger electron.

Avogadro's number (A_v): The number of entities in one mol of any substance, defined as the number of atoms in 0.012 kg of ^{12}C. $A_v = 6.02 \times 10^{23}$. N_A is an equivalent symbol for Avogadro's number.

azeotropic distillation: Purification of tritiated water by codistillation with an organic solvent. A fixed ratio of solvent and water are distilled at the corresponding boiling point; on condensing, the organic solvent and water separate into distinct phases, with minimal solubility of the organic solvent in water.

back extraction: Transfer of a substance that had previously been extracted into a second solvent back into the first solvent.

bandpass filter: A control for blocking all signals outside a selected range of characteristics such as frequency, energy, or mass to charge ratio.

bandwidth: The range of selected signal such as frequency, energy, or mass to charge ratio.

barn (b): Unit for cross section that equals 10^{-24} cm^2.

becquerel (Bq): The SI unit of activity of a radionuclide, equal to 1 s $^{-1}$ or 1 disintegration per second. One Bq $= 2.703 \times 10^{-11}$ curies (Ci).

berm: Man-made geographical feature; a mound of earth. In this context, a berm is usually an earthen wall that surrounds holding tanks to contain any liquid that might leak from them.

beta particle (β particle): Term given to an electron (β^- particle) or positron (β^+ particle) emitted from the nucleus.

bioassay: Analysis of biological material.

"blind" sample: Jargon; a sample of known content, but unknown to the analyst.

branching ratios: The fractional radiation decays that result in two or more products.

Bremsstrahlung radiation: "Braking radiation" (German). Electromagnetic radiation produced when one charged particle (say, an electron) is deflected or decelerated by another charged particle (say, a nucleus). Bremsstrahlung has a continuous spectrum that increases in intensity with decreasing energy.

carrier: A stable isotope in appreciable amount which, when thoroughly mixed with a trace of a specified radioelement, will carry the trace with it through a chemical or physical process. The stable and radioactive isotopes must behave identically.

chain reaction: A self-sustaining series of nuclear fissions that is propagated through the absorption of neutrons produced as the product of prior fissions.

Cherenkov radiation: Electromagnetic radiation resulting from charged particles traveling in a medium faster than light can travel in the same medium. The deceleration of the particle produces this radiation, which is sometimes seen as the characteristic "blue glow" at a nuclear reactor.

clear melt: The transparent form of melted solids.

Cockcroft-Walton accelerator: A high-voltage device inside which charged capacitors discharge to add electrons to hydrogen atoms to create negatively charged ions. The charged particles are then driven through an accelerating tube toward a target.

cold: Jargon; characterized by an absence of radiation or radionuclides.

colligative: A property that depends on the quantity of atoms, not their form.

colorimetry: Spectrophotometric analysis of visible electromagnetic radiation.

combined standard uncertainty: A standard uncertainty calculated by propagation of uncertainty. The combined standard uncertainty of a result y may be denoted by $u_c(y)$.

combustible liquid: Liquid with a flashpoint at or above 100°F. Class II combustible liquids have a flashpoint between 100°F and 140°F and Class III combustible liquids have a flashpoint above 140°F.

Compton scattering: Occurs when gamma rays interact with electrons in a material. The photon donates some of its energy to the electron. The electron is ejected from its atom and the reduction in energy of the photon results in an increase in its wavelength. The photon with the remaining energy is deflected.

continuum (energy): Range of particle energies, distributed in a continuous fashion from $E = 0$ to $E = $ maximum value.

conversion electron (CE): The inner orbital electron that receives the excess energy of a metastable nucleus during the course of an internal conversion. Because of its relative proximity to the nucleus, the conversion electron usually comes from the K shell.

coprecipitation: The simultaneous precipitation of a normally soluble component with a macro-component from the same solution by the formation of mixed crystals, by adsorption, occlusion, or mechanical entrapment.

coulombic barrier: Charge potential experienced when one electrically charged projectile approaches another charged object. For example, electrical charge repulsion when an alpha particle approaches the nucleus of an atom.

coverage factor: The factor k by which the combined standard uncertainty of a result, $u_c(y)$, of a measurement is multiplied to obtain the expanded uncertainty, U.

coverage probability: The approximate probability that the interval $y \pm U$ described by a measured result, y, and its expanded uncertainty, U, will contain the true value of the measurand.

critical value: Threshold value to which the result of a measurement is compared to make a detection decision (also sometimes called "critical level" or "decision level"). The critical value can be expressed as (for example) the critical gross count, critical net count, critical net count rate, or the critical activity.

cross section: Expression of the interaction probability for a nuclear reaction per unit of area, in cm^2 per atom. The term describes the area the nucleus presents for a particular projectile to strike.

cross-talk: Interfering signals among parallel measurements.

crystal lattice energy: The energy holding ions or atoms together in a crystal structure.

curie (Ci): A former unit of radioactivity, corresponding roughly to the acitivity of one gram of the radioactive isotope ^{226}Ra. 1 Ci $= 3.7 \times 10^{10}$ decays per second $= 37$ gigabequerels (GBq).

cyclotron: Circular charged particle accelerator invented by E.O. Lawrence in 1929. Pole diameters range in size from a few inches up to 236 inches with energies from several million electron volts (MeV) up to 700 MeV for protons. Heavier projectiles can also be accelerated.

DAC: Derived air concentration of a radionuclide that is related to an intake limit.

Dalton: Named after English scientist John Dalton, the Dalton is defined as exactly 1/12 of the relative atomic mass assigned to ^{12}C. It is functionally the same as an atomic mass unit (amu) and unified atomic mass unit (u).

daughter(s): Term given to the radioactive product of a radioactive parent, also progeny.

decay: A term describing the emission of energy from a radionuclide during the course of a nuclear transformation. Typically one decay mode, such as alpha particle emission, predominates for a particular nuclide. When more than one decay mode is evident, branching ratios are used to describe the fraction of the decay that occurs by each pathway.

decay chain: The succession of radioactive products (called daughters or progeny) to which a radionuclide decays.

decay constant (λ): A constant (in units of reciprocal time) that is characteristic of a particular radionuclide's decay.

decay fraction: The fraction of a particular mode of decay when a radionuclide decays by several modes; also termed branching ratio.

decontamination factor: The ratio of the proportion of contaminant to product before treatment to the proportion after treatment.

de-excitation: Process in which an nucleus, atom, or molecule releases energy, with the effect of reaching a less energetic state.

dewar: A metal or glass container designed to hold liquefied gases. The container was invented by Scottish chemist and physicist Sir James Dewar, who produced liquid hydrogen.

disintegration rate: Rate describing the number of nuclear transformations per unit time, in units of bequerels (Bq). Also referred to as the activity of a radionuclide sample.

doping: The process of adding a foreign material to an existing material in order to create a composite material that contains either an excess of electrons (an n-type material) or an excess of positive holes (a p-type material).

dosimetry: Relating to the practice of monitoring the degree of radiation exposure of individuals, whether externally or internally.

electrodeposition: Transfer of ions from solution to the surface of an electrode as part of an electric current.

electromagnetic radiation: Energy which has both an electrical and magnetic component; when characterized as a wave, identified by its wavelength or frequency; when characterized as an energy bundle, identified by its energy in electron volts.

electron: The negatively charged subatomic particle that is found in orbital spaces near the nucleus of the atom.

electron capture (EC): A decay mode characterized by the capture of an atomic electron from, most probably, an inner electron shell by the nucleus. Electron capture parallels positron emission. If the energy difference between the parent and the daughter is less than 1.022 MeV, electron capture is the sole decay mode.

electron multiplier: A device with multiple dynodes in series to increase an electron pulse for measurement. Electrons are accelerated as they move toward each dynode so that multiple electrons are freed at the dynode per arriving electron.

electron rest energy: Mass of an electron at rest expressed in units of energy: 0.511 MeV.

electron volt (eV): The unit of energy commonly used in quantifying nuclear processes, the electron volt is defined as the work done by an electron when it falls through a potential difference of 1 V stated another way, the quantity of energy that must be added to a standard electron (whose charge equals 1.602×10^{-19} C) to accelerate it through 1 V; potential difference. Mathematically, 1 eV = $(1.602 \times 10^{-19}$ C$)(1$ V$) = 1.602 \times 10^{-19}$ J.

elutriant: A solution that removes a previously retained substance from a sorbent, such as an ion-exchange column.

E_{max}: The maximum beta-particle energy for a beta-particle group.

excitation: Term used to describe the process in which an atom, molecule, or nucleus is elevated to an excited state.

exoergic: Releasing energy.

expanded uncertainty: An uncertainty, U, chosen so that the interval $y \pm U$ about the measured result y is believed to have a high, numerically defined, probability of containing the true value of the measurand.

extractant: A reagent that implements the transfer of a substance from one solvent to another.

extraction: Transfer of a substance from one solvent to another.

face velocity: The velocity at which air flows from the laboratory into the interior of the hood, measured at the face of a hood. This velocity must be high enough to prevent backspill of inhalants from the hood to the laboratory atmosphere, but not so high that it interferes with the stability of analytical reagent containers and equipment.

Fano factor: The observed variance relative to the calculated Poisson distribution variance, as observed in the peak width in spectral analysis.

Faraday cup: A sensitive device for collecting a stream of electrons or ions and measuring it in terms of current.

Federal Facility Compliance Act of 1992: Written to amend the Solid Waste Disposal Act, in order to clarify provisions concerning the application of requirements and sanctions to federal facilities.

fissile material: Any nuclide which is capable of fissioning, either spontaneously or induced, although the term mostly refers to heavy elements such as uranium or plutonium. The term "fissionable" may also be used.

fission: Process in which a nucleus is split into smaller nuclei. Fission may be spontaneous or induced, symmetric or asymmetric.

flammable liquid: Liquids with a flashpoint below 100°F, also known as Class I liquids. Class IA liquids have a flashpoint below 73°F and a boiling point below 100°F; Class IB liquids have a flashpoint below 73°F and a boiling point at or above 100°F; Class IC liquids have a flashpoint at or above 73°F but below 100°F.

flashpoint: The minimum temperature at which a liquid gives off vapor within a test vessel in sufficient concentration to form an ignitable mixture with air near the surface of the liquid.

flux (as a rate): The rate of transfer of particles or energy across a given barrier.

flux (chemical): A substance added to the analyte, with the goal of facilitating its dissolution.

footprint: The amount of floor space that an item requires.

Frisch grid: An ionization chamber used for alpha-particle spectroscopy. The size of an electronic pulse from the chamber is proportional to the energy of the alpha particle that produced the pulse. Because the energies of alpha particles are characteristic of the radionuclides that emit them, the Frisch grid chamber is able to distinguish between different radionuclides by measuring pulse height.

full width at half maximum (FWHM): Width of a peak at 1/2 of its maximum peak height. The FWHM measurement is a means of expressing the energy resolution of a spectrometer; also full peak at half maximum (FPHM).

functional group: A specific group of atoms within a molecule that predictably engages in characteristic chemical reactions.

fusion (nuclear): The process of two or more nuclei joining together to form a heavier nucleus.

fusion (chemical) : The process of liquifying solid materials by heating.

gamma ray (γ ray): Electromagnetic radiation emitted as a result of a nuclear transformation.

gas-filled detector: A chamber filled with gas that contains a cathode and an anode connected by electric circuit to a recording device to measure count rate or radiation exposure.

Gaussian: Normally distributed or shaped like the familiar "bell curve."

Gaussian distribution: A normal distribution of a population that is described by approximations of the Poisson distribution.

Gaussian peak: A spectral peak whose shape is Gaussian.

Geiger–Mueller (G-M) counter: A gas-filled ionization detector operating in the Geiger–Mueller applied voltage region.

getter: Jargon; term for reactive metals used to scavenge for impurities.

grab sample: A sample collected by hand.

gravimetric: Measurement by weight.

ground: A large conducting body such as the earth that is used as a common return for an electric circuit.

half-life ($t_{1/2}$): The time required for any given amount of a radionuclide to decay to one-half its value. $t_{1/2} = \ln 2/$decay constant.

HAZMAT: Term colloquially used to refer to hazardous materials and to functions of agents of the Office of Hazardous Materials Safety. The stated mission of the OHM is to "promulgate a national safety program that will minimize the risks to life and property inherent in commercial transportation of hazardous materials." See website at http://hazmat.dot.gov (01/15/2006).

Henry's Law: At contant temperature, the amount of a given gas dissolved in a given liquid is directly proportional to the partial pressure of the gas when it is in equilibrium with the liquid.

homogeneous: Uniform throughout.

homolog: In inorganic chemistry, an element in a similar position, such as a column, in the periodic table.

hoods: Devices designed to protect the analytical chemist from excess exposure to inhalants by continuously removing laboratory air to the outside, after the air has passed through a filtering system.

hopcalite: A catalyst for oxidizing hydrogen in hydrocarbons to water vapor.

hot particle: Jargon for an intensely radioactive particle.

hot swap: One of the features of a RAID system, whereby the drives are connected to the controller. A "hot swappable" RAID drive is one that can be changed out if it fails, without shutting down the system.

hot: Jargon; characterized by the presence of radiation or radionuclides.

hygroscopic: A characteristic of a material, wherein it absorbs water.

hyperpure: Extremely pure; describes purified germanium for a detector.

ICP/ MS: Inductively coupled plasma/mass spectrometer.

immiscible: Not able to be mixed together.

induced fission: The process in which neutrons are "fired" at a fissionable source material to cause it to split.

ingrowth: Describes the accumulation of product atoms from radioactive parent atoms as they decay.

input estimate: In a particular measurement, a measured or imported value of an input quantity.

input quantity: In a mathematical model of measurement, any of the quantities whose values are measured or imported and used to calculate the value of the output quantity Y.

interlaboratory testing: Program in which a number of laboratories participate to demonstrate competence.

internal conversion: A radioactive decay process in which an excited nucleus de-excites by transferring its energy to an inner orbital electron, which is consequently emitted. See *conversion electron.*

ionization (secondary): Ionization that results from the collision of ions or electrons through their interaction with media to form a second generation of ions.

isobar: Two or more atoms with the same mass number (A), but different atomic numbers (Z) and numbers of neutrons (N), are said to be isobaric.

isobaric interferences: In mass spectrometry, interferences from extraneous atoms or molecules that have masses very similar to that of the analyte atom.

isothermal: A process performed at a fixed energy.

isotope dilution: Addition and complete mixing of a second isotope with an isotope of the same element to measure the yield of a reaction.

laminar flow boxes: Similar to hoods, these boxes are designed to filter in-laboratory air continuously through the box, rather than venting to the outside, to isolate particulate matter inside the box.

lanthanides or lanthanide series: The fourteen elements with atomic numbers 58–71, inclusive, in which electrons are added to the 4f inner electron shell.

law of propagation of uncertainty: The general equation, based on local approximation of a mathematical model by a first-order Taylor polynomial, that may be used to calculate the combined standard uncertainty of the result of a measurement.

linear accelerator: A linear accelerator uses alternating electrodes to drive charged particles along a linear path toward a target.

loaded solvent: The solute-loaded result of an initial extraction. If several solutes are contained in this extract, one may be singled out and removed through back extraction.

Lucas cell: A cylindrical container for measuring radon gas that has a quartz window for transmitting scintillations at one end, has all other internal walls coated with ZnS(Ag) powder that responds with scintillations to alpha particles, and has a stopcock for connection to a pump to create a vacuum and then admitting a gas sample.

mass analyzer: A component of the mass spectrometer that separate ions based on mass, momentum, or velocity.

mass defect: Energy is either required or released when nucleons bind together to form a nucleus. The mass (or its equivalent energy) of the product nucleus then differs from the additive component masses. The mass defect, calculated as the total nucleon mass (M) minus the nuclear mass number (A), can be viewed as a reflection of the stablilty of the nucleus.

mass number (A): The sum of the number of protons and neutrons in the nucleus.

Material Safety Data Sheet (MSDS): A form that contains information about the physical and chemical properties of a substance. The MSDS is intended to provide personnel with safe methods of working with the substance in question.

matrix: The components of a sample mixture, other than the analyte.

measurand: The "particular quantity subject to measurement" (ISO 1993).

metastable nucleus: Nuclear isomer or isomeric state of an atom, wherein a proton or neutron in the nucleus is excited. In this state, a change in the spin of the nucleus is required before it can go to a lower state and release extra energy.

minimum detectable activity (MDA): A measure of the detection capability of a radioanalytical measurement process, defined as an estimate of the smallest true value of the activity (or massic activity, volumic activity, etc.), which gives a specified probability $1 - \beta$ of detection. The definition of the MDA presupposes that an appropriate detection criterion (i.e., the critical value), with a specified Type I error rate α, has been defined. It is a mistake to use the MDA itself as the detection criterion.

momentum: Mass of an object multiplied by the velocity of the object; must be conserved for interacting processes.

mono-energetic: Particles of a single energy, as in alpha particles.

Monte Carlo simulation: A numerical analysis technique that randomly generates values for uncertain variables. The simulation program then uses the results to construct a model that approximates the solution to some physical or mathematical problem.

muon: A subatomic particle.

National Institute of Standards and Technology (NIST): The federal organization that is officially responsible for standards in the United States.

negatron: A negatively charged electron emitted during a nuclear transformation or process.

network protocol: Designed to augment a secure channel by making it more secure, a network protocol requires that all communications by that channel adhere to a defined set of rules. The rules generally address such issues as data authentication and error detection.

neutrino: A subatomic particle with zero charge, near-zero mass, and 1/2 spin. The neutrino is emitted as a byproduct of positron particle emission, carrying away the appropriate amount of energy and momentum to conserve these quantities.

neutron: Fundamental subatomic particle of the atom that has no charge and a mass of approximately one dalton.

neutron activation: Process of activating stable nuclides by bombarding them with neutrons, thus forming radionuclides.

neutron flux: Measure of the neutron population per given cross section of space per unit time.

neutron moderator: A medium, such as deuterium or graphite, that slows down the fast neutrons produced in a nuclear reactor to reduce their kinetic energy to the thermal level. The thermal neutrons may then be absorbed by a new fuel atom and continue the chain reaction.

n-type: Describes a material whose structure contains an excess of electrons by which charge can move.

nuclear chemistry: The field of chemistry that studies and applies nuclear properties and reactions.

nuclear decay: See *decay*.

nuclear explosion: An explosion that results from nuclear processes by fission and fusion.

nuclear isomer: Nuclides having the same mass number and atomic number, but occupying different energy states.

nuclear reactions: Reactions taking place between the nucleus and an energetic particle.

Nuclear Regulatory Commission (NRC): The U.S. Nuclear Regulatory Commission is an independent agency established by the Energy Reorganization Act of 1974 to regulate civilian use of nuclear materials. Online at http://www.nrc.gov/.

nuclear transformation: Transformation of the nuclear state as a result of one or more nuclear processes.

nucleon: A term referring to either or both protons and neutrons.

nucleon number: The number of nucleons in the nucleus.

nucleus: The positively charged part of an atom in which most of the mass of the atom resides. The nucleus is composed of neutrons and protons and is the location of nuclear processes.

nuclide: Term given to denote any atomic species characterized by a specific atomic number (Z) and mass number (A). Nuclear isomers may also be recognized as distinct nuclides, provided they have half lives long enough to be observed, e.g., ^{234}Pa and ^{234m}Pa are nuclides distinct from each other.

output estimate: In a particular measurement, the calculated value of the output quantity.

output quantity: In a mathematical model of measurement, the quantity Y whose value is calculated from measured or imported values of the input quantities.

overvoltage: Voltage required for electrodeposition of a substance in excess of that predicted by half cell potentials due to extraneous reaction.

parent: In a decay chain, the parent is the immediate predecessor of some specified radionuclide; its progeny are usually referred to as *daughters*.

personal protective equipment (PPE): Any equipment worn in order to protect the wearer from the effects of an accident, whether chemical or radiological.

phosphor: A material that emits light (scintillates).

photomultiplier tube (PMT): Electronic tube that amplifies signal from incident light by multiplying (cascading) electrons, with more electrons from each successive plate.

photodisintegration: A nuclear reaction initiated by an energetic gamma ray.

photoelectric effect: The emission of a free electron from an atom that interacts with and totally absorbs a gamma ray.

photopeak: In gamma-ray spectral analysis, the photoelectric interaction peak, but more commonly defined as the full-energy peak.

planchet: A metal disk on which a sample is deposited or fixed in preparation for counting.

positron: Positively charged electron emitted in a nuclear transformation or process.

precipitate: The solid product of a chemical separation, left over after the supernatant liquid from which it came is removed.

principal quantum number (n): The energy of an electron in an atom is primarily dependent on this quantum number, as is the size of the atomic orbital in which it is contained. The greater the value of n for an electron, the higher the energy of that electron. The principal quantum number may have any positive integer value.

private key: An algorithm that allows a group or individual to enter a key agreement, wherein all communications will require that key before proceeding.

process stream: A part of an industrial process that transfers materials.

propagation of uncertainty: Mathematical operation of combining standard uncertainties (and estimated covariances) of input estimates to obtain the combined standard uncertainty of the output estimate.

proportional counter: A gas-filled ionization detector that operates in the proportional region of the applied voltage, in which the output pulse energy is proportional to the deposited energy.

proton: Fundamental subatomic particle of the atom that has a charge of $+1$ with a mass of approximately one dalton.

public key: A form of encryption that allows users to communicate securely without prior possession of a secret key.

p-type: Describes a material whose structure contains positive holes through which charge can move.

quality assurance (QA): A program that systematically monitors and evaluates laboratory operation to ensure that sample analysis meets standards of quality.

quality assurance plan (QAP): A planning tool that allows for the comprehensive management of laboratory operation to ensure that standards of quality are met.

quality assurance project plan (QAPP): A planning tool that allows for the comprehensive management of a data collection activity (DCA), from sample collection through data reporting, to ensure that standards of quality are met.

quality control (QC): System for ensuring maintenance of proper standards in the laboratory by use of standard and comparison samples.

quenching: In a scintillation detector, to extinguish light emanation from the sample, to a degree that depends on the quenching agent, thus reducing the measurement of radionuclide content. In a gas-filled detector, to terminate ionization promptly after it has been detected in order to avoid extended pulse tailing.

radioactivity: The property of an unstable nuclide that is characterized by the emission of particles or radiation from its atomic nucleus.

radioanalytical chemistry: Application of techniques, instrumentation and knowledge from radiochemistry and analytical chemistry for the quantitative analysis of a wide range of concentrations of both radioactive and non-radioactive species. Applicable to investigations ranging from geochemistry and age dating to diagnostic and therapeutic nuclear medicine procedures as well as to routine assay of radioactivity in the environment and surveillance of clandestine nuclear activities and terrorist threats.

radiochemistry: The subdiscipline of chemistry that studies radioactive materials and applies them in chemical processes.

radiocolloidal: Colloid-like behavior of a radionuclide at very low concentration in solution.

radioelements: Radionuclides of the same element.

radionuclide: Nuclide that is unstable and will de-excite spontaneously.

RAID drives: Redundant Array of Independent Discs. A RAID system is a collection of hard drives joined together, for goals that can be achieved. See online descriptions at http://bugclub.org/beginners/hardware/raid.html (01/15/2006).

range: The distance a particle travels through a specified material.

Raoult's Law: The vapor pressure of each component in an ideal solution is dependent on both the partial pressure and the mole fraction of each component in the solution.

recoil energy: Kinetic energy carried by the nucleus after it emits a particle.

recoil nucleus: Nucleus that results from the decay of a radionuclide. Also called the product nucleus, or daughter nucleus. Recoil balances the momentum associated with the emission of particles or radiation.

refractory: A solid that is stable and heat resistant at elevated temperatures.

remote access: The communication between a local and remote computer over a data link. Access to the data link is invariably limited to those with authorization.

resin: For ion-exchange processes, an organic structure for ion-exchange sites.

resonance structures: Two or more representations of the electronic structure of a molecule. The actual structure of the molecule may be viewed as a hybrid of all these structures. The more structures, the greater the stability of the molecule.

scavenger: A substance that reacts with (or otherwise removes) a trace component.

scintillation cocktail: A solution that consists of a scintillating substance and an organic solvent, which serves as the medium for counting a radionuclide sample in a liquid scintillation counter.

sector field analyzer: A spectrometer that separates ions by the curvature of their flight path, for which the radius is inversely proportional to the magnetic field and directly proportional to the square root of the mass to charge ratio.

secular equilibrium: A case where the activity (disintegration rate) of the parent and progeny becomes equal after a time. For this type of equilibrium, the half-life of the parent is at least 10 times greater than that of the daughter.

secure shell (ssh): A set of standards and accompanying network protocol that allows for the establishment of a secure channel between a local computer and one located remotely.

sequential analysis: Serial chemical analysis of several substances in the same solution.

shell (as in K, L, etc): The letter designation of the principal quantum number, n. Orbitals of the same n value are referred to as belonging to the same shell. Several subshells may be contained in a given shell; subshell designation depends on other quantum numbers

slurry: A watery mixture of insoluble material.

solubility product (K_{sp}): Mathematically, the product of the thermodynamic activities of the ionic components of a substance precipitated from a solution at equilibrium.

source: In the context of radionuclide analysis, a sample prepared for radionuclide measurement, i.e., the source of the radionuclides that are being measured.

specific activity: For a specified isotope, or mixture of isotopes, the activity (disintegration rate) of a material divided by the mass of the element.

spallation: To break off. In nuclear physics, the term refers to the emission of a large number of neutrons by a heavy radionuclide after it has been bombarded by charged particles.

spike: See *tracer*.

spiking: Jargon; addition to a sample of a tracer for the radionuclide of interest.

spontaneous fission: Natural decay process found in elements with Z greater than or equal to 90 (although it is unclear whether Th ($Z = 90$) can fission spontaneously).

stable nuclide: A nuclide that is not radioactive.

standard method: A procedure that has been selected by testing and agreement among skilled professionals.

standard reference material (SRM) or standard radioactive material (SRM): Material, often in liquid form, that is certified to have a known concentration or decay rate. SRM is used in instrument calibration and as primary standards for laboratory measurements.

standard uncertainty: Uncertainty of the result of a measurement expressed as a standard deviation (sometimes called a "one-sigma" uncertainty). The standard uncertainty of a measured value x may be denoted by $u(x)$.

stakeholder: In the context presented herein, a person or group that has an interest in the results produced by a laboratory or facility.

state-of-health: During the lifetime of an instrument, its performance or "health" tends to deteriorate. The degree of this decline can be inferred from measurements, specified for a particular system, that define its well being. Parameters may include battery power, calibration outputs and various hardware checks.

stopping power: The differential energy loss per travel distance, (dE/dx) of a particle.

STP (standard temperature and pressure): Zero degrees Celsius (273° Absolute) and one atmosphere pressure.

subatomic particle: Any fundamental particle that is smaller than the atom.

symmetric fission: Fission process in which product yields decrease on both sides of an intermediate mass that is produced with maximum yield.

Szilard–Chalmers effect: The rupture of the chemical bond between an atom and the molecule of which the atom was a component, as a result of a nuclear transformation of that atom.

thermoluminescent detector (TLD): Thermoluminescent material in powder form or small chips is placed in a badge for radiation monitoring. After a specified period of time, the badge is collected and "read" by heating it to emit light that is proportional to the energy deposition of the detected radiation.

thin target: Term applied to materials that are irradiated but do not significantly attenuate the projectile beam or flux.

traceability: Documentation for a standard that allows it to be historically linked to the laboratory of its formulation and testing, notably at NIST.

tracer: A substance that is added to a material to observe the behavior of one of its components.

transactinides: All elements with atomic number greater than lawrencium (103), the last of the actinide series.

transient equilibrium: A case where the ratio of the daughter activity to the parent activity becomes constant with a value that exceeds 1.0 after several half-lives. The half life of the parent is longer than that of the daughter.

transport velocity: The velocity at which air is flushed from the hood to the outside air. This velocity depends on fan speed and ductwork dimensions and is important to assure that airborne particles are transported from the hood and not deposited in the ductwork.

Type A evaluation: "Method of evaluation of uncertainty by the statistical analysis of series of observations" (ISO 1995).

Type B evaluation: "Method of evaluation of uncertainty by means other than the statistical analysis of series of observations" (ISO 1995).

uncertainty propagation: See *propagation of uncertainty*.

uncertainty: A "parameter, associated with the result of a measurement, that characterizes the dispersion of the values that could reasonably be attributed to the measurand" (ISO 1993, 1995).

VOC: Volatile organic chemicals that are emitted from certain liquids and solids.

volatilization: The conversion of a solid or liquid to a gas or vapor by application of heat, by reducing pressure, by chemical reaction or by a combination of these processes.

X-ray: One possible consequence of an internal conversion. Electromagnetic radiation that is emitted when an inner shell electron of an atom is removed and an electron from an orbital of higher energy drops into the position. It has a wavelength of 10 pm to 10 nm or an energy from electron volts to kiloelectron volts.

yield: The fraction of a substance that is recovered after separations are performed.

zeolite: One of hydrous aluminum silicates used as ion-exchange material.

References

ACGIH 2001. *Air Sampling Instruments for Evaluation of Atmospheric Contaminants, 9th ed.* Cincinnati OH: American Conference of Government Industrial Hygienists, Inc.

ACS 1990. American Chemical Society. *Informing Workers of Chemical Hazards: The OSHA Hazard Communication Standard.* 2nd ed. Washington, DC: ACS.

ACS 1998. American Chemical Society. *Living with the Laboratory Standard: A Guide for Chemical Hygiene Officers.* Washington, DC: ACS.

ACS 2000. American Chemical Society. *Safety Audit/Inspection Manual.* Washington, DC: ACS.

ACS 2003. American Chemical Society. *Safety in Academic Chemistry Laboratories.* 7th ed. Washington, DC: ACS.

Adloff, J.-P. and Guillaumont, R. 1993. *Fundamentals of Radiochemistry.* pp. 327–352. Boca Raton, FL: CRC Press.

Adriaens, A. G., Fassett, J. D., Kelly, W. R., Simons, D. S., and Adams, F. C. 1992. Determination of Uranium and Thorium Concentrations in Soils—Comparison of Isotope-Dilution Secondary Ion Mass-Spectrometry and Isotope-Dilution Thermal Ionization Mass-Spectrometry. *Anal Chem* 64(23), 2945–2950.

AIHA 1995. American Industrial Hygiene Association. *Laboratory Chemical Hygiene: An AIHA Protocol Guide.* Fairfax, VA: AIHA Press.

Alaimo, R. J. and Fivizzani, K. P. November/December 1996. Qualifications and training of chemical hygiene officers. *Chemical Health and Safety* 3(6), 10–13.

Aldstadt, J. H., Kuo, J. M., Smith, L. L., and Erickson, M. D. 1996. Determination of uranium by flow injection inductively coupled plasma mass spectrometry. *Anal Chim Acta* 319(1–2), 135–143.

Alonso, J. I. G., Babelot, J. F., Glatz, J. P., Cromboom, O., and Koch, L. 1993. Applications of a glove-box ICP-MS for the analysis of nuclear- materials. *Radioc Acta* 62(1–2), 71–79.

Alonso, J. I. G., Sena, F., Arbore, P., Betti, M., and Koch, L. 1995. Determination of fission-prroducts and actinides in spent nuclear-fuels by isotope-dilution ion chromatography inductively-coupled plasma-sass spectrometry. *J Anal Atom Spectromy* 10(5), 381–393.

Altshuler, B. and Pasternack, B. 1963. Statistical measures of the lower limit of detection of a radioactivity counter. *Health Phys* 9, 293–298.

Amphlett, C. B. 1964. *Inorganic Ion Exchangers.* New York, NY: Elsevier.

Anders, E. 1960. *The Radiochemistry of Technetium.* National Research Council Report NAS-NS 3021. Washington, DC: National Research Council.

ANSI 1995. American National Standards Institute, N42.22-1995. *Traceability of Radioactive Sources to NIST and Associated Instrument Quality Control*. Washington, DC: ANSI.

ANSI 2005. American National Standards Institute. Online at http://www.ansi.org/. N42 standards may be found by clicking "eStandards Store" then searching for N42. Viewed on September 20, 2005.

ANSI/IEEE 1995. American National Standards Institute—Institute of Electrical and Electronics Engineers Standard N42.22-1995. *American National Standard Traceability of Radioactive Sources to NIST and Associated Instrument Quality Control*. Online at www.ansi.org. Viewed on September 20, 2005.

APHA 1972. American Public Health Association, Intersociety Committee for a Manual of Methods for Ambient Air Sampling and Analysis. *Methods of Air Sampling and Analysis*. Washington, DC: APHA.

Armentrout, P. B. 2004. Fundamentals of ion-molecule chemistry. *J Anal Atom Spectrom* 19(5), 571–580.

ASME 1989. American Society of Mechanical Engineers, NQA-1. *Quality Aassurance Program Requirements for Nuclear Facilities*. New York, NY: ASME.

ASQC 1995. American Society for Quality Control, ANSI/ASQC E4-1994. *Specifications and Guidelines for Quality Systems for Environmental Data Collection and Environmental Technology Programs*. Milwaukee, WI: ASQC.

ASTM 2005. American Society for Testing and Materials. Standards may be found online at http://astm.org/. Click "Technical Committees" then search for the standard by the committee or sub-committee that developed it. Viewed on September 20, 2005.

Bachmann, K. 1982. Separation of trace elements in solid samples by formation of volatile inorganic compounds. *Talanta* 29, 1–29.

Barber, R. C., Greenwood, N. N., Hrynkiewicz, A. Z., Jeannin, Y. P., Lefort, M., Sakai, M., Ulehla, I., Wapstra, A. H., and Wilkinson, D. H. 1991. Criteria that must be satisfied for the discovery of a new chemical element to be recognized. *Pure Appl Chem*. 63, 879–886.

Barber, R. C., Greenwood, N. N., Hrynkiewicz, A. Z., Jeannin, Y. P., Lefort, M., Sakai, M., Ulehla, I., Wapstra, A. H., and Wilkinson, D. H. 1992. Discovery of the transfermium elements. *Prog Part Nucl Phys* 29, 453–530.

Barber, R. C., Greenwood, N. N., Hrynkiewicz, A. Z., Jeannin, Y. P., Lefort, M., Sakai, M., Ulehla, I., Wapstra, A. H., and Wilkinson, D. H. 1993. Naming of new elements. *Pure Appl Chem* 65, 1757–1813.

Barshick, C. M., Duckworth, D. C., and Smith, D. H. 2000. Inorganic mass spectrometry: fundamentals and applications in *Practical Spectroscopy*. Vol. 23, pp. 512. New York, NY: Marcel Dekker.

Baude, S., Broekaert, J. A. C., Delfosse, D., Jakubowski, N., Fuechtjohann, L., Orellana-Velado, N. G., Pereiro, R., and Sanz-Medel, A. 2000. Glow discharge atomic spectrometry for the analysis of environmental samples—a review. *J Anal Atom Spectrom* 15(11), 1516–1525.

Baumgartner, F. and Kim, M. A. 1990. Isotope effects in the equilibrium and non-equilibrium vaporization of tritiated water and ice. *Appl Radiat Isot* 41, 395–399.

Beals, D. M. and Hayes, D. W. 1995. Technetium-99, iodine-129 and tritium in the waters of the Savannah River Site. *Sci Total Environ* 173(1–6), 101–115.

Beasley, T., Cooper, L. W., Grebmeier, J. M., Aagaard, K., Kelley, J. M., and Kilius, L. R. 1998a. Np-237/I-129 atom ratios in the Arctic Ocean: Has Np-237 from western European and Russian fuel reprocessing facilities entered the Arctic Ocean? *J Environ Radioactiv* 39(3), 255–277.

Beasley, T. M., Kelley, J. M., Maiti, T. C., and Bond, L. A. 1998b. Np-237/Pu-239 atom ratios in integrated global fallout: a reassessment of the production of Np-237. *J Environ Radioactiv* 38(2), 133–146.

Beasley, T. M., Kelley, J. M., Orlandini, K. A., Bond, L. A., Aarkrog, A., Trapeznikov, A. P., and Pozolotina,V. N. 1998c. Isotopic Pu, U, and Np signatures in soils from Semipalatinsk-21, Kazakh Republic and the Southern Urals, Russia. *J Environ Radioactiv* 39(2), 215–230.

Becker, J. S. 2002. State-of-the-art and progress in precise and accurate isotope ratio measurements by ICP-MS and LA-ICP-MS—plenary lecture. *J Anal Atom Spectrom* 17(9), 1172–1185.

Becker, J. S. 2003. Mass spectrometry of long-lived radionuclides. *Spectrochim Acta B* 58(10), 1757–1784.

Becker, J. S. and Dietze, H. J. 1997. Double-focusing sector field inductively coupled plasma mass spectrometry for highly sensitive multi-element and isotopic analysis—Invited lecture. *J Anal Atom Spectrom* 12(9), 881–889.

Becker, J. S. and Dietze, H. J. 1998. Inorganic trace analysis by mass spectrometry. *Spectrochim Acta B* 53(11), 1475–1506.

Becker, J. S. and Dietze, H. J. 2000. Precise and accurate isotope ratio measurements by ICP-MS. *Fresen J Anal Chem* 368(1), 23–30.

Becker, J. S., Pickhardt, C., and Dietze, H. J. 2000. Laser ablation inductively coupled plasma mass spectrometry for the trace, ultratrace and isotope analysis of long-lived radionuclides in solid samples. *Int J Mass Spectrom* 202(1–3), 283–297.

Becker, J. S., Burow, M., Boulyga, S. F., Pickhardt, C., Hille, R., and Ostapczuk, P. 2002. ICP-MS determination of uranium and thorium concentrations and U-235/U-238 isotope ratios at trace and ultratrace levels in urine. *Atom Spectrosc* 23(6), 177–182.

Becker, J. S., Burow, M., Zoriy, M. V., Pickhardt, C., Ostapczuk, P., and Hille, R. 2004a. Determination of uranium and thorium at trace and ultratrace levels in urine by laser ablation ICP-MS. *Atom Spectrosc* 25(5), 197–202.

Becker, J. S., Zoriy, M., Halicz, L., Teplyakov, N., Muller, C., Segal, I., Pickhardt, C., and Platzner, I. T. 2004b. Environmental monitoring of plutonium at ultratrace level in natural water (Sea of Galilee-Israel) by ICP-SFMS and MC-ICP-MS. *J Anal Atom Spectrom* 19(9), 1257–1261.

Bellis, D., Ma, R., Bramall, N., McLeod, C. W., Chapman, N., and Satake, K. 2001. Airborne uranium contamination—as revealed through elemental and isotopic analysis of tree bark. *Environ Pollut* 114(3), 383–387.

Belloni, J., Haissinsky, M., and Salama, H. N. 1959. On the adsorption of some fission products on various surfaces. *J Phys Chem* 63, 881–887.

Bemis, C. E., Jr., Dittner, P. F., Silva, R. J., Hahn, R. L., Tarrant, J. R., Hunt, L. D., and Hensley, D. C. 1977. Production, L x-ray identification and decay of the nuclide (260)105. *Phys Rev C* 16, 1146–1157.

Bemis, C. E., Jr., Silva, R. J., Hensley, D. C., Keller, O. L., Jr., Tarrant, J. R., Hunt, L. D., Dittner, P. F., Hahn, R. L., and Goodman, C. D. 1973. X-Ray identification of element 104 *Phys Rev Lett* 31, 647–650.

Berkovits, D., Feldstein, H., Ghelberg, S., Hershkowitz, A., Navon, E., and Paul, M. 2000. U-236 in uranium minerals and standards.*Nuc Instrum Meth B* 172, 372–376.

Betti, M. 1996. Use of a direct current glow discharge mass spectrometer for the chemical characterization of samples of nuclear concern. *J Anal Atom Spectrom* 11(9), 855–860.

Bings, N. H., Bogaerts, A., and Broekaert, J. A. C. 2002. Atomic spectroscopy. *Anal Chem* 74(12), 2691–2711.

Blanchard, R. L., Kahn, B., and Birkhoff, R. D. 1960. The preparation of thin, uniform, radioactive sources by surface adsorption and electrodeposition. *Health Phys* 2, 246–255.

Bland, C. J. 1984. Tables of the geometrical factor for various source-detector configurations. *Nuc Instrum Methods* 223, 2–3, 602–606.

Blue, T. E. and Jarzemba, M. S. 1992. A model for radon gas adsorption on charcoal for open-faced canisters in an active environment. *Health Phys* 63(2), 226–232.

Bock, R. 1979. *A Handbook of Decomposition Methods in Analytical Chemistry.* New York: John Wiley.

Boni, A. L. 1966. Rapid ion exchange analysis of radiocesium in milk, urine, seawater and environmental samples. *Anal Chem* 38, 89–92.

Boulyga, S. F. and Becker, J. S. 2001. Determination of uranium isotopic composition and U-236 content of soil samples and hot particles using inductively coupled plasma mass spectrometry. *Fresen J Anal Chem* 370(5), 612–617.

Boulyga, S. F. and Becker, J. S. 2002. Isotopic analysis of uranium and plutonium using ICP-MS and estimation of burn-up of spent uranium in contaminated environmental samples. *J Anal Atom Spectrom* 17(9), 1143–1147.

Boulyga, S. F., Becker, J. S., Matusevitch, J. L., and Dietze, H. J. 2000. Isotope ratio measurements of spent reactor uranium in environmental samples by using inductively coupled plasma mass spectrometry. *Int J Mass Spectrom* 203(1–3), 143–154.

Boulyga, S. F., Desideri, D., Meli, M. A., Testa, C., and Becker, J. S. 2003. Plutonium and americium determination in mosses by laser ablation ICP-MS combined with isotope dilution technique. *Int J Mass Spectrom* 226(3), 329–339.

Bowyer, T. W., Abel, K. H., Hubbard, C. W., Panisko, M. E., Reeder, P. L., Thompson, R. C., and Warner, R. A. 1999. Field testing of collection and measurement of radioxenon for the comprehensive test ban treaty. *J Radioanal Nuc Ch* 240, 109–122.

B riesmeister, J. F. 1990. *MCNP-A general Monte Carlo code for neutron and photon transport.* Version 4.2. Technical Report LA-7396-M, Los Alamos National Laboratory.

Brown, T. A., Marchetti, A. A., Martinelli, R. E., Cox, C. C., Knezovich, J. P., and Hamilton, T. F. 2004. Actinide measurements by accelerator mass spectrometry at Lawrence Livermore National Laboratory. *Nuc Instrum Meth B* 223–224, 788–795.

Brown, E., Skougstad, M. W., and Fishman, M. J. 1970. Methods for collection and analysis of water samples for dissolved minerals and gases in *Techniques of Water-Resources Investigations, Book 5.* Chapter A1. Reston, VA: US Geological Survey.

Brown, R. M., Workman, W. J., and Kotzer, T. G. 1993. Tritium program at chalk river laboratories. Tritium monitoring in the environment in *Transactions of the American Nuclear Society.* Paper 1, pp. 20. Vol. 69. LaGrange Park, IL: American Nuclear Society.

Budnitz, R. J., Nero, A. V., Murphy, D. J., and Graven, R. 1983. *Instrumentation for Environmental Monitoring.* Vol. 1, Radiation. 2nd ed. New York, NY: John Wiley.

Burnett, W. C., Schultz, M. K., Corbett, D. R., Horwitz, E. P., Chiarizia, R., Dietz, M., Thakkar, A., and Fern, M. 1997. Preconcentration of actinide elements from soils and large volume water samples using extraction chromatography. *J Radioanal Nucl Chem* 226, 121–127.

Byrnes, M. E. 2001. *Sampling and Surveying Radiological Environments.* Boca Raton, FL: Lewis Publishers, CRC Press.

Carlton, W. H., Bauer, L. R., Evans, A. G., Geary, L. A., Murphy, C. E., Jr., Pinder, J. E., and Strom, R. N. 1992. *Cesium in the Savannah River Site Environment.* US Department of Energy Report WSRC-RP-92-250. Aiken, SC: Westinghouse Savannah River Company.

Cember, H. 1996. *Introduction to Health Physics.* New York, NY: McGraw-Hill.

Cerrai, E. and Ghersini, G. 1970. Reversed-phase extraction chromatography in inorganic chemistry, pp 3-189 in: *Advances in Chromatography*. Vol. 9. Giddings, J. C. and Keller, R. A., eds. New York, NY: Marcel Dekker.

Cesarano, C., Pugnetti, G., and Testa, C. 1965. Separation of cesium-137 from fission products by means of a Kel-F column supporting tetraphenylboron. *J Chromatog* 19, 589–593.

CFR 2004. Code of Federal Regulations. CFR is online at http://www.gpo.gov/nara/cfr/cfr-table-search.html#page1. (Viewed on December 2005)

CGIH 2001. *Air Sampling Instruments for Evaluation of Atmospheric Contaminants*, 9th ed. Cincinnati, OH: American Conference of Governmental Industrial Hygienists, Inc.

Chartier, F., Aubert, M., and Pilier, M. 1999. Determination of Am and Cm in spent nuclear fuels by isotope dilution inductively coupled plasma mass spectrometry and isotope dilution thermal ionization mass spectrometry after separation by high-performance liquid chromatography. *Fresen J Anal Chem* 364(4), 320–327.

Chatt, J. 1979. Recommendations for the naming of elements of atomic numbers greater than 100. *Pure Appl Chem.* 51, 381–384.

Chemweb 2005. Online through Champlain St. Lawrence College at http://www.slc.qc.ca/www/Chemweb/images/ET/sda.gif. Viewed on January 5, 2006.

Chiappini, R., Taillade, J. M., and Brebion, S. 1996. Development of a high-sensitivity inductively coupled plasma mass spectrometer for actinide measurement in the femtogram range. *J Anal Atom Spectrom* 11(7), 497–503.

Chiarappa-Zucca, M. L., Dingley, K. H., Roberts, M. L., Velsko, C. A., and Love, A. H. 2002. Sample preparation for quantitation of tritium by accelerator mass spectrometry. *Anal Chem* 74(24), 6285–6290.

Chieco, N. A. ed., 1997. *The Environmental Measurements Laboratory Procedures Manual*. U. S. Department of Energy Report HASL-300. New York, NY: Environmental Measurements Laboratory. Online at: http://www.eml.doe.gov/publications/Procman.cfm. Viewed on April 2, 2005.

Choppin, G. R., Liljenzin, J-O., and Rydberg, J. 1995. *Radiochemistry and Nuclear Chemistry*. 3rd ed. Oxford, London: Butterworth-Heinemann.

CNIC 1997. CNIC/IUPAC compromise recommendation approved August 30, 1997, Geneva, Switzerland. *Pure Appl Chem* 69, 2471–2473.

Coleman, G. H. 1965. *The Radiochemistry of Plutonium*. National Research Council Report NAS-NS 3058. Washington, DC: National Research Council.

Coomber, D. I. 1975. *Radiochemical Methods in Analysis*. New York, NY: Plenum Press.

Coplan, M. A., Moore, J. H., and Hoffman, R. A. 1984. Double-focusing ion mass-spectrometer of cylindrical symmetry. *Rev Sci Instrum* 55(4), 537–541.

Coplen, T. B. 2001. Atomic weights of the elements 1999. *J Phys Chem Ref Data* 30(3), 701–712.

Corish, J. and Rosenblatt, G. M. 2003. Name and symbol of the element with atomic number 110. *Pure Appl Chem* 75, 1613–1615.

Corish, J. and Rosenblatt, G. M. 2004. Name and symbol of the element with atomic number 111. *Pure Appl Chem* 76, 2101–2103.

Corley, J. P., Denham, D. H., Michels, D. E., Olsen, A. R., and Waite, D. A. 1977. *A Guide for Radiological Surveillance at ERDA Installations*. Energy Research and Development Administration, Division of Safety, Standards, & Compliance. ERDA77-24. Washington, DC: ERDA.

Corson, D. R., MacKenzie, K. R., Segrè, E. 1940. Artificially radioactive element 85. *Phys Rev* 58, 662.

Coryell, C. D. and Sugarman, N. 1951. Radiochemical studies: the fission products. *National Nuclear Energy Series*. 3 books, Vol. 9. New York, NY: McGraw-Hill.

Costa-Fernandez, J. M., Bings, N. H., Leach, A. M., and Hieftje, G. M. 2000. Rapid simultaneous multielemental speciation by capillary electrophoresis coupled to inductively coupled plasma time-of-flight mass spectrometry. *J Anal Atom Spectrom* 15(9), 1063–1067.

Cristy, S. S. 2000. Secondary ion mass spectrometry in *Inorganic Mass Spectrometry*. Vol. 23.. Barshick, C. M., Duckworth, D. C., and Smith, D. H., eds. pp. 159–221. New York, NY: Marcel Dekker.

Curie, P. and Sklodowska Curie, M. 1898. Sur une substance Nouvelle Radio-active, contenue dans la Pechblende. *Compt. Rend.* 127, 175.

Currie, L. A. 1968. Limits for qualitative detection and quantitative determination—application to radiochemistry. *Anal Chem* 40, 586–593.

Curtiss, L. F. and Davis, F. J. 1943. A counting method for the determination of small amounts of radium and radon. *J Res Nat Bur Stand.* 31, 181–195.

Dai, M. H., Buesseler, K. O., Kelley, J. M., Andrews, J. E., Pike, S., and Wacker, J. F. 2001. Size-fractionated plutonium isotopes in a coastal environment. *J Environ Radioactiv* 53(1), 9–25.

Daly, N. R. 1960. Scintillation type mass spectrometer ion detector. *Rev Sci Instrum* 31(3), 264–267.

Danesi, P. R., Bleise, A., Burkart, W., Cabianca, T., Campbell, M. J., Makarewicz, M., Moreno, J., Tuniz, C., and Hotchkis, M. 2003. Isotopic composition and origin of uranium and plutonium in selected soil samples collected in Kosovo. *J Environ Radioactiv* 64(2–3), 121–131.

Daubert v. Merrell Dow Pharmaceuticals 1993. 509 U.S. 579.

de Laeter, J. R. 1996. The role of off-line mass spectrometry in nuclear fission. *Mass Spectrom Rev* 15(4), 261–281.

de Laeter, J. R. 2001. *Applications of Inorganic Mass Spectrometry*. New York, NY: John Wiley.

de las Heras, L. A., Bocci, F., Betti, M., and Actis-Dato, L. O. 2000. Comparison between the use of direct current glow discharge mass spectrometry and inductively coupled plasma quadrupole mass spectrometry for the analysis of trace elements in nuclear samples. *Fresen J Anal Chem* 368(1), 95–102.

De, A. K., Khopkar, S. M., and Chalmers, R. A. 1970. *Solvent Extraction of Metals*. London: Van Nostrand Reinhold Co.

Dean, J. A. 1992. *Lange's Handbook of Chemistry*. 14th ed. Table 11.22, pp. 11–61. New York: McGraw-Hill.

Dean, J. A. 1999. *Lange's Handbook of Chemistry*. 15th ed. New York, NY: McGraw-Hill.

Delanghe, D., Bard, E., and Hamelin, B. 2002. New TIMS constraints on the uranium-238 and uranium-234 in seawaters from the main ocean basins and the Mediterranean Sea. *Mar Chem* 80(1), 79–93.

Delmore, J. E., Appelhans, A. D., and Peterson, E. S. 1995. A rare-earth-oxide matrix for emitting perrhenate anions. *Int J Mass Spectrom Ion Processes* 146, 15–20.

DeVoe, J. R. 1962. *Application of Distillation Techniques To Radiochemical Separations*. National Research Council Report NAS-NS-3108. Washington, DC: NRC.

Deyl, Z., ed. 1979. Electrophoresis: a survey of techniques and applications, Part A. Techniques. *J Chromatogr Libr* Vol. 18. New York, NY: Elsevier.

DHHS 1982. Department of Health and Human Services. *DHHS evaluation of results of environmental chemical testing by EPA in the vicinity of Love Canal—Implications for human health—Further considerations concerning habitability.* Washington, DC: DHHS.

Dienstbach, F. and Bachmann, K. 1980. Determination of plutonium in soil by gas chromatographic separation and alpha spectroscopy. *Anal Chem* 52, 62624.

DiNardi, S. R. ed. 2003. *The Occupational Environment: Its Evaluation, Control and Management.* Fairfax, VA: AIHA Press.

DOE 1987. Department of Energy Report DOE/EH-0053. *The Environmental Survey Manual*, Vol. 4. Appendices E to K. Washington, DC: DOE.

DOE/EM 1997. U.S. Department of Energy/Office of Environmental Management Report No. 0319. *Linking Legacies: Connecting Cold War Nuclear Weapons Production Processes to their Environmental Consequences.* Washington, DC: DOE/EM.

Douglas, D. J. and French, J. B. 1981. Elemental analysis with a microwave-induced plasma-quadrupole mass-spectrometer system. *Anal Chem* 53(1), 37–41.

Douglas, D. J. and Tanner, S. D. 1998. Fundamental considerations in ICP-MS in *Inductively Coupled Plasma Mass Spectrometry.* Chapter 8. Montaser, A., ed. Dordrecht: Wiley-VCH.

Duckworth, H. E., Barber, R. C., and Venkatasubramanian, V. S. 1986. *Mass Spectroscopy.* Cambridge, MA: Cambridge University Press.

Düllmann, C. E., Eichler, B., Eichler, R., Gäggeler, H. W., Jost, D. T., Piguet, D., and Türler, A. 2002a. IVO, a device for in situ volatilization and on-line detection of products from heavy ion reactions. *Nucl Instrum Meth A* 479, 631–639.

Düllmann, C. E., Brüchle, W., Dressler, R., Eberhardt, K., Eichler, B., Eichler, R., Gäggeler, H. W., Ginter, T. N., Glaus, F., Gregorich, K. E., Hoffman, D. C., Jäger, E., Jost, D. T., Kirbach, U. W., Lee, D. M., Nitsche, H., Patin, J. B., Pershina, V., Piguet, D., Qin, Z., Schädel, M., Schausten, B., Schimpf, E., Schött, H.-J., Soverna, S., Sudowe, R., Thörle, P., Timokhin, S. N., Trautmann, N., Türler, A., Vahle, A., Wirth, G., Yakushev, A. B., and Zielinski, P. M. 2002b. Chemical investigation of hassium (element 108). *Nature* 418, 859–862.

Durrant, S. F. 1999. Laser ablation-inductively coupled plasma mass spectrometry: achievements, problems, prospects. *J Anal Atom Spectrom* 14(9), 1385–1403.

Dushman, S. and Lafferty, J. M. 1962. *Scientific Foundations of Vacuum Technique.* New York, NY: John Wiley.

Eaton, G. F., Hudson, G. B., and Moran, J. E. 2004. Tritium-helium-3 age-dating of groundwater in the Livermore Valley of California.in *Radioanalytical Methods in Interdisciplinary Research.* Laue, C. A. and Nash, K. L., eds. pp. 235–245. Washington, DC: ACS.

Egorov, O. B., O'Hara, M. J., Addleman, S. R., Grate, J. W. 2003. Automation of radiochemical analysis: from groundwater monitoring to nuclear waste analysis in *Radioanalytical Methods in Interdisciplinary Research: Fundamental to Cutting Edge Applications.* ASC Symposium Series 868. Washington, DC: ACS.

Efurd, D. W., Rokop, D. J., Aguilar, R. D., Roensch, F. R., Banar, J. C., and Perrin, R. E. 1995. Identification and quantification of the source terms for uranium in surface waters collected at the rocky-flats facility. *Int J Mass Spectrom Ion Processes* 146, 109–117.

Ehmann, W. D. and Vance, D. E. 1991. *Radiochemistry and Nuclear Methods of Analysis.* New York, NY: Wiley-Interscience.

Eichholz, G. G. and Poston, J. W. 1979. *Principles of Nuclear Radiation Detection.* Ann Arbor, MI: Ann Arbor Science Publishers.

Eichholz, G. G., Nagel, A. E., and Hughes, R. B. 1965. Adsorption of ions in dilute aqueous solutions on glass and plastic surfaces. *Anal Chem* 37, 863–868.

Eichler, B., Türler, A., and Gäggeler, H. W. 1999. Thermochemical characterization of seaborgium compounds in gas adsorption chromatography. *J Phys Chem A* 103, 9296–9306.

Eichler, R., Brüchle, W., Dressler, C. E., Düllmann, C. E., Eichler, B., Gäggeler, H. W., Gregorich, K. E., Hoffman, D. C., Hübener, S., Jost, D. T., Kirbach, U. W., Laue, C. A., Lavanchy, V. M., Nitsche, H., Patin, J. B., Piguet, D., Schädel, M., Shaughnessy, D. A., Strellis, D. A., Taut, S., Tobler, L., Tsyganov, Y. S., Türler, A., Vahle, A., Wilk, P. A., and Yakushev, A. B. 2000. Chemical characterization of bohrium (element 107). *Nature (Letters)* 407, 63–65.

Eiden, G. C., Anderson, J. E., and Nogar, N. S. 1994. Resonant laser-ablation— semiquantitative aspects and threshold effects. *Microchem J* 50(3), 289–300.

Eisenbud, M. 1987. *Environmental Radioactivity.* 3rd ed. New York, NY: Academic Press.

Encinar, J. R., Alonso, J. I. G., Sanz-Medel, A., Main, S., and Turner, P. J. 2001. A comparison between quadrupole, double focusing and multicollector ICP-MS instruments Part I. Evaluation of total combined uncertainty for lead isotope ratio measurements. *J. Anal Atom Spectrom* 16, 315–321.

England, T. R. and Rider, B. F. 1994. Evaluation and compilation of fission product yields 1993. Los Alamos National Laboratory Report LA-UR-94-3106. ENDF (Evaluated Nuclear Data File) 349. Viewed on January, 2006.

EPA 1978. US Environmental Protection Agency Report EPA/EERF-78-1. *Radon in Water Sampling Program.* Washington, DC: EPA.

EPA 1980a. US Environmental Protection Agency Report 520/1-80-012. (Also Health Physics Committee Report HPSR-1). *Upprading Environmental Radiation Data.* Washington, DC: EPA.

EPA 1980b. US Environmental Protection Agency, Office of Monitoring Systems and Quality Assurance Publication QAMS 005/80. *Interim Guidelines and Specifications for Preparing Quality Assurance Project Plans.* Washington, DC: EPA.

EPA 1984. US Environmental Protection Agency Report EPA 520/5-884-006. *Eastern Environmental Radiation Facility Radiochemistry Procedures Manual.* Washington, DC: EPA. An index to the procedures contained in this manual is available online at www.epa.gov/region01/oarm/testmeth.pdf.

EPA 1986. US Environmental Protection Agency Report EPA/530/SW-86/055 (OSWER-9950.1). *RCRA Ground Water Monitoring Technical Enforcement Guidance Document.* Washington, DC: EPA.

EPA 2000a. US Environmental Protection Agency Manual 5360 A1. *EPA Quality Manual for Environmental Programs.* Washington, DC: EPA. This document can be found online at http://www.epa.gov/QUALITY/qa_docs.html.

EPA 2000b. US Environmental Protection Agency, Office of Environmental Information Report EPA QA/G-4. *Guidance for the Data Quality Objectives Process.* Washington, DC: EPA. This document can be found online at http://www.epa.gov/quality1/qs-docs/g4-final.pdf.

EPA 2000c. US Environmental Protection Agency Report 402-R-97-016 (also NRC Report NUREG-1575). *Multi-agency Radiation Survey and Site Investigation Manual (MARSSIM).* Washington, DC: EPA. May be downloaded from the EPA website at http://www.epa.gov/radiation/marssim/. (Viewed on January 2006)

EPA 2004. US Environmental Protection Agency, Report 402-B-04-001 (also NRC Report NUREG-1576, NTIS PB2004-105421). *Multi-Agency Radiological Laboratory Analytical Protocols Manual (MARLAP).* Washington, DC: EPA.

Erdmann, N., Nunnemann, M., Eberhardt, K., Herrmann, G., Huber, G., Kohler, S., Kratz, J. V., Passler, G., Peterson, J. R., Trautmann, N., and Waldek, A. 1998. Determination of the first ionization potential of nine actinide elements by resonance ionization mass spectroscopy (RIMS). *J Alloys Compd* 271, 837–840.

ETP 2005. Online at http://www.etpsci.com/htm/etp/tech_info/tech_articles/active_film.asp. Viewed on November 15, 2005.

Evans, R. D. 1955. *The Atomic Nucleus*. Pp. 470–510, 625 New York, NY: McGraw-Hill.

Faires, R. A. and Boswell, G. G. J. 1981. *Radioisotope Laboratory Techniques*. 4th ed. Oxford, London: Butterworth-Heinemann.

Fassel, V. A. 1971. Spectroscopic properties and analytical applications of induction-coupled plasmas. *Appl Spectrosc* 25(1), 148–&.

Fassel, V. A. and Kniseley, R. N. 1974 Inductively coupled plasmas. *Anal Chem* 46(13), 1155–&.

Faure, G. 1986. *Principles of Isotope Geology*. P. 399. New York, NY: John Wiley.

Fearey, B. L., Tissue, B. M., Olivares, J. A., Loge, G. W., Murrell, M. T., and Miller, C. M. 1992. High-precision thorium RIMS for geochemistry. *Inst Phys Conf Ser* (128), 209–212.

Fifield, L. K. 1999. Accelerator mass spectrometry and its applications. *Rep Prog Phys* 62(8), 1223–1274.

Fifield, L. K. 2000. Advances in accelerator mass spectrometry. *Nucl Instrum Meth B* 172, 134–143.

Fifield, L. K., Carling, R. S., Cresswell, R. G., Hausladen, P. A., di Tada M. L., and Day, J. P. 2000. Accelerator mass spectrometry of Tc-99. *Nucl Instrum Meth B* 168(3), 427–436.

Finney, G. D. and Evans, R. 1935. The radioactivity of solids determined by alpha-ray counting. *Phys Rev* 48, 503.

Finston, H. L. and Kinsley, M. T. 1961. *The Radiochemistry of Cesium*. National Research Council Report NAS-NS 3035. Washington, DC: National Research Council.

Firestone, R. B. 1990. *Table of Isotopes*. 8th ed. New York, NY: John Wiley.

Firestone, R. B. 1996. *Table of Isotopes*. 9th ed. New York, NY: John Wiley.

Friedlander, G. and Herrman, G. 2003. History of nuclear and radiochemistry in *Handbook of Nuclear Chemistry*. Vol. I. Chapter 1. Vertes, A., Nagy, S., and Klencsar, Z., eds. Dordrecht: Kluwer Academic.

Friedlander, G., Kennedy, J. W., Macias, E. S., and Miller, J. M. 1981. *Nuclear and Radiochemistry*. 3rd ed. New York, NY: John Wiley.

Fritz, J. S. and Umbreit, G. R. 1958. Ion exchange separation of metal ions. *Anal Chim Acta* 19, 509–516.

FRMAC 2002. Federal Radiological Monitoring and Assessment Center Report DOE/NV/11718—181. *Monitoring and Sampling Manual*. Vol. 1, Rev. 1. Washington DC: FRMAC.

Gäggeler, H. W. 1994. On-line gas chemistry experiments with transactinide elements. *J Radioanal Nucl Chem Articles* 183, 261.

Gäggeler, H. W., Jost, D. T., Kovacs, J., Scherer, U. W., Weber, A., Vermeulen, D., Türler, A., Gregorich, K. E., Henderson, R. A., Czerwinski, K. R., Kadkhodayan, B., Lee, D. M., Nurmia, M., Hoffman, D. C., Kratz, J. V., Gober, M. K., Zimmermann, H. P., Schädel, M., Brüchle, W., Schimpf, E., and Zvara, I. 1992. Gas phase chromatography experiments with bromides of tantalum and element-105. *Radiochim Acta* 57, 93–100.

Garcia, R. and Kahn, B. 2001. Total dissolution of environmental and biological samples for radiometric analysis by closed-vessel microwave digestion. *J Radiological Nucl Chem* 250, 85–91.

Gärtner, M., Boettger, M., Eichler, B., Gäggeler, H. W., Grantz, M., Hübener, S., Jost, D. T., Piguet, D., Dressler, R., Türler, A., and Yakushev, A. B. 1997. On-line gas chromatography of Mo, W and U (oxy) chlorides. *Radiochim Acta* 78, 59–68.

Gastel, M., Becker, J. S., Kuppers, G., and Dietze, H. J. 1997. Determination of long-lived radionuclides in concrete matrix by laser ablation inductively coupled plasma mass spectrometry. *Spectrochim Acta B* 52(14), 2051–2059.

Gerlich, D. 1992. Inhomogeneous Rf-fields—a versatile tool for the study of processes with slow ions. *Adv Chem Phys* 82, 1–176.

Ghiorso, A., Lee, D., Somerville, L. P., Loveland, W., Nitschke, J. M., Ghiorso, W., Seaborg, G. T., Wilmarth, P., Leres, R., Wydler, A., Nurmia, M., Gregorich, K., Czerwinski, K., Gaylord, R., Hamilton, T., Hannink, N. J., Hoffman, D. C., Jarzynski, C., Kacher, C., Kadkhodayan, B., Kreek, S., Lane, M., Lyon, A., McMahan, M. A., Neu, M., Sikkeland, T., Swiatecki, W. J., Türler, A., Walton, J. T., and Yashita, S. 1995a. Evidence for the synthesis of element $^{267}110$ produced by the ^{59}Co $+$ ^{209}Bi reaction. *Nucl Phys* 583, 861–866.

Ghiorso, A., Lee, D., Somerville, L. P., Loveland, W., Nitschke, J. M., Ghiorso, W., Seaborg, G. T., Wilmarth, P., Leres, R., Wydler, A., Nurmia, M., Gregorich, K., Czerwinski, K., Gaylord, R., Hamilton, T., Hannink, N. J., Hoffman, D. C., Jarzynski, C., Kacher, C., Kadkhodayan, B., Kreek, S., Lane, M., Lyon, A., McMahan, M. A., Neu, M., Sikkeland, T., Swiatecki, W. J., Türler, A., Walton, J. T., and Yashita, S. 1995b. Evidence for the possible synthesis of element $^{267}110$ produced by the ^{59}Co $+$ ^{209}Bi reaction. *Phys Rev C* 51, R2293–R2297.

Ghiorso, A., Nitschke, J. M., Alonso, J. R., Alonso, C. T., Nurmia, M., Seaborg, G. T., Hulet, E. K., and Lougheed, R. W. 1974. Element 106. *Phys Rev Lett* 33, 1490–1493.

Gibson, J. K. 1998. Laser ablation mass spectrometry of actinide dioxides: ThO_2, UO_2, NpO_2, PuO_2 and AmO_2. *Radiochim Acta* 81(2), 83–91.

Gill, C. G., Daigle, B., and Blades, M. W. 1991. Atomic mass-spectrometry of solid samples using laser ablation-ion trap mass-spectrometry (LAITMS). *Spectrochim Acta B* 46(8), 1227–1235.

Gilmore, G. and Hemingway, J. D. 1995. True coincidence summing in *Practical Gamma-Ray Spectrometry*. Chapter 7. West Sussex, England: John Wiley.

Gindler, J. E. 1962. *The Radiochemistry of Uranium*. National Research Council Report NAS-NS 3050. Washington, DC: National Research Council.

Gonzalez, J., Mao, X. L., Roy, J., Mao, S. S., and Russo, R. E. 2002. Comparison of 193, 213 and 266 nm laser ablation ICP-MS. *J Anal Atom Spectrom* 17(9), 1108–1113.

Gordon, L., Salutsky, M. L., and Willard, H. H. 1959. *Precipitation from Homogeneous Solution*. New York, NY: John Wiley.

Granstrom, M. L. and Kahn, B. 1955. The adsorption of cesium in low concentration by paper. *J Phys Chem* 59, 408–410.

Grate, J. W. and Egorov, O. 2003. Automated radiochemical separation, analysis, and sensing in *Handbook of Radioactivity Analysis*. pp. 1129–1164. San Diego, CA: Elsevier.

Gray, A. L. and Date, A. R. 1983. Inductively coupled plasma source-mass spectrometry using continuum flow ion extraction. *Analyst* 108(1290), 1033–1050.

Gregorich, K. E., Ginter, T., Loveland, W., Peterson, D., Patin, J. B., Folden, C. M., III, Hoffman, D. C., Lee, D. M., Nitsche, H., Omtvedt, J. P., Omtvedt, L. A., Stavsetra, L.,

Sudowe, R., Wilk, P. A., Zielinski, P., and Aleklett, K. 2003. Cross section limits for the $^{208}Pb(^{86}Kr,n)^{293}118$ reaction. *Eur Phys J A* 18, 633–638.

Gregorich, K. E., Henderson, R. A., Lee, D. M., Nurmia, M. J., Chasteler, R. M., Hall, H. L., Bennett, D. A., Gannett, C. M., Chadwick, R. B., Leyba, J. D., Hoffman, D. C., and Herrmann, G. 1988. Aqueous chemistry of element l05. *Radiochim Acta* 43, 223–231.

Groenewold, G. S., Delmore, J. E., Olson, J. E., Appelhans, A. D., Ingram, J. C., and Dahl, D. A. 1997. Secondary ion mass spectrometry of sodium nitrate: Comparison of ReO4- and Cs+ primary ions. *Int J Mass Spectrom Ion Processes* 163(3), 185–195.

Gruning, C., Huber, G., Klopp, P., Kratz, J. V., Kunz, P., Passler, G., Trautmann, N., Waldek, A., and Wendt, K. 2004. Resonance ionization mass spectrometry for ultratrace analysis of plutonium with a new solid state laser system. *Int J Mass Spectrom* 235(2), 171–178.

Guillaumont, R., Adloff, J.-P., and Peneloux, A. 1989. Kinetic and thermodynamic aspect of tracer-scale and single atom chemistry. *Radiochim Acta* 46, 169–176.

Guillaumont, R., Adloff, J.-P., Peneloux, A., and Delamoye, P. 1991. Sub-tracer scale behaviour of radionuclides. Application to actinide chemistry. *Radiochim Acta* 54, 1–15.

Gunnink, R. and Niday, J. B. 1972. Description of the GAMANAL program, UCRL-51061 in *Computerized Quantitative Analysis by Gamma-ray Spectrometry*. Vol. I. Livermore, CA: Lawrence Livermore Laboratory.

Gunther, D., Horn, I., and Hattendorf, B. 2000. Recent trends and developments in laser ablation-ICP-mass spectrometry. *Fresen J Anal Chem* 368(1), 4–14.

GV Instruments 2005. Application Report AN-20. Online at http://www.gvinstruments. co.uk/index.asp.

Gy, Pierre M. 1992. *Sampling of Heterogeneous and Dynamic Material Systems: Theories of Heterogeneity, Sampling and Homogenizing*. Amsterdam, The Netherlands: Elsevier.

Hahn, O. 1936. *Applied Radiochemistry*. Ithaca, NY: Cornell University Press.

Haissinsky, M. 1964. *Nuclear Chemistry and its Applications*. Reading, MA: Addison-Wesley.

Haldimann, M., Baduraux, M., Eastgate, A., Froidevaux, P., O'Donovan, S., Von Gunten, D., and Zoller, O. 2001. Determining picogram quantities of uranium in urine by isotope dilution inductively coupled plasma mass spectrometry. Comparison with alpha-spectrometry. *J Anal Atom Spectrom* 16(12), 1364–1369.

Harris, M. K., Herrington, P. B., Miley, H. S., Ellis, J. E., McKinnon, A. D., and St. Pierre, D. E. 1999. Data authentication demonstration radionuclide stations in *21st Seismic Research Symposium: Technologies for Monitoring the Comprehensive Nuclear-Test-Ban Treaty*. pp. 331–337. Los Alamos, NM: Los Alamos National Laboratory.

Heimbuch, A. M., Gee, H. Y., DeHaan, A. Jr., and Leventhal, L. 1965. The assay of alpha- and beta-emitters by means of scintillating ion-exchange resins in *Radioactive Sample Measurement Techniques in Medicine and Biology*. pp. 505–519. Vienna, Austria: International Atomic Energy Agency.

Heinrich, C. A., Pettke, T., Halter, W. E., Aigner-Torres, M., Audetat, A., Gunther, D., Hattendorf, B., Bleiner, D., Guillong, M., and Horn, I. 2003. Quantitative multi-element analysis of minerals, fluid and melt inclusions by laser-ablation inductively-coupled-plasma mass-spectrometry. *Geochim Cosmochim Ac* 67(18), 3473–3497.

Hellborg, R., Faarinen, M., Kiisk,M., Magnusson, C. E., Persson, P., Skog, G., and Stenstrom, K. 2003. Accelerator mass spectrometry—an overview. *Vacuum* 70(2–3), 365–372.

Henry, R., Koller, D., Liezers, M., Farmer, O. T., Barinaga, C., Koppenaal, D. W., and Wacker, J. 2001. New advances in inductively coupled plasma—mass spectrometry

(ICP-MS) for routine measurements in the nuclear industry. *J Radioanal Nucl Ch* 249(1), 103–108.

Hensley, J. W., Long, A. O., and Willard, J. E. 1949. Reactions of ions in aqueous solution with glass and metal surfaces. *Ind Eng Chem* 41, 1415–1421.

Herrmann, G., Riegel, J., Rimke, H., Sattelberger, P., Trautmann, N., Urban, F. J., Ames, F., Otten, E. W., Ruster, W., and Scheerer, F. 1991. Resonance ionization mass-spectroscopy of uranium. *Inst Phys Conf Ser* (114), 251–254.

Herzog, G. F. 1994. Applications of accelerator mass-spectrometry in extraterrestrial materials. *Nucl Instrum Meth B* 92(1–4), 492–499.

Heumann, K. G. 1992. Isotope-dilution mass-spectrometry. *Int J Mass Spectrom Ion Processes* 118, 575–592.

Heusser, G. 2003. *Low-background gamma spectroscopy of natural decay chain activity at the μBq/kg level*. Paper presented at the national meeting of the American Chemical Society, Division of Nuclear Chemistry and Technology; Paper No. 15. New York. Contact address for Heusser: Max Planck Institut fuer Kernphysik, POB 103 980 D-69029, Heidelberg, Germany.

Hidaka, H., Holliger, P., and Gauthier-Lafaye, F. 1999. Tc/Ru fractionation in the Oklo and Bangombe natural fission reactors, Gabon. *Chem Geol* 155(3–4), 323–333.

Hindman, F. D. 1986. Actinide separations for α spectrometry using neodymium fluoride coprecipitation. *Anal Chem* 58, 1238–1241.

Hoffman, D. C. 1994. The heaviest elements. *Chem & Eng News* 72, 24–34.

Hoffman, D. C. and Lee, D. M. 1999. Chemistry of the heaviest elements—One atom at a time. *J Chem Ed* 76, 331–347.

Hoffman, D. C. and Lee, D. M. 2003. Superheavy Elements in *Handbook of Nuclear Chemistry*, vol. 2. Chapter 10. pp. 397–416. Attila Vértes, S. N., Zoltán Klencsár, eds. Norwell, Massachusetts: Kluwer Academic Publishers.

Hoffman, D. C., Hamilton, T. M., and Lane, M. R. 1996. *Spontaneous Fission, in Nuclear Decay Modes*, Chapter 10. Poenaru, D. N. ed. Bristol: Institute of Physics Publishing.

Hoffman, D. C., Ghiorso, A., and Seaborg, G. T. 2000. *The Transuranium People: The Inside Story*. pp. 258–298, 379–396. London: Imperial College Press.

Hoffman, D. C., Lee, D. M., and Pershina, V. 2004. Transactinide Elements and Future Elements in: *The Chemistry of the Actinide and Transactinide Elements*. 3rd ed. vol. III. Chapter 14. Katz, J. J., Morss, L. R., Edelstein, N., Fuger, J. eds. London: Kluwer Academic Publishers.

Hofmann, S., Hessberger, F. P., Ackermann, D., Münzenberg, G., Antalic, S., Cagarda, P., Kindler, B., Kojouharova, J., Leino, M., Lommel, B., Mann, R., Popeko, A. G., Reshitko, S., Saro, S., Uusitalo, J., and Yeremin, A. V. 2002. New results on elements 111 and 112. *Eur Phys J A* 14, 147–157.

Hofmann, S., Ninov, V., Hessberger, F. P., Armbruster, P., Folger, H., Münzenberg, G., Schott, H. J., Popeko, A. G., Yeremin, A. V., Andreyev, A. N., Saro, S., Janik, R., and Leino, M. 1995a. Production and decay of ^{269}110. *Z Phys A* 350, 277–280.

Hofmann, S., Ninov, V., Hessberger, F. P., Armbruster, P., Folger, H., Münzenberg, G., Schott, H. J., Popeko, A. G., Yeremin, A. V., Andreyev, A. N., Saro, S., Janik, R., and Leino, M. 1995b. The new element 111. *Z Phys A* 350, 281.

Hofmann, S., Ninov, V., Hessberger, F. P., Armbruster, P., Folger, H., Münzenberg, G., Schott, H. J., Popeko, A. G., Yeremin, A. V., Saro, S., Janik, R., and Leino, M. 1996. *Z Phys A* 354, 229–230.

Hofstetter, K. J., Cable, P. R., and Beals, D. M. 1999. Field analyses of tritium at environmental levels. *Nucl Instr Methods in Phys Res* A422, 761–766.

Holden, N. E. and Hoffman, D. C. 2000. Spontaneous fission half-lives for ground-state nuclides. *Pure & Appl Chem* 72, 1525–1562.

Horowitz, E. P., Chiarizia, R., Dietz, M. L., Diamond, H., and Nelson, D. M. 1998. Separation and preconcentration of actinides from acidic media by extraction chromatography. *Anal Chim Acta* 281, 361–372.

Horrocks, D. L 1974. *Applications of Liquid Scintillation Counting*, New York, NY: Academic Press.

Hotchkis, M., Fink, D., Tuniz, C., and Vogt, S. 2000. Accelerator mass spectrometry analyses of environmental radionuclides: sensitivity, precision and standardisation. *Appl Radiat Isotopes* 53(1–2), 31–37.

Houk, R. S. and Thompson, J. J. 1982. Elemental and isotopic analysis of solutions by mass-spectrometry using a plasma ion-source. *Am Mineral* 67(3–4), 238–243.

Houk, R. S., Fassel, V. A., Flesch G. D., Svec H. J., Gray A. L., and Taylor C. E. 1980. Inductively coupled argon plasma as an ion-source for mass-spectrometric determination of trace-elements. *Anal Chem* 52(14), 2283–2289.

Houk, R. S., Svec, H. J., and Fassel, V. A. 1981. Mass-spectrometric evidence for suprathermal ionization in an inductively coupled argon plasma. *Appl Spectrosc* 35(4), 380–384.

Hulet, E. K., Lougheed, R. W., Wild, J. F., and Landrum, J. H. 1980. Chloride complexation of element 104. *J Inorg Nucl Chem* 42, 79–82.

Hyde, E. K., Hoffman, D. C., and Keller, O. L., Jr. 1987. A history and analysis of the discovery of elements 104 and 105. *Radiochim Acta* 42, 57–102.

IAEA 1989. International Atomic Energy Agency Technical Report Series No. 295. *Measurement of Radionuclides in Food and the Environment—A Guidebook.* Vienna, Austria: IAEA.

IAEA 2003. *Safeguards, Techniques and Equipment.* pp. 82. Vienna, Austria: International Atomic Energy Agency.

ICRU 1972. International Commission on Radiation Units and Measurements, Report 22. *Measurement of Low-level Radioactivity.* Washington, DC: ICRU.

ICRU 1979. International Commission on Radiation Units and Measurements, Report 31. *Average Energy Required to Produce an Ion Pair.* Washington, DC: ICRU.

ICRU 1984. International Commission on Radiation Units and Measurements, Report 37. *Stopping Powers for Electrons and Positrons.* Washington, DC: ICRU.

ICRU 1993. International Commission on Radiation Units and Measurments, Report 49. *Stopping Powers and Ranges of Protons and Alpha Particles.* Washington, DC: ICRU.

ICRU 1997. International Commission on Radiation Units and Measurments, Report 56. *Dosimetry of External Beta Rays for Radiation Protection.* Washington, DC: ICRU.

IEEE 1996. Institute of Electrical and Electronics Engineers. Report No. 325-1996. *IEEE Standard Test Procedures for Germanium Gamma-Ray Detectors.* Posted online at ieeexplore.ieee.org in 2002.

IEEE 2002. Institute of Electrical and Electronics Engineers. IEEE Standard for Software Quality Assurance Plans. IEEE (draft).

ISO 1987. International Organization for Standardization, ISO 9000/ASQC Q9000 Series. Milwaukee, WI: ASQC.

ISO 1993. International Organization for Standardization. *International Vocabulary of Basic and General Terms in Metrology.* Geneva, Switzerland: ISO.

ISO 1995. International Organization for Standardization. *Guide to the Expression of Uncertainty in Measurement.* Geneva, Switzerland: ISO.

ISO 2005. International Organization for Standardization online at http://www.iso.org/. Standards in Appendix A-4 may be found by clicking "ISO Store" and searching for the standards number. Viewed on January 10, 2006.

Jerome, S. M., Smith, D., Woods, M. J., and Woods, S. A. 1995. Metrology of plutonium for environmental measurements. *Appl Radiat Isotopes* 46(11), 1145–1150.

Kadkhodayan, B., Türler, A., Gregorich, K. E., Baisden, P. A., Czerwinski, K. R., Eichler, E., Gäggeler, H. W., Hamilton, T. M., Stoyer, N. J., Jost, D. T., Kacher, C. D., Kovacs, A., Kreek, S. A., Lane, M. R., Mohar, M. F., Neu, M. P., Sylwester, E. R., Lee, D. M., Nurmia, M. J., Seaborg, G. T., and Hoffman, D. C. 1996. On-line gas chromatographic studies of chlorides of rutherfordium and homologs Zr and Hf. *Radiochim Acta* 72, 169–178.

Kahn, B., Blanchard, R. L., Kolde, H. E., Krieger, H. L., Gold, S., Brinck, W. L., Averett, W. J., Smith, D. B., and Martin, A. 1971. *Radiological Surveillance Studies at a Pressurized Water Nuclear Power Reactor.* US Environmental Protection Agency Report RD71-1. Washington, DC: EPA.

Kahn, B., Rosson, R., and Cantrell, J. 1990. Analysis of ^{228}Ra and ^{226}Ra in public water supplies by gamma-ray spectrometer. *Health Phys* 59, 125–131.

Karam, L. R., Pibida, L., and McMahon, C. A. 2002. Use of resonance ionization mass spectrometry for determination of Cs ratios in solid samples. *Appl Radiat Isotopes* 56(1–2), 369–374.

Karge, H. G. and Weitkamp J. 1998. *Molecular Sieves: Science and Technology.* New York, NY: Springer-Verlag.

Karol, P. J., Nakahara, H., Petley, B. W., and Vogt, E. 2001. On the discovery of the elements 110–112. *Pure Appl Chem* 73, 959–967.

Karol, P. J., Nakahara, H., Petley, B. W., and Vogt, E. 2003. On the claims for discovery of elements 110, 111, 112, 114. *Pure Appl Chem* 75, 1601–1611.

Keller, O. L., Jr. 1984. Chemistry of heavy actinides and light transactinides. *Radiochim Acta* 37, 169–180.

Keller, J. H., Thomas, T. R., Spence, D. T., and Maech, W. J. 1973. An evaluation of materials and techniques used for monitoring airborne radioiodine species in *Proceedings of the Twelfth Air Cleaning Conference*, USA Department of Energy Report CONF-720823, Vol. 1, p. 322. Washington, DC: DOE.

Kelley, J. M. and Robertson, D. M. 1985. Plutonium Ion emission from carburized rhenium mass-spectrometer filaments. *Anal Chem* 57(1), 124–130.

Kelley, J. M., Bond, L. A., and Beasley, T. M. 1999. Global distribution of Pu isotopes and Np-237. *Sci Total Environ* 238, 483–500.

Kenna, T. C. 2002. Determination of plutonium isotopes and neptunium-237 in environmental samples by inductively coupled plasma mass spectrometry with total sample dissolution. *J Anal Atom Spectrom* 17(11), 1471–1479.

Kennedy, J. W., Seaborg, G. T., Segrè, E., and Wahl, A. C. 1946. Properties of ^{239}94. *Phys Rev* 70, 555–556.

Kepak, F. 1971. Adsorption and colloidal properties of radioactive elements in trace concentrations. *Chem Rev* 71, 357–369.

Kerl, W., Becker, J. S., Dietze, H. J., and Dannecker, W. 1997. Isotopic and ultratrace analysis of uranium by double-focusing sector field ICP mass spectrometry. *Fresen J Anal Chem* 359(4–5), 407–409.

Kersting, A. B., Efurd, D. W., Finnegan, D. L., Rokop, D. J., Smith, D. K., and Thompson, J. L. 1999. Migration of plutonium in ground water at the Nevada Test Site. *Nature* 397(6714), 56–59.

Kim, C. K., Seki, R., Morita, S., Yamasaki, S., Tsumura, A., Takaku, Y., Igarashi, Y., and Yamamoto, M. 1991. Application of a high-resolution inductively coupled plasma mass-spectrometer to the measurement of long-lived radionuclides. *J Anal Atom Spectrom* 6(3), 205–209.

Kim, C. S., Kim, C. K., Lee, J. I., and Lee, K. J. 2000. Rapid determination of Pu isotopes and atom ratios in small amounts of environmental samples by an on-line sample pretreatment system and isotope dilution high resolution inductively coupled plasma mass spectrometry. *J Anal Atom Spectrom* 15(3), 247–255.

King S. E., Phillips G. W., August R. A., Beach L. A., Cutchin J. H., and Castaneda C. 1987. Detection of tritium using accelerator mass-spectrometry. *Nucl Instrum Meth B* 29(1–2), 14–17.

Kingston, H. M. and Haswell, S., eds. 1997. *Microwave Enhanced Chemistry: Fundamentals, Sample Preparation, and Applications.* Washington, DC: ACS.

Kirbach, U. W., Folden, C. M., III, Ginter, T. N., Gregorich, K. E., Lee, D. M., Ninov, V., Omtvedt, J. P., Patin, J. B., Seward, N. K., Strellis, D. A., Sudowe, R., Türler, A., Wilk, P. A., Zielinski, P. M., Hoffman, D. C., and Nitsche, H. 2002. The Cryo-Thermochromatographic Separator (CTS): A new rapid separation and a detection system for on-line chemical studies of highly volatile osmium and hassium (Z = 108) tetroxides. *Nucl Instrum Meth A* 484, 587–594.

Kirby, H. W. and Salutsky, M. L. 1964. *The Radiochemistry of Radium.* National Research Council Report NAS-NS 3057. Washington, DC: National Research Council.

Kleinberg, J. and Cowan, G. A. 1960. *The Radiochemistry of Fluorine, Chlorine, Bromine and Iodine.* National Research Council Report NAS-NS 3005. Washington, DC: National Research Council.

Knoll, G. F. 1989. *Radiation Detection and Measurement.* 2nd ed. New York, NY: Wiley.

Knoll, G. F. 2000. *Radiation Detection and Measurement.* 3rd ed. New York, NY: John Wiley.

Kocher, D. C. 1977. *Nuclear Decay Data for Radionuclides Occurring in Routine Releases from Nuclear Fuel Cycle Facilities.* Oak Ridge National Laboratory Report ORNL/NUREG/TM-102. Oak Ridge, TN.

Kolthoff, I. M. 1932. Theory of coprecipitation: the formation and properties of crystalline precipitates. *J Phys Chem* 36, 860–881.

Kolthoff, I. M. and Elving P. J. 1971. *Treatise on Analytical Chemistry.* Pt. 1, Vol. 9. New York, NY: John Wiley.

Kolthoff, I. M. and Lingane, J. J. 1939. The fundamental principles and applications of electrolysis with the dropping mercury electrode and Heyrovsk's polarographic method of chemical analysis. *Chem Rev* 24, 1–94.

Koppenaal, D. W., Eiden, G. C., and Barinaga, C. J. 2004. Collision and reaction cells in atomic mass spectrometry: development, status, and applications. *J Anal Atom Spectrom* 19(5), 561–570.

Koppenol, W. H. 2002. Naming of new elements (IUPAC Recommendations 2002). *Pure Appl Chem* 74, 787–791.

Korenman, I. M. 1968. *Analytical Chemistry of Low Concentrations.* Jerusalem: Israel program for scientific translations ltd.

Korkisch, J. 1989. *Handbook of Ion Exchange Resins: Their Application to Inorganic Analytical Chemistry.* Boca Raton, FL: CRC Press.

Kovach, J. L. 1968. Adsorbents; review and projection in *Proceedings of the tenth AEC air cleaning conference.* First, M. W. and Morgan, J. M., Jr., eds. pp. 149–166.US DOE

Report CONF-680821. Springfield, VA: Clearinghouse for federal scientific and technical information.

Kratz, J. V. 1999a. Fast chemical separation procedures for transactinides in *Heavy Elements and Related New Phenomena*. Greiner, W. and Gupta, R. K. eds. pp. 43–63. Singapore: World Scientific.

Kratz, J. V. 1999b. Chemical properties of the transactinide elements in *Heavy Elements and Related New Phenomena*. Greiner, W. and Gupta, R. K. eds. pp. 129–193. Singapore: World Scientific.

Kraus, K. A. and Nelson, F. 1956. Anion exchange studies of the fission products in *Proceedings of the 1st International Conference on Peaceful Uses of Atomic Energy*. Vol. 7. A/CONF.8/7. pp. 113–125. New York, NY: United Nations Publishers.

Kusaka, Y. and Meinke, W. W. 1961. *Rapid Radiochemical Separations*. National Research Council Report NAS-NS 3104. Washington, DC: National Research Council.

Kutschera, W. 1994. Atom counting of long-lived radionuclides. *Nuclear Instruments & Methods in Physics Research Section a-Accelerators Spectrometers Detectors and Associated Equipment* 353(1–3), 562–562.

Ladd, M. F. C. and Lee, W. H. 1964. *Elementary Practical Radiochemistry*. London: Cleaver-Hume Press.

Laitinen, H. A. and Watkins, N. H. 1975. Cathodic stripping coulometry of lead. *Anal Chem* 47, 1352–1358.

Langmuir I. and Kingdon K. H. 1925. Thermionic effects caused by vapours of alkali metals. *P Roy Soc A* 107, 61–79.

Larsen, B. S. and McEwen, C. N. 1998. *Mass Spectrometry of Biological Materials*. pp. 469. New York, NY: Marcel Dekker.

Laue, C. A. and Nash, K. L., eds. 2003. *Radioanalytical Methods in Interdisciplinary Research, ACS Symposium Series 868*. Washington, DC: American Chemical Society.

Lazarev, Y. A., Lobanov, Y. V., Oganessian, Y. T., Utyonkov, V. K., Abdullin, F. S., Buklanov, G. V., Gikal, B. N., Iliev, S., Mezentsev, A. N., Polyakov, A. N., Sedykh, I. M., Shirokovsky, I. V., Subbotin, V. G., Sukhov, A. M., Tsyganov, Y. S., Zhuchko, V. E., Lougheed, R. W., Moody, K. J., Wild, J. F., Hulet, E. K., and McQuaid, J. H. 1994. Discovery of enhanced nuclear stability near the deformed shells N=162 and Z=108. *Phys Rev Lett* 73, 624–627.

Lazarev, Y. A., Lobanov, Y. V., Oganessian, Y. T., Utyonkov, V. K., Abdullin, F. S., Polyakov, A. N., Rigol, J., Shirokovsky, I. V., Tsyganov, Y. S., Iliev, S., Subbotin, V. G., Sukhov, A. M., Buklanov, G. V., Gikal, B. N., Kutner, V. B., Mezentsev, A. N., Subotic, K., Wild, J. F., Longheed, R. W., and Moody, K. J. 1996. Alpha decay of (273)110 - shell closure at N=162. *Phys Rev C* 54, 620–625.

Lederer, C. M., Hollander, J. M., and Perlman, I. 1967. *Table of Isotopes*, 6th ed. New York, NY: John Wiley.

Lee, S. H., Gastaud, J., La Rosa, J. J., Kwong, L. L. W., Povinec, P. P., Wyse, E., Fifield, L. K., Hausladen, P. A., Di Tada, L. M., and Santos, G. M. 2001. Analysis of plutonium isotopes in marine samples by radiometric, ICP-MS and AMS techniques. *J Radioanal Nucl Ch* 248(3), 757–764.

Lee, T., Ku, T. L., Lu, H. L., and Chen, J. C. 1993. 1st detection of fallout Cs-135 and potential applications of Cs-137/Cs-135 ratios. *Geochim Cosmochim Ac* 57(14), 3493–3497.

Lewis, D., Miller, G., Duffy, C. J., Efurd, D. W., Inkret, W. C., and Wagner, S. E. 2001. Los Alamos National Laboratory thermal ionization mass spectrometry results from intercomparison study of inductively coupled plasma mass spectrometry, thermal ionization mass spectrometry, and fission track analysis of mu Bq quantities of Pu-239 in synthetic urine (LA-UR-001698). *J Radioanal Nucl Ch* 249(1), 115–120.

Libby, W. F. 1940. Reactions of high energy atoms produced by slow neutron capture. *J Am Chem Soc* 62, 1930–1943.

Lieser, K. H. 2001. *Nuclear and Radiochemistry: Fundamentals and Applications.* 2nd ed. New York, NY: John Wiley.

Lindsey, A. J. 1964. Electrodeposition in: *Comprehensive analytical chemistry.* Vol. IIA. pp. 7–64. Wilson, C. L., Wilson, D. W., and Strouts, C. R. N., eds. New York, NY: Elsevier.

Litherland, A. E. and Kilius, L. R. 1997. Neutral injection for AMS. *Nucl Instrum Meth Phys B* 123(1–4), 18–21.

Liu, Y. F., Guo, Z. Y., Liu, X. Q., Qu, T., and Xie, J. L. 1994. Applications of accelerator mass-spectrometry in analysis of trace isotopes and elements. *Pure Appl Chem* 66(2), 305–334.

Liu, W. and van Wüllen, C. 1999. Spectroscopic constants of gold and eka-gold (element 111) diatomic compounds: The importance of spin–orbit coupling. *J Chem Phys* 110, 3730–3735.

Lockhart, L., Patterson, R., and Anderson, W. 1964. *Characteristics of Air Filter Media Used for Monitoring Airborne Radioactivity.* Naval Research Laboratory Report NRL-6054. Washington, DC: US Department of Defense.

Lougheed, R. W., Moody, K. J., Wild, J. F., Hulet, E. K., McQuaid, J. H., Lazarev, Y. A., Lobanov, Y. V., Oganessian, Y. T., Utyonkov, V. K., Abdullin, F. S., Buklanov, G. V., Gikal, B. N., Iliev, S., Mezentsev, A. N., Polyakov, A. N., Sedykh, I. M., Shirokovsky, I. V., Subbotin, V. G., Sukhov, A. M., Tsyganov, Y. S., and Zhuchko, V. E. 1994. Observation of enhanced nuclear stability near the 162 neutron shell. *J Alloys and Comp* 213, 61–66.

Love, A. H., Hunt, J. R., Roberts, M. L., Southon, J. R., Chiarappa-Zucca, M. L., and Dingle, K. H. 2002. Use of tritium accelerator mass spectrometry for tree ring analysis. *Environ Sci Tech* 36(13), 2848–2852.

Lucas, H. F. 1957. Improved low-level alpha scintillation counter for radon. *Rev Sci Instrum* 28, 680–683.

Magara, M., Hanzawa, Y., Esaka, F., Miyamoto, Y., Yasuda, K., Watanabe, K., Usuda, S., Nishimura, H., and Adachi, T. 2000. Development of analytical techniques for ultra trace amounts of nuclear materials in environmental samples using ICP-MS for safeguards. *Appl Radiat Isotopes* 53(1–2), 87–90.

Magara, M., Sakakibara, T., Kurosawa, S., Takahashi, M., Sakurai, S., Hanzawa, Y., Esaka, F., Watanabe, K., and Usuda, S. E. 2002. Isotope ratio measurement of uranium in safeguards environmental samples by inductively-coupled plasma mass spectrometry (ICP-MS). *J Nucl Sci Technol* 39(4), 308–311.

Malinovsky, D., Rodushkin, I., Baxter, D., and Ohlander, B. 2002. Simplified method for the Re-Os dating of molybdenite using acid digestion and isotope dilution ICP-MS. *Anal Chim Acta* 463(1), 111–124.

Maney, J. P. 2002. Optimizing data collection design. *Environ Sci Technol* 39 (19), 383A–389A.

Manninen, P. K. G. 1995. A rapid method for uranium isotope ratio measurement by inductively-coupled plasma-mass spectrometry. *J Radioan Nuc Ch Le* 201(2), 71–80.

March, R. E. and Hughes, R. J. 1989. Quadrupole storage mass spectrometry in *Chemical Analysis.* Vol. 102. Winefordner, J. D., ed. pp. 471. New York, NY: John Wiley.

Marcus, R. K. and Broekaert, J. A. C. 2002. *Glow Discharge Plasmas in Analytical Spectroscopy.* New York, NY: John Wiley.

Marcus, Y. and Kertes, A. S. 1969. *Ion Exchange and Solvent Extraction of Metal Complexes*. New York, NY: Wiley Interscience.

Marinsky, J. A., Glendenin, L. E., and Coryell, C. D. 1947. *J Am Chem Soc* 69, 2781.

Martin, B., Ockenden, D. W., and Foreman, J. K. 1961. The solvent extraction of plutonium and americium by tri-n-octylphosphine oxide. *J Inorg Nucl Chem* 21, 96–107.

Martin, M. J. and Blichert-Tolf, P. H. 1970. Radioactive atoms. *Nucl Data Tables* A8, 1–198, 1970.

McAninch, J. E., Hamilton, T. F., Brown, T. A., Jokela, T. A., Knezovich, J. P., Ognibene, T. J., Proctor, I. D., Roberts, M. L., Sideras-Haddad, E., Southon, J. R., and Vogel J. S. 2000. Plutonium measurements by accelerator mass spectrometry at LLNL. *Nucl Instrum Meth A* 172, 711–716.

McAninch, J. E., Marchetti, A. A., Bergquist, B. A., Stoyer, N. J., Nimz, G. J., Caffee, M. W., Finkel, R. C., Moody, K. J., Sideras-Haddad, E., Buchholz, B. A., Esser, B. K., and Proctor, I. D. 1998. Detection of Tc-99 by accelerator mass spectrometry: preliminary investigations. *J Radioan Nucl Ch* 234(1–2), 125–129.

McDowell, W. J. 1992. Photon/electron-rejecting alpha liquid scintillation (PERALS) spectrometry. A review. *Radioactivity and Radiochemistry* 3, 26–53.

McIntyre, J. I., Abel, K. H., Bowyer, T. W., Hayes, J. C., Heimbigner, T. R., Panisko, M. E., Reeder, P. L., and Thompson, R. C. 2001. Measurements of ambient radioxenon levels using the automated radioxenon sampler/analyzer (ARSA). *J Radioan Nucl Ch* 248 (3), 629–635.

McMillan, E. M. and Abelson, P. H. 1940. Radioactive element 93. *Phys Rev* 57, 1185.

Meeks, A. M., Giaquinto, J. M., and Keller, J. M. 1998. Application of ICP-MS radionuclide analysis to "Real World" samples of Department of Energy radioactive waste. *J Radioan Nucl Ch* 234(1–2), 131–135.

Miley, H. S., Bowyer, S. M., Hubbard, C. W., McKinnon, A. D., Perkins, R. W., Thompson, R. C., and Warner, R. A. 1998. A description of the DOE radionuclide aerosol sampler/analyzer for the Comprehensive Test Ban Treaty. *J Radioan Nucl Ch* 235 (1–2), 83–87.

Miller, D. R. 1988. Free jet sources in *Atomic and Molecular Beam Methods*. Vol. 1. Chapter 2. Scoles, G., ed. Oxford: Oxford University Press.

Moghissi, A. A., Bretthauer, E. W., and Compton, E. H. 1973. Separation of water from biological and environmental samples for tritium analysis. *Anal Chem* 45, 1565–1566.

Momyer, F., Jr. 1960. *The Radiochemistry of the Rare Gases*. National Academy of Sciences—National Research Council Report NAS-NS 3025. Washington, DC: National Research Council.

Montaser, A. 1992. *Inductively Coupled Plasmas in Analytical Atomic Spectrometry*. New York, NY: John Wiley.

Montaser, A. 1998. *Inductively Coupled Plasma Mass Spectrometry*. pp. 964. New York, NY: Wiley-VCH.

Moreno, J. M. B., Betti, M., and Alonso, J. I. G. 1997. Determination of neptunium and plutonium in the presence of high concentrations of uranium by ion chromatography inductively coupled plasma mass spectrometry. *J Anal Atom Spectrom* 12(3), 355–361.

Münzenberg, G., Armbruster, P., Folger, H., Hessberger, F. P., Hoffman, S., Keller, K., Poppensieker, K., Reisdorf, W., Schmidt, K.-H., and Schott, H.-J. 1984. The identification of element 108. *Z Phys A*, 317, 235–236.

Münzenberg, G., Armbruster, P., Hessberger, F. P., Hofmann, S., Poppensieker, K., Reisdorf, W., Schneider, J. H. R., Schneider, W. R. W., Schmidt, K.-H., Sahm, C.-C., and Vermeulen,

D. 1982. Evidence for element 109 from one correlated decay sequence following the fussion of ^{58}Fe with ^{209}Bi. *Z Phys A* 315, 145.

Münzenberg, G., Hofmann, S., Hessberger, F. P., Reisdorf, W., Schmidt, K. H., Schneider, J. H. R., Armbruster, P., Sahm, C.-C., and Thuma, B. 1981. Identification of element 107 by α correlation chains. *Z Phys A* 300, 107.

Muramatsu, Y., Hamilton, T., Uchida, S., Tagami, K., Yoshida, S., and Robison, W. 2001. Measurement of Pu-240/Pu-239 isotopic ratios in soils from the Marshall Islands using ICP-MS. *Sci Total Environ* 278(1–3), 151–159.

Muramatsu, Y., Uchida, S., Tagami, K., Yoshida, S., and Fujikawa, T. 1999. Determination of plutonium concentration and its isotopic ratio in environmental materials by ICP-MS after separation using ion-exchange and extraction chromatography. *J Anal Atom Spectrom* 14(5), 859–865.

Murthy, G. K. and Campbell, J. E. 1960. A simplified method for the determination of Iodine^{131} in milk. *J Dairy Science* 8, 1042–1049.

Murthy, G. K., Coakley, J. E., and Campbell, J. E. 1960. A method for the elimination of ashing in strontium-90 determination of milk. *Dairy Sci* 43, 151–154.

NAS-NRC 1960a. National Academy of Sciences—National Research Council Reports NAS-NS 30xx. *The Radiochemistry of ELEMENT*. Washington, DC: Volumes in the series published by the National Research Council, 1960–1988.

NAS-NRC 1960b. National Academy of Sciences—National Research Council Reports NAS-NS-31xx: Nuclear Science Series. *Radiochemistry Technique*. Available from NTIS, Springfield, VA. Volumes in the series published 1960–1988.

Navratil, O., Hala, J., Kopunec, R., Macasek, F., Mikulaj, V., and Leseticki, L. 1992. *Nuclear Chemistry*. New York, NY: Ellis Horwood, PTR Prentice Hall.

NCRP 1976a. National Council on Radiation Protection and Measurements, Report No. 47. *Tritium Measurement Techniques*. Bethesda, MD: NCRP.

NCRP 1976b. National Council on Radiation Protection and Measurements, Report No. 50. *Environmental Radiation Measurements*. Bethesda, MD: NCRP.

NCRP 1979. National Council on Radiation Protection and Measurements, Report No. 62. *Tritium in the Environment*. Bethesda, MD: NCRP.

NCRP 1983. National Council on Radiation Protection and Measurements Report No. 75. *I-129: Evaluation of Releases from Nuclear Power Generation*. Bethesda, MD: NCRP.

NCRP 1985a. National Council on Radiation Protection and Measurements, Report No. 81. *Carbon-14 in the Environment*. Bethesda, MD: NCRP.

NCRP 1985b. National Council of Radiation Protection and Measurements Report No. 58. *A Handbook of Radioactivity Measurement Procedures*. 2nd ed. Bethesda, MD: NCRP.

NCRP 1987a. National Council on Radiation Protection and Measurements, Report No. 94. *Exposure of the Population in the United States and Canada from Natural Background Radiation*. Bethesda, MD: NCRP.

NCRP 1987b. National Council on Radiation Protection and Measurements, Report No. 87. *Use of Bioassay Procedures for Assessment of Internal Radionuclide Deposition*, Bethesda, MD: NCRP, 1987.

Nelms S. 2005. *Inductively Coupled Plasma Mass Spectrometry Handbook*. Victoria, Australia: Blackwell Publishing.

Nichols, M. 2006. *Monte Carlo Simulation of Beta Particle Detection*. Ph.D. thesis. Atlanta, GA 30332: Georgia Institute of Technology.

Ninov, V., Gregorich, K. E., Loveland, W., Ghiorso, A., Hoffman, D. C., Lee, D. M., Nitsche, H., Swiatecki, W. J., Kirbach, U. W., Laue, C. A., Adams, J. L., Patin, J. B., Shaughnessy,

D. A., Strellis, D. A., and Wilk, P. A. 1999. Observation of superheavy nuclei produced in the reaction of ^{86}Kr with ^{208}Pb. *Phys Rev Lett* 83, 1104–1107.

Ninov, V., Gregorich, K. E., Loveland, W., Ghiorso, A., Hoffman, D. C., Lee, D. M., Nitsche, H., Swiatecki, W. J., Kirbach, U. W., Laue, C. A., Adams, J. L., Patin, J. B., Shaughnessy, D. A., Strellis, D. A., and Wilk, P. A. 2002. Retraction of 1999 published paper. *Phys Rev Lett* 89, 039901–1.

NIST 1994. National Institute of Standards and Technology, Technical Note 1297. *Guidelines for Evaluating and Expressing the Uncertainty of NIST Measurement Results.* Gaithersburg, MD: NIST.

Odom, R. W., Strathman, M. D., Buttrill, S. E., and Baumann, S. M. 1990. Nondestructive imaging detectors for energetic particle beams. *Nucl Instrum Meth B* 44(4), 465–472.

Oganessian, Y. T. 2001. The synthesis and decay properties of the heaviest elements. *Nucl Phys A* 685, 17c–36c.

Oganessian, Y. T. 2002. Synthesis and properties of even-even isotopes with Z = 110—116 in ^{48}Ca induced reactions. *J Nucl Radiochem Sci* 3, 5–8.

Oganessian, Y. T., Yeremin, A. V., Popeko, A. G., Bogomolov, S. L., Buklanov, G. V., Chelnokov, M. L., Chepigin, V. I., Gikal, B. N., Gorshkov, V. A., Gulbekian, G. G., Itkis, M. G., Kabachenko, A. P., Lavrentev, A. Y., Malyshev, O. N., Rohac, J., Sagaidak, R. N., Hofmann, S., Saro, S., Giardinas, G., and Morita, K. 1999a. Synthesis of nuclei of the superheavy element 114 in reactions induced by ^{48}Ca. *Nature* 400, 242–245.

Oganessian, Y. T., Utyonkov, V. K., Lobanov, Y. V., Abdullin, F. S., Polyakov, A. N., Shirokovsky, I. V., Tsyganov, Y. S., Gulbekian, G. G., Gobomolov, S. L., Gikal, B. N., Mezentsev, A. N., Iliev, S., Subbotin, V. G., Sukhov, A. M., Buklanov, G. V., Subotic, K., Itkis, M. G., Moody, K. J., Wild, J. F., Stoyer, N. J., Stoyer, M. A., and Lougheed, R. W. 1999b. Synthesis of superheavy nuclei in the ^{48}Ca + ^{244}Pu reaction. *Phys Rev Lett* 83, 3154–7.

Oganessian, Y. T., Yeremin, A. Y., Gulbekian, G. G., Bogomolov, S. L., Chepigin, V. I., Gikal, B. N., Gorshkov, V. A., Itkis, M. G., Kabachenko, A. P., Kutner, V. B., Lavrentev, A. Y., Malyshev, O. N., Popeko, A. G., Rohac, J., Sagaidak, R. N., Hofmann, S., Münzenberg, G., Veselsky, M., Saro, S., Iwasa, N., and Morita, K. 1999c. Search for new isotopes of element 112 by irradiation of ^{238}U with ^{48}Ca. *Eur Phys J A* 5, 63–68.

Oganessian, Y. T., Utyonkov, V. K., Lobanov, Y. V., Abdullin, F. S., Polyakov, A. N., Shirokovsky, I. V., Tsyganov, Y. S., Gulbekian, G. G., Bogomolov, S. L., Gikal, B. N., Mezentsev, A. N., Iliev, S., Subbotin, V. G., Sukhov, A. M., Ivanov, O. V., Buklanov, G. V., Subotic, K., Itkis, M. G., Moody, K. J., Wild, J. F., Stoyer, N. J., Stoyer, M. A., and Lougheed, R. W. 2000a. Synthesis of superheavy nuclei in the Ca-48+Pu-244 reaction: (288)114 - art. no. 041604. *Phys Rev C* 62, 041604 (R) 1–4.

Oganessian, Y. T., Utyonkov, V. K., Lobanov, Y. V., Abdulin, F. S., Polyakov, A. N., Shirokovsky, I. V., Tsyganov, Y. S., Gulbekian, G. G., Bogomolov, S. L., Gikal, B. N., Mezentsev, A. N., Iliev, S., Subbotin, V. G., Sukhov, A. M., Ivanov, O. V., Buklanov, G. V., Subotic, K., Itkis, M. G., Moody, K. J., Wild, J. F., Stoyer, N. J., Stoyer, M. A., Lougheed, R. W., Laue, C. A., Karelin, Y. A., and Tatarinov, A. N. 2000b. Observation of the decay of ^{292}116. *Phys Rev C* 63, 011301(R) 1–2.

Oganessian, Y. T., Utyonkov, V. K., Lobanov, Y. V., Abdullin, F. S., Polyakov, A. N., Shirokovsky, I. V., Tsyganov, Y. S., Gulbekian, G. G., Bogomolov, S. L., Gikal, B. N., Mezentsev, A. N., Iliev, S., Subbotin, V. G., Sukhov, A. M., Ivanov, O. V., Buklanov, G. V., Subotic, K., Voinov, A. A., Itkis, M. G., Moody, K. J., Wild, J. F., Stoyer, N. J., Stoyer, M. A., Lougheed, R. W., and Laue, C. A. 2002. *Eur Phys J A* 15, 201–204.

Oganessian, Y. T., Utyonkoy, V. K., Lobanov, Y. V., Abdullin, F. S., Polyakov, A. N., Shirokovsky, I. V., Tsyganov, Y. S., Gulbekian, G. G., Bogomolov, S. L., Mezentsev, A. N., Iliev, S., Subbotin, V. G., Sukhov, A. M., Voinov, A. A., Buklanov, G. V., Subotic, K., Zagrebaev, V. I., Itkis, M. G., Patin, J. B., Moody, K. J., Wild, J. F., Stoyer, M. A., Stoyer, N. J., Shaughnessy, D. A., Kenneally, J. M., and Lougheed, R. W. 2004a. *Phys Rev C* 69, 021601(R).

Oganessian, Y. T., Yeremin, A. V., Popeko, A. G., Malyshev, O. N., Belozerov, A. V., Buklanov, G. V., Chelnokov, M. L., Chepigin, V. I., Gorshkov, V. A., Hofmann, S., Itkis, M. G., Kabachenko, A. P., Kindler, B., Munzenberg, G., Sagaidak, R. N., Saro, S., Schott, H.-J., Streicher, B., Shutov, A. V., Svirikhin, A. I., and Vostokin, G. K. 2004b. *Eur Phys J A* 19, 3–6.

Oganessian, Y. T., Utyonkoy, V. K., Lobanov, Y. V., Abdullin, F. S., Polyakov, A. N., Shirokovsky, I. V., Tsyganov, Y. S., Gulbekian, G. G., Bogomolov, S. L., Gikal, B. N., Mezentsev, A. N., Iliev, S., Subbotin, V. G., Sukhov, A. M., Voinov, A. A., Buklanov, G. V., Subotic, K., Zagrebaev, V. I., Itkis, M. G., Patin, J. B., Moody, K. J., Wild, J. F., Stoyer, M. A., Stoyer, N. J., Shaughnessy, D. A., Kenneally, J. M., and Lougheed, R. W. 2004c. *Phys Rev C* 69, 054607.

Omtvedt, J. P., Alstad, J., Breivik, H., Dyve, J. E., Eberhardt, K., Folden III, C. M., Ginter, T., Gregorich, K. E., Hult, E. A., Johansson, M., Kirbach, U. W., Lee, D. M., Mendel, M., Nahler, A., Ninov, V., Omtvedt, L. A., Patin, J. B., Skarnemark, G., Stavsetra, L., Sudowe, R., Wiehl, N., Wierczinski, B., Wilk, P. A., Zielinski, P. M., Kratz, J. V., Trautmann, N., Nitsche, H., and Hoffman, D. C. 2002. SISAK liquid-liquid extraction experiments with pre-separated [257]Rf. *J Nucl Radiochem Sci* 3, 121–124.

Ostlund, H. G. and Mason, A. S. 1985. Atmospheric tritium in: *Tritium Laboratory Data Report No. 14*. Rosenstiel School of Marine and Atmospheric Science. Miami, FL: University of Miami.

OTA 1983. Office of Technology Assessment, NTIS order #PB84-114917. *Habitability of the Love Canal area: An analysis of the technical basis for the decision on the habitability of the emergency declaration area.* Washington, DC: U.S. Government Printing Office.

Oughton, D. H., Fifield, L. K., Day, J. P., Cresswell, R. C., Skipperud, L., Di Tada, M. L., Salbu, B., Strand, P., Drozcho, E., and Mokrov, Y. 2000. Plutonium from Mayak: Measurement of isotope ratios and activities using accelerator mass spectrometry. *Environ Sci Technol* 34(10), 1938–1945.

Oughton, D. H., Skipperud, L., Fifield, L. K., Cresswell, R. G., Salbu, B., and Day, P. 2004. Accelerator mass spectrometry measurement of Pu-240/Pu-239 isotope ratios in Novaya Zemlya and Kara Sea sediments. *Appl Radiat Isotopes* 61(2–3), 249–253.

Pajo, L., Schubert, A., Aldave, L., Koch, L., Bibilashvili, Y. K., Dolgov, Y. N., and Chorokhov, N. A. 2001a. Identification of unknown nuclear fuel by impurities and physical parameters. *J Radioanal Nucl Ch* 250(1), 79–84.

Pajo L., Tamborini G., Rasmussen G., Mayer K., and Koch L. 2001b. A novel isotope analysis of oxygen in uranium oxides: comparison of secondary ion mass spectrometry, glow discharge mass spectrometry and thermal ionization mass spectrometry. *Spectrochim Acta B* 56(5), 541–549.

Pappas, R. S., Ting, B. G., and Paschal, D. C. 2004. Rapid analysis for plutonium-239 in 1 ml of urine by magnetic sector inductively coupled plasma mass spectrometry with a desolvating introduction system. *J Anal Atom Spectrom* 19(6), 762–766.

Parrington, J. R., Knox, H. D., Breneman, S. L., Baum, E. M., and Feiner, F. 1996. *Chart of the Nuclides.* 15th ed. San Jose, CA: GE Nuclear Energy.

Pasternack, B. S. and Harley, N. H. 1971. Detection limits for radionuclides in the analysis of multi-component gamma ray spectrometer data. *Nucl Inst and Methods* 91, 533–540.

Penneman, R. A. and Keenan, T. K. 1960. *The Radiochemistry of Americium and Curium.* National Research Council Report NAS-NS 3006. Washington, DC: National Research Council.

Percival, D. R. and Martin, D. B. 1974. Sequential determination of radium-226, radium-228, actinium and thorium isotopes in environmental and process waste samples. *Anal Chem* 46, 1742–1749.

Perkins, R. W., Robinson, D. E., Thomas, C. W., and Young, J. A. 1989. IAEA-SM-306/125. Comparison of nuclear accident and nuclear test debris in *Proceedings of the International Symposium on Environmental Contamination Following a Major Nuclear Accident.* pp. 111–139. Vienna: IAEA, October 16–20, 1989.

Perrier, C. and Segrè, E. 1937. Some chemical properties of element 43. *J Chem Phys* 5, 712.

Perrin, R. E., Knobeloch, G. W., Armijo, V. M., and Efurd, D. W. 1985. Isotopic analysis of nanogram quantities of plutonium by using a sid ionization source. *Int J Mass Spectrom Ion Processes* 64(1), 17–24.

Perrin, D. D. 1964. *Organic Complexing Reagents: Structure, Behavior, and Application to Inorganic Analysis.* New York: Interscience Publishers.

Pershina, V. 1998a. Solution chemistry of element 105-Part I: Hydrolysis of group 5 cations: Nb, Ta, and Pa. *Radiochim Acta* 80, 65–73.

Pershina, V. 1998b. Solution chemistry of element 105-part II: Hydrolysis and complex formation of Nb, Ta, Ha and Pa in HCl solutions. *Radiochim Acta* 80, 75–84.

Pershina, V. and Bastug, T. 1999. Solution chemistry of element 105-Part III: Hydrolysis and complex fomation of Nb, Ta, Db and Pa in HF and HBr solutions. *Radiochim Acta* 84, 79–84.

Pershina, V. and Bastug, T. 2000. The electronic structure and properties of group 7 oxychlorides, MO_3Cl, where M=Tc, Re, and element 107, Bh. *J Chem Phys* 113, 1441–1446.

Pershina, V. and Fricke, B. 1996. Group 6 dioxydichlorides M_2Cl_2 (M=Cr, Mo, W, and element 106, Sg)—the electronic structure and thermochemical stability. *J Phys Chem* 100, 8748–8751.

Pershina, V. and Fricke, B. 1999. Electronic structure and chemistry of the heaviest elements in *Heavy Elements and Related New Phenomena.* vol. 1. pp. 184–262. Greiner, W. and Gupta, R. K., eds. Singapore: World Scientific.

Pershina, V. and Hoffman, D. C. 2003. The chemistry of the heaviest elements in *Theoretical chemistry and physics of heavy and superheavy elements in the series, "Progress in Theoretical Chemistry and Physics".* Chapter 3. pp. 55–114. Kaldor, U. and Wilson, S., eds. Dordrecht, The Netherlands: Kluwer Academic Publishers.

Pershina, V. and Kratz, J. V. 2001. Solution chemistry of element 106: Theoretical predictions of hydrolysis of group 6 cations Mo, W and Sg. *Inorg Chem* 40, 776–780.

Pershina, V., Sepp, W.-D., Bastug, T., Fricke, B., and Ionova, G. V. 1992. Relativistic effects in physics and chemistry of element 105. III. Electronic structure of hahnium oxyhalides as analogs of group 5 elements oxyhalides, *J Chem Phys* 97, 1123–1131.

Pershina, V., Bastug, T., Fricke, B., and Varga, S. 2001. The electronic structure and properties of group 8 oxides MO_4, where M=Ru, Os, and Element 108, Hs. *J Chem Phys* 115, 792–799.

Pershina, V., Bastug, T., Jacob, T., Fricke, B., and Varga, S. 2002. Intermetallic compounds of the heaviest elements: The electronic structure and bonding of dimers of element 112 and its homolog Hg. *Chem Phys Lett* 365, 176–183.

Pibida, L., Nortershauser, W., Hutchinson, J. M. R., and Bushaw, B. A. 2001. Evaluation of resonance ionization mass spectrometry for the determination of Cs-135/Cs-137 isotope ratios in low-level samples. *Radiochimica Acta* 89(3), 161–168.

Pitzer, K. S. 1975. Are elements 112, 114, and 118 relatively inert gases? *J Chem Phys* 63, 1032–1033.

Platzner, I. T. 1997. *Modern Isotope Ratio Mass Spectrometry*. New York, NY: John Wiley.

Platzner, I. T., Becker, J. S., and Dietze, H. J. 1999. Stability study of isotope ratio measurements for uranium and thorium by ICP-QMS. *Atomic Spectroscopy* 20(1), 6–12.

Porter, C. R. and Kahn, B. 1964. Improved determination of strontium-90 in milk by an ion-exchange method. *Anal Chem* 36, 676–678.

Porter, C. R., Kahn, B., Carter, M. W., Rehnberg, G. L., and Pepper, E. W. 1967. Determination of radiostrontium in food and other environmental media. *Env Sci Tech* 1, 745–750.

Poupard, D. and Jouniaux, B. 1990. Determination of picogram quantities of americium and curium by thermal ionization mass-spectrometry (Tims). *Radiochimica Acta* 49(1), 25–28.

Praphairaksit, N. and Houk, R. S. 2000. Reduction of mass bias and matrix effects in inductively coupled plasma mass spectrometry with a supplemental electron source in a negative extraction lens. *Anal Chem* 72, 4435–4440.

Priest, N. D., Pich, G. M., Fifield, L. K., and Cresswell, R. G. 1999. Accelerator mass spectrometry for the detection of ultra-low levels of plutonium in urine, including that excreted after the ingestion of Irish sea sediments. *Radiat Res* 152(6), S16–S18.

Probst, T. U. 1996. Studies on the long-term stabilities of the background of radionuclides in inductively coupled plasma mass spectrometry (ICP-MS)—A review of radionuclide determination by ICP-MS. *Fresen J Anal Chem* 354(7–8), 782–787.

Raisbeck, G. M. and Yiou, F. 1999. I-129 in the oceans: origins and applications. *Sci Total Environ* 238, 31–41.

Ratliff, T. A. 2003. *Laboratory Quality Assurance System*. 3rd ed. pp. 4, 31–32. Hoboken, NJ: John Wiley.

Rehkamper, M., Schonbachler, M., and Stirling, C. H. 2001. Multiple collector ICP-MS: Introduction to instrumentation, measurement techniques and analytical capabilities. *Geostandard Newslett* 25(1), 23–40.

Reyes, L. H., Gayon, J. M. M., Alonso, J. I. G., and Sanz-Medel, A. 2003. Determination of selenium in biological materials by isotope dilution analysis with an octopole reaction system ICP-MS. *J Anal Atom Spectrom* 18(1), 11–16.

Richardson, S. D. 2001. Mass spectrometry in environmental sciences. *Chem Rev* 101(2), 211–254.

Riegel, J., Deissenberger, R., Herrmann, G., Kohler, S., Sattelberger, P., Trautmann, N., Wendeler, H., Ames, F., Kluge, H. J., Scheerer, F., and Urban, F. J. 1993. Resonance ionization mass spectroscopy for trace analysis of neptunium. *Appl Phys B Laser O* 56(5), 275–280.

Rieman, W. and Walton, H. F. 1970. *Ion Exchange in Analytical Chemistry*. New York, NY: Pergamon Press.

Roberts, M. L. and Caffee, M. W. 2000. I-129 interlaboratory comparison: Phase II results. *Nucl Instrum Meth Phys B* 172, 388–394.

Rodushkin, I., Lindahl, P., Holm, E., and Roos, P. 1999. Determination of plutonium concentrations and isotope ratios in environmental samples with a double-focusing sector field ICP- MS. *Nucl Instrum Meth Phys A* 423(2–3), 472–479.

Rokop, D. J., Perrin, R. E., Knobeloch, G. W., Armijo, V. M., and Shields, W. R. 1982. Thermal ionization mass spectrometry of uranium with electrodeposition as a loading technique. *Anal Chem* 54(6), 957–960.

Rondinella, V. V., Betti, M., Bocci, F., Hiernaut, T., and Cobos, J. 2000. IC-ICP-MS applied to the separation and determination of traces of plutonium and uranium in aqueous leachates and acid rinse solutions of UO_2 doped with Pu-238. *Microchem J* 67(1–3), 301–304.

Rosen, A. 1998. Twenty to thirty years of DV-X α calculation: A survey of accuracy and application. *Adv Quant Chem* 29, 1–30.

Rosson, R., Jakiel, R., Klima, S., Kahn, B., and Fledderman, P. 2000. Correcting tritium concentrations in water vapor monitored with silica gel. *Health Phys* 78, 68–73.

Rucklidge, J. 1995. Accelerator mass-spectrometry in environmental geoscience—A review. *Analyst* 120(5), 1283–1290.

Russo, R. E., Mao, X. L., Liu, H. C., Gonzalez, J., and Mao, S. S. 2002. Laser ablation in analytical chemistry—A review. *Talanta* 57(3), 425–451.

Sahoo, S. K., Yonehara, H., Kurotaki, K., Fujimoto, K., and Nakamura, Y. 2002. Precise determination of U-235/U-238 isotope ratio in soil samples by using thermal ionisation mass spectrometry. *J Radioanal Nucl Ch* 252(2), 241–245.

Schädel, M. 1995. Chemistry of the transactinide elements. *Radiochim Acta* 70/71, 207–223.

Schädel, M., Brüchle, W., Dressler, R., Eichler, B., Gäggeler, H. W., Günther, R., Gregorich, K. E., Hoffman, D. C., Hübener, S., Jost, D. T., Kratz, J. V., Paulus, W., Schumann, D., Timokhin, S., Trautmann, N., Türler, A., Wirth, G., and Yakuschev, A. 1997a. Chemical properties of element 106 (seaborgium). *Nature Lett* 388, 55–57.

Schädel, M., Brüchle, W., Schausten, B., Schimpf, E., Jäger, E., Wirth, G., Günther, R., Kratz, J. V., Paulus, W., Seibert, A., Thorle, P., Trautmann, N., Zauner, S., Schumann, D., Andrassy, M., Misiak, R., Gregorich, K. E., Hoffman, D. C., Lee, D. M., Sylwester, E. R., Nagame, Y., and Oura, Y. 1997b. First aqueous chemistry with seaborgium (element 106). *Radiochim Acta* 77, 149–159.

Schädel, M., Brüchle, W., Jäger, E., Schausten, B., Wirth, G., Paulus, W., Günther, R., Eberhardt, K., Kratz, J. V., Seibert, A., Strub, E., Thörle, P., Trautmann, N., Waldek, W., Zauner, S., Schumann, D., Kirbach, U., Kubica, B., Misiak, R., Nagame, Y., and Gregorich, K. E. 1998. Aqueous chemistry of seaborgium (Z=106). *Radiochim Acta* 83, 163–165.

Schaumloffel, D., Giusti, P., Zoriy, M. V., Pickhardt, C., Szpunar, J., Lobinski, R., and Becker, J. S. 2005. Ultratrace determination of uranium and plutonium by nano-volume flow injection double-focusing sector field inductively coupled plasma mass spectrometry (nFI-ICP-SFMS). *J Anal Atom Spectrom* 20(1), 17–21.

Schmitt, G., Markell, C., and Hagen, D. 1990. Membrane approach to solid-phase extractions. *Anal Chim Acta* 236, 157–164.

Schneier, B. 2000. *Secrets & Lies: Digital Security in a Networked World*. New York, NY: John Wiley.

Schulze, J., Auer, M., and Werzi, R. 2000. Low level radioactivity measurement in support of the CTBTO. *Appl Radiat Isotopes* 53, 23–30.

Scoles, G. 1988. *Atomic and Molecular Beam Methods*. Vol. 1. Oxford: Oxford University Press.

Seaborg, G. T. 1945. The chemical and radioactive properties of the heavy elements. *Chem Eng News* 23, 2190.

Seaborg, G. T. 1967. Actinides reviews. *Actin Rev* 1, 3–38.

Seaborg, G. T., and Keller, O. L., Jr. 1986. Future elements in *The Chemistry of the Actinide Elements*. 2nd ed. Katz, J. J. Seaborg, G. T. and Morss, L. R. eds. pp. 1635–1643. London: Chapman and Hall.

Seaborg, G. T., Wahl, A. C., and Kennedy, J. W. 1946. Radioactive element 94 from deuterons on uranium. *Phys Rev* 69, 367.

Sekine, T. and Hasegaw, Y. 1977. *Solvent Extraction Chemistry: Fundamentals and Applications*. New York, NY: Marcel Dekker.

Seth, M., Schwerdtfeger, P., Dolg, M., Faegri, K., Hess, B. A., and Kaldor, U. 1996. Large relativistic effects in molecular properties of the hydride of superheavy element 111. *Chem Phys Lett* 250, 461–465.

Seth, M., Cooke, F., Schwerdtfeger, P., Heully, J.-L.,and Pelissier, M. 1998a. The chemistry of the superheavy elements. II. The stability of high oxidation states in group 11 elements: Relativistic coupled cluster calculations for the di-, tetra- and hexafluoro metallates of Cu, Ag, Au, and element 111. *J Chem Phys* 109, 3935–3943.

Seth, M., Faegri, K., and Schwerdtfeger, P. 1998b. The stability of the oxidation state +4 in group 14 compounds from carbon to element 114. *Angew Chem Int Ed Engl* 37, 2493–2496.

Seubert, A. 2001. On-line coupling of ion chromatography with ICP-AES and ICP-MS. *Trac-Trend in Anal Chem* 20(6–7), 274–287.

Shen, C. C., Cheng, H., Edwards, R. L., Moran, S. B., Edmonds, H. N., Hoff, J. A., and Thomas, R. B. 2003. Measurement of attogram quantities of Pa-231 in dissolved and particulate fractions of seawater by isotope dilution thermal ionization mass spectroscopy. *Anal Chem* 75(5), 1075–1079.

Sherwin, C. W. 1951.Vacuum evaporation of radioactive materials. *Rev Sci Instrum* 22 (5), 339–341.

Shleien, B. ed. 1992. *The Health Physics and Radiological Health Handbook*. Silver Springs, MD: Scinta.

Shleien, B., Slaback, L., Jr., and Birky, B. 1998. *Handbook of Health Physics and Radiological Health*. 3rd ed. Baltimore, MD: Lippincott.

Shuping, R. E., Philips, C. R., and Moghissi, A. A. 1970. Krypton-85 levels in the environment determined from dated krypton gas samples. *Radiol Health Data R* 11, 667–672.

Shuter, E. and Teasdale, W. E. 1989. Application of drilling coring, and sampling techniques to test holes and wells in *Techniques of Water-Resources Investigations*. Chapter F1. Book 2. Reston, VA: US Geological Survey.

Sill, C. W. and Olson, D. G. 1970. Sources and prevention of recoil contamination of solid-state alpha detectors. *Anal Chem* 42, 1596–1607.

Sill, C. W. and Williams, R. L. 1969. Radiochemical determination of uranium and the transuranium elements in process solutions and environmental samples. *Anal Chem* 41(12), 1624–1632.

Sill, C. W., Puphal, K. W., and Hindman, F. D. 1974. Simultaneous determination of alpha-emitting nuclides of radium through californium in soil. *Anal Chem* 46, 1725–1737.

Silva, R., Harris, J., Nurmia, M., Eskola, K., and Ghiorso, A. 1970. Chemical separation of rutherfordium. *Inorg Nucl Chem Lett* 6, 871–877.

Skipperud, L. and Oughton, D. H. 2004. Use of AMS in the marine environment. *Environ Int* 30(6), 815–825.

Smart, J. E. 1998. Real-time airborne radiation analysis and collection (RTARAC) searching for airborne species characteristic to nuclear proliferation. *J Radioanal Nucl Ch.* 235, 105–108.

Smith, D. H. 2000. Thermal ionization mass spectrometry in *Inorganic Mass Spectrometry: Fundamentals and Applications*. Vol. 23. Barshick, C. M., Duckworth, D. C., and Smith, D. H., eds. pp. 1–30. New York, NY: Marcel Dekker.

Smith, D. H. and Carter, J. A. 1981. A simple method to enhance thermal emission of metal ions. *Int J Mass Spectrom and Ion Processes* 40(2), 211–215.

Smolańczuk, R. 1997. Properties of the hypothetical spherical superheavy nuclei. *Phys Rev C* 56, 812–824.

Smolańczuk, R. 2001a. Production and decay of element 120. *Phys Lett B* 509, 227–230.

Smolańczuk, R. 2001b. Formation of superheavy elements in cold fusion reactions. *Phys Rev C* 63, 044607-1-8.

Song, M., Probst, T. U., and Berryman, N. G. 2001. Rapid and sensitive determination of radiocesium (Cs-135, Cs-137) in the presence of excess barium by electrothermal vaporization-inductively coupled plasma-mass spectrometry (ETV- ICP-MS) with potassium thiocyanate as modifier. *Fresen J Anal Chem* 370(6), 744–751.

Standard Methods 2005. Standard Methods for the Evaluation of Water and Wastewater. American Public Health Association (APHA), American Water Works Association (AWWA) and Water Environment Federation (WEF), 2005. Online at http://www.standardmethods.com/store/. Click "Browse Standard Methods." Viewed on January 2006.

Steinberg, E. P. 1962. Counting methods for the assay of radioactive samples in *Nuclear Instruments and Their Uses*. Chapter 5. Snell, A. H., ed. New York, NY: John Wiley.

Stevenson, P. C. and Nervik, W. E. 1961. *The Radiochemistry of the Rare Earths: Scandium, Yttrium, and Actinium*. National Research Council Report NAS-NS 3020. Washington, DC: National Research Council.

Stoffels, J. J., Wacker, J. F., Kelley, J. M., Bond, L. A., Kiddy, R. A., and Brauer, F. P. 1994. Environmental monitoring of Hanford nuclear facility effluents by thermal ionization mass-mpectrometry. *Appl Spectrosc* 48(11), 1326–1330.

Strub, E., Kratz, J. V., Kronenberg, A., Nähler, A., Thörle, P., Zauner, S., Brüchle, W., Jäger, E., Schädel, M., Schausten, B., Schimpf, E., Zongwei, L., Kirbach, U., Schumann, D., Jost, D., Türler, A., Asai, M., Nagame, Y., Sakara, M., Tsukada, K., Gäggeler, H. W., and Glanz, J. P. 2000. Fluoride complexation of rutherfordium (Rf, element 104). *Radiochim Acta* 88, 265.

Sturup, S., Dahlgaard, H., and Nielsen, S. C. 1998. High resolution inductively coupled plasma mass spectrometry for the trace determination of plutonium isotopes and isotope ratios in environmental samples. *J Anal Atom Spectrom* 13(12), 1321–1326.

Sulcek, Z. and Povondra, P. 1989. *Methods of Decomposition in Inorganic Analysis*. Boca Raton, FL: CRC Press.

Sunderman, D. N. and Townley, C. W. 1960. *The Radiochemistry of Barium, Calcium and Strontium*. National Research Council Report NAS-NS 3010. Washington, DC: National Research Council.

Sylwester, E. R., Gregorich, K. E., Lee, D. M., Kadkhodayan, B., Türler, A., Adams, J. L., Kacher, C. D., Lane, M. R., Laue, C. A., McGrath, C. A., Shaughnessy, D. A., Strellis, D. A., Wilk, P. A., and Hoffman, D. C. 2000. On-line gas chromatographic studies of Rf, Zr, and Hf bromides. *Radiochim Acta* 88, 837–843.

Takiue, M. and Ishikawa, H. 1978. Thermal neutron reaction cross section measurements for fourteen nuclides with a liquid scintillation spectrometer. *Nucl Instrum Methods* 148, 157–161.

Talvitie, N. A. 1972. Electrodeposition of actinides for alpha spectrometric determination. *Anal Chem* 44, 280–283.

Taylor, R. N., Croudace, I. W., Warwick, P. E., and Dee, S. J. 1998. Precise and rapid determination of U-238/U-235 and uranium concentration in soil samples using thermal ionisation mass spectrometry. *Chem Geol* 144(1–2), 73–80.

Taylor, R. N., Warneke, T., Milton, J. A., Croudace, I. W., Warwick, P. E., and Nesbitt, R. W. 2001. Plutonium isotope ratio analysis at femtogram to nanogram levels by multicollector ICP-MS. *J Anal Atom Spectrom* 16(3), 279–284.

Thiers, R. E. 1957. Separation, concentration, and contamination in *Trace Analysis*. Yoe, J. H. and Koch, H. J., eds. pp. 637–666. New York, NY: John Wiley.

Timokhin, S. N., Yakushev, A. B., Xu, H. G., Perelygin, V. P., and Zvara, I. 1996. Chemical identification of element 106 by thermochromatography. *J Radioanal Nucl Ch Lett* 212, 31–34.

Ting, B. G., Pappas, R. S., and Paschal, D. C. 2003. Rapid analysis for plutonium-239 in urine by magnetic sector inductively coupled plasma-mass spectrometry using Aridus desolvation introduction system. *J Anal Atom Spectrom* 18(7), 795–797.

Tolgyessy, J. and Bujdoso, E. 1991. *CRC Handbook of Radioanalytical Chemistry*. Boca Raton, FL: CRC Press.

Tolgyessy, J., Varga, S., and Krivan, V. 1971. *Nuclear Analytical Chemistry*. 5 books. Baltimore, MD: University Park Press.

Trautmann, N., Passler, G., and Wendt, K. D. A. 2004. Ultratrace analysis and isotope ratio measurements of long-lived radioisotopes by resonance ionization mass spectrometry (RIMS). *Anal Bioanal Chem* 378(2), 348–355.

Tuniz, C. 2001. Accelerator mass spectrometry: Ultra-sensitive analysis for global science. *Radiat Phys Chem* 61(3–6), 317–322.

Tuniz, C., Bird, J. R., Fink, D., and Herzog, G. F. 1998. *Accelerator Mass Spectrometry: Ultrasensitive Analysis for Global Science*. Baton Rouge, FL: CRC Press.

Türler, A., Gäggeler, H. W., Gregorich, K. E., Barth, H., Brüchle, W., Czerwinski, K. R., Gober, M. K., Hannink, N. J., Henderson, R. A., Hoffman, D. C., Jost, D. T., Kacher, C. D., Kadkhodayan, B., Kovacs, J., Kratz, J. V., Kreek, S. A., Lee, D. M., Leyba, J. D., Nurmia, M. J., Schädel, M., Scherer, U. W., Schimpf, E., Vermeulen, D., Weber, A., Zimmermann, H. P., and Zvara, I. 1992. Gas phase chromatography of halides of element-104 and Element-105. *J Radioanal Nucl Ch Ar* 160, 327–339.

Türler, A., Eichler, B., Jost, D. T., Piguet, D., Gäggeler, H. W., Gregorich, K. E., Kadkhodayan, B., Kreek, S. A., Lee, D. M., Mohar, M., Sylwester, E., Hoffman, D. C., and Hübener, S. 1996. On-line gas chromatography studies of chlorides of rutherfordium and homologs Zr and Hf. *Radiochim Acta* 73, 55–66.

Türler, A., Dressler, R., Eichler, B., Gäggeler, H. W., Jost, D. T., Schädel, M., Brüchle, W., Gregorich, K. E., Trautmann, N., and Taut, S. 1998. Decay properties of ^{265}Sg(Z=106) and ^{266}Sg(Z=106). *Phys Rev C* 57, 1648–1655.

Türler, A., Brüchle, W., Dressler, R., Eichler, B., Eichler, R., Gäggeler, H. W., Gärtner, M., Glatz, J. P., Gregorich, K. E., Hübener, S., Jost, D. T., Lebedev, V. Y., Pershina, V. G., Schädel, M., Taut, S., Timokhin, S. N., Trautmann, N., Vahle, A., and Yakushev, A. B. 1999. First measurement of a thermochemical property of a seaborgium compound. *Angew Chem Int Ed* 38, 2212–2213.

Uchida, S., Garcia-Tenorio, R., Tagami, K., and Garcia-Leon, M. 2000. Determination of U isotopic ratios in environmental samples by ICP-MS. *J Anal Atom Spectrom* 15(7), 889–892.

Underhill, D. W. 1996. The adsorption of argon, krypton and xenon on activated charcoal. *Health Phys* 71 (2), 160–166.

UNSCEAR 2000a. United Nations Scientific Committee on the Effects of Atomic Radiation, Annex B. *Exposures from Natural Radiation Sources*. Vienna: UN.

UNSCEAR 2000b. United Nations Scientific Committee on the Effects of Atomic Radiation, Annex C. *Exposures from Man-made Sources of Radiation*. Vienna: UN.

Vanhaecke, F. and Moens, L. 1999. Recent trends in trace element determination and speciation using inductively coupled plasma mass spectrometry. *Fresen J Anal Chem* 364(5), 440–451.

Vertes, A., Nagy, S., and Klencsar Z., eds. 2003. *Handbook of Nuclear Chemistry*. 5 books. Dordrecht: Kluwer Academic.

Vonderheide, A. P., Montes-Bayon, M., and Caruso, J. A. 2002. Development and application of a method for the analysis of brominated flame retardants by fast gas chromatography with inductively coupled plasma mass spectrometric detection. *J Anal Atom Spectrom* 17(11), 1480–1485.

Waber, J. T. and Averill, F. W. 1974. Molecular orbitals of PtF_6 and E110 F_6 calculated by the self-consistent multiple scattering X α method. *J Chem Phys* 60, 4460–4470.

Wahl, A. C. and Bonner, N. A. 1951. *Radioactivity Applied to Chemistry*. New York: John Wiley.

Wakat, M. A. 1971. Catalogue of γ-rays emitted by radionuclides. *Nucl Data Tables* 8, 445–666.

Walker, R. L., Carter, J. A., and Smith, D. H. 1981. A bulk resin bead procedure to obtain uranium and plutonium from radioactive solutions for mass-spectrometric analysis. *Anal Lett Pt A* 14(19), 1603–1612.

Walton, H. F. and Rocklin, R. D. 1990. *Ion Exchange in Analytical Chemistry*. Boca Raton, FL: CRC Press.

Wang, Z., Kahn, B., and Valentine, J. D. 2002. Efficiency calculation and coincidence summing correction for germanium detectors by Monte Carlo simulation. *IEEE T Nucl Sci* 49, 1925–1931.

Watanabe, K., Iguchi, T., Ogita, T., Uritani, A., and Harano, H. 2001. Development of failed fuel detection and location technique using resonance ionization mass spectrometry. *J Nucl Sci Technol* 38(10), 844–849.

Weast, R. C. 1985. *CRC Handbook of Chemistry and Physics*. Boca Raton, FL: CRC Press.

Wendt, K., Bhowmick, G. K., Bushaw, B. A., Herrmann, G., Kratz, J. V., Lantzsch, J., Muller, P., Nortershauser, W., Otten, E. W., Schwalbach, R., Seibert, U. A., Trautmann, N., and Waldek, A. 1997. Rapid trace analysis of Sr-89, Sr-90 in environmental samples by collinear laser resonance ionization mass spectrometry. *Radiochim Acta* 79(3), 183–190.

Wendt, K., Blaum, K., Bushaw, B. A., Gruning, C., Horn, R., Huber, G., Kratz, J. V., Kunz, P., Muller, P., Nortershauser, W., Nunnemann, M., Passler, G., Schmitt, A., Trautmann, N., and Waldek, A. 1999. Recent developments in and applications of resonance ionization mass spectrometry. *Fresen J Anal Chem* 364(5), 471–477.

Wendt, K., Trautmann, N., and Bushaw, B. A. 2000. Resonant laser ionization mass spectrometry: An alternative to AMS? *Nucl Instrum Meth B* 172, 162–169.

Wilk, P. A., Gregorich, K. E., Türler, A., Laue, C. A., Eichler, R., Ninov, V., Adams, J. L., Kirbach, U. W., Lane, M. R., Lee, D. M., Patin, J. B., Shaughnessy, D. A., Strellis, D. A., Nitsche, H., and Hoffman, D. C. 2000. Evidence for new isotopes of element 107: ^{266}Bh and ^{267}Bh. *Phys Rev Lett* 85, 2697–2700.

Williams, R. L. and Grothaus, G. E. 1984. Determination of actinides at the radiological and environmental sciences laboratory. *Nucl Instrum Meth A* 223 (2–3), 200–203.

Wilson, C. L., Wilson, D. W., and Strouts C. R. N. 1968. Physical separation methods in *Comprehensive Analytical Chemistry*. Volume IIB. New York, NY: Elsevier.

Winefordner, J. D., Gornushkin, I. B., Pappas, D., Matveev, O. I., and Smith, B. W. 2000. Novel uses of lasers in atomic spectroscopy. *J Anal Atom Spectrom* 15(9), 1161–1189.

Winn, W. G. 1993. A capability for monitoring effluent tritium with liquid scintillation/ robotics. Tritium monitoring in the environment.in *Transactions of the American Nuclear Society*. Vol. 69. Paper 8. pp. 26. LaGrange Park, IL: American Nuclear Society.

Wiza, J. L. 1979. Microchannel plate detectors. *Nucl Instrum Methods* 162(1–3), 587–601.

Wood, W. W. 1976. Guidelines for collection and field analysis of ground-water samples for selected unstable constituents in *Techniques of Water-Resources*. Book 1. Chapter D2. Reston, VA: US Geological Survey.

Wyse, E. J. and Fisher, D. R. 1994. Radionuclide bioassay by inductively-coupled plasma-mass spectrometry (ICP/MS). *Radiat Prot Dosim* 55(3), 199–206.

Wyse, E. J., MacLellan, J. A., Lindenmeier, C. W., Bramson, J. P., and Koppenaal, D. W. 1998. Actinide bioassays by ICPMS. *J Radioanal Nucl Ch* 234(1–2), 165–170.

Yakushev, A. B., Timokhin, S. N., Vedeneev, M. V., Xu, H. G., and Zvara, I. 1996. Comparative study of oxochlorides of molybdenum, tungsten and element 106. *J Radioan Nucl Ch Ar* 205, 63–67.

Yakushev, A. B., Buklanov, G. V., Chelnokov, M. L., Chepigin, V. I., Dmitriev, S. N., Gorshkov, V. A., Hübener, S., Lebedev, V. Y., Malyshev, O. N., Oganessian, Y. T., Popeko, A. G., Sokol, E. A., Timokhin, S. N., Türler, A., Vasko, V. M., Yeremin, A. V., and Zvara, I. 2001. First attempt to chemically identify element 112. *Radiochim Acta* 89, 743–745.

Yamamoto, M., Syarbaini, K., Kofuji, K., Tsumura A., Komura, K., Ueno, K., and Assinder, D. J. 1995. Determination of low-level Tc-99 in environmental samples by high-resolution ICP-MS. *J Radioan Nucl Ch Ar* 197(1), 185–194.

Yin, Q. Z., Jacobsen, S. B., Lee, C. T., McDonough, W. F., Rudnick, R. L., and Horn, I. 2001. A gravimetric K$_2$OsC^{16} standard: Application to precise and accurate Os spike calibration. *Geochim Cosmochim Ac* 65(13), 2113–2127.

Yiou, F., Raisbeck, G., and Imbaud, H. 2004. Extraction and AMS measurement of carrier free I-129/I-127 from seawater. *Nucl Instrum Meth B* 223–224, 412–415.

Yokoyama, T., Makishima, A., and Nakamura, E. 2001. Precise analysis of U-234/U-238 ratio using UO^{2+} ion with thermal ionization mass spectrometry for natural samples. *Chem Geol* 181(1–4), 1–12.

Yu, L. L., Fassett, J. D., and Guthrie, W. F. 2002. Detection limit of isotope dilution mass spectrometry. *Anal Chem* 74(15), 3887–3891.

Zhao, X. L., Kilius, L. R., Litherland, A. E., and Beasley, T. 1997. AMS measurement of environmental U-236 preliminary results and perspectives. *Nucl Instrum Meth B* 126(1–4), 297–300.

Zoriy, M. V., Pickhardt, C., Ostapczuk, P., Hille, R., and Becker, J. S. 2004. Determination of Pu in urine at ultratrace level by sector field inductively coupled plasma mass spectrometry. *Int J Mass Spectrom* 232(3), 217–224.

Zumwalt, L. R. 1950. Atomic Energy Commission Report AECU-567. *Absolute beta counting using end-window Geiger-Mueller counters and experimental data on beta-particle scattering effects*. Washington, DC: AEC.

Zvara, I., Belov, V. Z., Domanov, V. P., Korotkin, Y. S., Chelnokov, L. P., Shalayevsky, M. R., Shchegolev, V. A., and Hussonois, M. 1972. Chemical isolation of Kurchatovium. *Sov Radiochem* 14, 115–118.

Zvara, I., Aikhler, V., Belov, V. Z., Zvarova, T. S., Korotkin, Y. S., Shalayevsky, M. R., Shchegolev, V. A., and Hussonois, M. 1974. Gas chromatography and thermochromatography in the study of transuranium elements. *Sov Radiochem* 16, 709–715.

Zvara, I., Belov, V. Z., Domanov, V. P., and Shalayevsky, M. R. 1976. Chemical isolation of nielsbohrium as ekatantalum in the form of the anhydrous bromide; II experiments with a spontaneously fissioning isotope of nielsbohrium. *Sov Radiochem* 18, 328–334.

Index

Printed in the United States of America.